POROUS MEDIA

Applications in Biological Systems and Biotechnology

POROUS MEDIA

Applications in Biological Systems and Biotechnology

Edited by
KAMBIZ VAFAI

CRC Press
Taylor & Francis Group
Boca Raton London New York

CRC Press is an imprint of the
Taylor & Francis Group, an **informa** business

CRC Press
Taylor & Francis Group
6000 Broken Sound Parkway NW, Suite 300
Boca Raton, FL 33487-2742

First issued in paperback 2019

ISBN-13: 978-1-4200-6541-1 (hbk)
ISBN-13: 978-0-367-38367-1 (pbk)

Library of Congress Cataloging-in-Publication Data

Porous media : applications in biological systems and biotechnology / editor, Kambiz Vafai.
p. cm.
"A CRC title."
Includes bibliographical references and index.
ISBN 978-1-4200-6541-1 (alk. paper)
1. Biomedical materials. 2. Biotechnology--Materials. 3. Porous materials--Fluid dynamics. 4. Porous materials--Thermal properties. 5. Tissue engineering. 6. Biofilms. I. Vafai, K. (Kambiz)

R857.M3P67 2011
610.28'4--dc22 2010025878

Visit the Taylor & Francis Web site at
http://www.taylorandfrancis.com

and the CRC Press Web site at
http://www.crcpress.com

Contents

Preface

Fundamental and applied research in flow, heat, and mass transfer in porous media has received increased attention during the past several decades. This is due to the importance of this research area in many engineering applications, which can be modeled or approximated as transport through porous media such as thermal insulation, packed bed heat exchangers, drying technology, catalytic reactors, petroleum industries, geothermal systems, and electronic cooling. Other examples include computational biology, tissue replacement production, biofilms, drug delivery, advanced medical imaging, porous scaffolds for tissue engineering and effective tissue replacement to alleviate organ shortages, and transport in biological tissues. Another important application of porous media can be the diffusion process in the extracellular space (ECS), which is important for investigating central nervous system physiology. Significant advances have been made in modeling fluid flow, heat, and mass transfer through a porous medium including clarification of several important physical phenomena.

Despite some short treatises, this book, to the best of our knowledge, is the first to address, focus, and offer a comprehensive coverage of applications of porous media theory to biomedical and biological sciences. It covers various transport processes as well as mechanical behavior and material properties of biological tissues from a porous media point of view. This book will be of substantial help to researchers of various fields including but not limited to engineering, biophysics, biology, and medicine. Only the most outstanding contributors for each category in the field were selected and involved in the production of this book.

It is important to place this book in proper perspective relative to the first and second editions of *Handbook of Porous Media*. The material presented in these handbooks will enhance our understanding of what is presented here. As such, it is beneficial to briefly discuss the coverage within these two handbooks, as they may get referred to directly or indirectly during material discussion in this book. Both the first and the second editions of the *Handbook of Porous Media* were aimed at providing researchers with the most pertinent and up-to-date advances in modeling and analysis of flow, heat, and mass transfer. The first *Handbook of Porous Media* was arranged into 8 sections with a total of 19 chapters: section one covered fundamental topics of transport, which included theoretical models of fluid flow and the local volume averaging technique, capillary and viscous effects in porous media, and application of fractal and percolation concepts in characterizing porous materials; section two discussed basic aspects of conduction; in section three, various aspects of forced convection including numerical modeling were covered; section four concentrated on natural convection, thermal stability, and double diffusive convection; in section five, mixed convection was presented; section six was dedicated to the discussion of radiative transfer; and section seven was about turbulence; finally, the last section of the handbook covered several important applications of transport, which included packed bed chemical reactors, environmental applications (e.g., soil remediation), and drying and liquid composite molding (e.g., RTM and SRIM), which had received significant recent interest. Earlier chapters included other applications, such as forced convection heat transfer enhancement, which were covered along with other material.

The second edition of the *Handbook of Porous Media* addressed a substantially different set of topics as compared to the first edition. It included recent studies related to current and future challenges and advances in fundamental aspects of porous media, viscous dissipation, forced and double diffusive convection, turbulent flow, dispersion, particle migration and deposition, dynamic modeling of convective transport, and a number of other important topics related to porous media. This second edition handbook was arranged into seven parts with a total of 17 chapters: Part I covered fundamental topics of transport in porous media including theoretical models of fluid flow, the local volume averaging technique and viscous and dynamic modeling of convective heat transfer, and dispersion in porous media; Part II discussed various aspects of forced convection including numerical modeling, thermally developing flows and three-dimensional flow, and heat transfer within highly anisotropic medium; natural convection, double diffusive convection, and flows induced by both natural convection and vibrations were presented in Part III; Part IV looked at the effects of viscous dissipation for natural, mixed, and forced convection applications; Part V covered turbulence and in Part VI particle migration and deposition were discussed; finally, Part VII concentrated on several important applications including geothermal systems, liquid composite molding, combustion in inert porous media, and bioconvection applications. This final part also included the application of genetic algorithms (GAs) for identification of the hydraulic properties of porous materials in the context of petroleum, civil, and mining engineering. Whenever applicable, each of these handbooks discussed pertinent aspects of experimental works and techniques.

Understanding the physical, chemical, and biological processes governing an organism is important in biomedical engineering and physiology. Interactions involving mechanics, fluid mechanics, heat transfer, and mass transport in biology and medicine are crucial in understanding the causes of disease and in the development of new prophylactic, diagnostic, and therapeutic procedures for improving human health. Nearly all of the human tissues and organs can be categorized as porous media. Thus, this theory has found outstanding applications in biological and biomedical sciences, including but not limited to tissue engineering, biomaterials, biomechanics, biotransport phenomena, and biomedical imaging. Advances in numerical simulations and emergence of sophisticated porous models have significantly improved the study and analysis of biological systems. For instance, the most accurate description of initiation and growth of atherosclerosis is obtained via modeling the arterial wall as a multilayer porous medium. Another example is the application of porous scaffolds in tissue engineering. Developments in modeling transport phenomena in porous media have advanced the field of biology (see Khaled, A. -R. A. and Vafai, K., 2003, The Role of Porous Media in Modeling Flow and Heat Transfer in Biological Tissues, *International Journal of Heat Mass Transfer*, **46**, 4989–5003 and Khanafer, K., and Vafai, K., 2006, The Role of Porous Media in Biomedical Engineering as Related to Magnetic Resonance Imaging and Drug Delivery, *Heat and Mass Transfer*, **42**, 939–953). In these works, various biological areas such as diffusion in brain tissues and, during tissue-generation process, the use of magnetic resonance imaging (MRI) to characterize tissue properties, blood perfusion in human tissues, blood flow in tumors, bioheat transfer in tissues, and bioconvection that utilize different transport models in porous media have been synthesized. Pertinent works associated with MRI and drug delivery were reviewed to demonstrate the role of transport theory in advancing the progress in biomedical applications. Diffusion process is considered significant in many therapies such as delivering drugs to the brain. As such, progress in development of the diffusion equation using local volume averaging technique and evaluation of the applications associated with the diffusion equation was analyzed. Tortuosity and porosity have an impact on the diffusion transport. As such, different relevant models of tortuosity were presented and mathematical modeling of drug release from biodegradable delivery systems was analyzed. New models for the kinetics of drug release from porous biodegradable polymeric microspheres under bulk erosion and surface erosion of the polymer matrix were presented and diffusion of the dissolved drug, dissolution from the solid phase, and erosion of the polymer matrix were found to play a central role in controlling the overall drug release process. These studies paved the road for the researchers to develop comprehensive models based on porous media theory utilizing fewer assumptions as compared to other approaches.

Several pertinent areas of interest in which the porous media modeling is crucial are macromolecular transport in arterial walls, biofilms, tissue engineering, biodegradable porous drug delivery systems, diffusion-weighted magnetic resonance imaging (DW-MRI), and modeling and understanding heat transport and temperature variations within biological tissues and body organs, which are key issues in medical thermal therapeutic applications such as hyperthermia cancer treatment. In what follows, these cited areas are elaborated on. Macromolecular transport in arterial walls can affect the arteries of the brain, heart, kidneys, and the arms and legs. It is caused by the slow buildup of fatty substances, cholesterol, cellular waste products, calcium, and other substances found in the blood within the arterial walls. This buildup is called plaque. The transport of the low-density lipoprotein (LDL) from the blood into the arterial wall and its accumulation within the wall play an important role in the process of atherogenesis. This transport process is termed "arterial mass transport" and is influenced by blood flow in the lumen and transmural flow in the arterial wall. Four-layer model based on porous media can be used for the description of the mass transport in the arterial wall coupled with the mass transport in the arterial lumen. The endothelium, intima, internal elastic lamina (IEL), and media layers can all be treated as macroscopically homogeneous porous media and mathematically modeled using proper types of volume averaged porous media equations with the Staverman filtration and osmotic reflection coefficients employed to account for selective permeability of each porous layer to certain solutes (see Yang, N., and Vafai, K., 2006, Modeling of Low-Density Lipoprotein (LDL) Transport in the Artery-Effects of Hypertension, *International Journal of Heat and Mass Transfer*, **49**, 850–867; Ai, L., and Vafai, K., 2006, A Coupling Model for Macromolecule Transport in a Stenosed Arterial Wall, *International Journal of Heat and Mass Transfer*, **49**, 1568–1591; Khakpour, M., and Vafai, K., 2008, Critical Assessment of Arterial Transport Models, *International Journal of Heat and Mass Transfer*, **51**, 807–822; Yang, N., and Vafai, K., 2008 Low density Lipoprotein (LDL) Transport in an Artery-A Simplified Analytical Solution, *International Journal of Heat and Mass Transfer*, **51**, 497–505; Khakpour, M., and Vafai, K., 2008, A Comprehensive Analytical Solution for Macromolecular Transport within an Artery, *International Journal of Heat and Mass Transfer*, **51**, 2905–2913 and Khakpour, M., and Vafai, K., 2008, Effects of Gender-Related Geometrical Characteristics of Aorta-Iliac Bifurcation on Hemodynamics and Macromolecule Concentration Distribution, *International Journal of Heat and Mass Transfer*, **51**, 5542–5551).

Another area of focus is related to biofilms. A biofilm is a complex aggregation of microorganisms growing on a solid substrate. Structural heterogeneity, genetic diversity, complex community interactions, and an extracellular matrix of polymeric substances characterize biofilms. Biofilms are common in nature as bacteria possess mechanisms by which they can adhere to surfaces and to each other. Each year, microbial biofilm deposits on surfaces cost the global economy billions of dollars in equipment damage, product contamination, energy losses, and medical infections. In industrial environments, biofilms can develop on the interiors of pipes and lead to clogging and corrosion. Thus, it is important to develop new strategies on the basis of a better understanding of bacterial attachment, growth, and detachment as it is needed by many industries (see Shafahi, M., and Vafai, K., 2009, Biofilm Affected Characteristics of Porous Structures, *International Journal of Heat and Mass Transfer*, **52**, 574–581 and Shafahi, M., and Vafai, K., 2010, Synthesis of Biofilm Resistance Characteristics Against Antibiotics, *International Journal of Heat and Mass Transfer*, **53**, 2943–2950). Another area of focus is related to porous scaffolds for tissue engineering and its effective replacement to alleviate organ shortages. Tissue engineering is an emerging field bringing together chemical and material engineering, biology, and medicine. The major aim of this multidisciplinary field is to develop biological substitutes for the repair and regeneration of tissue or organ function. Examples of tissue-engineered substitutes that are currently being investigated throughout the world include skin, cartilage, bone, vascular, heart, breast, and liver. Porous scaffolds for tissue engineering serve to provide anatomical shape to the implant/repair, attach cells, direct cell growth/differentiation, and finally provide an environment for tissue formation. The success of tissue regeneration depends upon various factors of the functional design of the scaffold. This requires the scaffold to have optimum porosity and interconnectivity so as to accelerate the growth of tissues (see Khaled, A. -R. A. and Vafai, K., 2003, The Role of Porous Media in Modeling Flow and Heat Transfer in Biological Tissues, *International Journal of Heat Mass Transfer*, **46**, 4989–5003 and Khanafer, K., and Vafai, K., 2006. The Role of Porous Media in Biomedical Engineering as Related to Magnetic Resonance Imaging and Drug Delivery, *Heat and Mass Transfer*, **42**, 939–953).

Another application of a porous medium is it being used along with advanced fabrication techniques and materials to develop new technologies and computer models to enhance the performance of biodegradable porous drug-delivery devices. Controllably releasing a pharmacological agent to the site of action at a designed rate has numerous advantages over the conventional dosage forms. This interest stems from its importance in reducing dosing frequency, adverse side effects, and achieving enhanced pharmacological activity as well as maintaining constant and prolonged therapeutic effects. The basic formulation of a controlled release of a drug consists of an active agent and a carrier, which is usually made of polymeric materials. Biodegradable polymers have received considerable attention over the last decade, for controlling the drug delivery in the human body without the need to remove them after treatment. The biodegradable polymers can be used as either matrix devices or reservoirs. In matrix systems, the drug is dispersed or dissolved in the polymer and the release rate of the drug decreases as the time advances. While in reservoir, the drug is encapsulated in a biodegradable membrane. As such, the drug is released by diffusion through the membrane at a constant rate. The popularity of this technique has been improved by favorable intrinsic delivery properties of the microspheres (see Khanafer, K., and Vafai, K., 2006 The Role of Porous Media in Biomedical Engineering as Related to Magnetic Resonance Imaging and Drug Delivery, *Heat and Mass Transfer*, **42**, 939–953).

Yet another important area is the development of more effective imaging techniques such as DW-MRI for brain injuries, based on porous media modeling. Diffusion plays a crucial role in brain function: the spaces between cells can be likened to a foam and many substances move within this complicated region; besides delivering glucose and oxygen from the vascular system to brain cells, diffusion also moves informational substances between cells. Diffusion-weighted imaging, which is based on the molecular diffusion coefficient *in vivo*, is sensitive to cerebral ischemia within minutes of the onset of stroke. This technique shows superior capabilities in the early prediction of the brain stroke, compared to the conventional imaging techniques. In the neuroscience context, the ECS constitutes the microenvironment for brain cells. It is a conduit for cellular metabolites, a channel for chemical signaling mediated by volume transmission and a route for drug delivery. Therefore, the extracellular space represents a significant communication channel between neurons and glial cells. From a physical perspective, the extracellular space of the brain resembles that of a porous medium. Thus, theoretical approaches utilizing classical diffusion theory and porous media concepts can be used to measure diffusion properties in very small volumes of highly structured but delicate material (see Khanafer, K., Vafai, K. and A., Kangarlu, 2003, Water Diffusion in Biomedical Systems as Related to Magnetic Resonance Imaging *Magnetic Resonance Imaging Journal*, **21**, 17–31 and Khanafer, K., Vafai, K. and Kangarlu, A., 2003, Computational Modeling of Cerebral Diffusion-Application to Stroke Imaging, *Magnetic Resonance Imaging Journal*, **21**, 651–661).

The last cited focus area deals with an accurate description of the thermal interaction between vasculature and tissues, which is essential for the advancement of medical technology in treating fatal diseases such as tumors and breast cancer. Mathematical models have been used in the analysis of hyperthermia in treating tumors, cryosurgery, laser eye surgery, fetal-placental studies, and other applications. Thermal treatment has been demonstrated to be effective as a cancer therapy in recent years. The success of these types of treatments strongly depends on the knowledge of the heat-transfer processes in blood-perfused tissues. Thermal transport within living organisms and bioheat transfer is an important biological and therapeutic issue, which involves new aspects in thermal therapies, cryobiology, burn injury, disease diagnostics, and thermal comfort analysis. Understanding heat-transfer processes and temperature distributions within biological media are key issues in thermal therapy techniques such as in cancer treatment. The biological media can be treated as a blood saturated tissue represented by a porous matrix. Thermal side effects of various treatments are important issues in bioheat investigations such as bone drilling operations, frictional heating, and ophthalmology. Human eye is one of the most sensitive parts of the body when exposed to a thermal heat flux. Since there is no barrier (such as skin) to protect the eye against the absorption of an external thermal wave, the external flux can readily interact with cornea. The crucial role of blood-tissue in the eye thermal interactions subject to extreme thermal conditions has been established (see Shafahi, M., Vafai, K. Human Eye Response to Thermal Disturbances, to appear in *ASME Journal of Heat Transfer*). A principal issue in medical thermal therapeutic applications is modeling and understanding the heat transport and temperature variation within biological tissues and body organs. Thermal treatment also improves the efficiency of other cancer therapies such as chemotherapy and radiotherapy (see Mahjoob, S., and Vafai, K., 2009, Analytical Characterization of Heat Transport through Biological Media Incorporating Hyperthermia Treatment, *International Journal of Heat and Mass Transfer*, **52**, 1608–1618; Mahjoob, S., and Vafai, K., 2010, Analysis of Bioheat Transport Through a Dual Layer Biological Media, *ASME Journal of Heat Transfer*, **132**, 031101 pp. 1–14; Khanafer, K., and Vafai, K., 2009, Synthesis of Mathematical

Models Representing Bioheat Transport to appear in *Advances in Numerical Heat Transfer*, CRC Press, chapter 1, pp 1–28; Mahjoob, S., and Vafai, K., 2009, Analytical Characterization and Production of an Isothermal Surface for Biological and Electronic Applications, *ASME Journal of Heat Transfer*, **131**, 052604 pp. 1–12 and Mahjoob, S., Vafai, K., and Beer, N. R., 2008, Rapid Microfluidic Thermal Cycler for Polymerase Chain Reaction Nucleic Acid Amplification, *International Journal of Heat and Mass Transfer*, **51**, 2109–2122).

The core objective of this book is to explore innovative approaches and to discover ways to more effectively apply existing technologies to biomedical applications, which will be central to the vision and operation of the discussed research works. This book is targeted at researchers, practicing engineers, clinicians, as well as seasoned beginners in this field. A leading expert in the related subject area presents each topic. An attempt has been made to present the topics in a cohesive, concise, and yet complementary way with a common format. Nomenclature common to various sections was used as much as possible. This book will combine the efforts of world-class scientists and engineers to collaborate and respond to the problems of significance, along with providing porous media researchers with opportunities to pursue an exciting work related to biological systems.

The main goal here is to present the state-of-the-art research advancements related to the applications of porous media in biological systems and biotechnology. This book is arranged into 14 chapters and the subject matters presented in it are arranged as follows.

Chapter 1 takes a look at the general set of bioheat transfer equations for blood flows and its surrounding biological tissue using a volume averaging theory of porous media. This is a rigorous mathematical development, based on the volume averaging theory, so as to arrive at a set of the volume averaged governing equations for bioheat transfer and blood flow. A two-energy equation model for the blood and tissue temperatures is established for the case of isolated blood vessels and the surrounding tissue. Subsequently, the two-energy equation model is extended to the three-energy equation model, to account for the effect of countercurrent heat transfer between closely spaced arteries and veins in the blood circulatory system. In this model, three distinctive energy equations are derived for the arterial blood phase, venous blood phase, and tissue phase with three individual temperatures.

Chapter 2 presents electroporation, which is the use of electrical pulses to increase the cell membrane's permeability and is an important *in vivo* tool for clinical applications. The membrane plays a vital role in the life of a cell, and changing its properties may have far-reaching consequences on the cell itself, the tissue in which it is found, and in fact, on the entire body. A mass transfer model that associates some microscale phenomena on the cellular level with the macroscale conditions at the tissue level is introduced to help in explaining the process of drug uptake by cell.

Chapter 3 deals with hydrodynamics in porous media with applications to tissue engineering. Basic knowledge in biological processes is summarized from the cell's viewpoint and tissue growth with emphasis on the interactions between mechanical stimuli, nutrient transport, cell, and biomaterials. Moreover, different theoretical approaches that enable the modeling of the functional development of tissue-engineered material is presented in detail.

Chapter 4 focuses on interrelationship between the structure of microbial biofilms, with emphasis on porosity and their resistance to common modalities of antibiotic treatments. This resistance is currently a major problem in medicine, culminating in growing numbers of hospitalizations, amputations, sepsis, and death.

Chapter 5 centers on the influence of biofilms on porous media hydrodynamics. Microbial biofilms are formed in natural and engineered systems and can significantly affect the hydrodynamic properties of porous media. Biofilm growth influences porosity, permeability, dispersion, diffusion, and mass transport of reactive and nonreactive solutes. Understanding and controlling biofilm formation in porous media will maximize the potential benefit and minimize the detrimental effects of porous media biofilms.

Chapter 6 explores the application of porous media theory to the modeling of flow changes in cerebral aneurysms treated by endovascular coils. Conventional fluid mechanics modeling is not suitable for this setting because of the difficulty in describing the geometry of the random-shaped endovascular coil when solving the Navier–Stokes equations. Even if the geometry of the coils could be determined, the density and total number of nodal points required to capture the characteristics of the flow would represent a major limitation.

Chapter 7 deals with recent advances in Lagrangian particles methods and their applications for micro(pore)scale modeling of multiphase flow, biomass growth, and mineral precipitation in porous media. Contrary to Darcy-scale models that require phenomenological description of interactions between multiple phases, the pore-scale models are based on fundamental conservation laws and are able to provide an accurate description of complex nonlinear processes involved in biogeochemical transformations. A hybrid model for a multiscale representation of biochemical processes in porous media is also described. With the hybrid model a range of applications of (otherwise very computationally expensive) microscale models can be significantly extended.

Chapter 8 discusses passive mass transport processes in cellular membranes and their biophysical implications. Physical mechanisms of water and solute transport across biological and artificial membranes have been studied since the 1930s and yet they are not completely understood. This is partially due to a limited understanding of the structure of a membrane, which differs significantly from any bulk phase.

Chapter 9 is concerned with modeling and treatment of mass transport through biological tissues, especially that of the skin. The treatment of the skin as a porous media is addressed. A review of experimental findings and observations regarding the electrically induced creation of local transport regions (LTRs) is presented. Moreover, a description of various methods used to describe electroporation of the skin (both empirical and mechanistic) is then provided.

Chapter 10 highlights the biological applications of porous media in marine systems. The focus of this study is on demonstrating the importance and applicability of porous media theories in the emerging field of marine microbiology. Owing to the complex geometries that appear in such applications, a Lattice Boltzmann method approach has been developed and adopted to different applications. The examples include but are not limited to sinking marine aggregates, bioirrigation of macrozoobenthos larvae, oscillating flows near seabed topographies, tortuosity of marine sediments, and devices for generating uniform bottom shear stresses.

Chapter 11 deals with the transport of large biological molecules in deforming tissues and is an example of porous media theory applied to understand tissue homeostasis and repair. Specifically, the movement of growth factors and the synthesis and transport of extracellular matrix molecules through a cyclically loaded articular cartilage is described by combining reactive transport theory with poroelasticity within the framework of porous media theory. Through a series of models of increasing complexity, the interplay between cyclic deformation, interstitial fluid flow, reaction kinetics with cell receptors and a range of matrix molecules, and the mechanical- and chemical-induced biosynthesis is investigated. The outcome provides a framework for understanding the mechanical and chemical environment of a cartilage cell and some of the factors leading to a healthy cartilage or the optimal growth of new cartilage in tissue-engineered constructs.

Chapter 12 is devoted to a review of applications of magnetic stabilized beds as applied to the areas of biotechnology and biomedicine, with emphasis on its current status, historical, and future developments. This includes a description of the main principles and of the background theory.

Chapter 13 summarizes the potential *in situ* characterization techniques for studying porous media and conductive membranes, especially for investigations in solutions. Some of the techniques include spectroscopic imaging ellipsometry (SIE), quartz crystal microbalance (QCM), X-ray diffraction and reflection (XRD and XRR), and laser scanning confocal microscopy (LSCM), in combination with electrochemical techniques. These techniques are either surface or bulk sensitive, or both; and they can provide either spatial or temporal information simultaneously or separately to reveal the dynamic nature of the processes involved in the biofuel cell or the electrode and membrane. Combining the information obtained from these *in situ* techniques can intelligently encompass a wide spectrum of understanding of the cell behavior.

Chapter 14 is concentrated on the development of bioconvection patterns generated by populations of gravitactic microorganisms in porous media. A continuum model consisting of a coupled system of fluid flow and diffusion–convection equations describes the interactions between the microorganisms and the surrounding fluid.

Whenever applicable in each of these chapters, pertinent aspects of experimental work or numerical techniques are discussed. The experts in the field have reviewed each chapter of this handbook. Overall, there were many reviewers involved. As such, the authors and I are very thankful for the valuable and constructive comments received and, particularly, we would like to thank reviewers who performed multiple reviews.

Kambiz Vafai

Editor

Kambiz Vafai is professor of mechanical engineering at University of California, Riverside (UCR), where he started as the presidential chair in the department of Mechanical Engineering. He joined UCR from The Ohio State University, where he received outstanding research awards as assistant, associate, and full professor. Author of over 240 archival journal articles, book chapters, books (Ed.), and symposium volumes (Ed.), he has given numerous national and international invited lectures, keynote addresses, and presentations. He is a Fellow of American Association for Advancement of Science, American Society of Mechanical Engineers, World Innovation Foundation, and an Associate Fellow of the American Institute of Aeronautics and Astronautics. He is the editor-in-chief of the *Journal of Porous Media and Special Topics and Reviews in Porous Media*—an international journal—and serves on the editorial advisory board of the *International Journal of Heat and Mass Transfer*, *International Communications in Heat and Mass Transfer*, *Numerical Heat Transfer*, *International Journal of Numerical Methods for Heat and Fluid Flow*, *International Journal of Heat and Fluid Flow*, and *Experimental Heat Transfer*. He is the editor of the first and the second editions of the *Handbook of Porous Media*, which became best sellers and has been the Director/Chair of the First, Second, and Third International Conferences on Porous Media, all sponsored by ECI and NSF. He has supervised fifty doctoral and masters students, and has directed over twenty post docs and visiting scholars. He has worked on a multitude of fundamental research investigations, a number of which have addressed some pertinent concepts presented for the first time. He is among the very few engineering scientists within the prestigious ISI highly cited category with over 4,200 ISI citations covering a wide spectrum of disciplines and journals and an h-index of 33. He has carried out various sponsored research projects through companies, governmental funding agencies, and national labs. He has also consulted for various companies and national labs, and has been granted six U.S. patents. He was the recipient of the ASME Classic Paper Award in 1999 and has received the 2006 ASME Heat Transfer Memorial Award, which are amongst the most selective awards in the field of heat transfer. Dr. Vafai received his B.S. degree from the University of Minnesota (with the highest honors), Minneapolis, and the M.S. and Ph.D. degrees from the University of California, Berkeley.

Contributors

Audelino Álvaro
Department of Chemical Engineering and Textile
Faculty of Sciences, Chemistry
Universidad de Salamanca
Salamanca, Spain

Paulo A. Augusto
Department of Chemical Engineering and Textile
Faculty of Sciences, Chemistry
Universidad de Salamanca
Salamanca, Spain
and
Department of Chemical Engineering
Faculty of Engineering
Universidade do Porto
Porto, Portugal

Domingos Barbosa
Laboratory of Environmental Process Engineering and Energy (LEPAE)
Department of Chemical Engineering
Faculty of Engineering
Universidade do Porto
Porto, Portugal

S. M. Becker
Institute for Thermo-Fluid Dynamics
Hamburg University of Technology
Hamburg, Germany

H. Ben-Yoav
Department of Physical Electronics
School of Electrical Engineering
Faculty of Engineering
Tel Aviv University
Tel Aviv, Israel

Ramon Berguer
Vascular Mechanics Laboratory
Department of Biomedical Engineering and Section of Vascular Surgery
University of Michigan
Ann Arbor, MI

Teresa Castelo-Grande
Laboratory of Environmental Process Engineering and Energy (LEPAE)
Department of Chemical Engineering
Faculty of Engineering
Universidade do Porto
Porto, Portugal
and
Faculty of Natural Sciences, Engineering Technology (FCNET)
Universidade Lusófona do Porto
Porto, Portugal
and
Department of Chemical Engineering and Textile
Faculty of Sciences, Chemistry
Universidad de Salamanca
Salamanca, Spain

N. Cohen-Hadar
Department of Molecular Microbiology and Biotechnology
Faculty of Life Sciences
Tel Aviv University
Tel Aviv, Israel

Alfred B. Cunningham
Department of Civil Engineering
Center for Biofilm Engineering
Montana State University
Bozeman, MT

B. David
Laboratory of Mechanics of Soils, Structures and Materials (MSSMat CNRS-8579)
École Centrale Paris
Châtenay-Malabry Cedex, France

Angel M. Estevéz
Department of Chemical Engineering
Faculty of Engineering
Universidade do Porto
Porto, Portugal

Amihay Freeman
Department of Molecular Microbiology and Biotechnology
Faculty of Life Sciences
Tel Aviv University
Tel Aviv, Israel

Bruce S. Gardiner
Faculty of Engineering, Computing and Mathematics
The University of Western Australia
Western Australia, Australia

Robin Gerlach
Department of Chemical and Biological Engineering
Center for Biofilm Engineering
Montana State University
Bozeman, MT

Yair Granot
Graduate Group in Biophysics
Department of Mechanical Engineering
University of California
Berkeley, CA

Alan J. Grodzinsky
Center for Biomedical Engineering
Department of Electrical Engineering and Computer Science
Department of Mechanical Engineering
Massachusetts Institute of Technology
Cambridge, MA

Frederic Guichard
Department of Biology
McGill University
Montreal, Canada

The Hung Nguyen
Department of Mechanical Engineering
Polytechnic
University of Montreal
Montreal, Canada

Khodayar Javadi
Max Planck Institute for Marine Microbiology
Bremen, Germany

Armin Kargol
Department of Physics
Loyola University
New Orleans, LA

Marian Kargol
Department of Physics
The Jan Kochanowski University of Humanities and Sciences
Kielce, Poland

Arzhang Khalili
Max Planck Institute for Marine Microbiology
Bremen, Germany

Earth and Space Sciences
Jacobs University
Bremen, Germany

Khalil M. Khanafer
Vascular Mechanics Laboratory
Department of Biomedical Engineering and Section of Vascular Surgery
University of Michigan
Ann Arbor, MI

Kolja Kindler
Max Planck Institute for Marine Microbiology
Bremen, Germany

Zbigniew Koza
Institute of Theoretical Physics
University of Wrocław
Wrocław, Poland

Fujio Kuwahara
Department of Mechanical Engineering
Shizuoka University
Johoku, Hamamatsu, Japan

A. V. Kuznetsov
Department of Mechanical and Aerospace Engineering
North Carolina State University
Raleigh, NC

T. Lemaire
Laboratory of Multiscale Modelling and Simulation - Biomechanics,
Faculty of Sciences & Technology, (MSME CNRS-8208)
University Paris-Est
Créteil, France

Bo Liu
Max Planck Institute for Marine Microbiology
Bremen, Germany

Wei Liu
School of Energy and Power Engineering
Huazhong University of Science and Technology
Wuhan, P.R. China

Maciej Matyka
Max Planck Institute for Marine Microbiology
Bremen, Germany

Institute for Theoretical Physics
University of Wrocław
Wrocław, Poland

Paul Meakin
Idaho National Laboratory
Center for Advanced Modeling and Simulation
Idaho Falls, ID
and
Physics of Geological Processes
University of Oslo
Oslo, Norway
and
Multiphase Flow Assurance Innovation Center
Institute for Energy Technology
Kjeller, Norway

Mohammad R. Morad
Max Planck Institute for Marine Microbiology
Bremen, Germany

Akira Nakayama
Department of Mechanical Engineering
Shizuoka University
Johoku, Hamamatsu, Japan
and
School of Civil Engineering and Architecture
Wuhan Polytechnic University
Wuhan, P.R. China

Tri Nguyen-Quang
Department of Biology
McGill University
Montreal, Canada

C. Oddou
Laboratory of Multiscale Modelling and Simulation - Biomechanics
Faculty of Sciences & Technology, (MSME CNRS - 8208)
University Paris-Est
Créteil, France

J. Pierre
Laboratory of Osteo-Articular Biomechanics and Biomaterials (B2OA CNRS-7052)
University Paris 12
Faculty of Sciences & Technology
Créteil Cedex, France

Peter Pivonka
Faculty of Engineering, Computing and Mathematics
The University of Western Australia
Western Australia, Australia

Jesus M[a]. Rodríguez
Department of Chemical Engineering and Textile
Faculty of Sciences, Chemistry
Universidad de Salamanca
Salamanca, Spain

Boris Rubinsky
Graduate Group in Biophysics
Department of Mechanical Engineering
University of California
Berkeley, CA

Roman Stocker
Department of Civil and Environmental Engineering
Massachusetts Institute of Technology
Cambridge, MA

Alexandre M. Tartakovsky
Pacific Northwest National Laboratory
Computational Mathematics Group
Richland, WA

David W. Smith
Faculty of Engineering, Computing and Mathematics
The University of Western Australia
Western Australia, Australia

Zhijie Xu
Idaho National Laboratory
Energy Resource Recovery & Management
Idaho Falls, ID

Bor Yann Liaw
Hawaii Natural Energy Institute, SOEST
University of Hawaii at Manoa
Honolulu, HI

Lihai Zhang
Department of Civil and Environmental Engineering
The University of Melbourne
Melbourne, Australia

1

A General Set of Bioheat Transfer Equations Based on the Volume Averaging Theory

Akira Nakayama

Department of Mechanical Engineering, Shizuoka University,
Johoku, Hamamatsu, Japan
School of Civil Engineering and Architecture,
Wuhan Polytechnic University, Wuhan, P.R. China

Fujio Kuwahara

Department of Mechanical Engineering, Shizuoka University,
Johoku, Hamamatsu, Japan

Wei Liu

School of Energy and Power Engineering, Huazhong University
of Science and Technology, Wuhan, P.R. China

CONTENTS

1.1 Introduction

There has been considerable interest in developing sound and accurate thermal models that describe heat transfer within a living tissue with blood perfusion. Since the landmark paper by Pennes (1948), a number of bioheat transfer equations for living tissue have been proposed to remedy possible shortcomings in his equation. Although Pennes' model is often adequate for roughly describing the effect of blood flow on the tissue temperature, there exist some serious shortcomings in his model due to its inherent simplicity, as pointed out by Wulff (1974), namely, assuming uniform perfusion rate without accounting for blood flow direction, neglecting the important anatomical features of the circulatory network system such as countercurrent arrangement of the system, and choosing only the venous blood stream as the fluid stream equilibrated with the tissue.

To overcome these shortcomings, a considerable number of modifications have been proposed by various researchers. Wulff (1974) and Klinger (1978) considered the local blood mass flux to account the blood flow direction, while Chen and Holmes (1980) examined the effect of thermal equilibration length on the blood temperature and added the dispersion and microcirculatory perfusion terms to the Klinger equation.

All foregoing papers concerned mainly with the cases of isolated vessels and the surrounding tissue. The effect of countercurrent heat transfer between closely spaced arteries and veins in the tissue must be taken into full consideration when the anatomical configuration of the main supply artery and vein in the limbs is treated. Following the experimental study conducted by Bazett and his colleagues (1948a,b), Scholander and Krog (1957), and Mitchell and Myers (1968) investigated such an effect and successfully demonstrated that the countercurrent heat exchange reduces heat loss from the extremity to the surroundings, which could be quite significant because of a large surface to volume ratio. Keller and Seiler (1971) established a bioheat transfer model equation to include the countercurrent heat transfer, using a one-dimensional configuration for the subcutaneous tissue region with arteries, veins, and capillaries. Weinbaum and Jiji (1979) proposed a new model, which is based on some anatomical understanding, considering the countercurrent arteriovenous vessels. As pointed out by Roetzel and Xuan (1998), the model may be useful in describing a temperature field in a single organ, but would not be convenient to apply to the whole thermoregulation system. Excellent reviews on these bioheat transfer equations may be found in Chato (1980) and Charny (1992).

Khaled and Vafai (2003) and Khanafer and Vafai (2006) stress that the theory of porous media is most appropriate for treating heat transfer in biological tissues since it contains fewer assumptions as compared to different bioheat transfer equations. Roetzel and Xuan (1998) and Xuan and Roetzel (1997) exploited the volume averaging theory (VAT) previously established for the study of porous media (e.g., Cheng 1978, Nakayama 1995), to formulate a two-energy equation model accounting for the thermal nonequilibrium between the blood and peripheral tissue. In their model, the perfusion term is replaced by the interfacial convective heat transfer term. This point should be examined since the interfacial convective heat transfer is different from perfusion heat transfer. Naturally, the former takes place even in the absence of the latter.

In this chapter, we present a rigorous mathematical development based on VAT so as to achieve a complete set of the volume averaged governing equations for bioheat transfer and blood flow. Most shortcomings in existing models will be overcome. We start with the case of isolated blood vessels and the surrounding tissue, to establish a two-energy equation model for the blood and tissue temperatures. We shall identify the terms describing the blood perfusion and dispersion in the resulting equation and revisit the Pennes model, the Wulff model, and their modifications.

Subsequently, the two-energy equation model is extended to the three-energy equation model, so as to account for the effect of countercurrent heat transfer between closely spaced arteries and veins in the blood circulatory system. In this model, three distinctive energy equations are derived for the arterial blood phase, venous blood phase, and tissue phase with three individual temperatures. Capillaries providing a continuous connection between the countercurrent terminal arteries and veins are modeled introducing the perfusion bleed-off rate. It will be shown that the resulting model, under appropriate conditions, naturally reduces to those introduced by Chato (1980), Bejan (1979), Weinbaum and Jiji (1985), and others for countercurrent heat transfer for the case of closely aligned pairs of vessels. A useful expression for the longitudinal effective thermal conductivity for the tissue can be obtained without dropping the perfusion source terms. The expression turns out to be quite similar to Bejan's and Wienbaum and Jiji's expressions. Furthermore, the effect of spatial distribution of perfusion bleed-off rate on total countercurrent heat transfer is discussed in depth exploiting the present bioheat transfer model.

As for an application of a bioheat equation, the freezing process within a tumor during cryoablation therapy is investigated both analytically and numerically. The freezing front in a tumor during percutaneous cryoablation can be traced exploiting a bioheat equation. It will be shown that there exists a limiting size of the tumor that one single cryoprobe can freeze at the maximum. The freezing front moves radially outward from the cryoprobe and reaches the end, where the heat from the surrounding tissue to the frozen tissue balances with the heat being absorbed by the cryoprobe. An excellent agreement between the analytical and numerical results is achieved for the time required to freeze the tumor using the cryoprobe of one single needle. The resulting analytical expression for estimating the limiting radius provides useful information for cryotherapy treatment plans.

1.2 Volume Averaging Procedure

In an anatomical view, three compartments are identified in the biological tissues, namely, blood vessels, cells and interstitium, as illustrated in Figure 1.1. The interstitial space can be further divided into the extracellular matrix and the interstitial fluid. However, for sake of simplicity, we divide the biological tissue into two distinctive regions, namely, the vascular region and the extravascular region (i.e., cells and the interstitium) and treat the whole anatomical structure as a fluid-saturated porous medium, through which the blood infiltrates. The extravascular region is regarded as a solid matrix (although the extravascular fluid is present), and will be simply referred to as the "tissue" region to differentiate it from the "blood" region.

Thus, we shall try to apply the principle of heat and fluid flow in a fluid-saturated porous medium to derive a set of the volume averaged governing equations for the bioheat transfer and blood flow. For the volume averaging (smoothing process) to be meaningful, we consider a control volume V in a fluid-saturated porous medium, as shown in Figure 1.2, whose length scale $V^{1/3}$ is much smaller than the macroscopic characteristic length $V_c^{1/3}$, but, at the same time, much greater than the microscopic (anatomical structure) characteristic length (see e.g., Nakayama [1995]). Under this condition, the volume average of a certain variable ϕ is defined as

$$\langle \phi \rangle \equiv \frac{1}{V} \int_{V_f} \phi dV \tag{1.1}$$

Another average, namely, intrinsic average, is given by

$$\langle \phi \rangle^f \equiv \frac{1}{V_f} \int_{V_f} \phi dV \tag{1.2}$$

where V_f is the volume space that the fluid (blood) occupies. Obviously, two averages are related as

$$\langle \phi \rangle = \varepsilon \langle \phi \rangle^f \tag{1.3}$$

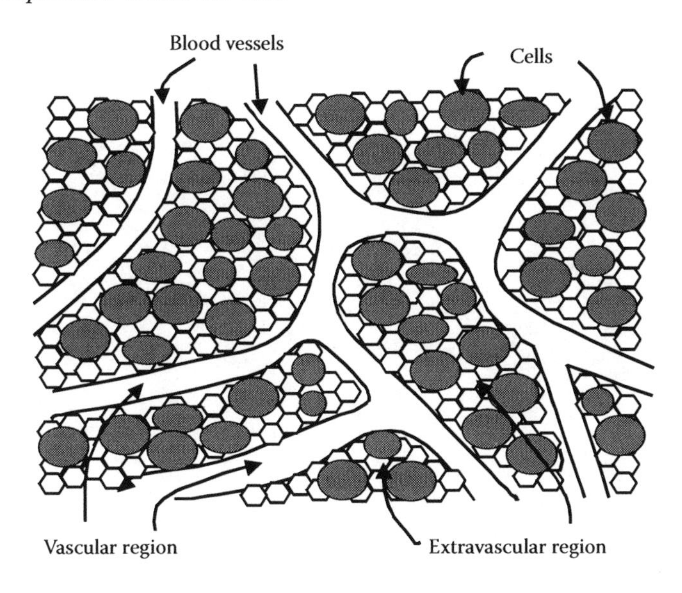

FIGURE 1.1
Schematic view of biological tissue.

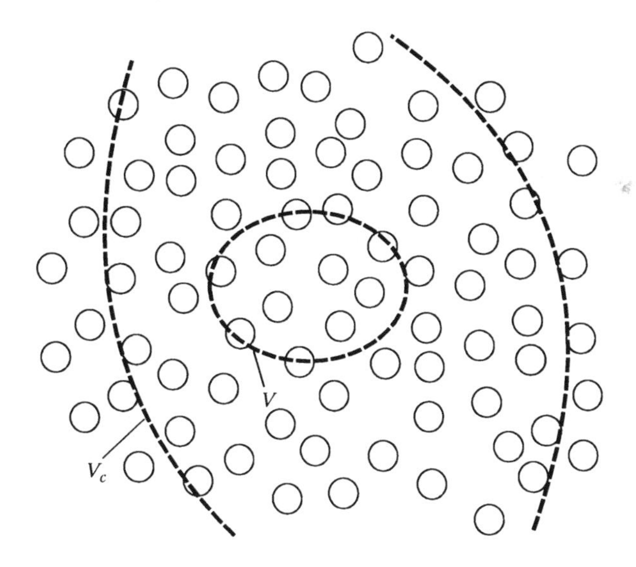

FIGURE 1.2
Control volume in a porous medium.

where $\varepsilon \equiv V_f/V$ is the local porosity, namely, the volume fraction of the vascular space, which is generally less than 0.1. Naturally, anatomical data are required to find the porosity. Following Cheng (1978), Nakayama (1995),

Quintard and Whitaker (1993), and many others, we decompose a variable into its intrinsic average and the spatial deviation from it:

$$\phi = \langle \phi \rangle^f + \tilde{\phi} \tag{1.4}$$

We shall exploit the following spatial average relationships:

$$\langle \phi_1 \phi_2 \rangle^f = \langle \phi_1 \rangle^f \langle \phi_2 \rangle^f + \langle \tilde{\phi}_1 \tilde{\phi}_2 \rangle^f \tag{1.5}$$

$$\left\langle \frac{\partial \phi}{\partial x_i} \right\rangle = \frac{\partial \langle \phi \rangle}{\partial x_i} + \frac{1}{V} \int_{A_{\mathrm{int}}} \phi n_i dA \quad \text{or} \quad \left\langle \frac{\partial \phi}{\partial x_i} \right\rangle^f = \frac{1}{\varepsilon} \frac{\partial \varepsilon \langle \phi \rangle^f}{\partial x_i} + \frac{1}{V_f} \int_{A_{\mathrm{int}}} \phi n_i dA \tag{1.6a,b}$$

and

$$\left\langle \frac{\partial \phi}{\partial t} \right\rangle = \frac{\partial \langle \phi \rangle}{\partial t} \tag{1.7}$$

where A_{int} is the local interface between the blood and solid matrix, while n_i is the unit vector pointing outward from the fluid side to the solid side. The similarity between the volume averaging and the Reynolds averaging used in the study of turbulence is quite obvious. However, it should be noted that the present volume averaging procedure is somewhat more complex than the Reynolds averaging procedure, since it involves surface integrals, as clearly seen from equation (1.6). It should also be noted that biological tissues in reality are highly compliant. To include the compliance of the tissues, the foregoing spatial averaging relationships must be modified accordingly to account for the deformation of the elementary control volume. In the present study, we neglect such effects for simplicity.

We subdivide the anatomic structure into the blood phase (fluid phase) and the tissue and other solid tissue phase (solid matrix phase), in which metabolic reactions may take place. We shall consider the microscopic governing equations, namely, the continuity equation, the Navier–Stokes equation, and the energy equation for the blood phase and the heat conduction equation for the solid matrix phase.

For the blood phase:

$$\frac{\partial u_j}{\partial x_j} = 0 \tag{1.8}$$

$$\frac{\partial u_i}{\partial t} + \frac{\partial}{\partial x_j} u_j u_i = -\frac{1}{\rho} \frac{\partial p}{\partial x_i} + \frac{\partial}{\partial x_j} \nu_f \left(\frac{\partial u_i}{\partial x_j} + \frac{\partial u_j}{\partial x_i} \right) \tag{1.9}$$

$$\rho_f c_{p_f} \left(\frac{\partial T}{\partial t} + \frac{\partial}{\partial x_j} u_j T \right) = \frac{\partial}{\partial x_j} \left(k_f \frac{\partial T}{\partial x_j} \right) \tag{1.10}$$

For the solid matrix phase:

$$\rho_s c_s \frac{\partial T}{\partial t} = \frac{\partial}{\partial x_j} \left(k_s \frac{\partial T}{\partial x_j} \right) + S_m \tag{1.11}$$

where the subscripts f and s stand for the fluid and solid, respectively. It is assumed that the fluid (blood) is incompressible and Newtonian, and all properties are constant.

1.3 Governing Equation for Blood Flow

Let us integrate the continuity equation (1.8) over a local control volume using formula (1.6b) as

$$\frac{\partial \varepsilon \langle u_j \rangle^f}{\partial x_j} + \frac{1}{V} \int_{A_{\text{int}}} u_j n_j dA = 0 \tag{1.12}$$

where A_{int} is the local interface between the blood and solid matrix within the control volume V, while n_j is the unit vector pointing outward from the fluid side to solid side. For sake of simplicity, the porosity ε is assumed to vary moderately within a porous medium.

The second term describes the volume rate of the fluid bleeding off to the solid matrix through the interfacial vascular wall, as illustrated in Figure 1.3. In most microcirculatory systems of the body, there is a net filtration of fluid from the intravascular to the extravascular compartment, such that capillary fluid filtration exceeds reabsorption. However, this would not cause fluid to accumulate within the interstitium since the lymphatic system removes excess fluid from the interstitium and returns it back to the intravascular compartment, as indicated in the figure. Thus, the second term describing the net filtration is negligibly small, such that equation (1.12) reduces to

$$\frac{\partial \langle u_j \rangle}{\partial x_j} = 0 \tag{1.13}$$

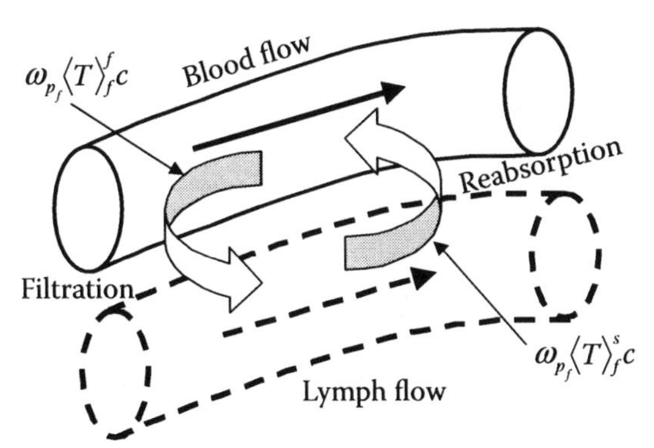

FIGURE 1.3
Capillary blood flow and extravascular flow.

Accordingly, the Navier–Stokes equation (1.9) may be integrated to give

$$\frac{\partial \langle u_i \rangle^f}{\partial t} + \frac{\partial}{\partial x_j} \langle u_j \rangle^f \langle u_i \rangle^f = -\frac{1}{\rho_f} \frac{\partial \langle p \rangle^f}{\partial x_i} + \frac{\partial}{\partial x_j} \nu_f \left(\frac{\partial \langle u_i \rangle}{\partial x_j} + \frac{\partial \langle u_j \rangle}{\partial x_i} \right) + \frac{1}{V_f} \int_{A_{\text{int}}} \left(-\frac{p}{\rho} + \nu_f \left(\frac{\partial u_i}{\partial x_j} + \frac{\partial u_j}{\partial x_i} \right) \right) n_j dA - \frac{\partial}{\partial x_j} \langle u_j u_i \rangle^f \tag{1.14}$$

To close the foregoing macroscopic momentum equations (1.14), the terms associated with the surface integral are modeled according to Vafai and Tien (1981) as

$$\frac{1}{V_f} \int_{A_{\text{int}}} \left(-\frac{p}{\rho_f} + \nu_f \left(\frac{\partial u_i}{\partial x_j} + \frac{\partial u_j}{\partial x_i} \right) \right) n_j dA - \frac{\partial}{\partial x_j} \langle u_j u_i \rangle^f = -\frac{\nu_f}{K} \varepsilon \langle u_i \rangle^f - b\varepsilon^2 \left(\langle u_k \rangle^f \langle u_k \rangle^f \right)^{1/2} \langle u_i \rangle^f \tag{1.15}$$

such that

$$\frac{\partial \langle u_i \rangle^f}{\partial t} + \frac{\partial}{\partial x_j} \langle u_j \rangle^f \langle u_i \rangle^f = -\frac{1}{\rho} \frac{\partial \langle p \rangle^f}{\partial x_i} + \frac{\partial}{\partial x_j} \nu_f \left(\frac{\partial \langle u_i \rangle^f}{\partial x_j} + \frac{\partial \langle u_j \rangle^f}{\partial x_i} \right) - \frac{\nu_f}{K_{ij}} \varepsilon \langle u_j \rangle^f - b_{ij} \varepsilon^2 \left(\langle u_k \rangle^f \langle u_k \rangle^f \right)^{1/2} \langle u_j \rangle^f \tag{1.16}$$

where K_{ij} and b_{ij} are the permeability and Forchheimer tensors, respectively. These tensors, which depend on the anatomical structure, can be determined following the procedure established for anisotropic porous structure (Nakayama et al. 2004), as sufficient information on the anatomical structure and properties is provided. For the vessels of sufficiently small diameter, the foregoing equation reduces to Darcy's law:

$$-\frac{1}{\rho} \frac{\partial \langle p \rangle^f}{\partial x_i} - \frac{\nu_f}{K_{ij}} \langle u_j \rangle = 0 \tag{1.17}$$

where $\langle u_j \rangle = \varepsilon \langle u_j \rangle^f$ is the Darcian velocity (i.e., apparent velocity). We may use the Darcy law for most tissue regions except for the regions where large arteries or veins are located.

1.4 Two-Energy Equation Model for Blood Flow and Tissue

1.4.1 Related Work

Pennes (1948) carried out temperature measurements in the limb and found that the maximum muscle temperature is located very close to the axis of the limb. Using his experimental data, Pennes proposed what is known today as the Pennes bioheat equation. In his model, he assumed that the net heat transferred from the blood to tissue is proportional to the temperature difference between the arterial blood entering the tissue and the venous blood leaving from the tissue, and introduced the Pennes perfusion heat source.

The Pennes bioheat equation has been used for various bioheat transfer problems and found satisfactory for roughly describing the effect of blood flow on the tissue temperature. However, a number of researchers including Wulff (1974) and Klinger (1978) pointed out serious shortcomings in his model due to its inherent simplicity, namely, assuming uniform perfusion rate without accounting for blood flow direction, neglecting the important anatomical features of the circulatory network system such as countercurrent arrangement of the system, and choosing only the venous blood stream as the fluid stream equilibrated with the tissue.

Possible modifications have been proposed by some researchers, so as to remedy these shortcomings. Wulff (1974) and Klinger (1978) considered the local blood mass flux to account the blood flow direction, whereas Chen and Holmes (1980) examined the effect of thermal equilibration length on the blood temperature and added the dispersion and microcirculatory perfusion terms to the Klinger equation.

In this section, we exploit VAT described in the foregoing sections to obtain a complete set of the volume averaged governing equations for bioheat transfer and blood flow. Most shortcomings in existing models can be overcome.

1.4.2 Two-Energy Equation Model Based on VAT

Before actually integrating the energy equation (1.10), it may be quite instructive to focus our attention on the volume average of the convection term. Using equations (1.5) and (1.6), it is straightforward to show

$$\varepsilon\left\langle \frac{\partial}{\partial x_j}\rho_f c_{p_f} u_j T\right\rangle^f = \frac{\partial}{\partial x_j}\rho_f c_{p_f}\langle u_j\rangle\langle T\rangle^f + \frac{\partial}{\partial x_j}\varepsilon\rho_f c_{p_f}\left\langle \tilde{u}_j\tilde{T}\right\rangle^f + \frac{1}{V}\int_{A_{\mathrm{int}}}\rho_f c_{p_f} u_j T n_j dA \tag{1.18}$$

where the first term on the right-hand side describes the macroscopic convection, while the second term on the right-hand side takes account of the thermal dispersion (Nakayama et al. 2004). It is the last term on the right-hand side that corresponds to the blood "perfusion" heat source. Thus, the blood perfusion heat source term is identified as an extra surface integral term resulting from changing the sequence of integration and derivation, as we obtain the macroscopic energy equation by integrating the microscopic convection term over a local control volume.

Having expanded the integrated convection term, we may readily transform both the energy equation (1.10) for the blood flow and the conduction equation (1.11) for the solid matrix into the corresponding volume averaged equations as follows:

For the blood phase:

$$\varepsilon\rho_f c_{p_f}\frac{\partial\langle T\rangle^f}{\partial t}+\rho_f c_{p_f}\frac{\partial}{\partial x_j}\langle u_j\rangle\langle T\rangle^f$$
$$=\frac{\partial}{\partial x_j}\left(\varepsilon k_f\frac{\partial\langle T\rangle^f}{\partial x_j}+\frac{k_f}{V}\int_{A_{\mathrm{int}}}Tn_j dA-\varepsilon\rho_f c_{p_f}\left\langle\tilde{u}_j\tilde{T}\right\rangle^f\right)$$
$$+\frac{1}{V}\int_{A_{\mathrm{int}}}k_f\frac{\partial T}{\partial x_j}n_j dA-\frac{1}{V}\int_{A_{\mathrm{int}}}\rho_f c_{p_f}u_j Tn_j dA \qquad (1.19)$$

For the solid matrix phase:

$$(1-\varepsilon)\rho_s c_s\frac{\partial\langle T\rangle^s}{\partial t}=\frac{\partial}{\partial x_j}\left((1-\varepsilon)k_s\frac{\partial\langle T\rangle^s}{\partial x_j}-\frac{k_s}{V}\int_{A_{\mathrm{int}}}Tn_j dA\right)$$
$$-\frac{1}{V}\int_{A_{\mathrm{int}}}k_f\frac{\partial T}{\partial x_j}n_j dA+\frac{1}{V}\int_{A_{\mathrm{int}}}\rho_f c_{p_f}u_j Tn_j dA+(1-\varepsilon)S_m \qquad (1.20)$$

where $\langle T\rangle^s$ is the intrinsic average of the solid matrix temperature. Note that the dispersion heat flux $\rho_f c_{pf}\langle\tilde{u}_j\tilde{T}\rangle=\varepsilon\rho_f c_{pf}\langle\tilde{u}_j\tilde{T}\rangle^f$ appears in the volume averaged energy equation (1.19) for the blood phase, which may well be modeled under the gradient diffusion hypothesis:

$$-\varepsilon\rho_f c_{pf}\langle\tilde{u}_j\tilde{T}\rangle^f=\varepsilon k_{dis_{kj}}\frac{\partial\langle T\rangle^f}{\partial x_k} \qquad (1.21)$$

A number of expressions have been proposed for the thermal dispersion thermal conductivity $k_{dis_{kj}}$. Nakayama et al. (2006) obtained a transport equation for the dispersion heat flux vector, which naturally reduces to the foregoing gradient diffusion form. For a bundle of vessels of radius R, they obtained the following expression for the predominant axial component of $k_{dis_{kj}}$:

$$k_{dis_{xx}}=\frac{1}{48}\left(\frac{\rho_f c_{pf}\langle u\rangle^f R}{k_f}\right)^2 k_f \quad \frac{\rho_f c_{pf}\langle u\rangle^f R}{k_f}<1 \quad \text{(capillary blood vessels)} \qquad (1.22a)$$

$$k_{dis_{xx}}=2.55\left(\frac{\rho_f c_{pf}\langle u\rangle^f R}{k_f}\right)^{7/8} Pr^{1/8}k_f \quad \frac{\rho_f c_{pf}\langle u\rangle^f R}{k_f}>1 \quad \text{(large arteries and veins)} \qquad (1.22b)$$

To close the foregoing macroscopic energy equations (1.19) and (1.20), the terms associated with the surface integral, describing the interfacial heat transfer and perfusion between the fluid and solid, must be modeled. For the interfacial heat transfer, Newton's cooling law may be adopted as

$$\frac{1}{V}\int_{A_{\text{int}}} k_f \frac{\partial T}{\partial x_j} n_j dA = a_f h_f \left(\langle T\rangle^s - \langle T\rangle^f\right) \tag{1.23}$$

where a_f and h_f are the specific surface area and interfacial heat transfer coefficient, respectively. For the bundle of vascular tubes of radius R, we have $a_f = 2\varepsilon/R$ and $h_f = Nu(k_f/2R)$, such that $a_f h_f = Nu\left(\varepsilon k_f/R^2\right)$, where Nu is the Nusselt number based on the local diameter of the vascular tube. If the local porosity ε and specific surface area a_f are provided for the complex tissue-vascular structure, we may estimate the interfacial heat transfer coefficient using $h_f = Nu(k_f a_f/4\varepsilon)$. Roetzel and Xuan (1998) set $Nu = 4.93$ for both arterial and venous blood vessels. We may appeal to a numerical experiment proposed by Nakayama et al. (2002) for complex porous structures.

As for modeling the blood perfusion term, we may refer back to Figure 1.3 and note that the transcapillary fluid exchange takes place between the blood and the surrounding tissue. However, the fluid lost from the vascular space will be compensated by the flow of extravascular fluids and lymph from the tissue to vascular space. It is quite reasonable to assume that extravascular fluids and all lymph in the tissue space have the same temperature as the tissue itself. Thus, we assume that the transcapillary fluid exchange takes place at the rate of ω (m^3/sm^3) and model the blood perfusion term as

$$\frac{1}{V}\int_{A_{\text{int}}} \rho_f c_{p_f} u_j T n_j dA = \rho_f c_{p_f} \omega \left(\langle T\rangle^f - \langle T\rangle^s\right) \tag{1.24}$$

Note that the perfusion rate ω, unlike that of Pennes, varies locally, and we assume that its local value is provided everywhere. Pennes found that his model fits the experimental data for $\omega = 2 \times 10^{-4}$ to 5×10^{-4} (m^3/sm^3). The perfusion rate varies spatially. In general, it is not an easy task to do *in vivo* measurements for living tissues. Charny (1992) in his review describes how to measure the perfusion rate in terms of the effective thermal conductivity.

Furthermore, the surface integral terms $\frac{k_f}{V}\int_{A_{\text{int}}} T n_j dA$ and $-\frac{k_s}{V}\int_{A_{\text{int}}} T n_j dA$ present the tortuosity heat fluxes, which are usually small, as convection dominates over conduction (see e.g., Nakayama et al. 2001). Therefore, their effects may well be absorbed in effective thermal conductivities, as done by Xuan and Roetzel (1997). Having modeled the terms associated with dispersion, interfacial heat transfer, blood perfusion, and tortuosity, the individual macroscopic energy equations may finally be written for the blood and tissue phases as follows.

For the blood phase:

$$\varepsilon \rho_f c_{p_f} \frac{\partial \langle T\rangle^f}{\partial t} + \rho_f c_{p_f} \frac{\partial}{\partial x_j} \langle u_j\rangle \langle T\rangle^f = \frac{\partial}{\partial x_j}\left(\varepsilon k_f \frac{\partial \langle T\rangle^f}{\partial x_j} + \varepsilon k_{dis_{jk}} \frac{\partial \langle T\rangle^f}{\partial x_k}\right)$$
$$- a_f h_f \left(\langle T\rangle^f - \langle T\rangle^s\right) - \rho_f c_{p_f} \omega \left(\langle T\rangle^f - \langle T\rangle^s\right) \tag{1.25}$$

in which the left-hand side term denotes the macroscopic convection term, while the four terms on the right-hand side correspond to the macroscopic conduction, thermal dispersion, interfacial convective heat transfer, and blood perfusion, respectively.

For the solid tissue phase:

$$(1-\varepsilon)\rho_s c_s \frac{\partial \langle T\rangle^s}{\partial t} = \frac{\partial}{\partial x_j}\left((1-\varepsilon)k_s \frac{\partial \langle T\rangle^s}{\partial x_j}\right) + a_f h_f \left(\langle T\rangle^f - \langle T\rangle^s\right) + \rho_f c_{pf}\omega\left(\langle T\rangle^f - \langle T\rangle^s\right) + (1-\varepsilon)S_m \quad (1.26)$$

in which the left-hand side term denotes the thermal inertia term, while the four terms on the right-hand side correspond to the macroscopic conduction, interfacial convective heat transfer, blood perfusion heat source, and metabolic heat source, respectively.

The resulting equations (1.25) and (1.26) appear to be a correct form for the case of thermal nonequilibrium and are expected to clear up possible confusions associated with the blood perfusion term. The continuity equation (1.13), Darcy's law (1.17), and the two-energy equations (1.25) and (1.26) form a closed set of the macroscopic governing equations. The present model in a multidimensional and anisotropic form is quite general and can be applied to find both velocity and temperature fields, as we prescribe the spatial distributions of permeability tensor, porosity, interfacial heat transfer coefficient, metabolic reaction rate, and perfusion rate. It is interesting to note that, when the velocity field, porosity, and metabolic reaction are prescribed, we only need to know the local value of the lumped convection-perfusion parameter, namely, $(a_f h_f + \rho_f c_{p_f}\omega)$ (in addition to appropriate thermal boundary conditions) to solve the two-energy equations (1.25) and (1.26) for the blood and tissue temperatures, $\langle T\rangle^f$ and $\langle T\rangle^s$.

1.4.3 Pennes Model

It should be noted that most existing bioheat transfer models already reside in the present model based on the theory of porous media. We shall revisit some of the existing models and try to generate them from the present general model, starting with the Pennes model (1948), which in our notation runs as

$$(1-\varepsilon)\rho_s c_s \frac{\partial \langle T\rangle^s}{\partial t} = \frac{\partial}{\partial x_j}\left((1-\varepsilon)k_s \frac{\partial \langle T\rangle^s}{\partial x_j}\right) + \rho_f c_{p_f}\omega_{\text{Pennes}}\left(T_{a0} - \langle T\rangle^s\right) + (1-\varepsilon)S_m \quad (1.27)$$

where ω_{Pennes} is the mean blood perfusion rate, while T_{a0} is the mean brachial artery temperature. We compare the Pennes model against the energy equation (1.26) for the solid tissue phase and find the following relationship:

$$\rho_f c_{pf} \omega_{\text{Pennes}} \left(T_{a0} - \langle T \rangle^s\right) = a_f h_f \left(\langle T \rangle^f - \langle T \rangle^s\right) + \rho_f c_{p_f} \omega \left(\langle T \rangle^f - \langle T \rangle^s\right) \tag{1.28}$$

Perhaps, Pennes considered that the blood perfusion is the predominant heat source for the tissue, and did not bother to describe the interfacial convective heat transfer between the blood and tissue via the vascular wall. Instead, he introduced T_{a0} to adjust the total heat transfer, which takes place as the blood enters and leaves the tissue. We may assume $T_{a0} \simeq \langle T \rangle^f$ for small vessels, and find

$$\omega_{\text{Pennes}} = \omega + \frac{a_f h_f}{\rho_f c_{pf}} \tag{1.29}$$

Thus, Pennes' perfusion rate may be regarded as an effective one that includes interfacial convective heat transfer as well. Pennes assumed that blood enters the smallest vessels of the microcirculation at T_{a0}, where all heat transfer between the blood and tissue takes place. The assumption of the complete thermal equilibration with the surrounding tissue is valid only when Peclet number is sufficiently small.

1.4.4 Wulff Model and Klinger Model

Wulff (1974) criticized the Pennes model, pointing out that the moving blood through a tissue convects heat in any direction, not just in the direction of the local tissue temperature gradient. He assumed that the blood temperature $\langle T \rangle^f$ is equivalent to the tissue temperature within a tissue control volume and proposed a new bioheat transfer equation. The equation later generalized by Klinger (1978) runs in our notation as

$$(1-\varepsilon)\rho_s c_s \frac{\partial \langle T \rangle^s}{\partial t} = \frac{\partial}{\partial x_j}\left((1-\varepsilon) k_s \frac{\partial \langle T \rangle^s}{\partial x_j}\right) - \rho_f c_{pf} \frac{\partial \langle u_j \rangle \langle T \rangle^s}{\partial x_j} + (1-\varepsilon) S_m \tag{1.30}$$

We can obtain a similar equation by combining equations (1.25) and (1.26) setting $\langle T \rangle^f = \langle T \rangle^s$ as follows:

$$\begin{aligned}
&\left(\varepsilon \rho_f c_{p_f} + (1-\varepsilon)\rho_s c_s\right) \frac{\partial \langle T \rangle^s}{\partial t} + \rho_f c_{p_f} \frac{\partial}{\partial x_j} \langle u_j \rangle \langle T \rangle^s \\
&= \frac{\partial}{\partial x_j}\left(\left(\varepsilon k_f + (1-\varepsilon) k_s\right) \frac{\partial \langle T \rangle^s}{\partial x_j} + \varepsilon k_{dis_{jk}} \frac{\partial \langle T \rangle^s}{\partial x_k}\right) + (1-\varepsilon) S_m
\end{aligned} \tag{1.31}$$

We can easily see that the foregoing equation reduces to the Klinger equation when the ratio of vascular volume to total volume (i.e., porosity ε) is sufficiently small. Since the porosity is generally less than 0.1, the foregoing two equations are quite close to each other.

Another interpretation on the directional effect on the tissue temperature field is possible. When the blood flow is strong enough to neglect the macroscopic diffusion, the energy equation (1.25) for the blood flow reduces to

$$\rho_f c_{pf} \frac{\partial}{\partial x_j} \langle u_j \rangle \langle T \rangle^f = -a_f h_f \left(\langle T \rangle^f - \langle T \rangle^s \right) - \rho_f c_{p_f} \omega \left(\langle T \rangle^f - \langle T \rangle^s \right) \quad (1.32)$$

Substitution of the foregoing equation into the energy equation for the tissue (1.26) yields the Klinger equation (1.30). The assumption implicit here is that the blood flow velocity is sufficiently high that the ratio of the bulk convection heat transfer to conduction heat transfer, namely, the Peclet number, is much greater than unity. Thus, the Klinger model applies to the tissue with comparatively large vessels.

1.4.5 Chen and Holmes Model

Chen and Holmes (1980) assumed that all tissue-arterial blood heat exchange occurs along the circulatory network after the blood flows through the terminal arteries and before it reaches the level of the arterioles, which prompted them to propose the following bioheat transfer model:

$$\begin{aligned} &\rho c \frac{\partial T_t}{\partial t} + \rho_f c_{pf} \frac{\partial}{\partial x_j} \langle u_j \rangle T_t \\ &= \frac{\partial}{\partial x_j} \left(\left(\varepsilon k_f + (1-\varepsilon) k_s \right) \frac{\partial T_t}{\partial x_j} + k_p \frac{\partial T_t}{\partial x_j} \right) + \rho_f c_{p_f} \omega_j^* \left(T_a^* - T_t \right) + (1-\varepsilon) S_m \end{aligned} \quad (1.33)$$

where

$$\rho = \varepsilon \rho_f + (1-\varepsilon) \rho_s \quad (1.34a)$$

$$c = \left(\varepsilon \rho_f c_{pf} + (1-\varepsilon) \rho_s c_s \right) / \rho \quad (1.34b)$$

and

$$T_t = \left(\varepsilon \rho_f c_{p_f} \langle T \rangle^f + (1-\varepsilon) \rho_s c_s \langle T \rangle^s \right) \Big/ \rho c \quad (1.34c)$$

is the temperature of the continuum based on a volume average. Moreover, ω_j^* is the perfusion bleed-off to the tissue only from the microvessels past the jth generation of branching, while T_a^* is the blood temperature at the jth generation of branching. Both ω_j^* and T_a^* require the anatomical data. Chen and Holmes (1980) also took account of the "eddy" conduction due to the random flow of blood, by introducing the thermal conductivity k_p, which corresponds to our dispersion thermal conductivity k_{dis}. The energy equation similar to their equation (1.33) may be obtained by combining the two-energy equations (1.25) and (1.26) in the present model as

$$\rho c \frac{\partial T_t}{\partial t} + \rho_f c_{pf} \frac{\partial}{\partial x_j} \langle u_j \rangle \langle T \rangle^f = \frac{\partial}{\partial x_j} \left(\varepsilon k_f \frac{\partial \langle T \rangle^f}{\partial x_j} + (1-\varepsilon) k_s \frac{\partial \langle T \rangle^s}{\partial x_j} + \varepsilon k_{dis_{jk}} \frac{\partial \langle T \rangle^f}{\partial x_k} \right) + (1-\varepsilon) S_m \tag{1.35}$$

When the three temperature gradients on the right-hand side are close and $\varepsilon k_{dis_{jk}} = k_p \delta_{jk}$, the foregoing equation reduces to

$$\rho c \frac{\partial T_t}{\partial t} + \rho_f c_{p_f} \frac{\partial}{\partial x_j} \langle u_j \rangle \langle T \rangle^f = \frac{\partial}{\partial x_j} \left((\varepsilon k_f + (1-\varepsilon) k_s) \frac{\partial T_t}{\partial x_j} + k_p \frac{\partial T_t}{\partial x_j} \right) + (1-\varepsilon) S_m \tag{1.36}$$

which is close to the equation of Chen and Holmes, except that $\rho_f c_{p_f} \omega_j^* (T_a^* - T_t)$ is missing, as in the models of Wulff and Klinger, since it should vanish, as we add equations (1.25) and (1.26).

1.5 Three-Energy Equation Model for Countercurrent Heat Transfer in a Circulatory System

1.5.1 Related Work

Bazett and his colleagues (1948a,b) conducted a series of experimental studies on countercurrent heat exchange in the circulatory system. They found that the axial temperature gradient in the limb artery of human, under conditions of very low ambient temperature, is an order of magnitude higher than that under normal ambient conditions. From these experimental observations, they proposed the concept of venous shunting to the periphery, namely, that the countercurrent heat transfer takes place in the deep vasculature at the same time the blood is directed to the cutaneous circulation in close proximity to the surroundings. Their experimental finding brought attention to the important role of countercurrent heat exchange in bioheat transfer. Especially when the anatomical configuration of the main supply artery and vein in the limbs is treated, the effect of countercurrent heat transfer between closely spaced arteries and veins in the tissue must be taken into full consideration.

Following the experimental studies conducted by Bazett and his colleagues (1948a,b), Scholander and Krog (1957), and Mitchell and Myers (1968) investigated such an effect and successfully demonstrated that the countercurrent heat exchange reduces heat loss from the extremity to the surroundings, which could be quite significant owing to a large surface to volume ratio. These models, however, were not able to take account of either metabolic reaction or perfusion bleed-off from the artery to vein. Keller and Seiler (1971) established a one-dimensional bioheat transfer model to include the countercurrent heat transfer for the subcutaneous tissue region with arteries, veins, and capillaries. Weinbaum and Jiji (1979, 1985) proposed a model, which is based on some anatomical understanding, considering the countercurrent arterio-venous vessels. Roetzel and Xuan (1998) pointed out that the model may be useful in describing a temperature field in a single organ but would not be convenient to apply to the whole thermoregulation system. The foregoing survey prompts us to establish a multidimensional model that can be applied to the regions of extremity, where the countercurrent heat transfer happens between closely spaced arteries and veins in the blood circulatory system. Excellent reviews on these bioheat transfer equations may be found in Chato (1980), Charny (1992), and Khaled and Vafai (2003).

In this section, we shall extend the volume averaging procedure described for the heat transfer between the isolated vessels and the surrounding tissue to the case of countercurrent bioheat transfer in a blood circulatory system. The set of macroscopic governing equations consists of continuity and momentum equations for both arterial and venous blood phases and three individual energy equations for the two blood phases and the surrounding tissue phase. It will be shown that most shortcomings in existing models are overcome in the present model. Capillaries providing a continuous connection between the countercurrent terminal arteries and veins are modeled introducing the perfusion bleed-off rate, originally introduced in the pioneering paper by Pennes (1948). It has been found that the resulting model under certain conditions reduces to existing models for countercurrent heat transfer such as Chato (1980), Bejan (1979), Keller and Seiler (1971), Roetzel and Xuan (1998), and Weinbaum and Jiji (1985) for the case of closely aligned pairs of artery and vein. A general expression has been presented for the longitudinal effective thermal conductivity in the energy equation for the tissue. To examine the present model, we shall apply it to the countercurrent blood vessel configuration examined by Chato (1980). While Chato assumed the constancy of the perfusion bleed-off rate, we shall allow the spatial distribution of perfusion bleed-off rate and investigate its effect on the total countercurrent heat transfer.

1.5.2 Three-Energy Equation Model Based on the Volume Averaging Theory

A schematic view of the tissue layer close to the skin surface is shown in Figure 1.4, in which the arteries and veins are paired, such that the countercurrent heat transfer takes place. Thus, we assign individual dependent variables such as temperature to the arterial blood, venous blood, and tissue, which leads us to propose a three-energy equation model.

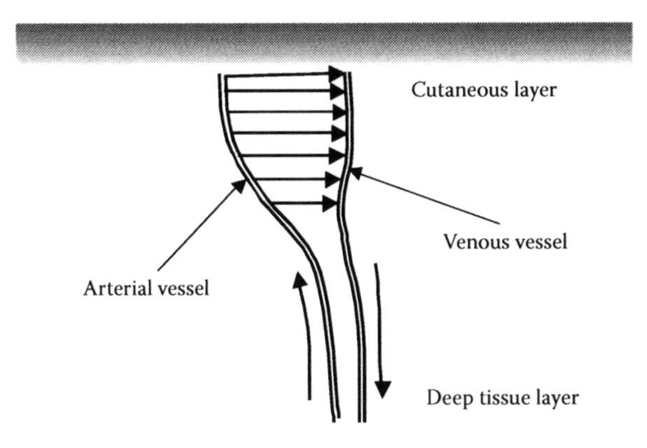

FIGURE 1.4
Schematic view of countercurrent heat exchange near the skin surface.

All dependent variables in the microscopic governing equations for the arterial blood, venous blood, and tissue phases are decomposed in this manner, and then these governing equations are integrated over the local control volume. After some manipulations following Nakayama and Kuwahara (2008) and Nakayama et al. (2008), we obtain the volume averaged set of the governing equations, which can be written assigning the subscripts a, v, and s to arterial blood vessels (arteries and arterioles), venous blood vessels (veins and venules), and tissue, as follows:

For the arterial blood phase:

$$\frac{\partial \varepsilon_a \langle u_j \rangle^a}{\partial x_j} + \omega_a' = 0 \tag{1.37}$$

$$-\frac{1}{\rho}\frac{\partial \langle p \rangle^a}{\partial x_i} - \frac{\nu}{K_{aij}} \varepsilon_a \langle u_j \rangle^a - \omega_a' u_{i_{\text{int}}} = 0 \tag{1.38}$$

$$\begin{aligned} &\varepsilon_a \rho_f c_{p_f} \frac{\partial \langle T \rangle^a}{\partial t} + \rho_f c_{p_f} \frac{\partial}{\partial x_j} \varepsilon_a \langle u_j \rangle^a \langle T \rangle^a \\ &= \frac{\partial}{\partial x_j}\left(\varepsilon_a k_a \frac{\partial \langle T \rangle^a}{\partial x_j} + \varepsilon_a k_{disa_{jk}} \frac{\partial \langle T \rangle^a}{\partial x_k}\right) - a_a h_a \left(\langle T \rangle^a - \langle T \rangle^s\right) - \rho_f c_{p_f} \omega_a' \langle T \rangle^a \end{aligned} \tag{1.39}$$

For the venous blood phase:

$$\frac{\partial \varepsilon_v \langle u_j \rangle^v}{\partial x_j} + \omega_v' = 0 \tag{1.40}$$

$$-\frac{1}{\rho}\frac{\partial \langle p \rangle^v}{\partial x_i} - \frac{\nu}{K_{vij}} \varepsilon_v \langle u_j \rangle^v - \omega_v' u_{i_{\text{int}}} = 0 \tag{1.41}$$

$$\varepsilon_v \rho_f c_{p_f} \frac{\partial \langle T \rangle^v}{\partial t} + \rho_f c_{p_f} \frac{\partial}{\partial x_j} \varepsilon_v \langle u_j \rangle^v \langle T \rangle^v$$
$$= \frac{\partial}{\partial x_j}\left(\varepsilon_v k_v \frac{\partial \langle T \rangle^v}{\partial x_j} + \varepsilon_v k_{disv_{jk}} \frac{\partial \langle T \rangle^v}{\partial x_k}\right) - a_v h_v \left(\langle T \rangle^v - \langle T \rangle^s\right) - \rho_f c_{p_f} \omega'_v \langle T \rangle^v \tag{1.42}$$

For the solid tissue phase:

$$(1-\varepsilon)\rho_s c_s \frac{\partial \langle T \rangle^s}{\partial t}$$
$$= \frac{\partial}{\partial x_j}\left((1-\varepsilon) k_s \frac{\partial \langle T \rangle^s}{\partial x_j}\right) + a_a h_a \left(\langle T \rangle^a - \langle T \rangle^s\right) + \rho_f c_{p_f} \omega'_a \langle T \rangle^a$$
$$+ a_v h_v \left(\langle T \rangle^v - \langle T \rangle^s\right) + \rho_f c_{p_f} \omega'_v \langle T \rangle^v + (1-\varepsilon) S_m \tag{1.43}$$

where ε_a and ε_v are the volume fractions of the arterial blood and that of the venous blood, respectively, such that $\varepsilon = \varepsilon_a + \varepsilon_v$. The terms associated with the surface integral are modeled as

$$\frac{1}{V_f}\int_{A_{\text{int}}}\left(-\frac{p}{\rho_f} + \nu_f\left(\frac{\partial u_i}{\partial x_j} + \frac{\partial u_j}{\partial x_i}\right)\right) n_j dA = -\frac{\nu_f}{K_{ij}} \varepsilon \langle u_j \rangle^f \tag{1.44}$$

which is simply Darcy's law and

$$\int_{A_{\text{int}}} \rho_f u_j n_j dA/V = \rho_f \omega' \tag{1.45}$$

is the mass flow rate per unit volume through the interface A_{int}, modeled in terms of the perfusion bleed-off rate ω' (1/sec). The perfusion bleed-off rate ω' describes the volume rate of the fluid per unit volume, bleeding off to the solid matrix through the interfacial vascular wall. Thus, the momentum bleed-off rate is modeled as

$$\int_{A_{\text{int}}} \rho_f u_i u_j n_j dA/V = \rho_f \omega' u_{i_{\text{int}}} \tag{1.46a}$$

where $u_{i_{\text{int}}}$ is the velocity vector averaged over the interface. Likewise, the enthalpy bleed-off rate is modeled as

$$\int_{A_{\text{int}}} \rho_f c_{p_f} u_j T n_j dA/V = \rho_f c_{p_f} \omega' \langle T \rangle^f \tag{1.46b}$$

For the interfacial heat transfer, Newton's cooling law is adopted as

$$\frac{1}{V}\int_{A_{\text{int}}} k_f \frac{\partial T}{\partial x_j} n_j dA = a_f h_f \left(\langle T \rangle^s - \langle T \rangle^f\right) \tag{1.47}$$

where a_f and h_f are the specific surface area and interfacial heat transfer coefficient, respectively. Furthermore, $k_{dis_{jk}}$ is the thermal dispersion conductivity tensor, as introduced in Nakayama et al. (2006).

For the microcirculation of peripheral tissue in which capillaries provide a continuous connection between the terminal artery and vein (i.e., arterial-venous anastomoses), as shown in Figure 1.4, we may readily set $\omega_a' = -\omega_v'$ such that the present energy equation (1.43) for the solid tissue phase reduces to

$$
\begin{aligned}
(1-\varepsilon)\rho_s c_s \frac{\partial \langle T \rangle^s}{\partial t} &= \frac{\partial}{\partial x_j}\left((1-\varepsilon)k_s \frac{\partial \langle T \rangle^s}{\partial x_j}\right) \\
&\quad + a_a h_a \left(\langle T \rangle^a - \langle T \rangle^s\right) + a_v h_v \left(\langle T \rangle^v - \langle T \rangle^s\right) \\
&\quad + \rho_f c_{p_f} \omega_a' \left(\langle T \rangle^a - \langle T \rangle^v\right) + (1-\varepsilon) S_m
\end{aligned} \tag{1.48}
$$

1.5.3 Keller and Seiler Model

Most existing bioheat transfer models for countercurrent bioheat transfer already reside in the present model based on the theory of porous media. Let us revisit some of the existing models and try to generate them from the present general model.

Keller and Seiler (1971) noted that the axial temperature gradient in the limb is much higher than the transverse one and considered an energy balance within a control volume for the idealized one-dimensional steady case, as illustrated in Figure 1.5, for which they proposed

$$
\begin{aligned}
&(1-\varepsilon)k_s \frac{d^2 \langle T \rangle^s}{dx^2} + a_a h_a \left(\langle T \rangle^a - \langle T \rangle^s\right) + a_v h_v \left(\langle T \rangle^v - \langle T \rangle^s\right) \\
&\quad + \rho_f c_{p_f} \omega' \left(\langle T \rangle^a - \langle T \rangle^s\right) + (1-\varepsilon) S_m = 0
\end{aligned} \tag{1.49}
$$

which is almost identical to what we would get for the one-dimensional case from our multidimensional expression (1.48), except that the temperature difference in the perfusion term somewhat differs from ours. Keller and

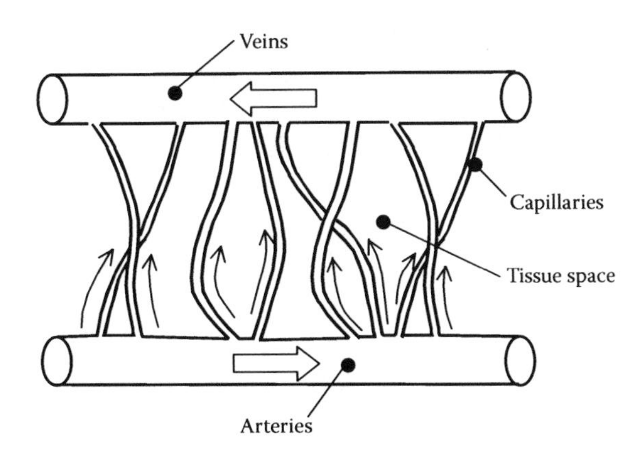

FIGURE 1.5
One-dimensional model for countercurrent heat exchange.

Seiler (1971) obtained solutions assuming that the arterial blood enters the peripheral region at the isothermal core temperature and that the venous blood is completely equilibrated with the tissue at the cutaneous layer.

1.5.4 Chato Model

Chato's countercurrent heat transfer model (1980) differs from Keller and Seiler's (1971) model in its neglect of heat transfer between the blood and the tissue. In this way, he was able to concentrate on the two temperatures instead of three as in Keller and Seiler's model. Chato assumed that the flow rate decreases linearly, which corresponds with the case of constant perfusion bleed-off rate. His one-dimensional model can easily be generated from our general expressions (1.39) and (1.42) along with (1.37) and (1.40), dropping the transient and conduction terms as

$$\rho_f c_{p_f} \frac{d}{dx} \varepsilon_a \langle u \rangle^a \langle T \rangle^a = -a_f h_f \left(\langle T \rangle^a - \langle T \rangle^v \right) - \rho_f c_{p_f} \omega_a' \langle T \rangle^a \tag{1.50}$$

$$\rho_f c_{p_f} \frac{d}{dx} \varepsilon_v \langle u \rangle^v \langle T \rangle^v = -a_f h_f \left(\langle T \rangle^v - \langle T \rangle^a \right) + \rho_f c_{p_f} \omega_a' \langle T \rangle^a \tag{1.51}$$

where the interfacial heat transfer coefficients are assumed to be the same. The continuity equations (1.37) and (1.40) readily provide

$$\varepsilon_a \langle u \rangle^a = u_0 - \omega_a' x \tag{1.52}$$

and

$$\varepsilon_v \langle u \rangle^v = -u_0 + \omega_a' x \tag{1.53}$$

Note that u_0 is the apparent velocity at $x = 0$ and that the right-hand side terms in the two equations (1.50) and (1.51) cancels out each other, as they should for this "perfect" heat exchange system. Chato (1980) obtained arterial and venous temperature profiles along the length of the vessels and demonstrated that the effect of perfusion bleed-off is to increase the heat transfer between the vessels as compared with the case of constant mass flow rate (i.e., $\omega_a' = 0$).

1.5.5 Roetzel and Xuan Model

Roetzel and Xuan (1998) used the theory of porous media to simulate a transient response of the limb to external stimulus, in which the effect of the countercurrent heat exchange on the temperature response is expected to be significant. Their energy equation for the tissue in our notation runs as

$$(1-\varepsilon) \rho_s c_s \frac{\partial \langle T \rangle^s}{\partial t} = \frac{\partial}{\partial x_j} \left((1-\varepsilon) k_s \frac{\partial \langle T \rangle^s}{\partial x_j} \right) + a_a h_a \left(\langle T \rangle^a - \langle T \rangle^s \right) + a_v h_v \left(\langle T \rangle^v - \langle T \rangle^s \right) + (1-\varepsilon) S_m \tag{1.54}$$

Comparison of the foregoing equation against our expression (1.48) for the tissue reveals that the perfusion term $\rho_f c_{p_f} \omega'_a (\langle T \rangle^a - \langle T \rangle^v)$ is missing. Obviously, they did not retain the term describing the transcapillary fluid exchange via arterial-venous anastomoses, namely, $\int_{A_{\text{int}}} \rho_f c_{p_f} u_j T n_j dA/V = \rho_f c_{p_f} \omega' \langle T \rangle^f$. If they did, they would have obtained our expression (48), which may be rearranged in their form as

$$(1-\varepsilon)\rho_s c_s \frac{\partial \langle T \rangle^s}{\partial t} = \frac{\partial}{\partial x_j}\left((1-\varepsilon) k_s \frac{\partial \langle T \rangle^s}{\partial x_j}\right) + \left(a_a h_a + \rho_f c_{p_f} \omega'_a\right)\left(\langle T \rangle^a - \langle T \rangle^s\right) + \left(a_v h_v - \rho_f c_{p_f} \omega'_a\right)\left(\langle T \rangle^v - \langle T \rangle^s\right) + (1-\varepsilon) S_m \tag{1.55}$$

In their model, the convection-perfusion parameters, namely, $(a_f h_f \pm \rho_f c_{p_f} \omega')$, are replaced by the interfacial convective heat transfer coefficients, $a_f h_f$. This difference should not be overlooked since the perfusion heat sources could be quite significant for the bioheat transfer in the extremities, as Chato (1980) demonstrated using his model.

1.5.6 Weinbaum–Jiji Model and Bejan Model

Weinbaum and Jiji (1979) considered bioheat transfer between a paired countercurrent terminal artery and vein. They took account of the vascular structure in which vessel number density, velocity, and diameter vary significantly from the deep tissue layer toward the skin layer. Later, Weinbaum and Jiji (1985) proposed a simplified model in which an effective thermal conductivity tensor is introduced as a function of the local blood velocity. They claimed that the perfusion heat source vanishes within the capillary bed and derived a single equation to describe the steady-state tissue temperature variations, which, when the vessels are in parallel to the temperature gradient, reduces to

$$\frac{d}{dx}\left(\left((1-\varepsilon) k_s + \frac{\pi \varepsilon_a}{2\sigma} \frac{\left(\rho_f c_{p_f} \langle u \rangle^a R\right)^2}{(1-\varepsilon) k_s}\right) \frac{d \langle T \rangle^s}{dx}\right) + (1-\varepsilon) S_m = 0 \tag{1.56}$$

where σ is a geometrical factor of the vessel structure, whereas R is the local radius of the vessel. It is seen that the longitudinal effective thermal conductivity due to countercurrent flow is proportional to the square of blood mass flow rate. It is also interesting to note that the concept of the longitudinal effective thermal conductivity in countercurrent heat transfer was already explicit in Bejan (1979) in which he presented a novel method for thermal insulation system optimization. Bejan (1979) seems to be the first to point out the relationship associated with the square of the mass flow rate and the longitudinal effective thermal conductivity by convection. His expression is a simple one:

$$Q = -\frac{\left(m_f c_{pf}\right)^2}{UP} \frac{d \langle T \rangle^s}{dx} \tag{1.57}$$

where Q, $\dot{m}_f$, U, and P are the heat flow from the warm end to the cold end, the mass flow rate of the hot (or cold) fluid, the overall heat transfer coefficient, and the wetted perimeter, respectively. The group $(m_f c_{pf})^2/(UP)$ plays the same role as Ak_{eff}in the one-dimensional insulation system. Upon noting that $m_f = A\rho_f \varepsilon_a \langle u \rangle^a$and $P = Aa_f$, Bejan's equation (1.57) may be translated in the present bioheat transfer problem as

$$\frac{d}{dx}\left(\left((1-\varepsilon)k_s + \frac{\left(\rho_f c_{p_f} \varepsilon_a \langle u \rangle^a\right)^2}{a_f U}\right)\frac{d\langle T \rangle^s}{dx}\right) + (1-\varepsilon)S_m = 0 \qquad (1.58)$$

In these countercurrent heat transfer models, namely, Bejan's and Wienbaum and Jiji's, the perfusion heat sources are ignored. Thus, in what follows, we shall attempt to reduce the present set of governing equations to a single equation for the tissue temperature variations, without neglecting these perfusion heat source terms.

When the blood flow is strong enough to neglect the macroscopic diffusion, the energy equations (1.39) and (1.42) for arterial and venous blood flows for the one-dimensional steady state reduce to

$$\rho_f c_{p_f} \frac{d}{dx}\varepsilon_a \langle u \rangle^a \langle T \rangle^a = -a_f h_f \left(\langle T \rangle^a - \langle T \rangle^s\right) - \rho_f c_{p_f} \omega_a' \langle T \rangle^a \qquad (1.59)$$

$$\rho_f c_{p_f} \frac{d}{dx}\varepsilon_v \langle u \rangle^v \langle T \rangle^v = -a_f h_f \left(\langle T \rangle^v - \langle T \rangle^s\right) + \rho_f c_{p_f} \omega_a' \langle T \rangle^v \qquad (1.60)$$

where the interfacial heat transfer coefficients are assumed to be the same as in the case of Chato. However, the foregoing equations are different from Chato's equations (1.50) and (1.51), since we do take account of the heat transfer between the bloods and tissue. Upon noting the continuity relationship $\varepsilon_a \langle u \rangle^a = -\varepsilon_v \langle u \rangle^v$as given by the continuity equations (1.37) and (1.40) with $\omega_a' = -\omega_v'$, we subtract equation (1.60) from (1.59) to obtain

$$\rho_f c_{p_f} \frac{d}{dx}\varepsilon_a \langle u \rangle^a \left(\langle T \rangle^a + \langle T \rangle^v\right) = -a_f h_f \left(\langle T \rangle^a - \langle T \rangle^v\right) - \rho_f c_{p_f} \omega_a' \left(\langle T \rangle^a + \langle T \rangle^v\right) \qquad (1.61a)$$

or

$$\rho_f c_{p_f} \varepsilon_a \langle u \rangle^a \frac{d}{dx}\left(\langle T \rangle^a + \langle T \rangle^v\right) = -a_f h_f \left(\langle T \rangle^a - \langle T \rangle^v\right) \qquad (1.61b)$$

as we note the continuity relationship, namely, $d(\varepsilon_a \langle u \rangle^a)/dx = -\omega_a'$. Weinbaum and Jiji (1985) proposed that the mean tissue temperature around an artery–vein pair can be approximated as

$$\langle T \rangle^s = \frac{\langle T \rangle^a + \langle T \rangle^v}{2} \qquad (1.62)$$

Following their approximation, we obtain

$$\langle T \rangle^a - \langle T \rangle^v = -2\frac{\rho_f c_{p_f} \varepsilon_a \langle u \rangle^a}{a_f h_f}\frac{d\langle T \rangle^s}{dx} \qquad (1.63)$$

from (1.61b). Using equations (1.59) and (1.60), we may replace both interfacial and perfusion heat source terms in the energy equation (1.48) for the tissue by the blood convection terms as

$$\begin{aligned}
&\frac{d}{dx}\left((1-\varepsilon)k_s\frac{d\langle T\rangle^s}{dx}\right)+a_a h_a\left(\langle T\rangle^a-\langle T\rangle^s\right)+a_v h_v\left(\langle T\rangle^v-\langle T\rangle^s\right)\\
&\quad+\rho_f c_{p_f}\omega_a'\left(\langle T\rangle^a-\langle T\rangle^v\right)+(1-\varepsilon)S_m\\
&\quad=\frac{d}{dx}\left((1-\varepsilon)k_s\frac{d\langle T\rangle^s}{dx}\right)-\rho_f c_{p_f}\frac{d}{dx}\left(\varepsilon_a\langle u\rangle^a\langle T\rangle^a+\varepsilon_v\langle u\rangle^v\langle T\rangle^v\right)\\
&\quad+(1-\varepsilon)S_m=0
\end{aligned}\tag{1.64}$$

As we note the continuity relationship $\varepsilon_a \langle u\rangle^a = -\varepsilon_v \langle u\rangle^v$ and use equation (1.63) for the last expression in (1.64), we finally have

$$\frac{d}{dx}\left(\left((1-\varepsilon)k_s+2\frac{\left(\rho_f c_{p_f}\varepsilon_a\langle u\rangle^a\right)^2}{a_f h_f}\right)\frac{d\langle T\rangle^s}{dx}\right)+(1-\varepsilon)S_m=0 \tag{1.65}$$

which we find almost identical to Bejan's equation (1.58), as we note the overall heat transfer coefficient corresponds to

$$U=\frac{1}{\frac{1}{h_a}+\frac{1}{h_v}}=\frac{h_f}{2} \tag{1.66}$$

It is most interesting to find that the foregoing relationship for the longitudinal effective thermal conductivity holds for all cases, with or without perfusion bleed-off sources, as long as the local values are used to evaluate the effective thermal conductivity by convection.

1.6 Effect of Spatial Distribution of Perfusion Bleed-Off Rate on Total Countercurrent Heat Transfer

As an example for illustration, we shall consider Chato's one-dimensional problem of countercurrent heat transfer as schematically shown in Figure 1.5. Chato (1980) assumed the constancy of the perfusion bleed-off rate ω_a', namely, a linear decrease in the arterial flow rate, and that all of the bleed-off fluid that leaves the artery reenters the vein at the same location. We shall relax his assumption, allowing the spatial variation of ω_a' so as to investigate its effect on the total countercurrent heat transfer. Let us assume that the perfusion bleed-off rate ω_a' follows:

$$\omega_a'=(1+n)\overline{\omega}_a'\left(\frac{x}{L}\right)^n \tag{1.67}$$

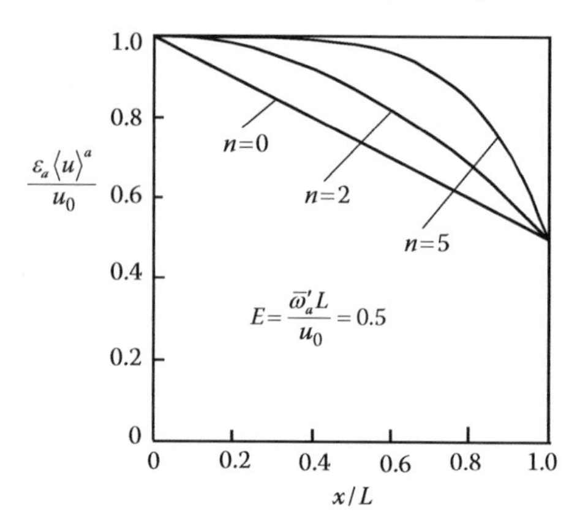

FIGURE 1.6
Effect of the exponent n on perfusion rate.

along the blood vessel of length L, where $\overline{\omega'}_a$ is the average perfusion rate such that the total amount of perfusion is given by $\overline{\omega'}_a L$, irrespective of the value of n. The exponent n may take any value equal to zero (i.e., Chato's case) or greater than zero, such that we can compare the results against Chato's and elucidate the effect of blood pressure on the bioheat transfer for fixed total amount of perfusion. As we substitute the foregoing equation into the continuity equations (1.37) and (1.40), we readily obtain

$$\varepsilon_a \langle u \rangle^a = u_0 - \overline{\omega}'_a L \left(\frac{x}{L}\right)^{1+n} \tag{1.68}$$

$$\varepsilon_v \langle u \rangle^v = -u_0 + \overline{\omega}'_a L \left(\frac{x}{L}\right)^{1+n} \tag{1.69}$$

where u_0 is the apparent blood velocity at $x = 0$. As illustrated in Figure 1.6, the exponent n controls the distribution of the perfusion rate. For a large exponent n, the perfusion bleed-off takes place rather suddenly toward the end of the vessel, indicating poor blood circulation.

Upon substituting these velocity distributions into the momentum equations (1.38) and (1.41), we obtain

$$\langle p \rangle\big|_{x=0} - \langle p \rangle\big|_{x=L} = \left(\varepsilon_a \langle p \rangle^a + \varepsilon_v \langle p \rangle^v\right)\Big|_L^0 = \mu \left(\frac{\varepsilon_a}{K_a} - \frac{\varepsilon_v}{K_v}\right) u_0 L \left(1 - \frac{E}{2+n}\right) \tag{1.70}$$

where

$$E = \frac{\overline{\omega'}_a L}{u_0} \tag{1.71}$$

is the dimensionless perfusion bleed-off rate, while μ is the viscosity. Thus, the pressure difference within the body may never be large since

$\varepsilon_a/K_a \cong \varepsilon_v/K_v$. Equation (1.70), however, indicates that the blood pressure difference increases for either small perfusion bleed-off rate E or large exponent n, which may result from aging.

Following Chato (1980), we note that the axial conduction terms in the blood energy equations are negligibly small as compared to the convection and perfusion terms. Then, the energy equations (1.50) and (1.51) along with the foregoing velocity distributions reduce to

$$\frac{d\langle T\rangle^a}{d(x/L)} = -\frac{N}{1-E\left(\frac{x}{L}\right)^{1+n}}\left(\langle T\rangle^a - \langle T\rangle^v\right) \tag{1.72}$$

$$\frac{d\langle T\rangle^v}{d(x/L)} = \frac{N+(1+n)E\left(\dfrac{x}{L}\right)^n}{1-E\left(\dfrac{x}{L}\right)^{1+n}}\left(\langle T\rangle^v - \langle T\rangle^a\right) \tag{1.73}$$

where

$$N = \frac{a_f h_f L}{\rho_f c_{p_f} u_0} \tag{1.74}$$

is the number of heat transfer units. The boundary conditions are given by

$$x/L = 0\text{: } \langle T\rangle^a = \langle T\rangle^a_0 \tag{1.75}$$

$$x/L = 1\text{: } \langle T\rangle^v = \langle T\rangle^v_L \tag{1.76}$$

A series of numerical integrations were carried out for various sets of three important dimensionless parameters, namely, the dimensionless perfusion rate E, the number of heat transfer units N, and the exponent n. Thus, the temperature profiles along the vessel axes are obtained for the case of $n = 0$ and presented in Figures 1.7(a) and 1.7(b) for a physiological range of E and N values. The results appear to be in perfect agreement with the exact expressions reported by Chato (1980). The difference between the present curves for $n = 0$ and those based on Chato's solution is indiscernible in the figure. Naturally, a better blood circulation (i.e., larger E) results in warming the venous blood efficiently. The figures show that its efficiency as a heat exchanging system increases with N.

The temperature profiles along the vessel axes for the case of $n = 5$ are presented in Figures 1.8(a) and 1.8(b). It is interesting to note that the arterial blood temperature for the case of nonzero n always stays higher than that for the case of $E = 0$ (i.e., without perfusion) even at the end of the vessel. Following Chato (1980), we shall evaluate the total heat transfer from the artery to vein in terms of

$$q_{a-v} = \rho c_{p_f} u_0 \left(\langle T\rangle^v_0 - \langle T\rangle^v_L\right) \tag{1.77}$$

or its dimensionless form, namely,

$$\frac{q_{a-v}}{\rho c_{p_f} u_0 \left(\langle T\rangle^a_0 - \langle T\rangle^v_L\right)} = \frac{\langle T\rangle^v_0 - \langle T\rangle^v_L}{\langle T\rangle^a_0 - \langle T\rangle^v_L} \tag{1.78}$$

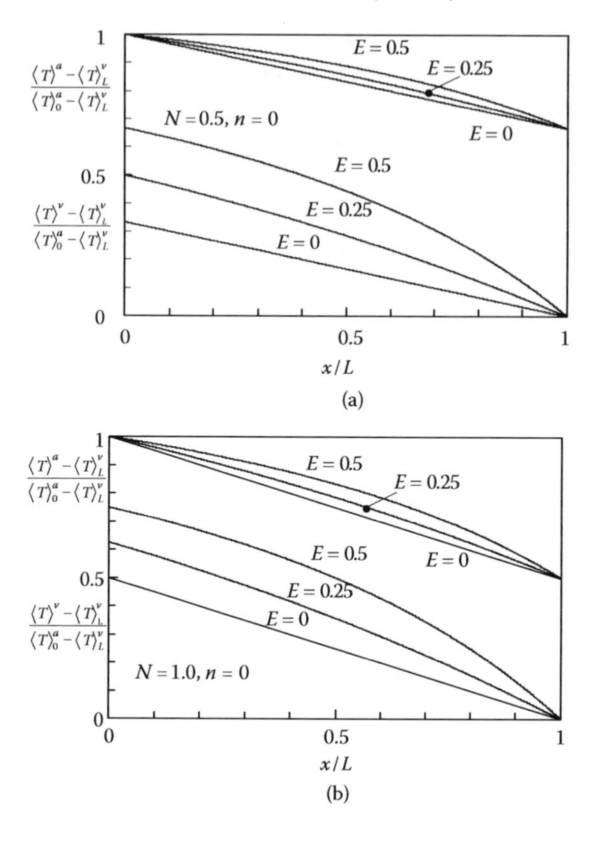

FIGURE 1.7
Arterial and venous blood temperature profiles along the vessel axes for the case of $n = 0$ (a) $N = 0.5$, $n = 0$ and (b) $N = 1.0$, $n = 0$.

The total heat transfer from the artery to the vein is plotted against the exponent n for various sets of N and E values in Figure 1.9. The total heat transfer decreases as increasing n (i.e., worsening the blood circulation), while it increases with E (i.e., increasing the perfusion rate).

1.7 Application of Bioheat Equation to Cryoablation Therapy

1.7.1 Related Work

Cryotherapy is often preferred to more traditional kinds of surgical therapy because of its minimal pain, scarring, and cost. The therapy has been gaining

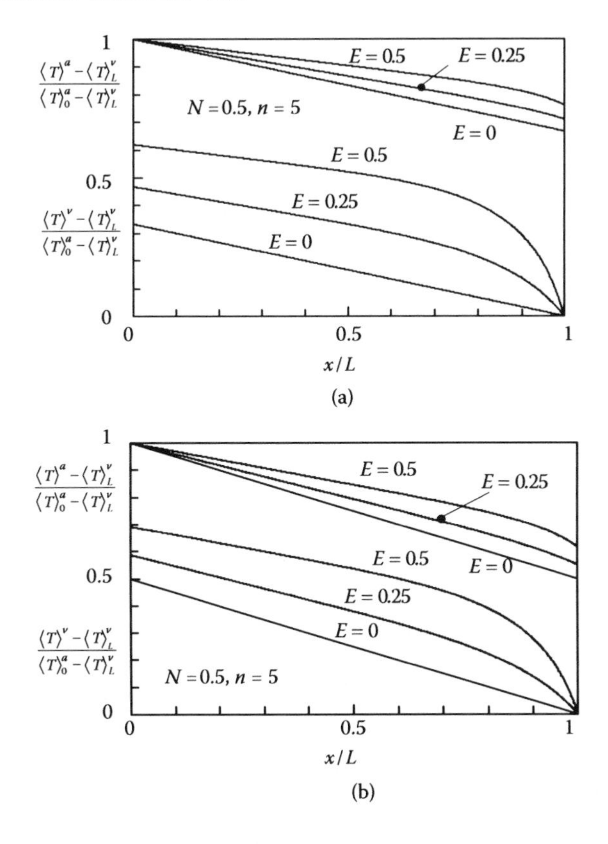

FIGURE 1.8
Arterial and venous blood temperature profiles along the vessel axes for the case of $n = 5$ (a) $N = 0.5$, $n = 5$ and (b) $N = 1.0$, $n = 5$.

significant acceptance as minimally invasive therapy for treatments of various malignant cancers. In a cryosurgical treatment, a single or multiprobe metal system is placed in contact with the target tissue through the skin. We have placed the emphasis of this paper upon the treatment of malignant lung tumor, since its application to lung cancer has been practiced on a trial basis for some years in Japanese medical schools (Nakatsuka et al. 2004).

The cryoprobe in consideration houses a small coaxial nozzle internally. A high-pressure gas supply line is connected to the probe so as to supply Argon gas, which expands through the nozzle to the probe tip and then flows backward through the internal channel leading to the cryoprobe outlet. Owing to the Joule–Thompson effect, the outer surface temperature of the probe decreases below −135°C. As the tissue temperature is lowered, an ellipsoidal ice ball forms around each probe increasing in size, eventually encompassing and invading the entire tumor. This freezing process continues for 5–15 min. Then, the thawing process takes place as supplying Helium gas. Because of

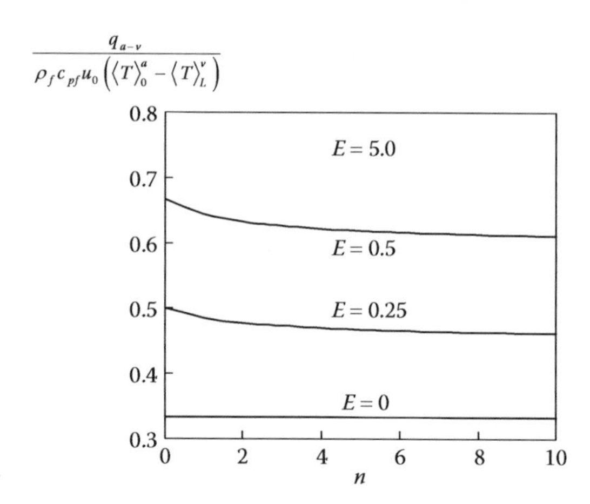

FIGURE 1.9
Effect of the exponent n on the total heat transfer from artery to vein.

the difference in the inversion temperature, the probe temperature during this process goes up to about 20°C to thaw the frozen tissue. This freezing-thawing sequence is repeated several times to kill abnormal cells or tissues such as are found in malignant tumors. Cryoinjury is believed to be due to two primary mechanisms: one is the direct injury to the cells from the freeze–thaw cycle, and the other is the indirect injury that results from the biological response to the damage caused by freezing, primarily the vasculature of the tumor.

As with any medical treatment, there are risks involved, primarily that of damage to nearby healthy tissue (Butz et al. 2000). We must know the exact time required to freeze the entire cancer without damaging its surrounding healthy tissue. However, some standards for setting clinical parameters such as freezing rate and time are quite empirical today. Therefore, improvements in cryosurgery depend upon developing reliable mathematical models and pre-operational simulation tools based on them.

Perhaps, Bischof et al. (1997) is the first to predict ice ball formation around a single cryosurgical probe. They used a cylindrical model to predict the interface location and temperature profile. Rewcastle et al. (1998) proposed a finite difference model for single probe freezing and generated isotherms within the ice ball during its growth. Keanini and Rubinsky (1992) and Baissalov et al. (2001) dealt with the problem of optimization in cryosurgery by regarding the placement of cryoprobes and freezing protocol design. Wan et al. (2003) appealed to finite element methodology to simulate ice ball formation in a multiprobe cryosurgery. Rabin and Shitzer (1998) and Rossi et al. (2007) introduced fairly sophisticated numerical techniques for freezing an angioma, while Rossi and Rabin (2007) developed an elegant experimental technique to create a two-dimensional freezing problem associated with prostate cryosurgery with urethral warming. However, none of them considered the case of lung cancer nor were concerned with the effects of the blood perfusion on the temporal evolution of ice formation, which leads to the fact, namely, that there exists the limiting size of the tumor that one single cryoprobe can freeze at the maximum. No attempts were made to estimate the limiting radius for freezing tumors.

In this section, we shall appeal to the bioheat equation derived using the volume averaging theory and solve it both numerically and analytically to simulate the ice ball evolution and to locate the freezing front as time goes by. The analytical results based on the integral method agree very well with the numerical results based on the enthalpy method. Thus, the resulting analytical expression may be exploited for estimating the time for freezing a cancer of a given size. It will also be pointed out that there exists the limiting size of the cancer that one single cryoprobe can freeze at the maximum. It is believed that the present results lend quantitative support to the current empirical standards for cryosurgical clinical applications.

1.7.2 Bioheat Equation for Cryoablation

The general bioheat equation (1.26) for the solid tissue phase may be used to attack this problem. When the ratio of blood to total lung volume is small, equation (1.26) reduces to

$$\rho_s c_s \frac{\partial T}{\partial t} = \frac{\partial}{\partial x_j}\left(k_s \frac{\partial T}{\partial x_j}\right) + \rho_f c_{p_f} \omega (T_f - T) + a_f h (T_f - T) + S_m \tag{1.79}$$

where the second time and third term on the right-hand side correspond to the blood perfusion to the tissue and the interfacial heat transfer from the blood to tissue through the vessel wall, respectively. Similarity between our equation and Pennes' equation is obvious as we rewrite the foregoing equation as

$$\rho_s c_s \frac{\partial T}{\partial t} = \frac{\partial}{\partial x_j}\left(k_s \frac{\partial T}{\partial x_j}\right) + \rho_f c_{p_f} \omega_{\text{eff}} (T_f - T) + S_m \tag{1.80}$$

where

$$\omega_{\text{eff}} = \omega + \frac{a_f h}{\rho_f c_{p_f}} \tag{1.81}$$

is the effective perfusion rate. However, ω_{eff} conceptually differs from Pennes' perfusion rate ω_{Pennes} in the Pennes equation (1.27), which is purely empirical. It should also be noted that T_f in $(T_f - T)$ is the local blood temperature, whereas T_{a0} in equation (1.27) is the mean brachial artery temperature. The interfacial convective heat transfer between the blood and tissue can never be insignificant for countercurrent bioheat transfer. Even when there is no perfusion, that is, $\omega_{\text{Pennes}} = 0$, the effective perfusion rate never vanishes since $\omega_{\text{eff}} = a_f h/\rho_f c_{p_f}$. Thus, equation (1.81) must always be used for countercurrent bioheat transfer for the case of closely aligned pairs of vessels.

1.7.3 Numerical Analysis Based on Enthalpy Method

The enthalpy method is often used for locating an interface in phase change problems since it allows us to use a fixed mesh. An easy approach to implement the method is to include the latent heat by artificially increasing the specific heat capacity around the freezing point, thus making it a function of temperature as illustrated in Figure 1.10.

This simplest temperature function satisfies the obvious relationship among the latent heat of solidification h_{sf}, artificial maximum heat capacity $c_{\max}$, and artificial temperature band ΔT, namely, $h_{sf} = c_{\max}(2\Delta T)$. The temperature band ΔT should be set according to the mesh resolution. Naturally, a finer grid system allows us to use a smaller ΔT, which provides us a sharper freezing front where $T = T_i$. In this study, ΔT was set to from $1°$ to $5°$. The temporal development of the freezing front is found fairly insensitive to ΔT in this range.

Any standard scheme may be used to discretize the governing equation (1.80). We shall use a finite volume method as proposed by Patankar (1980) to obtain a two-dimensional finite volume expression. We consider a control volume of size $\Delta x \Delta y$ centering the node P (pole), as shown in Figure 1.11, and

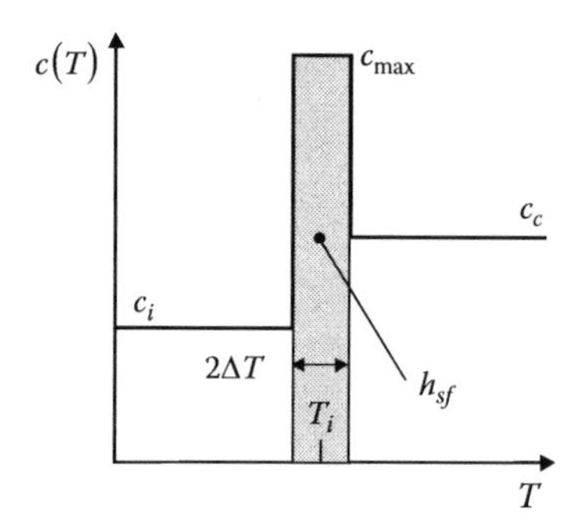

FIGURE 1.10
Effective specific heat capacity.

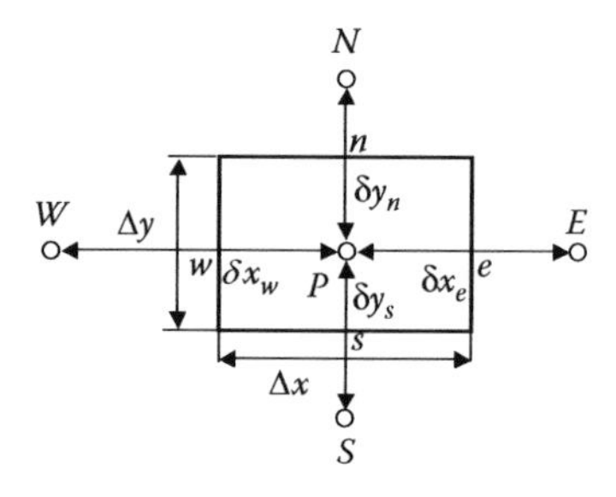

FIGURE 1.11
Grid nomenclatures.

let the upper-case letters E (East), W (West), N (North), and S (South) denote its neighboring nodes. Furthermore, we let the lowercase versions of the same letters e, w, n, and s denote four faces of the control volume, and $(\delta x)_e$, $(\delta x)_w$, $(\delta y)_n$, and $(\delta y)_s$ denote the distances between the nodes. Then, the discretized version of the bioheat equation may be written as follows (see e.g., Nakayama [1995] for details):

$$a_P T_P = a_E T_E + a_W T_W + a_N T_N + a_S T_S + b \tag{1.82}$$

where

$$a_E = \left(\frac{\Delta y}{\delta x_e}\right) \frac{1}{T_E - T_P} \int_{T_P}^{T_E} k(T)dT \tag{1.83a}$$

$$a_W = \left(\frac{\Delta y}{\delta x_w}\right) \frac{1}{T_P - T_W} \int_{T_W}^{T_P} k(T)dT \tag{1.83b}$$

$$a_N = \left(\frac{\Delta x}{\delta y_n}\right) \frac{1}{T_N - T_P} \int_{T_P}^{T_N} k(T)dT \tag{1.83c}$$

$$a_S = \left(\frac{\Delta x}{\delta y_s}\right) \frac{1}{T_P - T_S} \int_{T_S}^{T_P} k(T)dT \tag{1.83d}$$

$$a_P = \left(\frac{\Delta x \Delta y}{\Delta t}\right) \overline{\rho c} + a_E + a_W + a_N + a_S + \rho_f c_{p_f} \omega_{\text{eff}} \Delta x \Delta y \tag{1.83e}$$

$$b = \left(\frac{\Delta x \Delta y}{\Delta t}\right) \overline{\rho c} T_P^o + \left(\rho_f c_{p_f} \omega_{\text{eff}} T_f + S_m\right) \Delta x \Delta y \tag{1.83f}$$

$$\overline{\rho c} = \frac{1}{T_P - T_P^o} \int_{T_P^o}^{T_P} \rho c(T)dT \tag{1.83g}$$

The superscript o indicates the value at the *old* time t, whereas no superscript is assigned for the value at the new time $t + \Delta t$.

The present computer code is capable of dealing with arbitrary two-dimensional shapes of cryoprobe and tumor as illustrated in Figure 1.12. The initial and boundary conditions for the freezing process using the cryoprobe of outer radius R_p are given as follows:

$$t = 0\text{: } T = T_f \quad \text{(everywhere)} \tag{1.84}$$

$$t > 0 : T|_{x^2+y^2=R_P^2} = T_p \quad \text{(cryoprobe outer surface)} \tag{1.85a}$$

$$T|_{x^2+y^2\to\infty} = T_f \quad \text{(deep tissue region)} \tag{1.85b}$$

Computations were carried out using highly nonunform grid systems, namely, (250 × 500) to cover the right-half domain 30 mm × 60 mm for the case of the longitudinal tumor of 20 mm × 27 mm, and (350 × 700) to cover the right-half domain 160 mm ×160 mm for the case of determining the limiting radius. The results associated with the limiting radius are found to be independent of any additional expansion of the calculation domain. Grid

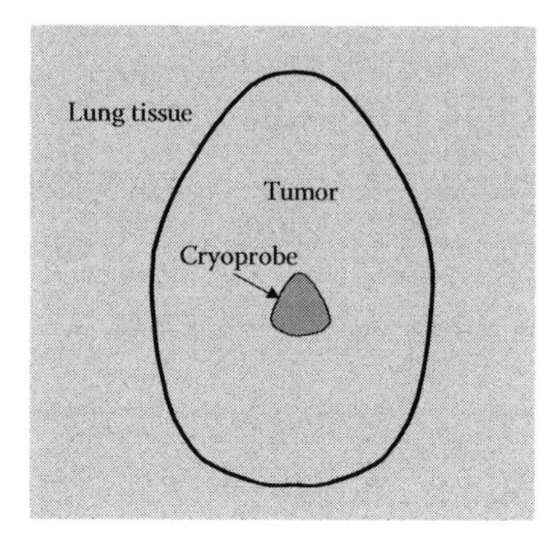

FIGURE 1.12
Numerical model for cryoablation.

nodes are laid out densely around the probe. Grid refinement tests were carried out to ensure that the results are independent of grid systems. Convergence was measured in terms of the maximum change in temperature during an iteration, which was set to 10^{-5}.

1.7.4 Analytical Treatment Based on Integral Method

In what follows, we shall exploit an integral method to derive an analytical expression for the limiting radius of the tumor that one single cryoprobe can freeze at the maximum (Nakayama et al. 2008). For the sake of simplicity in this analytical treatment, we shall assume that the probe is a circular cylinder and that the tumor is so large that heat transfer to the healthy lung tissue is negligible.

The temperature around the cryoprobe is schematically shown in Figure 1.13, where T_p and T_i are the temperatures of the probe and freezing front, respectively, while T_0 is the body temperature. Upon referring to the figure, we may introduce the energy balance relationship at the freezing front at $r = R_i$ as follows:

$$\rho_i h_{sf} R_i \frac{dR_i}{dt} = R_i k_i \left.\frac{\partial T}{\partial r}\right|_{r=R_i} - R_i k_c \left.\frac{\partial T}{\partial r}\right|_{r=R_i} - \rho_c c_c \left(T_0 - T_i\right) R_i \frac{dR_i}{dt} : r = R_i \tag{1.86}$$

where the subscripts i and c refer as to the frozen and unfrozen regions, respectively. The first, second, and third terms on the right-hand side correspond to the conduction heat flux evaluated at the ice side, the conduction heat flux evaluated at the unfrozen side, and the sensible heat entering to the interface as the interface (freezing front) at $r = R_i(t)$ moves radially outward from the cryoprobe.

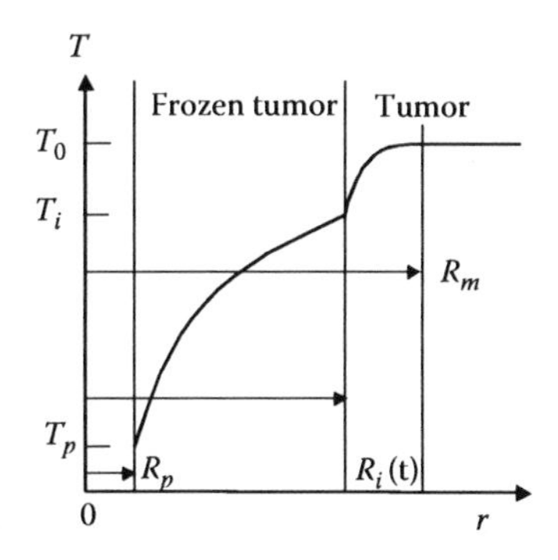

FIGURE 1.13
Temperature profile around a cryoprobe.

The freezing front moves so slowly that quasi-steady approximation may be valid. Thus, assuming that the temperature profile within the frozen region follows that obtained at the steady state, namely,

$$\frac{T - T_p}{T_i - T_p} = \frac{\ln(r/R_p)}{\ln(R_i/R_p)} : R_p \leq r \leq R_i \tag{1.87}$$

where we may estimate the first term on the right-hand side as

$$R_i k_i \left. \frac{\partial T}{\partial r} \right|_{r=R_i} = k_i \frac{T_i - T_p}{\ln(R_i/R_p)} \tag{1.88}$$

To estimate the second term on the right-hand side (representing the heat flux from the unfrozen tumor to the interface), we write the bioheat equation (1.79) for the unfrozen tumor region using the cylindrical coordinate system, which, under the quasi-steady approximation, may be integrated to give

$$-k_c R_i \left. \frac{dT}{dr} \right|_{r=R_i} - \rho_c c_c \omega_{\text{eff}} \int_{R_i}^{R_m} r(T - T_0)dr + S_m \int_{R_i}^{R_m} r dr = 0 \tag{1.89}$$

Let us assume that the temperature in this unfrozen region follows:

$$\frac{T - T_0}{T_i - T_0} = \left(1 - \frac{r - R_i}{R_m - R_i}\right)^2 : R_i \leq r \leq R_m \tag{1.90}$$

The equation satisfies $T = T_i$ at $r = R_i$ and $T = T_0$ and $\partial T/\partial r = 0$ at $r = R_m$ such that the boundary condition given by (1.85b) is satisfied in an approximate sense. Then substituting this temperature profile into equation (1.89), we have

$$2k_c R_i \frac{T_i - T_0}{R_m - R_i} - \rho_c c_c \omega_{\text{eff}} (T_i - T_0) \left(\frac{1}{3}(R_m - R_i) R_i + \frac{1}{12}(R_m - R_i)^2\right)$$
$$+ S_m \frac{R_m^2 - R_i^2}{2} = 0 \tag{1.91}$$

which forms a cubic equation for $R_i/(R_m - R_i)$. The root of the cubic equation is quite complex. However, it is found that the following explicit expression based on Newton's shooting method gives a quite accurate value for the root:

$$\frac{R_i}{R_m - R_i} = R_i\sqrt{\frac{1}{6}\left(\frac{\omega_{\text{eff}}}{\alpha_c} + 3\frac{S_m}{k_c(T_0 - T_i)}\right)} + \frac{1}{8}\left(\frac{\frac{\omega_{\text{eff}}}{\alpha_c} + 6\frac{S_m}{k_c(T_0 - T_i)}}{\frac{\omega_{\text{eff}}}{\alpha_c} + 3\frac{S_m}{k_c(T_0 - T_i)}}\right) \tag{1.92}$$

where $\alpha_c = k_c/\rho_c c_c$ is the thermal diffusivity of the unfrozen tumor. Thus, the second on the right-hand side of equation (1.86) may be estimated as

$$R_i k_c \left.\frac{\partial T}{\partial r}\right|_{r=R_i} = 2k_c(T_0 - T_i)\left(R_i\sqrt{\frac{1}{6}\left(\frac{\omega_{\text{eff}}}{\alpha_c} + 3\frac{S_m}{k_c(T_0 - T_i)}\right)} + \frac{1}{8}\left(\frac{\frac{\omega_{\text{eff}}}{\alpha_c} + 6\frac{S_m}{k_c(T_0 - T_i)}}{\frac{\omega_{\text{eff}}}{\alpha_c} + 3\frac{S_m}{k_c(T_0 - T_i)}}\right)\right) \tag{1.93}$$

Upon substituting (1.88) and (1.93) into (1.86), we have

$$(\rho_i h_{sf} + \rho_c c_c(T_0 - T_i))R_i\frac{dR_i}{dt} = k_i\frac{T_i - T_p}{\ln(\frac{R_i}{R})} - 2k_c(T_0 - T_i)\left(R_i\sqrt{\frac{1}{6}\left(\frac{\omega_{\text{eff}}}{\alpha_c} + 3\frac{S_m}{k_c(T_0 - T_i)}\right)} + \frac{1}{8}\left(\frac{\frac{\omega_{\text{eff}}}{\alpha_c} + 6\frac{S_m}{k_c(T_0 - T_i)}}{\frac{\omega_{\text{eff}}}{\alpha_c} + 3\frac{S_m}{k_c(T_0 - T_i)}}\right)\right) \tag{1.94}$$

which reduces to

$$dt^* = \frac{1 + Sr}{Ste}\frac{R_i^* \ln R_i^*}{1 - \frac{\ln R_i^*}{Cr}\left(\sqrt{\frac{2}{3}}(\omega^* + 3Met)^{1/2}R_i^* + \frac{\omega^* + 6Met}{4(\omega^* + 3Met)}\right)}dR_i^* \tag{1.95}$$

where

$$R_i^* = R_i/R_p \tag{1.96a}$$

and

$$t^* = \alpha_i t/R_p^2 \tag{1.96b}$$

is Fourier number, where $\alpha_i = k_i/\rho_i c_i$ is the thermal diffusivity of the ice. Moreover, the following dimensionless parameters are introduced:

$$Ste_i = \frac{c_i (T_i - T_p)}{h_{sf}}: \text{ Stefan number} \tag{1.97a}$$

$$Sr = \frac{\rho_c c_c (T_0 - T_i)}{\rho_i h_{sf}} \tag{1.97b}$$

$$\omega^* = \frac{\omega_{\text{eff}} R_p^2}{\alpha_c} \tag{1.97c}$$

$$Cr = \frac{k_i (T_i - T_p)}{k_c (T_0 - T_i)} \tag{1.97d}$$

$$Met = \frac{S_m R_p^2}{k_c (T_0 - T_i)} \tag{1.97e}$$

The foregoing ordinary differential equation (1.95) may readily be integrated using any standard integration scheme such as Runge–Kutta–Gill, to find the dimensionless time $t^* = \alpha_c t/R_p^2$ required for freezing the tumor of a given dimensionless radius $R_i^* = R_i/R_p$. Obviously, the quasi–steady assumption is valid when $t^* Ste_i/(1+Sr) > 1$, which roughly gives $t > 1$ sec. Thus, the assumption holds most part of the freezing process except its initial short period.

It is interesting to note that there exists the limiting radius R_{lim} of the tumor that one single cryoprobe can freeze at the maximum. Its dimensionless value $R_{\text{lim}}^* = R_{\text{lim}}/R_p$ may be obtained setting $dR_{\text{lim}}^*/dt^* = 0$, for which equation (1.95) yields

$$\frac{\ln R_{\text{lim}}^*}{Cr} \left(\sqrt{\frac{2}{3}} (\omega^* + 3Met)^{1/2} R_{\text{lim}}^* + \frac{\omega^* + 6Met}{4(\omega^* + 3Met)} \right) = 1 \tag{1.98}$$

This implicit equation gives the dimensionless limiting radius R_{lim}^* for a given set of the dimensionless values, Met, Cr, and ω^*. Usually, ω^* is much larger than Met. For such cases, the following explicit expression based on Newton's shooting method may be used to give a reasonably accurate value for R_{lim}^*:

$$R_{\text{lim}}^* = \frac{\left(\sqrt{\frac{3}{2\omega^*}} Cr + R_0^* \right) R_0^* + \sqrt{\frac{3}{32\omega^*}} R_0^* (1 - \ln R_0^*)}{R_0^* (1 + \ln R_0^*) + \sqrt{\frac{3}{32\omega^*}}} \tag{1.99}$$

where

$$R_0^* = \frac{\frac{3Cr^2}{\omega^*} + \frac{3Cr}{8\omega^*} \left(1 - \ln\left(\sqrt{\frac{3}{2\omega^*}} Cr \right) \right)}{\sqrt{\frac{3}{2\omega^*}} Cr \left(1 + \ln\left(\sqrt{\frac{3}{2\omega^*}} Cr \right) \right) + \sqrt{\frac{3}{32\omega^*}}} \tag{1.100}$$

For the present case of $Cr = 15.4$, equation (1.99) along with (1.100) gives $R_{\text{lim}}^* = 29.9$ and 12.9 for $\omega^* = 0.031$ ($\omega = 0.004$/sec) and 0.310 ($\omega = 0.04$/sec), respectively.

TABLE 1.1
Thermophysical Properties

	Frozen Tumor	Tumor	Lung
Subscript	i	c	l
k [W/mK]	2.20	0.52	0.281
ρ $[\text{kg/m}^3]$	1,000	1,000	550
c [J/kgK]	2,000	4,000	3,710
α $[\text{m}^2/\text{s}]$	1.10×10^{-6}	1.30×10^{-7}	1.38×10^{-7}
$h_{sf} = 3.34 \times 10^5$ [J/kg]			

1.7.5 Limiting Radius for Freezing a Tumor during Cryoablation

Some tissue freezes over a fairly large range of temperatures. However, for the case of lung cancer, the blood comes out from the vessels during the freezing–thawing sequence. The subsequent freezing takes place around the probe surrounded by the blood as a conducting medium. To a first approximation, we may use a single temperature for the phase change. Numerical calculations based on the enthalpy method were carried out for the case, in which the cryosurgical and biological parameters are given by

$$R_p = 1 \text{ mm}, \quad T_p = -135°\text{C}, \quad T_i = -0°\text{C}$$
$$T_f = T_0 \text{ (body temperature)} = 37°\text{C}, \quad S_m = 1,200 \text{ W/m}^3$$

The effective perfusion rate ω_{eff} within the tumor can be quite high since some blood vessels are connected to the tumor. Here, we assume the effective perfusion rate in the range of $\omega_{\text{eff}} = 0.004$ to 0.04/sec. Moreover, the thermophysical properties for frozen and unfrozen tissues in the lung are listed in Table 1.1, according to Yokoyama (1993).

For the case in which $T_p = -135°\text{C}$, $T_i = 0°\text{C}$, $T_0 = 37°\text{C}$, $S_m = 1{,}200$ W/m^3, $\omega_{\text{eff}} = 0.004/\text{s}$, $R_p = 1$ mm, we have $Ste_i = 0.808$, $Sr = 0.443$, $Met = 6.24 \times 10^{-5}$, $Cr = 15.4$, and $\omega^* = 0.031$. A typical evolution of the isotherms obtained for a longitudinal tumor of 20 mm × 27 mm is presented in Figures 1.14(a)–(c). The outermost isotherm in each figure corresponds to the freezing front (i.e., $T = T_i = 0°$). Figure 1.14(c) clearly indicates that ill placement of the probe may result in a substantial damage to the surrounding healthy tissue.

Let us consider the freezing process when the probe is placed in a large tumor. The temporal evolutions of the freezing front for the cases of

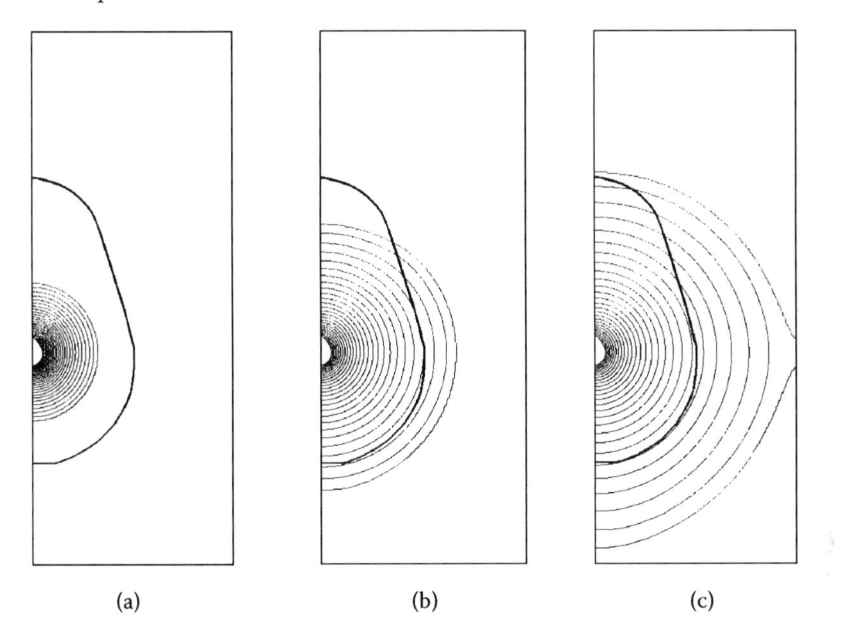

FIGURE 1.14
Temporal evolution of isotherms (interval 5°): (a) 60 sec, (b) 300 sec, and (c) 600 sec.

$\omega_{\text{eff}} = 0.004/\text{sec}$ (low perfusion) and $0.040/\text{sec}$ (high perfusion) are illustrated in Figures 1.15(a) and 1.15(b), respectively, along with the curve analytically obtained by integrating the ordinary differential equation (1.95). The figures may also be used to know the time required to kill the circular tumor of radius R_i. The numerical results obtained for these two cases in the figures clearly show that the limiting radii R_{lim} for $\omega_{\text{eff}} = 0.004/\text{sec}$ and $0.040/\text{sec}$ are around 29.9 mm and 12.8 mm, respectively, which are estimated on the basis of the analytical expression (1.99).

Finally, the curve representing the limiting radius is generated from equation (1.99) and plotted against the effective perfusion rate in Figure 1.16. We learn from the figure that a single probe, even when placed in the center of the target, is capable of freezing only the size of a tumor whose equivalent radius is less than the limiting radius, R_{lim}. The figure indicates that, for the case of comparatively high perfusion rate, a single probe of radius 1 mm can freeze a tumor only within the radius of 20 mm or less. This is consistent with the fact reported by Nakatsuka et al. (2004). In practice, we may introduce a factor λ and estimate the range of the killed tissue by $r \leq \lambda R_{\text{lim}}$. The factor λ has to be chosen carefully, depending on the specific clinical and surgical constraints, such as the number of cryoprobes available, the time set for a single freezing process, and the level of malignancy.

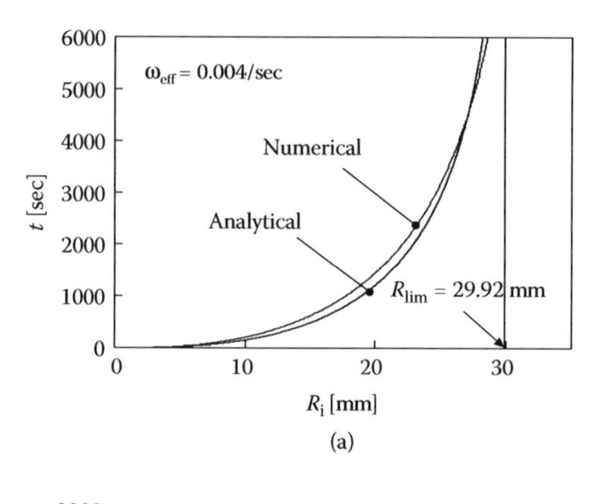

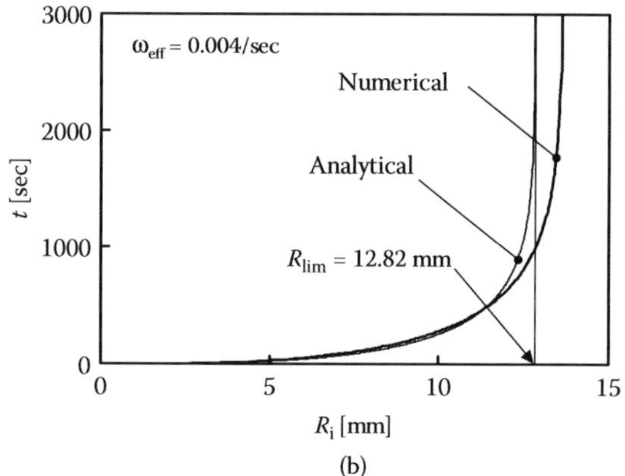

FIGURE 1.15
Time required for freezing the tumor: (a) low perfusion rate, (b) high perfusion rate.

1.8 Conclusions

In this chapter, a general set of bioheat transfer equations for blood flows and its surrounding biological tissue was derived using a VAT established in the field of fluid-saturated porous media. Unknown correlations were modeled in terms of macroscopic determinable quantities. It has been shown that the resulting two-energy equation model reduces to existing empirical models such as the Pennes model, the Wulff model, and their modifications, under appropriate conditions. Subsequently, the two-energy equation model has been extended to the three-energy equation model, so as to account for the effect of countercurrent heat transfer between closely spaced arteries and veins in the

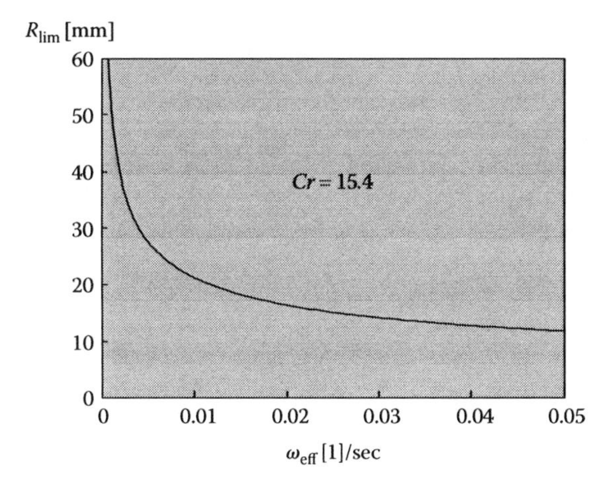

FIGURE 1.16
Effects of the perfusion rate on the limiting radius.

blood circulatory system. The resulting model, under appropriate conditions, naturally reduces to those introduced by Chato, Bejan, Keller, and Seiler and Weinbaum and Jiji for countercurrent heat transfer for the case of closely aligned pairs of vessels.

As for an application of the bioheat equation, the freezing process within a tumor during cryoablation therapy was investigated both analytically and numerically. The freezing front in a tumor during percutaneous cryoablation was traced exploiting the bioheat equation. It has been found that there exists a limiting size of the tumor that one single cryoprobe can freeze at the maximum. An excellent agreement between the analytical and numerical results has been achieved for the time required to freeze the tumor using the cryoprobe of one single needle. The resulting analytical expression for estimating the limiting radius provides useful information for cryotherapy treatment plans.

1.9 Nomenclature

A	Surface area (m^2)
a_f	Specific surface area (1/m)
A_{int}	Interface between the fluid and solid (m^2)
b_{ij}	Forchheimer tensor (1/m)
c_p	Specific heat at constant pressure (J/kg K)
Cr	Dimensionless parameter associated with the thermal conductivity ratio

h_f	Interfacial heat transfer coefficient (W/m^{2}K)
h_{sf}	Latent heat of solidification (W/kg)
k	Thermal conductivity (W/mK)
K_{ij}	Permeability tensor (m^2)
Met	Dimensionless number associated with metabolic reaction rate
n_j	Unit vector pointing outward from the fluid side to solid side (–)
p	Pressure (Pa)
r	Radial coordinate (m)
R_i	Radius of the freezing front (m)
$R_{\lim}$	Limiting radius (m)
S_m	Metabolic reaction rate (W/m^3)
Ste	Stephan number (–)
T	Temperature (K)
t^*	Fourier number (–)
T_0	Body temperature (K)
T_i	Phase change tempearture (K)
T_p	Probe temperature (K)
u_i	Velocity vector (m/sec)
V	Representative elementary volume (m^3)
x, y	Cartesian coordinates (m)
α	Thermal diffusivity (m^2/sec)
ε	Porosity (–)
ν	Kinematic viscosity (m^2/sec)
P	Density (kg/m^2)
ω	Perfusion rate (1/sec)
ω'	Net filtration rate (1/sec)
ω^*	Dimensionless perfusion rate (–)
ω_{eff}	Effective perfusion rate (1/sec)

Special symbols

$\tilde{\phi}$	Deviation from intrinsic average
$\langle\phi\rangle$	Volume average
$\langle\phi\rangle^{f,s,a,v}$	Intrinsic average

Subscripts and superscripts

a	Artery
c	Unfrozen tumor
dis	Dispersion
f	Fluid
i	Ice interface
p	Cryoprobe
s	Solid
v	Vein
$*$	Dimensionless

1.10 References

Baissalov, R., Sandison, G. A., Reynolds, D., and Muldrew, K. (2001). Simultaneous optimization of cryoprobe placement and thermal protocol for cryosurgery. *Physics in Medicine and Biology,* **46**:1799–1814.

Bazett, H. C., Love, L., Eisenberg, L., Day, R., and Forster, R. E. (1948a). Temperature change in blood flowing in arteries and veins in man. *Journal of Applied Physiology,* **1**:3–19.

Bazett, H. C., Mendelson, E. S., Love, L., and Libet, B. (1948b). Precooling of blood in the arteries, effective heat capacity and evaporative cooling as factors modifying cooloing of the extremities. *Journal of Applied Physiology,* **1**:169–182.

Bejan, A. (1979). A general variational principle for thermal insulation system design. *International Journal of Heat and Mass Transfer,* **22**:219–228.

Bischof, J. C., Smith, D., Pazhayannur, P. V., Manivel, C., Hulbert, J., and Roberts, K. P. (1997). Cryosurgery of dunning AT-1 rat prostate tumor: thermal, biophysical, and viability response at the cellular and tissue level. *Cryobiology,* **34**:42–69.

Butz, T., Warfield, S. K., Tuncali, K., Silverman, S. G., van Sonnenberg, E., Jolesz, F. A., and Kikinis, R. (2000). Pre- and intra-operative planning and simulation of percutaneous tumor ablation. *Proceedings of the Medical Image Computing and Computer Assisted Intervention, MICCAI,* pp. 317–326.

Charny, C. K. (1992). *Mathematical Models of Bioheat Transfer. Advances in Heat Transfer,* **22**:19–155, Academic Press, New York.

Chato, J. C. (1980). Heat transfer to blood vessels. *ASME Journal of Biomechanical Engineering,* **102**:110–118.

Chen, M. M. and Holmes, K. R. (1980). Microvascular contributions in tissue heat transfer. *Annals of the New York Academy of Sciences,* **335**:137–150.

Cheng, P. (1978). Heat transfer in geothermal systems. *Advances in Heat Transfer,* **14**:1–105, Academic Press, New York.

Keanini, R. G. and Rubinsky, B. (1992). Optimization of multi-probe cryosurgery. *ASME Transactions Journal of Heat Transfer,* **114**:796–802.

Keller, K. H. and Seilder, L. (1971). An analysis of peripheral heat transfer in man. *Journal of Applied Physiology,* **30**:779–789.

Khaled, A.-R. A. and Vafai, K. (2003). The role of porous media in modeling flow and heat transfer in biological tissues. *International Journal of Heat Mass Transfer*, **46**:4989–5003.

Khanafer, K. and Vafai, K. (2006). The role of porous media in biomedical engineering as related to magnetic resonance imaging and drug delivery. *Heat Mass Transfer*, **42**:939–953.

Klinger, H. G. (1978). Heat transfer in perfused biological tissue. II. The 'macroscopic' temperature distribution in tissue. *Bulletin of Mathematical Biology*, **40**:183–199.

Mitchell, J. W. and Myers, G. E. (1968). An analytical model of the counter-current heat exchange phenomena. *Biophysics Journal*, **8**:897–911.

Nakatsuka, S., Kawamura, M., Sugiura, H., Nakano, K., Izumi, Y., Kobayashi, K., Jinzaki, M., Hashimoto, S., Kuribayashi, S., Wakabayashi, G., and Kitajima, M. (2004). Preliminary experience with percutaneous cryoablation for malignant lung tumors under CT fluoroscopic guidance. *Low Temperature Medicine*, **30**(1):9–15.

Nakayama, A. (1995). *PC-Aided Numerical Heat Transfer and Convective Flow.* CRC Press, Boca Raton, FL.

Nakayama, A. and Kuwahara, F. (2008). A general bioheat transfer model based on the theory of porous media. *International Journal of Heat Mass Transfer*, **51**:3190–3199.

Nakayama, A., Kuwahara, F., and Hayashi, T. (2004). Numerical modelling for three-dimensional heat and fluid flow through a bank of cylinders in yaw. *Journal of Fluid Mechanics*, **498**:139–159.

Nakayama, A., Kuwahara, Y., Iwata, K., and Kawamura, M. (2008). The limiting radius for freezing a tumor during percutaneous cryoablation. *Journal of Heat Transfer*, **130**(11):111101.

Nakayama, A., Kuwahara, F., and Kodama, Y. (2006). A thermal dispersion flux transport equation and its mathematical modelling for heat and fluid flow in a porous medium. *Journal of Fluid Mechanics*, **563**:81–96.

Nakayama, A., Kuwahara, F., and Liu, W. (2008). A macroscopic model for countercurrent bioheat transfer in a circulatory system. *Journal of Porous Media*, **12**:289–300.

Nakayama, A., Kuwahara, F., Sugiyama, M., and Xu, G. (2001). A two-energy equation model for conduction and convection in porous media. *International Journal of Heat Mass Transfer*, **44**:4375–4379.

Nakayama, A., Kuwahara, F., Umemoto, T., and Hayashi, T. (2002). Heat and fluid flow within anisotropic porous medium. *Journal of Heat Transfer*, **124**:746–753.

Patankar, S. V. (1980). *Numerical Heat Transfer and Fluid Flow.* Hemisphere Publishing Corp., Washington D.C.

Pennes, H. H. (1948). Analysis of tissue and arterial blood temperature blood temperature in the resting human forearm. *Journal of Applied Physiology*, **1**:93–122.

Quintard, M. and Whitaker, S. (1993). One and two equation models for transient diffusion processes in two-phase systems. *Advances in Heat Transfer*, **23**:369–465.

Rabin, Y. and Shitzer, A. (1998). Numerical solution of the multidimensional freezing problem during cryosurgery. *ASME Transactions Journal of Heat Transfer*, **120**:32–37.

Rewcastle, J., Sandison, G., Hahn, L., Saliken, J., McKinnon, J., and Donnelly, B. (1998). A model for the time-dependent thermal distribution within an ice ball surrounding a cryoprobe. *Physics in Medicine and Biology,* **43**: 3519–3534.

Roetzel, W. and Xuan, Y. (1998). Transient response of the human limb to an external stimulus. *International Journal of Heat and Mass Transfer*, **41**:229–239.

Rossi, M. R. and Rabin, Y. (2007). Experimental verification of numerical simulations of cryosurgery with application to computerized planning. *Physics in Medicine and Biology*, **52**:4553–4567.

Rossi, M. R., Tanaka, D., Shimada, K., and Rabin, Y. (2007). An efficient numerical technique for bioheat simulations and its application to computerized cryosurgery planning. *Computer Methods and Programs in Biomedicine*, **85**:41–50.

Scholander, P. F. and Krog, J. (1957). Countercurrent heat exchange and vascular bundles in sloths. *Journal of Applied Physiology,* **10**: 405–411.

Vafai, K. and Tien, C. L. (1981). Boundary and inertia effects on flow and heat transfer in porous media. *International Journal of Heat and Mass Transfer*, **24**:195–203.

Wan, R., Liu, Z., Muldrew, K., and Rewcastle, J. (2003). A finite element model for ice ball evolution in a multi-probe cryosurgery. *Computer Methods in Biomechanics and Biomedical Engineering*, **6(3)**:197–208.

Weinbaum, S. and Jiji, L. M. (1979). A two phase theory for the influence of circulation on the heat transfer in surface tissue. In *Advances in Bioengineering*, ed. M. K. Wells. ASME, New York, NY, pp. 179–182.

Weinbaum, S. and Jiji, L. M. (1985). A new simplified bioheat equation for the effect of blood flow on local average tissue temperature. *ASME Journal of Biomechanical Engineering*, **107**:131–139.

Wulff, W. (1974). The energy conservation equation for living tissue. *IEEE Transactions on Biomedical Engineering*, **BME-21**:494–495.

Xuan, Y. and Roetzel, W. (1997). Bioheat equation of the human thermal system. *Chemical Engineering & Technology*, **20**:268–276.

Yokoyama, S. (1993). *Bioheat Transfer Phenomena (in Japanese)*. Hokkaido University Press, Japan.

2

Mathematical Models of Mass Transfer in Tissue for Molecular Medicine with Reversible Electroporation

Yair Granot, Boris Rubinsky

Graduate Group in Biophysics, Department of Mechanical Engineering, University of California, Berkeley, CA

CONTENTS

2.1 Introduction

Every cell is distinguished from its environment by means of a semipermeable membrane. This membrane plays a crucial role in the ability of the cell to function; the loss of membrane integrity, except for brief periods, would almost certainly lead to the cell's death. Manipulation of the cell membrane permeability has been long sough after by many researchers as a means of controlling cells and tissues. A controlled method of inserting or extracting certain substances from the cell without compromising its viability may be at the foundation of molecular medicine where drugs or genes can be directed

Acknowledgment
B.R. was supported by The Israel Science Foundation (grant no. 403/06).

toward a cell to alter its behavior. One prominent method of modifying the permeability of the membrane is the use of electric fields to create a large transmembrane potential in a process known as electroporation.

Interactions between electric fields and tissues have been studied for centuries long before the detailed structures of the cell and its membrane were discovered (Rubinsky 2007). Several studies from the mid-twentieth century and later (Coster 1965; Sale and Hamilton 1967; Neumann and Rosenheck 1972; Crowley 1973; Lindner et al. 1977; Abidor et al. 1979) began to explore in a more detailed manner the specific interactions between transmembrane potential and the membrane's characteristics, such as its permeability and its mechanical structure. Although we still lack complete knowledge about the exact mechanism that causes the perceived changes to the membrane under large electric fields, electroporation is widely regarded today as the process that is in the heart of the measured phenomena. The explanation is based on the assumption that a large transmembrane potential causes a physical change to the mechanical structure of the membrane, creating miniature pores. These pores may expand in size if the transmembrane potential persists and usually take one of the two configurations: stable pores, which may stay open even after the transmembrane potential returns to its normal value at rest, or unstable pores, which reseal very rapidly when the high electric field is terminated. Another outcome of electroporation may be the complete disintegration of the membrane due to numerous pores that have grown so large as to prevent the membrane from resealing.

When the membrane does not reseal after the application of high electric field pulses, the process is called *irreversible electroporation,* and one of its obvious results is that the cell will not be able to survive. It is important to note that in some cases even cells that managed to reseal their membrane may ultimately die as a direct result of the electroporation. For example, this may occur when the membrane was in a high-permeability state for a long time and the concentration of various ions reached levels that prevented the cell from recovering. The advantages of tissue ablation using irreversible electroporation have gained attention recently (Davalos et al. 2005; Esser et al. 2007; Rubinsky et al. 2007) after years of being considered an unwanted result in the process of reversible electroporation.

Since its discovery, reversible electroporation has become one of the most useful lab techniques for introducing proteins, genes, and other molecules into cells. It is one of the prominent methods used in laboratories today for transfection, for example, for creating knockout mice. *In vitro* applications of reversible electroporation are usually carried out by placing a suspension of cells in a cuvette, a small tube with two metallic plates that act as electrodes. A voltage is applied to the plates of the cuvette and thus a high-magnitude electric field is created inside the cuvette, where the cells are found. The field is relatively uniform with values close to the applied voltage divided by the distance between the plates. This is a reasonably controlled environment, and many protocols have been devised over the years for various cell types

(Nickoloff 1995). Many of the protocols are empirical and describe routines that have worked well under certain conditions.

Considerably less attention has been focused on gaining more insight into the actual mechanism that is responsible for the increase in the membrane permeability. Research in this direction has several subcategories: attempts to image or otherwise determine the mechanical structure of the membrane (Chang and Reese 1990; Teissie et al. 2005), development of biophysical models that explain the time evolution of electroporation (Pastushenko et al. 1979; Weaver and Chizmadzhev 1996; Pliquett et al. 2004; Krassowska and Filev 2007), and building devices that may monitor the electroporation process, preferably in real time (Kinosita et al. 1988; Hibino et al. 1993; Pavlin et al. 2005; Ivorra and Rubinsky 2007).

An important tool in estimating the change of membrane permeability as a result of electroporation is monitoring the electrical conductivity (Davalos et al. 2002). In contrast with high-specificity mechanisms such as ion channels for passive transport or ion pumps for active transport, which allow only certain ions to flow through them, the pores that are created during electroporation are relatively nonspecific. This allows numerous particles, including ions, to flow quite freely through the membrane, which ordinarily functions as a high-capacitance element and does not permit low-frequency electrical currents to pass easily through the cell. Measuring the change in the electrical conductivity of a cell can thus reveal changes in the membrane permeability due to electroporation. For tissue electroporation, and particularly in the case of *in vivo* applications, this may be the only practical method of controlling electroporation in real time.

The changes in tissue conductivity, however, are not easily translated into permeability values. The tissue is a complex and nonuniform structure with cells, extracellular matrix, blood vessels, and extracellular medium. Low-frequency electric currents through the tissue are mostly carried by ions that usually flow between the cells. The tissue conductivity when the cells are not electroporated is determined by several factors such as the amount of extracellular medium and the density with which the cells are packed. When the cells in a certain part of the tissue become electroporated, ions start to flow *through* the cells in that region, increasing the tissue conductivity there. At the same time, the number of pores as well as their size depends on local electrical fields, which may vary considerably. The creation of pores immediately affects the transmembrane potential so a delicate equilibrium must be reached for the pores to become stable. Otherwise, they may shrink and disappear or, alternatively, expand to such a degree that the membrane may be destroyed. The function linking the permeability of the cells in the tissue and the electrical conductivity is quite difficult to model and it is not an injective function, that is, there are numerous permeability values and geometrical configurations in a tissue that yield the same tissue conductivity value.

The electroporation process is difficult to isolate from other processes that occur in the cell and its membrane. Many parameters affect the triggering and amplitude of electroporation, so under different conditions the behavior of the cell may be difficult to analyze. An effort to model electroporation as an isolated phenomenon would require strict calibration and control of many environmental parameters during electroporation tests and measurements. For example, when hydraulic stress is applied, the transmembrane potential threshold is decreased so pores start to form under conditions that would not cause normal cells to go through electroporation. Furthermore, electroporation is not unique in the sense that similar phenomena due to other causes can be observed, unrelated to transmembrane potential. This is true for both creating pores in the membrane as well as sealing them. The ability of the cell to reseal, for example, after a mechanical puncture of the membrane has been studied carefully (Steinhardt et al. 1994; McNeil and Steinhardt 1997). Similar mechanisms may be involved in the resealing of the membrane after electroporation. This makes the isolated study of this process quite challenging.

In this chapter, we will review several aspects of the electroporation process and how it is used to introduce drugs into cells in a specific part of a tissue. We begin by describing the fundamental mechanism of electroporation and continue with some mathematical models of ion transport during the process. We then describe a method of monitoring and controlling *in vivo* electroporation and relate that to mass transfer in tissue, which gives rise to a hierarchical model for drug delivery using electroporation.

2.2 Fundamental Aspects of Reversible Electroporation

The electroporation process can be divided into five steps (Teissie et al. 2005):

1. Excitation
2. Pore expansion
3. Stabilization
4. Resealing
5. Long-term recovery

In the excitation stage, the membrane is excited by some high voltage pulse and becomes more permeable due to the creation of pores, roughly the size of 1 nm (Weaver and Chizmadzhev 1996). This is perhaps the most elusive part of the electroporation process, and different theories exist on how exactly these initial "aqueous pathways" are created. One of the leading assumptions describes this initial stage as the creation of hydrophobic pores. A bilayer lipid membrane at rest that may look like the schematic diagram shown in

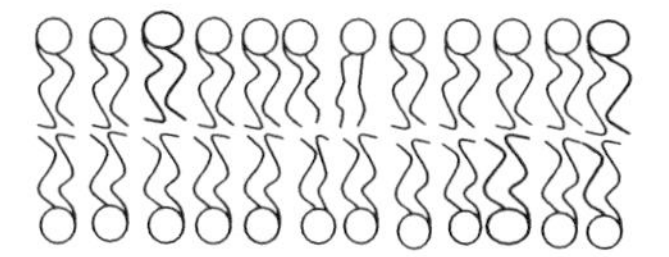

FIGURE 2.1
Schematic diagram of a bilayer lipid membrane.

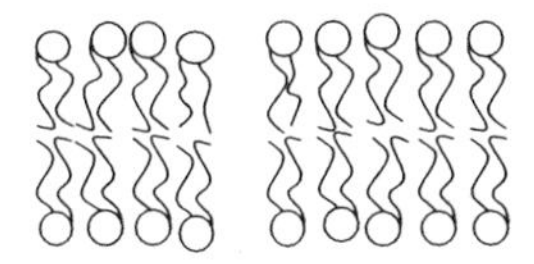

FIGURE 2.2
Schematic diagram of a membrane with a hydrophobic pore.

Figure 2.1 will deform to create a small pore, which at this stage is referred to as a hydrophobic pore, such as the one shown in Figure 2.2. The pore is hydrophobic because the fatty hydrophobic chains are exposed. This is not a thermodynamically preferred configuration, and the pore is not stable.

There are substantial experimental data that show that lipid bilayers experience electrical breakdown when the transmembrane voltage exceeds some threshold, usually around 200 mV. This may occur after as little as 1 μsec, or up to several milliseconds, depending on different attributes of the membrane and the environment (Chen et al. 2006). The fact that this initial step can be detected in both artificial planar bilayer lipid membranes as well as the cell's membrane indicates that this phenomenon does not depend on some active processes in the cell and cannot be attributed to unique components of the cell membrane such as ion channels or other membrane proteins. This does not mean that such elements do not play a role in the overall electroporation process. In fact, many experiments suggest they do, but the fundamental phenomenon does not depend on them.

In the second stage, the pores expand as the excitation pulse continues (Prausnitz et al. 1995). Two simultaneous processes occur. The existing pores expand in size, increasing the initial radius from less than 1 nm to several nanometers up to 100 nm. At the same time, new pores continue to form in those parts of the membrane that are above the electroporation voltage threshold. The bilayer membrane becomes leaky, and significant molecular transport occurs. This is a complicated scenario since the increase in permeability also has a great effect on the transmembrane potential. The increase in membrane conductivity lowers the transmembrane potential, which has an attenuating effect on the process itself. When the radius of a pore increases, it turns from a hydrophobic into a hydrophilic pore, such as the one depicted in Figure 2.3. This kind of pore is much more stable than the hydrophobic pore that was first created. Molecular dynamics simulations (Tieleman 2004; Tarek 2005) have

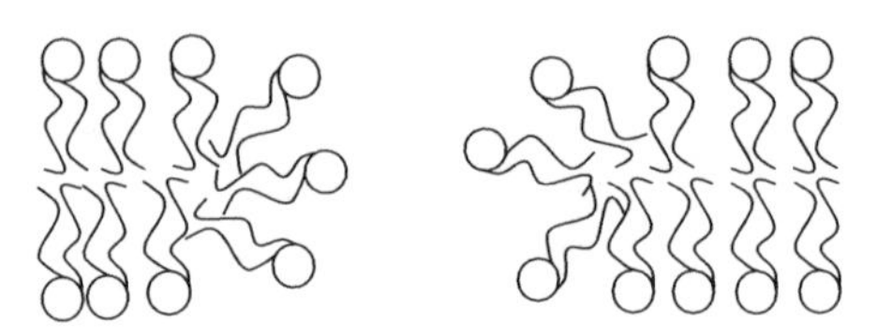

FIGURE 2.3
Schematic of a membrane with a hydrophilic pore.

been able to show very similar structures to the pores described here. This sounds promising in terms of confirming the theories regarding the creation of pores and the mechanisms of electroporation in general, but some of the details of the results of these simulations do not fully agree with the data recorded in experiments. For instance, the fact that only relatively large transmembrane voltages are able to simulate these pores raises concerns about the accuracy of such simulations. A possible explanation to this specific problem may be found in some impurities that usually exist in membranes but have not been included in these simulations. Forces exerted on such impurities may enhance the effect of water molecules that are being forced into the membrane due to some nonuniformity in the electric field near the membrane–cytosol boundary.

Direct imaging of electroporation pores could possibly provide a convincing proof to the theory of electroporation. Many attempts to do that have failed, and in one case where the authors have claimed to have imaged pore using freeze fracture electron microscopy techniques (Chang and Reese 1990) results are inconclusive. The pores that are seen in this case may have been experimental artifacts (Teissie et al. 2005). These images were not reproducible under different circumstances and so the ultimate proof remains elusive.

The next stage is the stabilization. After the electric pulse has ended the transmembrane potential decreases and the pores begin to reseal. This process occurs very rapidly in the order of 1 msec or less. The permeability of the membrane does not return to its original value. Although many of the pores reseal with the removal of the external electric field, some of the larger pores remain open for very long. This implies that, while most of the pores were kept open because of the forces generated by the external electric field, some of them took on a more stable configuration and they remain open and allow general transport of molecules across the membrane.

The fourth stage of the process is resealing where the stable pores begin to reseal and the membrane returns to a state very similar to its initial configuration. This process may take several seconds and even minutes after the termination of the pulse. Its duration is influenced by many factors such as temperature, mechanical stress on the membrane, membrane composition, and the chemical environment. For example, drugs that interfere with the cytoskeleton of the cell, as well as other physical treatments that change its organization, have a considerable effect on the resealing process. This may be similar to the mechanism of pore resealing that the cell applies regardless of the pore's origin (Steinhardt et al. 1994). Although the number of pores in this stage is much smaller than, during the pulse and the molecular transport has also decreased considerably, most of the important mass transfer occurs during this stage. The short duration of the pulses, which are required to prevent the cells membrane from completely disintegrating, make it very difficult to obtain substantial molecular and ionic transport during the pulse duration. The existence of the very large electric field also has an effect on the charge molecules that may pass during the pulse duration, although this effect may sometimes be constructive and sometime destructive depending on the charge and the geometrical configuration.

The final stage of the process, the long-term recovery, may linger for several hours, but it is probably the less important part of electroporation. Some studies have shown that even 4 h after electroporation, when the pores have long been resealed and the original permeability has been restored, the cell may not return to its exact original state (Rols et al. 1995). For instance, endocytosis and macropinocytosis were observed in electroporated cells even though cells that were not treated with electrical pulses did not exhibit similar behavior. Another long-lasting effect, which is not really a part of the electroporation process but rather a consequence of this procedure, is the terminal effect that electroporation may have on some of the cells. This is what we have referred to as irreversible electroporation, and it is sometimes the ultimate goal of the procedure.

In many cases irreversible electroporation is considered as an instance of unresealable pores. This is certainly the case in several scenarios where the applied voltage is very high or when the expansion of the pores persists for a very long time to the point that the membrane ruptures and it is not able to reseal. Sometimes the damage is too great even for the cytoskeletal mechanisms of the cell to fix. The loss of membrane can only mean one thing—cell death. However, another mechanism of irreversible electroporation exists, one which may be just as important. In this case, the cell manages to keep the membrane in tact and the pores do reseal, either spontaneously or with the aid of the cytoskeleton, but homeostasis is lost and the cell is not able to recover from this shock. This will lead to cell death but on a much longer time scale and possibly with very different characteristics.

2.3 Mathematical Models of Ion Transport during Electroporation

Many of the models developed for electroporation focused on the membrane itself and often referred to a single cell (Abidor et al. 1979; Lewis 2003; Krassowska and Filev 2007), but an important thing to consider in the case of *in vivo* electroporation is what cells go through when they are packed in a tissue (Esser et al. 2007). In some solid tissues the cells may be very tightly packed but not so much in others, so there are considerable extracellular paths with conductive medium to allow most of the electric current at low frequency to flow between the cells. We consider two important effects of the packing of cells: first, the local transmembrane potential at various parts of each cell's membrane, which changes considerably with the geometry of the cell, and, second, the volumetric ratio of intracellular to extracellular medium. This ratio is much smaller in tissue than in common *in vitro* setups so there is less effect of "chemical stress" due to longer periods of open pores. When pores remain open for a long time and the amount of extracellular medium is relatively large, concentrations of essential ions may drop below critical levels, while unwanted molecules may invade the cell in non-negligible quantities, exerting chemical stress on the cell. In a tightly packed tissue on the other hand, even when the cell's membrane is permeable for a relatively long period of time and basically uncontrolled exchange occurs freely between the cell and its surroundings, the cell may still keep its homeostasis because of the limited amount of extracellular medium in its immediate vicinity.

As for the first effect, Figure 2.4 depicts a schematic diagram of several cells of various shapes packed together. Some cells in this example are very close and some have larger extracellular paths between them. The colors on the cell membranes represent the simulated transmembrane potential due to

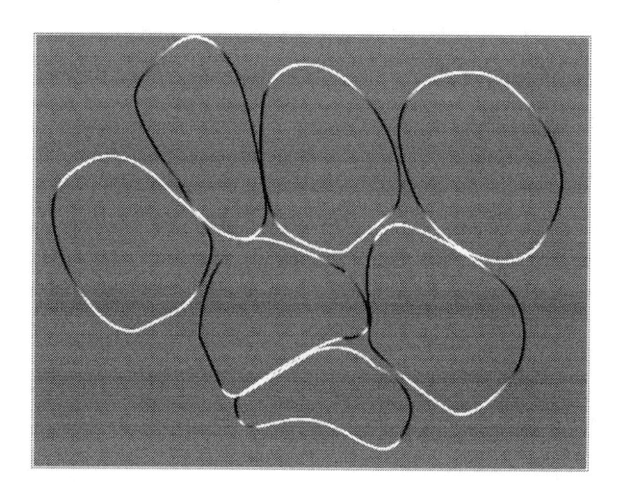

FIGURE 2.4
Schematic diagram of packed cells in a solid tissue and the simulated transmembrane potential because of a horizontal electric field that is applied externally. The warm colors represent large transmembrane potentials and cool colored regions are where the potential is close to zero. The resting potential is ignored in this case.

a horizontal electric field that is applied to this small sample. The value of the transmembrane potential strongly depends on the geometrical shape of the cells. While the peaks tend to be aligned with the external field, like as in the case of spherical cells, the breadth and amplitude of the high transmembrane potential region varies significantly from cell to cell. Realistic biophysical models of electroporation should take into account the variation of the transmembrane potential in solid tissue, but since even theoretical models regarding a single spherical cell are currently less than satisfying, a more comprehensive treatment of this point would probably have to be postponed.

2.4 Electrical Impedance Tomography of *in vivo* Electroporation

Medical treatment procedures that are performed deep inside the body generally require some sort of a feedback mechanism to monitor and control the procedure. Relying on the change in tissue conductivity may be an effective method of monitoring *in vivo* electroporation. One modality of imaging tissue conductivity is known as electrical impedance tomography, or EIT (Cheney et al. 1999; Brown 2003; Bayford 2006). Electric currents are injected into the tissue using external electrodes. These currents flow through the tissue using the path of least resistance, which depends, amongst other things, on the geometrical configuration of the tissue, its internal structure, the location of the electrodes, and also on the specific conductivity distribution. By measuring the electric potential on the boundary of the tissue using other electrodes it is possible to estimate the path of the current and deduce the internal conductivity distribution.

Some of the advantages of using EIT for medical imaging include the low cost and simplicity of the hardware components of the imaging device. The basic system only requires a low-power current source, several electrodes, a voltage measurement device, and a signal processing unit. All of these components are standard, low-cost devices, with the possible exception of the signal processing unit, which can actually be separated from the other components to simplify the system and lower its cost (Granot et al. 2008). Other medical imaging systems such as magnetic resonance imaging, computerized tomography, or positron emission tomography rely on special equipment with strict requirements for several features, which make them expensive. However, the image quality of EIT is inferior to the quality of the images produced by these modalities particularly in terms of image resolution and contrast. These disadvantages can sometimes be partially compensated for by the high frame rate of EIT, which can image the target at 30 images or more per second.

To assess the ability of EIT to image electroporated tissue we can use a mathematical model to simulate the changes in tissue conductivity due to electroporation and then simulate the measured voltages in the EIT measurement

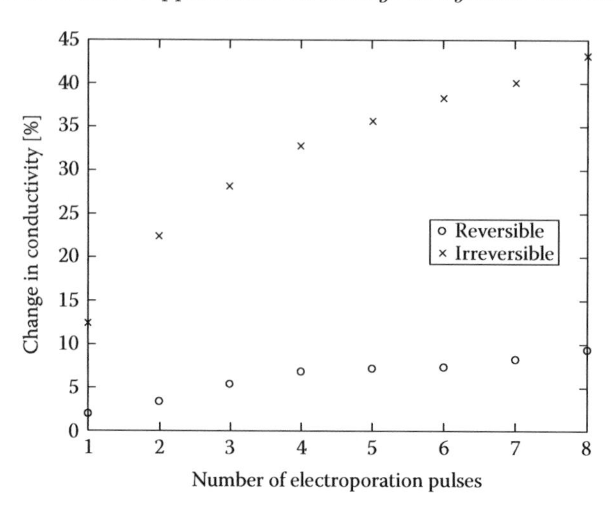

FIGURE 2.5
Change in rat liver conductivity during a series of eight electroporation pulses. The change is expressed in terms of the mean change in tissue conductivity as percentage of the initial conductivity. Two electric fields were used: one for reversible electroporation at 450 V/cm and one for irreversible electroporation at 1,500 V/cm.

phase for various current injections. The measurements would then be used to calculate the conductivity distribution in the tissue using the conventional EIT algorithms.

Currently there are no reliable models to predict the increase in conductivity as a function of the electroporation parameters. As a first estimate we rely on empirical data that were published for reversible and irreversible electroporation in rat liver (Ivorra and Rubinsky 2007). In this study, rat liver tissue were exposed *in vivo* to nearly uniform electric fields and the changes in conductivity were measured. The researchers used eight electroporation pulses and recorded the measurements after every pulse. Two groups of rats were tested, one with electroporation pulses that were designed to cause reversible electroporation and were set to 450 V/cm and the other group designed to cause irreversible electroporation with the electric field set to 1,500 V/cm. Figure 2.5 depicts the results of these tests showing the mean change in conductivity for each group after every pulse from the electroporation protocol. The pulses were applied at intervals of 0.5 sec, and each pulse lasted 100 μsec.

These results give some idea about the extent of changes in conductivity we may expect from a similar electroporation protocol. Nevertheless, an *in vivo* medical application would ordinarily make use of nonuniform electric fields because of the difficulty in creating uniform fields with a minimally invasive device and the need to shape the effective field in such a way as to treat certain parts of the tissue without affecting other neighboring regions. Therefore, we need to expand the experimental data to interpolate the conductivity changes for other electric field values. We assume, for the purpose of this model, that below 450 V/cm there are no significant changes in conductivity because the transmembrane potential of the cells would be below the electroporation threshold. We further assume that above 1,500 V/cm the conductivity changes reach a saturated value and will not increase any further. This is a reasonable assumption because tests have shown (Weaver and Chizmadzhev 1996) that above a certain threshold the effect of the membrane on conductivity is negligible and the conductivity value is similar to that of a configuration with no membrane at all.

A further assumption is that the conductivity increases linearly with electric fields between 450 and 1,500 V/cm. An increase in the conductivity has been observed in several studies (Pavlin et al. 2005; Pavlin and Miklavcic 2008) although not enough data has been studied to validate the linearity assumption. However, this simple model seems close enough to available experimental results to make it a good basis for a first model. Figure 2.6 depicts the results of combining these assumptions with the empirical results of electroporation in rat liver tissue. Every line represents the percent of change in conductivity relative to the initial value for a different number of pulses. The first pulse represents the changes after a single pulse and the last line the changes after the entire set of eight pulses. For nonuniform electric fields, different parts of the tissue experience electroporation in a different manner, and the changes in conductivity at each region would depend on the local electric field.

This model can be used to calculate the effects of electroporation on tissue conductivity for various configurations. To obtain the electric field for each point in the model we need to solve the Laplace equation:

$$\nabla(\sigma \nabla u) = 0 \tag{2.1}$$

where σ is the electrical conductivity and u is the electric potential. An analytic solution is possible only for a limited number of cases so usually numerical methods are applied. One of the most useful tools to solve the Laplace equation is the finite element method (FEM) in which the domain is divided into very small homogenous elements for which the equation can be easily solved (Vauhkonen et al. 2001). The boundary conditions are obtained by assuming that the boundaries of the tissue are electrically insulated and that the electrodes have a known and constant potential as determined by the voltage we apply. The initial conductivity of the tissue at every point σ is assumed to be known and usually taken to be homogeneous.

The solution of the Laplace equation is u for every point, and by taking the gradient of u we obtain the electric field. Using the functions of Figure 2.6 we have the conductivity after the first pulse of electroporation. This stage in the process may be refined by using an iterative procedure (Pavselj et al. 2005)

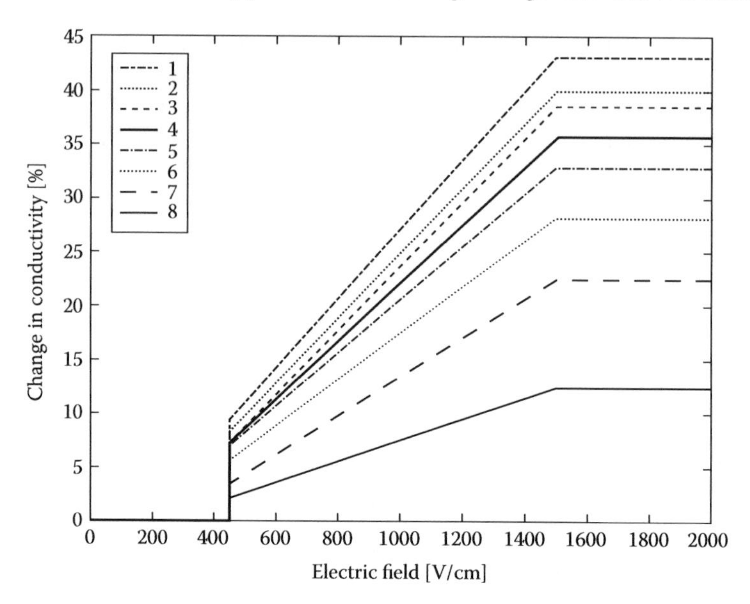

FIGURE 2.6
Change in tissue conductivity as a function of the electrical field. A simple piecewise linear model is used to interpolate the changes in conductivity of rat liver tissue for various electric fields. Results are shown after each electroporation pulse, which are represented here by different lines numbered 1 through 8.

since the electric field and the conductivity have a mutual effect. An electric field above the electroporation threshold for some minimal duration will create pores in the membranes of cells in that part of the tissue and cause the tissue conductivity to increase. The increase in conductivity will have a negative effect on the local electric field, which will decrease, thus changing the electroporation process. To accurately solve this iterative process we require a reliable model of electroporation mechanism that will precisely describe the time evolution of pores as a function of the changing electric field. For many purposes, the first order solution without additional iterations will suffice.

In general, those regions of the tissue where the electric field during the electroporation pulses was high will display an increased conductivity over areas of lower electrical fields. Imaging these regions using EIT will enable real time monitoring and control of the electroporation process. Using additional electrodes and possibly the electroporation electrodes as well (Granot and Rubinsky 2007) we can reconstruct the conductivity map of the electroporated tissue. We rely again on the Laplace equation and use FEM to solve it albeit this time the solution is much more difficult. In the simulation part we have calculated the electric potential u with known conductivity and boundary conditions. This time the conductivity σ is the unknown and the potential u is known only on the boundary, where the voltage measurement electrodes are located. An additional boundary condition exists since we know the current density on the boundary of the tissue under the current injecting electrodes.

Calculating the conductivity, given the currents when the potential on the boundary is known, which is often referred to as the inverse problem, turns out to be mathematically ill-posed (Lionheart 2004). This problem does not have a unique solution for every input, which means that different internal conductivity distributions result in the same boundary potentials. Furthermore, solutions are usually not stable so very small changes in the input can result in very large changes in the output. The principal way to overcome this problem is to use some a priori knowledge in the form of a regularization matrix (Borsic et al. 2002) that is introduced to the solution. A very common regularization matrix relies on the a priori "knowledge" that the internal impedance is a slowly varying spatial function, or in other words, that the impedance does not change very much between adjacent parts of the tissue. This assumption is very useful in solving the equations, but it obviously restricts the solution in terms of spatial resolution. Small inhomogeneities are difficult to detect, and it is essentially impossible to determine sharp borders between two tissue types, even when their impedance is different.

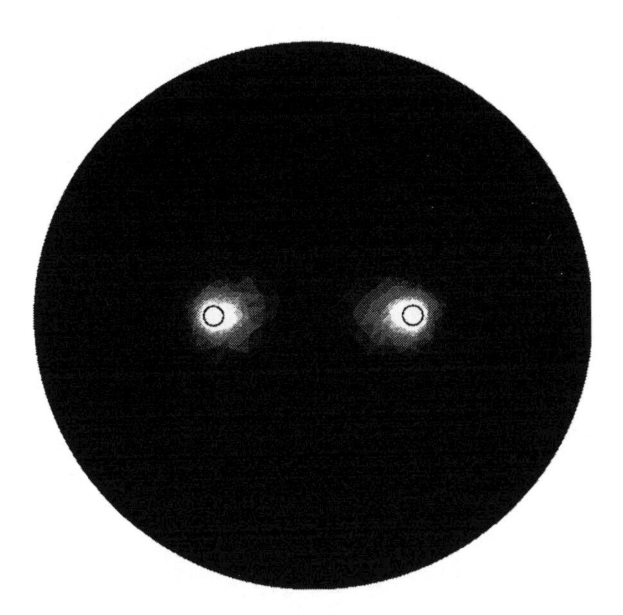

FIGURE 2.7
Example of a reconstructed conductivity map after electroporation. Electroporation was performed using the two needle electrodes, which are the two white discs surrounded by black circles. The conductivity is shown as different shades where bright areas represent higher conductivity.

Most of the algorithms for solving the inverse problem seek a solution in which the conductivity distribution is such that when it is used to solve the *forward* problem the electric potential obtained is close to the measured voltage on the electrodes (Cheney et al. 1990). The search for this conductivity distribution can be performed iteratively, in a single step or using search techniques such as genetic algorithms, simulated annealing, and even exhaustive search (Olmi et al. 2000; Lionheart 2004). Regardless of the search method, once an acceptable solution has been reached, the conductivity distribution depicts the structure of the tissue, or, in our case, the location of electroporated regions. In Figure 2.7, we show an example of a common reconstructed two-dimensional image of simulated electroporation. A homogeneous tissue with two needles that are used for electroporation serves as a model. Additional electrodes are placed around the tissue and are used for the EIT current injections and voltage measurements. The final results of the entire simulation procedure described earlier are shown in Figure 2.7.

2.5 Mass Transfer in Tissue with Reversible Electroporation

A mathematical model for mass transfer of drug molecules in tissue is an important tool for *in vivo* reversible electroporation applications even with monitoring techniques such as real-time imaging using EIT. Such a model can serve as a basic design tool for piecing together a treatment protocol on one hand and for real-time adjustments of the treatment using information supplied by feedback mechanisms such as EIT. Here we describe the basic design of a class of multiscale models that may be used for these purposes.

From the large-scale perspective, looking at the tissue, the model describes the generally nonuniform electric field in the treated region. The electric field is affected by the tissue geometry, the local conductivity at different regions of the tissue if the tissue is not homogenous, and by the electroporation electrodes configuration, both in terms of the electrodes' location and the applied voltages. The diffusion of drug molecules across the tissue is also modeled at large scale to track how the initial concentration changes as the drug enters the electroporated cells. From the small-scale perspective, looking at individual cells, the model describes the electroporation effects on the cell membrane. We need to consider the membrane's permeability due to the creation of pores, their expansion and the resealing process as well as the mass transfer of drug molecules across the membrane.

Beginning with the microscale part of the model we examine a single cell in a uniform electric field and study the effects of electroporation on its membrane. This model is a simplified view for two main reasons, but may be extended to a more realistic model following similar arguments. The first

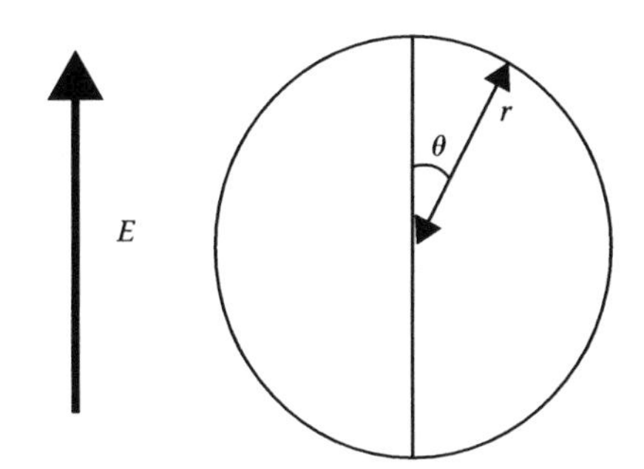

FIGURE 2.8
A two-dimensional model of a single spherical cell in a uniform electric field. E is the electrical field, r is the cell's radius, and θ is the angle between the direction of the electric field and the specific location on the cell membrane.

assumption of a uniform electric field does not usually hold, as we have just stated, but even with a nonuniform field on the large-scale tissue level, the changes in the electric field at the small-scale cell level are generally negligible. We can therefore assume that the electric field in the vicinity of the cell we are about to examine is uniform, although the field strength will vary for different regions of the tissue. The second assumption, treating the cell as a single cell is more delicate. As Figure 2.4 illustrates, the cells in a solid tissue do not resemble an ensemble of independent single cells, but have mutual effects on each other in terms of geometrical shape, transmembrane potential, and more. Nevertheless, we make this simplifying assumption for the sake of clarity in the description of a general class of models. More accurate models can rely on similar methods and treat various scenarios of packed cells in tissue using models that are described in the literature (Esser et al. 2007).

As we have seen, the determining factor in cell membrane electroporation is the transmembrane potential. To estimate this potential we examine a two-dimensional model of a spherical cell, as shown in Figure 2.8. The transmembrane potential, V_m, for such a cell with a radius r, found in a uniform electric field E, will vary for different regions of the membrane. Specifically, it will depend on the angle θ between the direction of the electric field and the membrane region (Neumann et al. 1989).

$$V_m = 1.5rE\cos(\theta) \tag{2.2}$$

This translates to large transmembrane potentials along the electric field axis, where θ is close to 0° or 180°, and to a negligible transmembrane potential near the equator of the cell, where θ is close to 90° or 270°. We can use one of the several biophysical models (Neu and Krassowska 2003; Smith et al. 2004) to estimate the size of pores at different regions of the cell and their time evolution during the application of the electroporation pulse or pulses and after the external field has been removed. Such models estimate the number and size of the pores, usually by solving several differential equations. Solving the equations is often a difficult task because many of the variables depend on each other. For instance, the local transmembrane potential changes dramatically as pores are created owing to the change in membrane permeability to ions. As an example we shall take a single-cell model (Krassowska and Filev 2007) and show how it may be used to determine the cell membrane permeability. The number of pores, according to this model, is computed as a function of V_m using the pore density N:

$$\frac{dN}{dt} = \alpha e^{(V_m/V_{ep})^2}\left(1 - \frac{N}{N_0 e^{q(V_m/V_{ep})^2}}\right) \tag{2.3}$$

where α is the pore creation rate coefficient (10^9 m^{-2}sec^{-1}), V_{ep} is the characteristic voltage of electroporation (0.258 V), N_0 is the equilibrium pore density for the membrane area at $V_m = 0$ ($1.5 \cdot 10^9$ m^{-2}), and q is an electroporation constant ($q = 2.46$). For a known transmembrane voltage such as that of equation (2.2) we can compute the pore density as a function of time.

The size of pores also plays an important role in determining the membrane permeability for two reasons. First, larger pores will allow large molecules to cross the membrane and effectively render the cell permeable to molecules that would not be able to enter the cell if only smaller pores existed, regardless of the number of small pores. Second, the area of a pore depends on its radius, and, thus, larger pores contribute more effective area through which molecules can travel than smaller sized pores. In the current example, we assume that all of the pores have the same size, in agreement with the model we are using, where soon after the end of the pulse the pores shrink to the minimal pore radius R_p, and begin to reseal. This model does not adequately describe reversible electroporation for transfection or other processes where large molecules are introduced into cells over extended periods of time, but for the sake of clarity we continue with this example and only note that more elaborate models for long-term electroporation may be used in a similar manner.

The amount of molecules that enter the cell under these assumptions depends on the number of pores, which decrease rapidly after the electroporation pulses is over since the pores start to reseal. During the pulse the pores are already open and some of them are even larger than the minimal pore radius we consider, but since the pulse duration is very short compared to the phase of stable pores after the pulse, we will only take into account the mass transfer that occurs after the pulse. We calculate the total area of the pores, A_p, from the number of pores per cell and the area of each pore:

$$A_p = \pi R_p^2 \cdot N_p \tag{2.4}$$

where N_p is the number of pores at the end of the electroporation pulse and is computed by integrating the pore density over the entire cell membrane surface:

$$N_p = \oint N dS \tag{2.5}$$

Shortly after the electroporation pulse ends, the pores start to reseal and their number decreases exponentially (Neu and Krassowska 1999) with a time constant, τ, of a few seconds. Therefore when considering the postpulse mass transfer, we need to add a correction term that makes the area an exponentially decaying function of time t where $t = 0$ at the end of the electroporation pulse:

$$A_p = \pi R_p^2 \cdot N_p e^{-t/\tau} \tag{2.6}$$

Equation (2.6) is the final result of this stage and gives the total area of the cell envelope for which the membrane is open and where ions and other small molecules may pass quite freely. As we have mentioned earlier, the case for large molecules is somewhat more delicate, and there only the number of cells above a certain radius should be considered. If we focus our attention on small molecules we can assume that the role of electroporation is completed at this stage and we may start to consider the mass transfer mechanisms that form the second component of the small-scale part of the model.

The membrane constitutes the principal barrier obstructing molecules from entering or leaving the cell. However, the fact that some pores exist and that the membrane no longer surrounds the entire perimeter of the cell does not mean that molecules are completely free to travel. The details of molecular motion through the pores are beyond the scope of this chapter. We will only briefly state that interactions between the pore and the various molecules that pass through it affect the nature of mass transfer through the pore and may depend on the size and shape of the pore, the charge or polarity of the molecules, and more. Molecules may sometimes pass through a relatively small pore in a single-file fashion, which makes the process very different from that of free-flowing molecules. Mechanisms of molecular uptake into electroporated cells may include convection, diffusion, and even electrophoresis for charged molecules when an electric field is applied. Convection is an interesting mechanism to consider since cell swelling has been noticed following electroporation due to water rushing into the electroporated cells (Abidor et al. 1994; Ivorra and Rubinsky 2007). Electrophoresis has been suggested as the reason for the increased uptake of charged molecules when low-voltage pulses are applied following electroporation (Klenchin et al. 1991). Several studies have shown that applying low-voltage pulses after the electroporation pulses increases the amount of molecules such as DNA, which are introduced into the target cells. Electrophoresis may be involved in this process (Satkauskas et al. 2002) in a manner that resembles gel electrophoresis where DNA segments are moved using an electric field. However, other studies (Liu et al. 2006) have questioned whether this mechanism is indeed the reason for the experimental results. We will not go into more details here but rather focus on the third and mechanism mentioned earlier—diffusion. In the following paragraphs, we assume that the driving force of molecular uptake of molecules into cells is the diffusion process in which ions, for example, travel along their concentration gradient where the membrane barrier does not exist.

The electroporation process has left the membrane unsealed and an area described by equation (2.6) is now a part of the cell boundary where an aqueous pathway for molecules exists. With no obstructions to impede their motion, molecules that are found in large concentrations in the extracellular medium will now diffuse into the cell. When no molecules are added or removed from the system the diffusion equation describes the concentration of our molecules, c, as a function of space and time:

$$\frac{\partial c}{\partial t} = \nabla(D\nabla c) \tag{2.7}$$

where D is the diffusion coefficient. Although we shall use a similar version of this equation later in this chapter for the large-scale diffusion of molecules in the tissue, for the small-scale part of the model we will simplify our analysis a little further. We are interested in the number of molecules that enter a cell by diffusing through the aqueous pores across the cell membrane. To obtain that, we take the diffusion flux J, which is given in dimensions of amount of molecules per unit area per unit time and multiply by the total electroporated area A_p. This results in the amount of molecules that enter the cell per unit time. According to Fick's first law of diffusion for an external molecule concentration c_{ex} and an internal concentration c_{in}, the flux is given by

$$J = -P \cdot (c_{in} - c_{ex}) \tag{2.8}$$

where P is the permeability of the molecules through the membrane pores. The permeability is often determined experimentally and is a measure of how well molecules may flow through the pore under certain conditions. For very large pores and very small molecules it could be approximated by the ratio of D, the diffusion coefficient, and δ the membrane thickness $P = D/\delta$. This corresponds to a more common version of Fick's first law for the flow in one dimension where x is in the direction normal to the pore area:

$$J = -D\frac{\partial c}{\partial x} \tag{2.9}$$

For the limit of δ going to zero, in the case of a very thin membrane, $P\Delta c$ in equation (2.8) approaches D times the partial derivative of c along x in equation (2.9). Nevertheless, actual values of the permeability depend on a mixture of parameters such as the size of the molecule and the interaction between the molecule and the pore so this estimate may serve as an upper limit but in realistic scenarios the permeability may be much lower.

In many instances of reversible electroporation the concentration of molecules inside the cell, c_{in}, is much lower compared to that of the external concentration. The goal of the process is in fact to increase the internal concentration. In certain cases c_{in} may be taken to be zero if, for example, molecules in the cytoplasm bind very rapidly to some cellular compartment and are not free to diffuse inside the cell (Granot and Rubinsky 2008).

The final stage we examine in the small-scale model is the process of mass transfer, which happens in every cell in the tissue. This stage will serve as the link between the cellular-scale part of the model and the tissue-scale part. When developing the model of large-scale mass transport in the tissue, we assume that the cells are infinitesimally small, and model them as a distributed reaction rate. This means that at the cell level molecules that enter the cell will never flow out and in a certain sense may be treated as if they have been removed from the system. From the large-scale model point of view, it is as if we have a mass sink that removes the molecules from the system, but this sink is not a point sink, but rather a distributed sink through which molecules leave the system in every region where electroporation occurred. This essentially means that c_{in} is zero at the beginning of the process and remains zero throughout the electroporation procedure although molecules are entering the cell.

The amount of molecules that disappear in each region depends on the extent of electroporation in that specific location. A convenient way of describing this at the cellular scale is by modeling the cells as uniformly packed in the tissue so that each spherical cell is contained in a cube. The edges of the cube are equal to the diameter of the spherical cell, $2r$. It is a deviation from our earlier assumption of a single cell for the electroporation analysis, but it is sufficiently reliable for the purpose of this illustrative example. The reaction rate that describes how much matter in terms of the molecules we are interested in is removed from the system depends on the volume around the cells where the molecules are initially found. A cube with edges as described earlier has a volume of $V_0 = (2r)^3$ for which the reaction rate is the flow of molecules per unit time divided by the surrounding volume:

$$R = JA_p/V_0 \tag{2.10}$$

To sum up the cellular-scale part of the model we have a single spherical cell surrounded by an extracellular medium with a high concentration of some molecule whose internal concentration is zero, and which we would like to introduce into the cell. By inducing an above-threshold transmembrane potential pores are created in the cell's membrane and some of the molecules diffuse into the cytoplasm. These molecules bind to internal structures in the cell and are effectively removed from the system.

From the tissue-scale perspective we are dealing with two aspects. First the electric field values in the tissue are determined by the large-scale configuration of the electrodes and the tissue's properties. This analysis is the basis for the local electric field in the surroundings of each cell in the cellular-scale model. Once we have calculated the reaction rate as detailed earlier, we return to the large-scale model for the tissue-scale diffusion equation. The concentration throughout the extracellular space is calculated using the spatially dependent variable R:

$$\frac{\partial c}{\partial t} - \nabla(D\nabla c) = R \tag{2.11}$$

Equation (2.11) is solved using the diffusion coefficient D and is subject to initial conditions as well as boundary conditions. The initial concentration of molecules in the extracellular space is assumed to be uniform throughout the tissue, neglecting the volume of cells in which there are no molecules. The boundary conditions can often be modeled as no mass flux at the boundary of the tissue.

Using the FEM we can solve the spatially dependent time evolution of c to determine how the molecules diffuse around the tissue and also to obtain an estimate of the amount of molecules that enter cells in different regions. In general, more molecules will disappear into cells in areas where electroporation has created more pores, and molecules will diffuse from higher concentration regions where electroporation is less effective to regions of more intense electroporation. In the following section we shall work out a detailed example of such a process for delivering chemotherapeutical drugs into a cancerous tumor.

2.6 Studies on Molecular Medicine with Drug Delivery in Tissue by Electroporation

A common example of using electroporation for delivering drug molecules is a method called *electrochemotherapy* (Mir et al. 1991; Mir and Orlowski 1999). Drugs such as bleomycin are effective in destroying malignant cells, but are not able to efficiently penetrate the membrane under normal conditions. Using reversible electroporation the tissue around a tumor becomes permeable due to the pores that are created in the cell membrane and bleomycin can enter the cytoplasm. Once inside the cell the drug can act and destroy it. To find out whether the cell will be harmed, we shall estimate the number of bleomycin molecules that are expected to enter the cell.

The configuration we have chosen for this worked-out example is a two-dimensional analysis of a homogeneous tissue where electroporation is induced using two needle electrodes. When two long needles are inserted into the tissue to be used as electrodes, the electric field between and around them may be analyzed in two dimensions, neglecting the boundary effects at the edges of the needles. We therefore use a cross-section in the middle of the electrodes as depicted in Figure 2.7.

We solve equation (2.1) to obtain the potential at every point in the model and thus the electric field. The equation was solved using the finite element method with the code written in Comsol Multiphysics (version 3.3a, www.comsol.com). The boundary conditions were defined as a potential of 1,000 V on one electrode and ground on the other electrode. The outer edges of the tissue, which are far from the electrodes and the region of interest, are assumed to be electrically insulating. When using finite elements ones needs

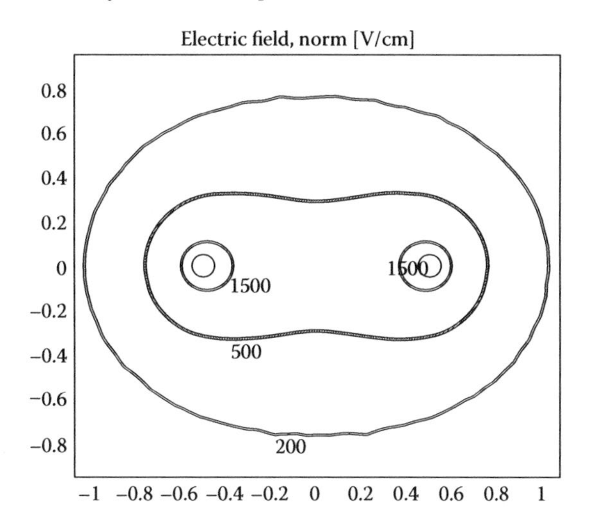

FIGURE 2.9
Electric field for a 1,000 V pulse with two needle electrodes that are 1 cm apart. The electrodes are marked by the two black circles at the center part of the tissue. The contours depict the isopotential lines at 200 V/cm, 500 V/cm, and 1,500 V/cm.

to verify that the solution for a certain mesh does not depend on the specific mesh that was chosen. Obtaining two different solutions using different meshes and comparing them is a simple way of checking that the solution is stable.

An example of a typical solution to equation (2.1) with a 1,000 V voltage difference over a 1 cm gap between the electrodes is shown in Figure 2.9. The electric field for this configuration is such that the largest field is around the electrodes, reaching values well above the ratio of voltage over distance. When plate electrodes are used, the electric field is more uniform and may be estimated by dividing the voltage by the distance between the plates. In our example, the field is not uniform at all so care must be taken to avoid fields that are too large near the electrodes, or that are too small in the treated region. The electrodes in such a procedure are usually inserted into the treated tissue so the field should be adjusted in such a way that the entire area between the electrodes shall have an electric field above the electroporation threshold. The example shown in Figure 2.9 may have irreversible electroporation close to the needles, for example, where the electric field exceeds 1,500 V/cm, but most of the tissue is expected to experience reversible electroporation. The reversible electroporated region may include the entire area where the electric field is above 200 V/cm.

Choosing a specific protocol for electroporation will determine the results of the process since many more parameters need to be considered except for the electric field. The number of pulses, their duration, and the intervals between pulses are some of the dominant factors. The model may use theoretical or experimental data to determine the results of the electroporation protocol and use them in the following stages of the model. We shall proceed with this example using the parameters of a single-cell electroporation model (Krassowska and Filev 2007) and a detailed example of bleomycin mass transfer with electroporation (Granot and Rubinsky 2008).

In this model, there is a single pulse, lasting 1 msec. The threshold for electroporation is an electrical field 130 V/cm, and values of 1,000 V/cm and above are assumed to cause irreversible electroporation. In this specific case, we may notice that irreversible electroporation, which leads to cell death, is not necessarily an unwanted effect, as the purpose of the procedure is to introduce a lethal drug into cells with the intent of killing them. In fact, using irreversible electroporation without any drugs for tissue ablation has been suggested and successfully tested in several studies (Davalos et al. 2005; Miller et al. 2005; Edd et al. 2006; Al-Sakere et al. 2007; Rubinsky 2007; Rubinsky et al. 2007). Still, care must be taken when choosing the electroporation parameters to bring to a minimum the thermal effects caused by Joule heating (Davalos et al. 2003).

During the pulse pores are created, and by solving equations (2.3) and (2.5) we can obtain the number of pores in each cell as a function of the electric field in every region of the tissue. Shortly after the electroporation pulse ends, the pores start to reseal and their number decreases exponentially with a time constant of approximately 1.5 sec, as shown in equation (2.6). The diffusion coefficient D depends on the specific type of molecule we are interested in. For bleomycin, D is assumed to be 10^{-4} mm^2/sec, which is also a reasonable value for many other molecules that have similar traits (Neumann et al. 1998).

Before the electroporation procedure begins, the drug is injected into the tissue and allowed to diffuse and reach a uniform concentration. We recall that very few molecules will be able to enter the cells at this stage. The initial concentration of the drug is taken to be $c\ (t = 0) = 5\ \mu\text{M} = 5 \cdot 10^{-12}\ \text{mol} \cdot \text{mm}^{-3}$.

The next step is solving equation (2.10) to obtain the reaction rate and determine how rapidly the drug is depleted from the extracellular medium and absorbed by the cells. With this solution we turn to equation (2.11), which gives the drug concentration at every point in the tissue and shows how it evolves over time. To solve this equation we return to the finite elements method and obtain a numeric solution. This analysis also allows us to calculate the amount of drug that enters each cell. If we take bleomycin, for example, it is highly toxic and a few hundred molecules that enter the cell are sufficient to cause cell death. The change in local concentration is mostly due to the drug entering the cells with only a small effect by the diffusion of molecules from other parts of the tissue. Thus, for each location we calculate the average number of molecules that enter a cell in that small region, based on the change

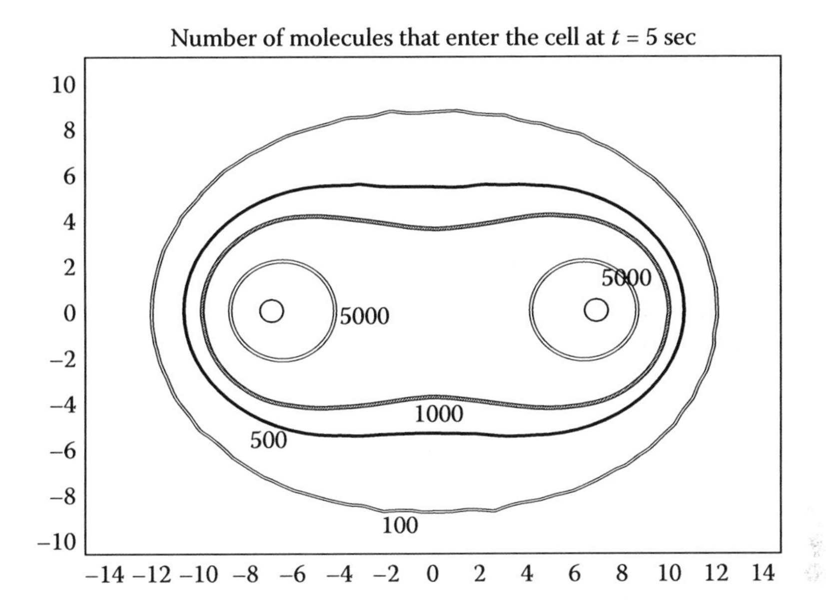

FIGURE 2.10
An example of the number of molecules that are expected to enter cells at various parts of the tissue.

in concentration from the initial value. The drop in drug concentration in the cube with volume V_0 around a cell is attributed to molecules flowing inside the cell. Since the concentration is given in mol/mm^3, N the number of molecules per cell is

$$N = N_A(c(t=0) - c)V_0 \tag{2.12}$$

where N_A is Avogadro's number.

For the model and parameters in this example a simulation of 5 sec is enough to illustrate the effects of reversible electroporation on the uptake of bleomycin. Under different circumstances diffusion may be effective for longer periods, depending on the resealing rate of the cells. Figure 2.10 shows the average number of bleomycin molecules that are expected to enter cells at different areas of the tissue 5 sec after electroporation. In the case of bleomycin, no effect on the cell is expected when fewer than 100 molecules enter the cytoplasm (Poddevin et al. 1991). For this reason the tissue outside of the outer contour in Figure 2.10 will not be affected by the procedure. Roughly 70% of the cells that have been penetrated by 500 or more molecules per cell were expected to die (Tounekti et al. 1993), and the probability of cell death increases further with the increase in the amount of bleomycin absorbed by the cell.

2.7 Future Research Needs in Mathematical Modeling of the Field of Electroporation

As reversible electroporation gains momentum as a clinical method for *in vivo* application the need for mathematical models is expected to rise. The combination of increasingly accurate models that are expected to improve, for instance, as molecular dynamics simulations (Tarek 2005) become more available together with real time monitoring techniques such as the one presented in this chapter will probably aid in understanding the nature of the electroporation process. Several issues still remain as a topic of active scientific research both on the academic level and for developing practical clinical tools.

Today, electroporation research, and particularly those studies that use conductivity measurements to monitor the process, is conducted mostly using one of three approaches:

1. Electroporation of membranes—usually artificial membranes that are grown specifically for this purpose in a chamber specifically designed for these experiments.
2. Electroporation of cells in suspension—*in vitro* procedure where cells are suspended in some medium and usually placed in an electroporation cuvette.
3. Electroporation of tissue—usually an *in vivo* procedure where either needle electrodes are inserted into the tissue or plate electrodes are placed on both sides of a treated tissue.

Every one of these methods has some benefits and some disadvantages. The most controlled environment is when electroporation of artificial membranes is performed. But, although electroporation can be observed in artificial bilayers, it does not always reveal the same characteristics as cell membranes, which by themselves also demonstrate a variety of behaviors.

The electroporation of cells in suspensions is also very important, and although it is somewhat less controlled than the electroporation of the membrane by itself, it is more closely related to *in vivo* electroporation. It also has many practical applications as a method, for example, for introducing genes into cells or producing knockout mice. Measuring the conductivity of the suspension is not a simple procedure, and results may vary significantly depending on the type and concentration of the medium for instance.

Conducting *in vivo* electroporation experiments is obviously the most direct approach to study this procedure. Nevertheless, nonuniform fields that are created by needle electrodes and to some extent by plate electrodes as well, combined with tissue inhomogeneities create a very complex environment for analysis. If the EIT technique matures into a reliable tool for monitoring electroporation, it may be used very efficiently for further research into the effects of *in vivo* electroporation, both in the reversible and irreversible modes.

A key ingredient in the path toward reliable imaging using EIT is the study of the conductivity changes as a function of the desired electroporation result in tissue. The effects of electroporation on the conductivity are very different from one case to another and depend on many factors such as the size of the cells, the extracellular paths between cells, the extracellular medium, and the equilibrium in the presence of a membrane with or without pores. A robust model that will be able to predict the change in tissue conductivity given the key parameters of the electroporation protocol and the tissue will be an essential part of any method of monitoring electroporation by measuring conductivity.

As for longer-term goals, a possible intermediate step on the way to a more complete understanding of electroporation may be a mathematical model that predicts the results of a certain electroporation protocol (e.g., no effect, reversible electroporation or irreversible electroporation) as a function of the electroporation parameters. These parameters may include the applied field, the number of pulses, the duration of each pulse, the pulse repetition rate, the shape of the pulse (square, trapezoidal, sine), and the carrier frequency of the applied pulse, which does not necessarily need to be a direct current pulse. A good model will need to take into account the behavior of a cell after it has been electroporated since it is not unlikely that, even if the cell manages to reseal, it may not be able to maintain its homeostasis and would die shortly after the procedure.

Currently most of the studies in this field rely on past experience and educated guesses or trial-and-error methods. This is useful for certain scenarios, but when we need to consider other parameters as well, it may be too cumbersome. For instance, when we plan a new electroporation protocol in a temperature-sensitive environment, we can use the bioheat equation to estimate the local increase in temperature because of the electroporation pulses. Although the bioheat models are not perfected, they provide a very good base for making an initial plan. What we need to complement that is a model that can predict the results of the electroporation protocol. Such a tool would be useful to find an optimal working point in which we obtain the required results from the electroporation model while keeping the temperature within the defined boundaries.

Pursuing these research directions may increase our knowledge of the basic mechanisms of electroporation and open the way for numerous applications of reversible and irreversible electroporation.

2.8 References

Abidor, I. G., Arakelyan, V. B., Chernomordik, L. V., Chizmadzhev, Y. A., Pastushenko, V. F., and Tarasevich, M. R. (1979). Electric breakdown of bilayer lipid membranes I. The main experimental facts and their qualitative discussion. *Bioelectrochemistry and Bioenergetics,* **6**(1):37–52.

Abidor, I. G., Li, L. H., and Hui, S. W. (1994). Studies of cell pellets: II. Osmotic properties, electroporation, and related phenomena: membrane interactions. *Biophysics Journal,* **67**(1):427–435.

Al-Sakere, B., Andre, F., Bernat, C., Connault, E., Opolon, P., Davalos, R. V., Rubinsky, B., and Mir, L. M. (2007). Tumor ablation with irreversible electroporation. *PLoS ONE,* **2**(11):e1135.

Bayford, R. H. (2006). Bioimpedance tomography (electrical impedance tomography). *Annual Review of Biomedical Engineering,* **8**(1):63–91.

Borsic, A., Lionheart, W. R. B., and McLeod, C. N. (2002). Generation of anisotropic-smoothness regularization filters for EIT. *IEEE Transactions on Medical Imaging,* **21**(6):579–587.

Brown, B. H. (2003). Electrical impedance tomography (EIT): a review. *Journal of Medical Engineering & Technology,* **27**(3): 97–108.

Chang, D. C. and Reese, T. S. (1990). Changes in membrane structure induced by electroporation as revealed by rapid-freezing electron microscopy. *Biophysics Journal,* **58**(1):1–12.

Chen, C., Smye, S. W., Robinson, M. P., and Evans, J. A. (2006). Membrane electroporation theories: a review. *Medical and Biological Engineering and Computing,* **V44**(1):5–14.

Cheney, M., Isaacson, D., and Newell, J. C. (1999). Electrical Impedance Tomography. *SIAM Review,* **41**(1):85–101.

Cheney, M., Isaacson, D. Newell, J. C., Simske, S., and Goble, J. (1990). NOSER: an algorithm for solving the inverse conductivity problem. *International Journal of Imaging Systems and Technology,* **2**(2): 66–75.

Coster, H. G. L. (1965). A quantitative analysis of the voltage-current relationship of fixed charge membranes and the associated property of "punch-through." *Biophysical Journal,* **5**:669–686.

Crowley, J. M. (1973). Electrical breakdown of biomolecular lipid membranes as an electromechanical instability. *Biophysical Journal,* **13**: 711–724.

Davalos, R. V., Mir, L. M., and Rubinsky, B. (2005). Tissue ablation with irreversible electroporation. *Annals of Biomedical Engineering,* **V33**(2): 223–231.

Davalos, R. V., Rubinsky, B., and Mir, L. M. (2003). Theoretical analysis of the thermal effects during in vivo tissue electroporation. *Bioelectrochemistry,* **61**(1–2):99–107.

Davalos, R. V., Rubinsky, B., and Otten, D. M. (2002). A feasibility study for electrical impedance tomography as a means to monitor tissue electroporation for molecular medicine. *IEEE Transactions on Biomedical Engineering,* **49**(4):400–403.

Edd, J., Horowitz, L., Davalos, R. V., Mir, L. M., and Rubinsky, B. (2006). In-vivo results of a new focal tissue ablation technique: irreversible electroporation. *IEEE Transactions on Biomedical Engineering,* **53**(5):1409–1415.

Esser, A. T., Smith, K. C., Gowrishankar, T. R., and Weaver, J. C. (2007). Towards solid tumor treatment by irreversible electroporation: intrinsic redistribution of fields and currents in tissue. *Technology in Cancer Research and Treatment,* **6**(4):261–274.

Granot, Y. and Rubinsky, B. (2007). Methods of optimization of electrical impedance tomography for imaging tissue electroporation. *Physiological Measurement,* (10):1135–1147.

Granot, Y. and Rubinsky, B. (2008). Mass transfer model for drug delivery in tissue cells with reversible electroporation. *International Journal of Heat and Mass Transfer,* **51**(23–24):5610–5616.

Granot, Y., Ivorra, A., and Rubinsky, B. (2008). A new concept for medical imaging centered on cellular phone technology. *PLoS ONE,* **3**(4):e2075.

Hibino, M., Itoh, H., and Kinosita, K. J. (1993). Time courses of cell electroporation as revealed by submicrosecond imaging of transmembrane potential. *Biophysics Journal,* **64**(6):1789–1800.

Ivorra, A. and Rubinsky, B. (2007). In vivo electrical impedance measurements during and after electroporation of rat liver. *Bioelectrochemistry,* **70**(2): 287–295.

Kinosita, K. J., Ashikawa, I., Saita, N., Yoshimura, H., Itoh, H., Nagayama, K., and Ikegami, A. (1988). Electroporation of cell membrane visualized under a pulsed-laser fluorescence microscope. *Biophysics Journal,* **53**(6):1015–1019.

Klenchin, V. A., Sukharev, S., Serov, S. M., Chernomordik, L. V., and Chizmadzhev, Yu. A., (1991). Electrically induced DNA uptake by cells is a fast process involving DNA electrophoresis. *Biophysics Journal,* **60**: 804–811.

Krassowska, W. and Filev, P. D. (2007). Modeling electroporation in a single cell. *Biophysics Journal,* **92**(2):404–417.

Lewis, T. J. (2003). A model for bilayer membrane electroporation based on resultant electromechanical stress. *IEEE Transactions on Dielectrics and Electrical Insulation,* **10**(5):769–777.

Lindner, P., Neumann, E., and Rosenheck, K. (1977). Kinetics of permeability changes induced by electric impulses in chromaffin granules. *Journal of Membrane Biology,* **32**(1):231–254.

Lionheart, W. R. B. (2004). EIT reconstruction algorithms: pitfalls, challenges and recent developments. *Physiological Measurement,* (1):125.

Liu, F., Heston, S., Shollenberger, L. M., Sun, B., Mickle, M., Lovell, M., and Huang, L. (2006). Mechanism of in vivo DNA transport into cells by electroporation: electrophoresis across the plasma membrane may not be involved. *The Journal of Gene Medicine,* **8**:353–361.

McNeil, P. L. and Steinhardt, R. A. (1997). Loss, Restoration, and Maintenance of Plasma Membrane Integrity. *Journal of Cell Biology,* **137**(1):1–4.

Miller, L., Leor, J., and Rubinsky, B. (2005). Cancer cells ablation with irreversible electroporation. *Technology in Cancer Research and Treatment,* **4**(6):699–706.

Mir, L. M. and Orlowski, S. (1999). Mechanisms of electrochemotherapy. *Advanced Drug Delivery Reviews,* **35**:107–118.

Mir, L. M., Belehradek, M., Domenge, C., Orlowski, S., Poddevin, B., Belehradek, J. J., Schwaab, G., Luboinski, B., and Paoletti, C. (1991). Electrochemotherapy, a new antitumor treatment: first clinical trial. *Comptes Rendus de l'Academie des Sciences Serie III Sciences de la Vie,* **313**: 613–618.

Neu, J. C. and Krassowska, W. (1999). Asymptotic model of electroporation. *Physical Review E,* **59**(3):3471.

Neu, J. C. and Krassowska, W. (2003). Modeling postshock evolution of large electropores. *Physical Review E,* **67**(2):021915.

Neumann, E. and Rosenheck K. (1972). Permeability changes induced by electric impulses in vesicular membranes. *Journal of Membrane Biology,* **10**(1):279–290.

Neumann, E., Sowers, A. E., and Jordan, C. A. (1989). *Electroporation and Electrofusion in Cell Biology.* Springer-Verlag, Berlin.

Neumann, E., Toensing, K., Kakorin, S., Budde, P., and Frey, J. (1998). Mechanism of electroporative dye uptake by mouse B cells. *Biophysics Journal,* **74**(1):98–108.

Nickoloff, J. A., ed. (1995). *Animal cell electroporation and electrofusion protocols. Methods in Molecular Biology.* Totowa, NJ, Humana Press.

Olmi, R., Bini, M., and Priori, S. (2000). A genetic algorithm approach to image reconstruction in electrical impedance tomography. *IEEE Transactions on Evolutionary Computation,* **4**(1):83–88.

Pastushenko, V. F., Chizmadzhev, Y. A., and Arakelyan, V. B. (1979). Electric breakdown of bilayer lipid membranes II. Calculation of the membrane lifetime in the steady-state diffusion approximation. *Bioelectrochemistry and Bioenergetics,* **6**(1):53–62.

Pavlin, M. and Miklavcic, D. (2008). Theoretical and experimental analysis of conductivity, ion diffusion and molecular transport during cell electroporation—relation between short-lived and long-lived pores. *Bioelectrochemistry,* **74**(1):38–46.

Pavlin, M., Kanduser, M., Rebersek, M., Pucihar, G., Art, F. X., Magjarevic, R., and Miklavcic, D. (2005). Effect of cell electroporation on the conductivity of a cell suspension. *Biophysics Journal,* **88**(6):4378–4390.

Pavselj, N., Bregar, Z., Cukjati, D., Batiuskaite, D., Mir, L. M., and Miklavcic, D. (2005). The course of tissue permeabilization studied on a mathematical model of a subcutaneous tumor in small animals. *IEEE Transactions on Biomedical Engineering,* **52**(8):1373.

Pliquett, U., Elez, R., Piiper, A., and Neumann, E. (2004). Electroporation of subcutaneous mouse tumors by rectangular and trapezium high voltage pulses. *Bioelectrochemistry,* **62**(1):83–93.

Poddevin, B., Orlowski, S., Belehradek, J., and Mir, L. M. (1991). Very high cytotoxicity of bleomycin introduced into the cytosol of cells in culture. *Biochemical Pharmacology,* **42**(Suppl. 1):S67–S75.

Prausnitz, M. R., Corbett, J. D., Gimm, J. A., Golan, D. E., Langer, R., and Weaver, J. C. (1995). Millisecond measurement of transport during and after an electroporation pulse. *Biophysics Journal,* **68**(5):1864–1870.

Rols, M. P., Femenia, P., and Teissie, J. (1995). Long-lived macropinocytosis takes place in electropermeabilized mammalian cells. *Biochemical and Biophysical Research Communications,* **208**(1):26–35.

Rubinsky, B. (2007). Irreversible electroporation in medicine. *Technology in Cancer Research and Treatment,* **6**(4):255–259.

Rubinsky, B., Onik, G., and Mikus, P. (2007). Irreversible electroporation: a new ablation modality—clinical implications. *Technology in Cancer Research and Treatment,* **6**(1):37–48.

Sale, A. J. H. and Hamilton, W. A. (1967). Effects of high electric fields on microorganisms. 1. Killing of bacteria and yeasts. *Biochimica et Biophysica Acta,* **148**:781–788.

Satkauskas, S., Bureau, M. F., Puc, M., Mahfoudi, A., Scherman, D., Miklavcic, D., and M. Mir, L. (2002). Mechanisms of in vivo DNA electrotransfer: respective contributions of cell electropermeabilization and DNA electrophoresis. *Molecular Therapy,* **5**:133–140.

Smith, K. C., Neu, J. C., and Krassowska, W. (2004). Model of creation and evolution of stable electropores for DNA delivery. *Biophysics Journal,* **86**:2813–2826.

Steinhardt, R. A., Bi, G., and Alderton, J. M. (1994). Cell membrane resealing by a vesicular mechanism similar to neurotransmitter release. *Science,* **263**(5145):390–393.

Tarek, M. (2005). Membrane electroporation: a molecular dynamics simulation. *Biophysics Journal,* **88**(6):4045–4053.

Teissie, J., Golzio, M., and Rols, M. P. (2005). Mechanisms of cell membrane electropermeabilization: a minireview of our present (lack of?) knowledge. *Biochimica et Biophysica Acta,* **1724**(3):270–280.

Tieleman, D. P. (2004). The molecular basis of electroporation. *BMC Biochemistry,* **5**(1):10.

Tounekti, O., Pron, G., Belehradek, J. Jr., and Mir, L. M. (1993). Bleomycin, an apoptosis-mimetic drug that induces two types of cell death depending on the number of molecules internalized. *Cancer Research,* **53**(22):5462–5469.

Vauhkonen, M., Lionheart, W. R. B., Heikkinen, L. M., Vauhkonen, P. J., and Kaipio, J. P. (2001). A Matlab package for the EIDORS project to reconstruct two-dimensional EIT images. *Physiological Measurement,* **22**(1): 107–111.

Weaver, J. C. and Chizmadzhev, Y. A. (1996). Theory of electroporation: a review. *Bioelectrochemistry and Bioenergetics,* **41**:135–160.

3

Hydrodynamics in Porous Media with Applications to Tissue Engineering

C. Oddou, T. Lemaire

Laboratory of Multiscale Modelling and Simulation – Biomechanics, Faculty of Sciences & Technology, (MSME CNRS-8208), University Paris-Est, Créteil, France

J. Pierre

Laboratory of Osteo-Articular Biomechanics and Biomaterials (B2OA CNRS-7052), University Paris 12, Faculty of Sciences & Technology, Créteil Cedex, France

B. David

Laboratory of Mechanics of Soils, Structures and Materials (MSSMat CNRS-8579), École Centrale Paris, Châtenay-Malabry Cedex, France

CONTENTS

3.1 Nomenclature

Parameter	Notation	Unit
Substrate overall length scale	L	m
Porosity of the medium	ϕ	
Length scale of the pore	a	m
Specific pore wall area	$S_V \approx \dfrac{2\phi}{a}$	m^{-1}
Effective pore length of the sample	L_P	m
Tortuosity	$T = L_P^2 / L^2$	
Porous medium permeability	$K = \dfrac{\phi\,(2a)^2}{96} \times \dfrac{1}{T} = \dfrac{1}{6} \times \dfrac{\phi^3}{S_V^2 \times T}$	m^2
Concentration of the nutrient (oxygen, ...) molecules	C	$\mathrm{mol} \times \mathrm{m}^{-3}$
Michaelis–Menten constant	K_M	$\mathrm{mol} \times \mathrm{m}^{-3}$
Cellular nutrient maximal consumption rate	V_{Max}	$\mathrm{mol} \times \mathrm{cel}^{-1} \times \mathrm{sec}^{-1}$
Rate of nutrient (oxygen, ...) consumption by unit area	R_S	$\mathrm{mol} \times \mathrm{m}^{-2} \times \mathrm{sec}^{-1}$
Surface density of the cells	σ_{cel}	$\mathrm{cel} \times \mathrm{m}^{-2}$
Maximum ratio of cell oxygen consumption	$R_{\max} = \sigma_{cel} \times V_{\max}$	$\mathrm{mol} \times \mathrm{m}^{-2} \times \mathrm{sec}^{-1}$
Rate of nutrient (oxygen, ...) consumption by unit volume	$R_V = R_S \times S_V$	$\mathrm{mol} \times \mathrm{m}^{-3} \times \mathrm{sec}^{-1}$

Parameter	Notation	Unit
Reference concentration of the nutrient	c_0	$\mathrm{mol} \times \mathrm{m}^{-3}$
Diffusion coefficient of the nutrient solute	D	$\mathrm{m}^2 \times \mathrm{s}^{-1}$
Damköhler number	$Da = \dfrac{\sigma_{\mathrm{cel}} \times V_{\mathrm{Max}} \times a}{D \times c_0}$	
Perfusion mean velocity	u_0	$\mathrm{m} \times \mathrm{s}^{-1}$
Density of the culture fluid	ρ	$\mathrm{kg} \times \mathrm{m}^{-3}$
Dynamic viscosity of the culture fluid	η	$\mathrm{Pa} \times \mathrm{sec}$
Reynolds number	$Re = \dfrac{\rho \times a \times {}^{u_0}/_{\phi}}{\eta}$	
Péclet number	$Pe = \dfrac{a \times {}^{u_0}/_{\phi}}{D}$	
Viscous stresses	$\tau = \eta \times \dfrac{u_0}{a}$	Pa
Debye length	L_D	m
Ionic force	C_i	$\mathrm{mol} \times \mathrm{m}^{-3}$
Vacuum permittivity	ε_0	$\mathrm{C} \times \mathrm{V}^{-1} \times \mathrm{m}^{-1}$
Relative dielectric constant of the solvent	ε	
Faraday constant	F	$\mathrm{C} \times \mathrm{mol}^{-1}$
Gas constant	R	$\mathrm{J} \times \mathrm{K}^{-1} \times \mathrm{mol}^{-1}$
Absolute temperature	T	K
Electric double-layer potential	φ	V
Donnan osmotic swelling pressure	π_D	Pa
Oxygen molar concentration	c_{O_2}	$\mathrm{mol} \times \mathrm{m}^{-3}$
Oxygen binary diffusivity in water	D_{O_2}	$\mathrm{m}^2 \times \mathrm{sec}^{-1}$
Perfusion velocity	u	$\mathrm{m} \times \mathrm{sec}^{-1}$
Typical size of the solid heterogeneities in the porous medium	h	m
Fluid kinematic viscosity	$\nu = \frac{n}{\rho}$	$\mathrm{m}^2 \times \mathrm{sec}^{-1}$
Dimensionless Michaelis constant	λ_M	
Dimensionless oxygen flux at the frontier	$\vec{N}'_{O_2}$	

(*Continued*)

Parameter	Notation	Unit
Normal unit vector of a considered surface	$\vec{n}$	
Tangent unit vector to a considered surface	$\vec{t}$	
Identity tensor	$\overline{\overline{I}}$	
Streaming potential	ψ	V
Ionic binary diffusion coefficient in water	$D_{\pm}$	$\mathrm{m}^2 \times \mathrm{sec}^{-1}$
Ionic flux density	$\overrightarrow{J_{\pm}}$	$\mathrm{mol} \times \mathrm{m}^{-2} \times \mathrm{sec}^{-1}$
Poiseuille permeability tensor	$\boldsymbol{\kappa}_P$	$\mathrm{m}^2 \times \mathrm{Pa}^{-1} \times \mathrm{sec}^{-1}$
Osmotic permeability tensor	$\boldsymbol{\kappa}_C$	$\mathrm{m}^5 \times \mathrm{mol}^{-1} \times \mathrm{sec}^{-1}$
Electroosmotic permeability tensor	$\boldsymbol{\kappa}_E$	$\mathrm{m}^2 \times \mathrm{sec}^{-1}$
Nondimensional number comparing electrical current and diffusion	$J_d = \frac{I_0 L}{D_0 F c_0}$	
Reference electric current density	I_0	$\mathrm{C} \times \mathrm{m}^{-2} \times \mathrm{sec}^{-1}$
Length of the representative elementary volume	l_{Ch}	m
Reference diffusion coefficient	D_0	$\mathrm{m}^2 \times \mathrm{sec}^{-1}$
Dean number	$Dn = Re \times \left({}^{a}/_{r_c}\right)^{\frac{1}{2}}$	
Curvature radius of the flow streamlines	r_c	m
Secondary transverse fluid velocity	u_t	$\mathrm{m} \times \mathrm{sec}^{-1}$
Characteristic oxygen diffusion velocity within a pore	$u_D \approx \frac{D_{O_2}}{\mathrm{a}}$	$\mathrm{m} \times \mathrm{sec}^{-1}$

3.2 Introduction

Interaction between fluid flow and living media is a complex matter, far from being completely understood: it is clear that not only cell and tissue metabolism is likely to be influenced by the transport phenomena of nutrient and waste products that are regulated by the flow, but it may also be directly affected by the various stresses generated by the fluid motion. Fluid flow within natural tissues plays important roles in morphogenesis, metabolism function, and pathogenesis. In the design of new biomaterials mimicking biological tissues, it is well recognized today that three-dimensional *in vitro* culture better recapitulates physiological cell environment. Indeed, fluid flow and solid strain that are imposed within tissue cultured in bioreactors not only affect cell nourishment but also exert on cells mechanical actions such as pressure effects, drag interactions, and viscous shear stresses (Chen 2008).

This fact is largely exploited in tissue engineering, a pluridisciplinary research field based on the employment of biological cell systems to develop therapeutic strategies aiming at the replacement, repair, maintenance, and enhancement of tissue function (Lanza et al. 2002). The general strategy for tissue engineering then involves seeding cells into a biocompatible porous scaffold and culturing this seeded structure in a bioreactor (Martin et al. 2004). Thus, tissue engineering presents sound promise for the next future, providing a large number of transplantable organoids. Indeed, it requires proposing challenging development in effective bioreactors to overcome nutrient transport limitations and to subject cells to optimal mechanical stresses.

Current strategies generally take advantage of the porous media morphology of the scaffolds generally displaying a high porosity $\phi \geq 0.5$ (volume of pores divided by the total volume) and large pore length scales ($a \approx 100\,\mu\text{m}$), that is to say an order of magnitude larger than the cell length scale. The use of perfusion bioreactors to force culture fluid to flow through the medium allows enhancement of nutrient transport and generation of mechanical *stimuli* upon the cells (Abousleiman and Sikavitsas 2006; Cimetta et al. 2007; Kim et al. 2007). For small tissue-engineered construct these methods have been shown to be successful in comparison to "static" culture (Glowacki et al. 1998; Goldstein et al. 2001; Bancroft et al. 2002; Cartmell et al. 2003). However, problems arise when the tissue size is scaled up. Cells residing away from the inlets and outlets may sit in almost stagnant regions, where both nutrient delivery and shear stress are compromised. If the center of the construct is to receive adequate flow, then regions near the inlet and outlet may suffer by receiving too much shear stress. The nonuniformity of the flow and shear-stress distributions is problematic (see Figure 3.1).

It avers then very critical to elaborate a comprehensive analysis of the fluid mechanics and related transport phenomena inside such porous media and the way they control the different life processes. As presented here, within such multiscale natural or artificial materials, the dynamic interactions between the fluid phase and the solid structure as well as the diffusive, convective transport of nutrient by fluid and oxygen consumption by immersed cells are phenomena not always completely understood and controlled. Some pertinent predictions about the microfluid dynamic environment imposed on three-dimensional engineered cell systems within bioreactors is expected from such an analysis.

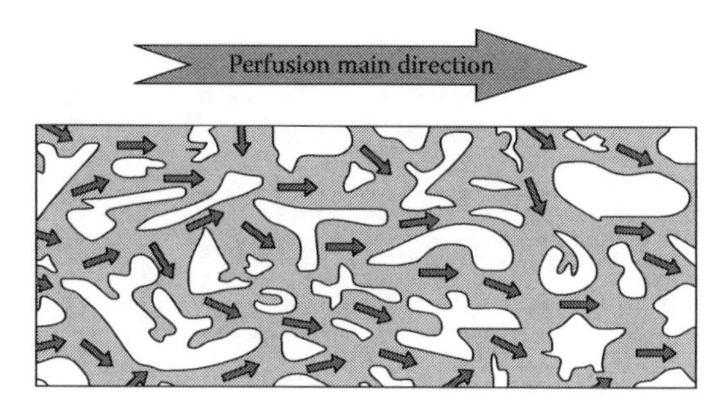

FIGURE 3.1
Flow perfusion culture can help the nutrient and waste transport within the porous scaffold but it is often characterized by a heterogeneous distribution of the hydromechanical effects, such as nutrient or oxygen convective transport and viscous shear stresses acting on cells. Illustrated by an arrow, the macroscopic perfusion direction is horizontal.

Experimental facts show that the scaffolds generally used in tissue engineering are not optimally designed to facilitate cell migration and to provide appropriate microenvironment for cell proliferation and differentiation. In the framework of new computational and theoretical studies in tissue engineering, we review here some physical and mathematical concepts that describe fluid and mass transport within porous substrates seeded with cells. On the basis of our own experience, some examples of model-driven experimental approaches are presented mainly dealing with bone and cartilage tissue, with emphasis upon the multiscale and multiphysics aspects of the implied processes. Analyses of some factors that affect cell biology and tissue genesis are presented. The aim of this review is thus to recap some of the most relevant advances, with emphasis on compilation of physical laws, substrate material properties, modeling, and experimentation in relation with the fluid mechanics inside such porous media. In relation to some comments about emerging biological ideas on how the flow environment of the cells is "sensed" to drive morphogenetic processes and responses, the viewpoint developed here is relevant to the engineering sciences applied at the mesoscopic and microscopic scales including pore and cell lengths.

3.3 Cell and Tissue Engineering: Physicochemical Determinants of the Development

3.3.1 Cell Metabolism—Nutrient and Oxygen Consumption: The Michaelis–Menten Formulation

Normal tissue cells are anchorage dependant, that is to say that they are generally not viable when suspended in a fluid alone and must adhere to a more or less solid-phase substratum. The feedback between the physicochemical properties of this phase and the cell metabolism is strongly implied in their development, differentiation, disease, and regeneration states (Discher et al. 2005). In tissue engineering, new bioactive, degradable and porous, composite substrates associated with bone marrow stem cells may serve as basic materials for preparing synthetic tissue. Indeed, such porous media are generally offering large-pore open space, enabling tissue progenitor and endothelial cells to migrate into the overall structure and hence contribute to the long-term development and irrigation of the newly formed tissue.

Flow perfusion culture within three-dimensional porous scaffolds is an efficient way of fostering cell population growth and matrix production throughout these seeded media owing to the enhancement of nutrient delivery and mechanical action genesis (Bancroft et al. 2003). In this framework, three-dimensional porous media with sufficiently high porosity promoting tissue formation are required to have specific internal microarchitectural features. These features characterizing highly porous interconnected structures include large surface-to-volume ratios of the pore network favoring cell in growth and cell distribution throughout the matrix. A very important parameter of these porous structures is then the specific surface area of the porous medium, S_V. This parameter represents the surface per unit of apparent volume, which is available for cell attachment and tissue deposition, and can be approximated knowing the pore size and the porosity, owing to the approximate formula: $S_V \cong {}^{2\phi}/_{a}$. Higher specific surface areas are expected to favor higher cell attachment and proliferation inside the scaffold structures. In this respect, the use of porous substrates in tissue engineering is seen as a convenient way of considerably increasing cell adhesion and exchange surface within a given volume.

During the culture phase, cells have to proliferate, colonize homogeneously the porous scaffold, and synthesize extracellular matrix. Nutrient supply of the cells is a key factor to obtain a successful culture, especially in the case of "large" implants (with volume around few cubic centimeters). Different classes of molecules can interact with cells (Lanza et al. 2002). Among the soluble nutrient elements, a large number of studies were focused on oxygen, because this chemical element has a major impact on tissue growth, particularly for osteoarticular systems (Tuncay et al. 1994; Arnett et al. 2003). For instance, in the case of *in vitro* bone cells culture, osteoblast metabolism is influenced by the local oxygen concentration, an effect also known for cartilage and liver cells. Furthermore, the magnitude of cell oxygen local consumption could be affected by both temperature and cell concentration. Moreover, cell oxygen need could evolve during culture time and increase during cell division phase.

For small nutrient components, such as oxygen molecules, that pass directly across the cell membrane and are subject to enzymatic chemical reactions, the kinetics of their uptake generally follows a Michaelis–Menten law considered as fundamental in enzymology. This law stipulates that at low concentration ($C \leq K_M$, K_M Michaelis–Menten constant) the chemical

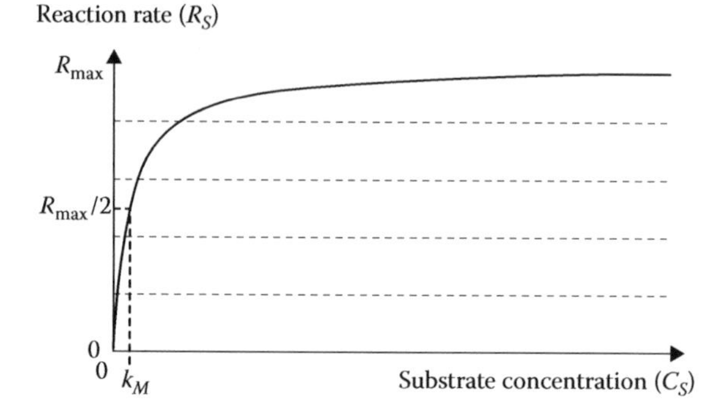

FIGURE 3.2
The Michaelis–Menten law displaying a nonlinear relation between the nutrient or oxygen consumption rate, R_S, by unit area of cell layer versus the local concentration of the nutrient or oxygen solute molecule.

reaction rate of the cell uptake is of first-order type in concentration, whereas at increasing concentration, this rate asymptotically approaches a constant value, $V_{\max}$. Then, the rate R_S of oxygen consumption by unit area of cell layer with a surface density σ_{cel} takes the following expression (Figure 3.2):

$$R_S = -\sigma_{cel} \times V_{\max} \times \frac{C}{C + K_M} = -R_{\max} \times \frac{C}{C + K_M} \tag{3.1}$$

where the minus sign indicates that the presence of a cell layer has a sink effect on the nutrient molecules concentration C.

Note that the oxygen consumption by unit volume within such porous substrates is then

$$R_V = S_V \times R_S \tag{3.2}$$

Owing to the complexity of the biological and biochemical phenomena taking place in nutrient uptake by cells, simplified models are often used for the description of these enzymatic processes. As a result, and by extension, this law is also used for the description of other biological phenomena such as cell population growth, absorption of biochemical molecules within kidneys, or consumption of drugs by tumoral cells. On the basis of experimental studies, it has been shown that the maximal oxygen consumption rate, $V_{\max}$, depends on the cell type and may vary by several orders of magnitude. This variability could partially be explained by the variety of the experimental setups (culture protocol, scaffold type, cell concentration, state of the cells, etc.) as well as by the diversity of the measurement methods (oxygen electrode or measurement by fluorescence, for example). However, regarding the Michaelis constant K_M, literature values are close to each other for a given type of given nutrient.

3.3.2 Effects of Nutrient Transport

A key problem, often encountered in tissue engineering with large-size constructs, is the lack of spatial homogeneity in the distribution of cells and extracellular matrix. Indeed, according to Martin et al. (2004), the supply of oxygen and soluble nutrients becomes critically limiting for the *in vitro* culture of three-dimensional tissues. The consequence of such a limitation is exemplified by early studies showing that cellular spheroids larger than 1 mm in diameter generally contain a hypoxic, necrotic center, surrounded by a rim of viable cells (Sutherland et al. 1986). Similar observations were reported for different cell types cultured on three-dimensional scaffolds under static conditions. For example, glycosaminoglycan (GAG) deposition by chondrocytes cultured on poly(glycolic acid) meshes was poor in the central part of the constructs (~400 μm from the outer surface) (Martin et al. 1999), and deposition of mineralized matrix by stromal osteoblasts cultured into poly(DL-lactic-*co*-glycolic acid) foams reached a maximum penetration depth of 240 μm from the top surface (Ishaug et al. 1997).

As the matrix density determines eventual mechanical functionality of engineered tissues, such an inhomogeneous spatial distribution can result in inadequate overall mechanical properties (Vunjak-Novakovic et al. 1999; Kelly and Prendergast 2005; Sengers et al. 2007). Because engineered constructs should be at least a few millimeters in size to serve as grafts for tissue replacement, mass-transfer limitations represent one of the greatest challenges to be addressed. The causes of these restrictions are diverse, implying supply of nutrient and soluble biochemical factors, removal of waste products, and non-homogeneity in consumption sites due to cell seeding or migration. The main control parameter that determines whether solute gradient will occur is the required surface density of the cells (corresponding to σ_{cel}) lining on the pore wall associated with the cellular nutrient consumption rate V_{Max}. Depending on the tissue-engineering application, porous scaffold properties and cultivation processes can be different, as explained furthermore in Section 3.3, but the overall transport restrictions have to be overcome by controlling the physical and biochemical environments in "the heart of the bioreactor." Thus it is necessary to downscale the description of the phenomena to reach the length scale a of the medium pore. Furthermore, the nondimensional parameter, named Damköhler number Da, defined as the ratio between the consumption flux of nutrient at the pore wall and the diffusion flux inside the pore, avers fundamental in the study of transport processes of nutrients such as oxygen or glucose. This parameter can also be viewed as a good evaluation, at the pore scale a, of the ratio between the consumption and diffusive rates of change in nutrient concentration and is so expressed by

$$Da = \frac{R_V}{D \times c_0/a^2} = \frac{\sigma_{cel} \times V_{Max} \times a}{D \times c_0} \tag{3.3}$$

In these expressions, c_0 stands for the reference concentration of the oxygen or nutrient, whereas D is its diffusion coefficient. As will be shown later on, transport can be enhanced by means of perfusion through the scaffold with a mean velocity u_0. At the pore scale the dynamics of the perfusion fluid flow (solvent fluid momentum transfer) is then characterized by the Reynolds number defined by

$$Re = \frac{\rho \times a \times {}^{u_0}/_{\phi}}{\eta} \tag{3.4}$$

Here ρ and η are, respectively, the density and the dynamic viscosity of the culture fluid. Moreover, for problems in solute nutrient mass transport, it is pertinent to similarly introduce the Péclet number Pe, which represents the ratio of the convective and diffusive effects implied at the length scale a of the pores and given by

$$Pe = \frac{a \times {}^{u_0}/_{\phi}}{D} \tag{3.5}$$

3.3.3 Effects of Mechanical Loading: Cell and Tissue Mechanobiology

Increasing evidence suggests that mechanical forces, which are known to be important modulators of cell physiology, might increase the biosynthetic activity of cells in bioartificial matrices and, thus, possibly improve or accelerate tissue regeneration *in vitro* (Butler et al. 2000). Various studies have demonstrated the validity of this principle, particularly in the context of musculoskeletal tissue engineering. For example, cyclical mechanical stretch was found to (1) enhance proliferation and matrix organization by human heart cells seeded on gelatin-matrix scaffolds (Akhyari et al. 2002), (2) improve the mechanical properties of tissues generated by skeletal muscle cells suspended in collagen or Matrigel (Powell et al. 2002), and (3) increase tissue organization and expression of elastin by smooth muscle cells seeded in polymeric scaffolds (Kim et al. 1999). Pulsatile radial stress of tubular scaffolds seeded with smooth muscle cells improved structural organization and suture retention of the resulting engineered blood vessels, and enabled the vessels to remain open for 4 weeks following *in vivo* grafting (Niklason et al. 1999). Dynamic deformational loading or shear of chondrocytes embedded in a three-dimensional environment stimulated GAG synthesis (Davisson et al. 2002) and increased the mechanical properties of the resulting tissues (Mauck et al. 2000; Waldman et al. 2003). Strains in elongation and torsion on collagen gels embedding mesenchymal progenitor cells induced cell alignment, formation of orientated collagen fibers, and upregulation of ligament-specific genes (Altman et al. 2002). This study provided evidence that specific mechanical forces applied to three-dimensional cellular constructs might not only enhance the development of an engineered tissue but also direct the differentiation of multipotent cells along specific lineages.

Despite numerous proof-of-principle studies showing that mechanical conditioning can improve the structural and functional properties of engineered tissues, little is known about the specific mechanical forces or regimes of application (i.e., type of applied stresses, magnitude, frequency, continuous or intermittent, duty cycle, etc.) that are stimulatory for a particular tissue. In addition, engineered tissues at different stages of development might require different regimes of mechanical conditioning owing to the increasing accumulation of extracellular matrix (ECM) and developing structural organization. In this highly complex field, a comprehensive understanding can only be achieved through hypothesis-driven experiments aimed at elucidating the mechanisms of downstream processes of cellular responses to well-defined and specific mechanical stimuli. In this context, bioreactors can play an important role since they provide controlled environments for reproducible and accurate application of specific regimes of mechanical forces to three-dimensional constructs (Démarteau et al. 2003a). This must be coupled with quantitative analysis and computational modeling of the physical forces experienced by cells within the engineered tissues, including mechanically induced fluid flows and changes in mass transport.

The role of bioreactors in applying mechanical forces to three-dimensional constructs could be broadened beyond the conventional approach of enhancing cell differentiation and/or ECM deposition in engineered tissues. For example, they could also serve as valuable *in vitro* models to study the pathophysiological effects of physical forces on developing tissues and to predict the responses of an engineered tissue to physiological forces on surgical implantation. Together with biomechanical characterization, bioreactors could thus help in defining when engineered tissues have a sufficient mechanical integrity and biological responsiveness to be implanted (Démarteau et al. 2003b). Moreover, quantitative analysis and computational modeling of stresses and strains experienced both by normal tissues *in vivo* for a variety of activities and by engineered tissues in bioreactors could lead to more precise comparisons of *in vivo* and *in vitro* mechanical conditioning, and help in determining potential regimes of physical rehabilitation that are most appropriate for the patient receiving the tissue.

Despite these results, the mechanisms whereby cells are sensing mechanical actions produced by their environment have not been well established. This is particularly true in the case of the viscous shear stresses, τ, generated by slow interstitial flow within porous tissue (Swartz and Fleury 2007), which can be evaluated by the following expression:

$$\tau = \eta \times \frac{u_0}{a} \tag{3.6}$$

As it is clearly recognized now that the cell biochemical transduction depends upon its mechanical environment, thorough studies of the flow behavior within three-dimensional matrices in relation with biophysical and biochemical signaling of the cells have to be undertaken.

3.3.4 Other Physicochemical Factors Affecting Cell Metabolism

In addition to mechanical *stimuli*, cultivated tissues are known to react to other physical signals. For instance, electromagnetic phenomena are important from cell biology to medicinal applications. Indeed, modern molecular biology tends to correlate the action of ion transporters and ion channels to the "electric" action of cells and tissues. Also, cell proliferation is improved by applying adequate electric fields. Thus, the triggers exerted by ion concentrations and concomitant electric field gradients have been traced along signaling cascades till gene expression changes in the nucleus (Funk et al. 2009). Moreover, at the small scale, the living tissues can present a "membrane behaviour." They are impermeable to organic solutes with large molecules, such as polysaccharides, while permeable to water and small, uncharged solutes. Permeability may depend on solubility properties, charge, or chemistry as well as solute size. Osmosis provides the primary means by which water is transported into and out of cells. The turgor pressure of a cell is largely maintained by osmosis, across the cell membrane, between the cell interior and its relatively hypotonic environment (Maton et al. 1997). Moreover the swelling properties of connective biological tissues such as cartilage can be explained by the osmotic disjoining pressure (Huyghe and Janssen 1997).

To take into account these electrochemical effects, it is necessary to combine the transport equations with equations governing the electrolyte movement coupled with local electrodynamical field evolutions.

An important property inherent in many biological charged porous media is the negative charge of their surface, which is a consequence of the presence of some chemical negative sites such as hydroxyl complex. This negative charge is compensated by the adsorption of cations on the surface forming the inner compact layer commonly referred to as the immobile stern layer. Nevertheless the majority of the excess of positively charged counterions are located in the electrolyte aqueous solution externally to the solid phase forming an outer diffuse layer composed of mobile charges. Together with the fixed charged groups of the solid matrix these ions form the so-called electrical double layer (see Figure 3.3).

The thickness of this double layer is characterized by the Debye length $L_D = \sqrt{\varepsilon\varepsilon_0 RT/2F^2C_i}$,which inversely depends on the ionic concentrations, that is to say on the ionic force C_i (Hunter 2001). Here ε_0 is the vacuum permittivity, ε is the relative dielectric constant of the solvent, R is the gas constant, T is the absolute temperature, and F is the Faraday constant. These electrical phenomena are generally purely microscopical since the Debye length is classically of a few nanometers. The dimensionless electric double-layer potential $\overline{\varphi} = {}^{F\varphi}/_{RT}$ obeys the well-known nonlinear Poisson–Boltzmann equation:

$$\Delta(\bar{\varphi}) = \frac{1}{L_D^2}\sinh\bar{\varphi} \tag{3.7}$$

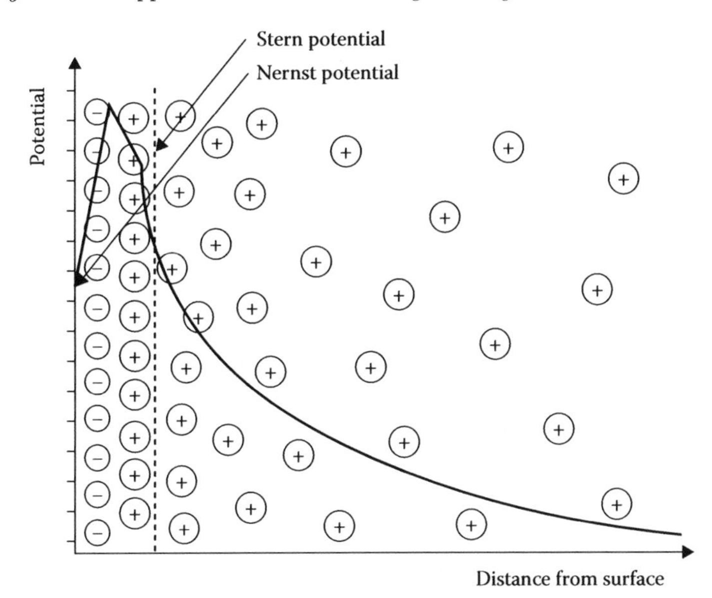

FIGURE 3.3
Equilibrium electrostatic potential in an electrolyte solution bordered by a plane negative surface. The microstructure of the ionic distribution at the interface induces the phenomenon of Debye shielding by the ion cloud of the opposite sign.

Owing to the difficulty in solving this equation in a three-dimensional configuration, the hyperbolic sine is often linearized following the Debye–Hueckel approximation, which is valid for small double-layer potentials ($\overline{\varphi} \ll 1$). On the basis of a multiscale description of multiphysical flow in porous media, Moyne and Murad (2002) prove that this equation applies to phenomena at purely microscopic scale. However, this microscopic effect can have significant consequences at the macroscale. For instance, when considering symmetric Cartesian pores, the macroscopical swelling effects observed for cartilagenous tissues can be explained by Donnan osmotic swelling pressure π_D (Donnan 1924) and are governed by the value of this double-layer potential in the symmetry plane (Langmuir 1938; Israelachvili 1991).

When advected by the streaming velocity of the fluid, the excess in mobile charge population in the counterion atmosphere leads to macroscopic observed electrokinetic phenomena such as streaming currents, resulting from the influence of fluid movement upon charge flow. In addition, to counterbalance this apparent charge accumulation and to conserve charge, the movement of the net charge generates an electric potential, often referred to as streaming potential, which gives rise to other macroscopic electrokinetic phenomena.

The spatial variability of this potential generates electrophoretic movement of the mobile charges, inducing a conduction Ohmic current, which opposes the streaming current and consequently slows down the counterions of the diffuse double layer. Owing to the viscous drag interaction, the ions pull the solvent, resulting in a concomitant electroosmotic seepage flow opposing the pressure-gradient driven flow. This electrokinetic coupling has been commonly referred to as the electro-viscous effect as its overall influence upon the flow is usually treated through an increase in the viscosity of the liquid (Hunter 1981).

Many efforts have been made to better understand the role of coupled electro-hydraulic phenomena on the stimulation of cell activity. For instance, Lemaire et al. (2006, 2008) carried out a multiscale approach to quantify the viscous shear stresses, τ, generated by interstitial flow by taking into account the electrokinetic phenomena occurring at the scale of the cell membrane. Dealing with cortical tissue, these studies proposed a coupled Darcy law to describe the bone interstitial fluid flow that develops not only because of the pressure gradient effect but also in response to streaming potential and chemical gradients. Indeed, this description includes chemical-osmotic driven effects (gradient of the Nernst potential) that are also manifested particularly when the salinity or the pH varies spatially (Gu et al. 1998). This example indicates the potentiality of stimulating cell culture using electrokinetic effects (Funk et al. 2009).

3.4 Bioreactors and Implants

Tissue engineering not only promises for the future development of a new generation of artificial organs, but also provides a basis for quantitative *in vitro* studies of tissue genesis by culturing cells on three-dimensional substrates in the presence of specific biochemical and physical factors. Bioreactors and substrates are designed to maintain ad hoc levels of physiological parameters in the cell environment, including enhancement of mass transport rates and exposure to specific mechanical stimuli. Thus, functional tissue engineering not only requires cellular components capable of differentiating in appropriate lineages, but also necessitates the use of specific structural templates whose material nature and structural design foster the tissue growth. Moreover, it requires the development of bioreactors providing necessary biochemical and physical regulatory signals predisposing cell population growth, guiding differentiation, and inducing extracellular matrix production.

3.4.1 Different Types of Bioreactors

Bioreactors are classically used in the food-processing industry (fermentation and water treatment) as well as in the pharmaceutical industry (proteins manufacturing). They are generally defined as devices in which biological or biochemical processes occur in a well-controlled environment (pH temperature, pressure, nutrient supply, etc., [Martin et al. 2004]). As regards tissue engineering, bioreactors are used to obtain a culture environment adapted to the implant development.

As summarized by Martin and Vermette (2005), the bioreactor functions should be ideally the same as those performed by the uterus during embryo development. Even if this objective is still unreachable in the near future, a few researchers use the human body as a bioreactor (those applications are based on the body's self regeneration, [Service 2005]). For *in vitro* tissue-engineering applications, different kind of bioreactors are classically used (see Figure 3.4) not only to cultivate implants but also to seed biomaterials with living cells.

However, "traditional" cell cultures within Petri dishes or wells seem to be unsuitable when the implant volume is too large because the diffusion flux carrying oxygen or nutrient is too weak compared to the nutrient requirements of cells. To obtain a tissue volume of clinical interest, the use of a flowing culture media could be beneficial to the cell culture, as shown by numerous experimental and also numerical experiments. In this case, the flow induces an increase of the nutrient flux, of the oxygen supply, and of the waste removal by supplementing diffusion transport processes with convection transport processes (generally more intense). Moreover, the flow generates a mechanical loading on the cells, which can be beneficial to the implant development under definite conditions (Nauman et al. 2001; Bancroft et al. 2002; Raimondi et al. 2002; Cartmell et al. 2003; Sikavitsas et al. 2003; Healy et al. 2005; Gemmiti and Guldberg 2006; Leclerc et al. 2006; Zhao et al. 2007).

In contrast to the "static" feature of the traditional cell cultures (culture media without movement), "dynamic" culture processes using a flowing culture media have been widely developed. Depending on the bioreactor characteristics and the biomaterial's geometry, the culture media flows at the implant periphery or flows throughout the implant. For example, in a "hollow fiber bioreactor" (Dulong and Legallais 2005) cells are confined within the fiber where no flow occurs (the culture media flows around the fiber). Convection allows transporting a large amount of oxygen or nutrient to the outer surface of the fiber, but the transport between the outer surface of the fiber and the cells is only achieved by diffusion.

At the opposite, in a perfusion bioreactor (Goldstein et al. 2001; Raimondi et al. 2002; Cartmell et al. 2003; Sikavitsas et al. 2003) due to the imposed perfusion flow rate, convection occurs within the pores of the implant.

Between these two last configurations are placed the "stirring flask bioreactors" (Malda et al. 2004; Sucosky et al. 2004; Lewis et al. 2005; Bilgen et al.

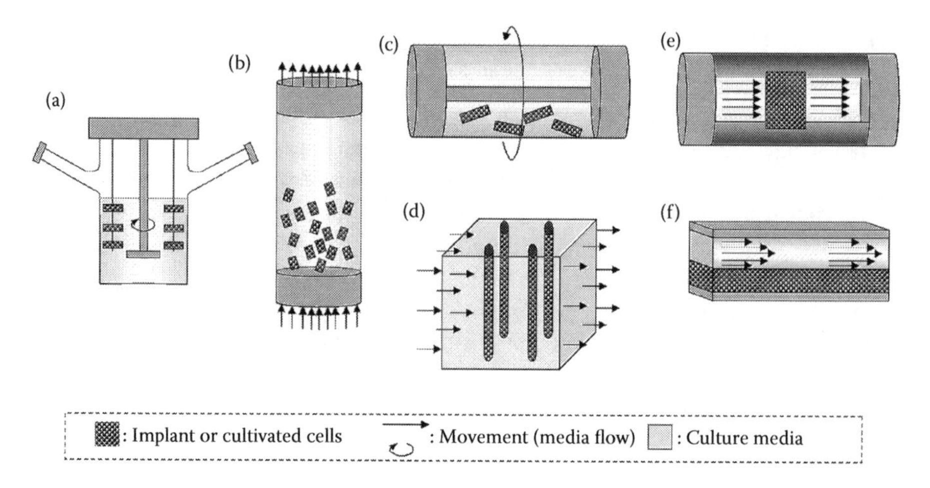

FIGURE 3.4
Diagram of bioreactors classically used in tissue engineering. (a) Stirred flask bioreactor: implants are attached to a fixed wire, a stirring rod mixing the culture media (Malda et al. 2004; Sucosky et al. 2004; Lewis et al. 2005; Bilgen et al. 2005; 2006); (b) fluidized bed bioreactor: implants are placed within the cylindrical body of the bioreactor, the culture media flows from the bottom to the top of the bioreactor (imposed flow rate) (David 2002; Janssen et al. 2006); (c) rotating bioreactor: implants are placed within the cylindrical body of the bioreactor, which rotates around its axis (Botchwey et al. 2001, 2003, 2004; Singh et al. 2005); (d) hollow fiber bioreactor: implants or cells are confined within the fibers, the culture media flows around the fibers (Dulong and Legallais 2005); (e) perfusion bioreactor: the implant is attached to the body of the bioreactor and is perfused by the culture media (imposed flow rate) (Goldstein et al. 2001; Raimondi et al. 2002; Cartmell et al. 2003; Sikavitsas et al. 2003); (f) parallel-plate bioreactor (or microfluidic bioreactor): the implant is growing on the lower surface of the bioreactor, the culture media flows between the upper surfaces of the implant and the bioreactor (imposed flow rate) (Gemmiti and Guldberg 2006; Leclerc et al. 2006; Zhao et al. 2007).

2005, 2006), the "fluidized bed bioreactors" (David 2002; Janssen et al. 2006), or the "rotating bioreactors" (Botchwey et al. 2001, 2003, 2004; Singh et al. 2005). For those bioreactors, the flow (and, as a result, the associated convective transport) will occur or not within the implant depending on the value of the pressure gradient imposed by the flow at the scale of the implant, on the one hand, and the hydraulic permeability of the implant, on the other. Such a characteristic parameter of the medium micro architecture has dimension of length square and an order of magnitude approaching the pore radius squared.

This order of magnitude can be obtained from a simple three-dimensional model of an isotropic network of cylindrical pores, leading to (Guyon et al. 2001):

$$K = \frac{\phi(2a)^2}{96} \times \frac{1}{T} = \frac{1}{6} \times \frac{\phi^3}{S_V^2 \times T} \tag{3.8}$$

where T is the tortuosity of the pores. In relation to this simple model of twisted and crooked pores, channels crossing a thickness L of the medium in the wise stream perfusion direction display an "effective pore length" L_P such that

$$L_P^2 = T \times L^2 \quad (T \geq 1) \tag{3.9}$$

The reality in such media is much more complex, with dead-ending or connecting pores. This means there is a need to scale the permeability from the capillary tube model to include increased path length owing to the crookedness of the path (tortuosity) or lack of connection between points in the medium (connectivity). In theory, the two parameters should be inversely related, with highly connected (or highly porous) substrates having low tortuosity and vice versa. For the purpose of this chapter we will consider the two parameters interchangeable and propose an order of magnitude of 10 for the tortuosity parameter of highly connected porous biomaterial (corresponding to what is found in biological porous tissue such as trabecular bone). These concepts and formula were applied in Table 3.1 to evaluate the specific permeability of some porous substrates currently used in osteoarticular tissue engineering.

As a result, in various culture processes, the choice of the implant scaffold (chemical nature and physical properties) is a key parameter for the success of the culture.

3.4.2 Microarchitectural Design of Substrates

Tissue formation requires a physical support on which the cells can adhere, migrate, proliferate, and differentiate. As these processes are often *in fine* associated with neovascularization, the supporting material should be porous. Ideally, the pores should be interconnected to favor rapid and complete angiogenesis within the implant. The pore size should be greater than 150 μm to allow formation of new tissue (Shors 1999). Since the ultimate goal is to completely replace the defective site, owing to a new organoid, a resorbable vehicle should be preferred: the ideal material should resorb in parallel to the new tissue formation so that the mechanical support provided by the vehicle can be progressively and completely replaced by new tissue. One of the greatest difficulties with cell therapy lies in obtaining such an optimal vehicle.

Even if three-dimensional cultures of tissues can be achieved without using any artificial scaffold (see e.g., Gemmiti and Guldberg 2006), the majority of cell cultures in the framework of tissue engineering need the use of a porous

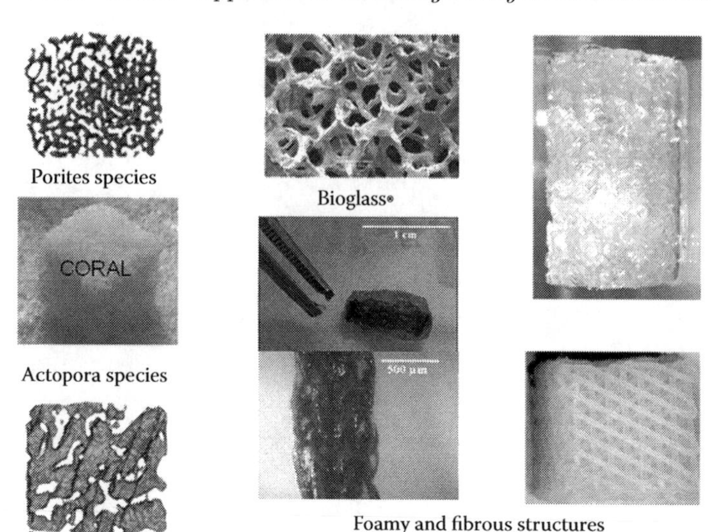

FIGURE 3.5
Various scaffolding porous materials and structures, either natural or artificial, used in osteoarticular tissue engineering. (Courtesy of D. Hutmacher, H. Petite, and X. Wang.)

biomaterial as substratum. In such a case the nature of the constitutive biomaterials and the local architecture of the substrate are fundamental parameters for the culture's success. The ideal biomaterials or scaffold must be sterilizable, biocompatible (noncytotoxic and at least biotolerated), and ideally bioresorbable (at mid or long term). The scaffold must also be a physical substratum giving the opportunity to the cells to bind with its surface, to proliferate, to differentiate, and to synthesize ECM. Moreover, its mechanical properties have to be as close as possible to those of the natural organ. This last function is especially important when the implant must have a holding function after implantation (bony or cartilaginous implants for example).

In the framework of bone and cartilage tissue engineering, various biomaterials have been used. Examples are illustrated in Figure 3.5 and some of their physicochemical properties are given in Table 3.1. Even if metallic porous scaffolds have been utilized (titanium [van den Dolder et al. 2003], tantalum and nitinol [Maurin et al. 2005]), the majority of three-dimensional cell cultures have been achieved using polymeric or ceramic scaffolds. Among the polymeric biomaterials (for a review see [Jagur-Grodzinski 2006]) the most popular are nonexhaustively the following: the polylactic acid or polylactide (PLA) (Saini and Wick 2003), the polyglycolic acid (PGA) (Obradovic et al. 2000; Wilson et al. 2002), polylactic-co glycolic-acid (PLAGA) (Botchwey et al. 2001, 2003, 2004), hyaluronic acid (HA) (Raimondi et al. 2002), or polycaprolactone (PCL) (Hutmacher et al. 2001). Scaffolds made of hydrogel can also

Local Architectural Properties of Some Porous Substrates Used in Osteoarticular Tissue Engineering

Properties → Scaffolds ↓	**Order of Magnitude in Porosity (%)**	**Mean Pore Diameters (μm)**	**Connectivity (Tortuosity)**	**Specific Pore Surface Area ($\times 10^3 m^{-1}$)**	**Permeability (μm^2)**	**Production Method**	**References**
Natural exoskeleton:							
Porites	50	300	High	7	47	Harvesting, drying, and cleaning	Guillemin et al. 1989
Acropora	15	500		1	39		
Copolymer Scaffolds:							
Collagen	99	100	High	40	10	Freezing and drying process	Tierney et al. 2009
Glycoaminoglycan							
Structures made of polymeric Materials:							
Polycaprolactone	60	400	High	6	100	Rapid prototyping: fused deposition modeling	Hutmacher et al. 2001
Synthetic β-tricalcium	80	250	High	13	52	Polymeric sponge method	Liu et al. 2007, 2008
phosphate β-TCP							
Porous bioglass	70	250	High	11	46	Glass melting procedure	Zenati et al. 2006

be interesting because this material has high water content and can offer an internal structure close to ECM structure of several biological tissues (Drury and Mooney 2003). It is also of importance to mention that some ceramics (synthetic: hydroxyapatite [Rose et al. 2004], tricalcic phosphate [Pothuaud et al. 2005; Janssen et al. 2006] or natural: coral [Petite et al. 2000] and nacre [Atlan et al. 1999]) show a chemical composition rich in calcium, which makes them specific to the culture of bone implants.

Furthermore, the interaction between biomaterials and the biological environment has to be carefully considered. Some ceramics (hydroxyapatite and other bioactive calcium phosphates) are known for their bioactivity because they interact positively with the body by accelerating the tissue-regeneration processes. Other biomaterials are referred to as "bioresorbable" because their resorption takes place after implantation within a period of time varying from a few weeks to several years. Bioresorbable materials may be natural (alginate, hyaluronic acid, collagen, fibrin gels, or coral) or synthetic (PCL, bioglass, tricalcium phosphate, etc.). As regards the interaction between the biomaterials and the biological environment, another key factor resides in the materials surface because its nature may or may not favor the adsorption of biological molecules acting on the cell behavior. For example, an experimental study has shown that by coating a biomaterial's surface with hydroxyapatite, the behavior of bone cells can be modified (El-Ghannam et al. 1997). In that way, numerous biological molecules (ligands increasing the cell adhesion, growth factors, hormones, enzymes, etc.) can be grafted on the biomaterial's surface (Jagur-Grodzinski 2006) to control or to adjust the cell behavior.

At last, but not at least, the microstructure of scaffolds plays a key role. As regards the porous biomaterials with interconnected pores, it is generally considered that a pore size smaller than 100–150 μm does not allow a satisfying tissue in-growth within the implant (Shors 1999; Rose et al. 2004; Rezwan et al. 2006) and that a mean pore size between 200 and 400 μm is optimal to favor not only the cell proliferation and differentiation but also the generation of a vascular network after implantation (Brekke and Toth 1998).

Since a careful design of the scaffold microstructure is essential to the culture success, various techniques such as "rapid prototyping" have been used to obtain well-defined and controllable microarchitectural properties (Hutmacher et al. 2001; Zein et al. 2002). Among the rapid prototyping techniques, the "Fused Deposition Modeling" based on the principle of extrusion allows the manufacturing of polymeric scaffolds (Hutmacher et al. 2004). Scaffolds are first virtually designed using a computer and then manufactured in three dimensions by superposing flat fibrous layers. With such a technique, pores are fully interconnected and the scaffold geometry is controllable and reproducible (for more details see Zein et al. 2002). Other techniques such as "jet based methods" are also developed to "print" directly the cells within the scaffold (Ringeisen et al. 2006).

3.5 Theoretical Models of Active Porous Media

The term "model" can have different meanings according to the scientific domain. For instance, in biology, an animal model is used, in a study of pathology on a given animal species, to understand the same type of pathology in a human being. In physics or mathematics, a model is a simplified representation of a process that can be treated, owing to mathematical expressions. In this chapter, the term model is used in its mathematical acceptation.

In general, a model can be very useful to understand experimental phenomena and promote new ideas in experimental investigations. In tissue engineering, many different physical parameters participate in tissue growth. The modeling tool can be useful for independently studying the principal physical mechanisms and thus identifying those that mainly govern the culture processes.

3.5.1 Length and Time Scales of the Different Physicochemical Phenomena

Various physicochemical phenomena are associated to the biological steps of the implant development, and they interact together at very different time and length scales (see Figure 3.6).

Physicochemical phenomena occurring during the tissue development are (this list being nonexhaustive) transport and consumption of nutrient and oxygen, generation of waste by cells, mechanical loading of tissue and/or cells, osmosis phenomena (transport trough the cell membrane), electrochemical phenomena (Debye length), electromechanical phenomena (piezoelectricity), or chemomechanical phenomena (swelling).

The time and length scales related to these physicochemical phenomena are showing important variability in their extent: the characteristic time for cell adhesion is about few hours while cells synthesize ECM over several weeks. As regards the length scales, few hundreds of nanometers have to be considered by speaking about focal adhesion points of cells while few centimeters are the characteristic length scale of the substrate and bioreactor. As a result, because of the difference in magnitude of both time and length scales as well as the plurality of the implied phenomena, a good understanding of the *in vitro* three-dimensional cell culture is still very complex.

3.5.2 Convection–Diffusion–Reaction Phenomena: Basic Equations and Characteristic Nondimensional Parameters

Porous medium numerical simulations have been designed to study physical and biochemical factors that control the functional development of

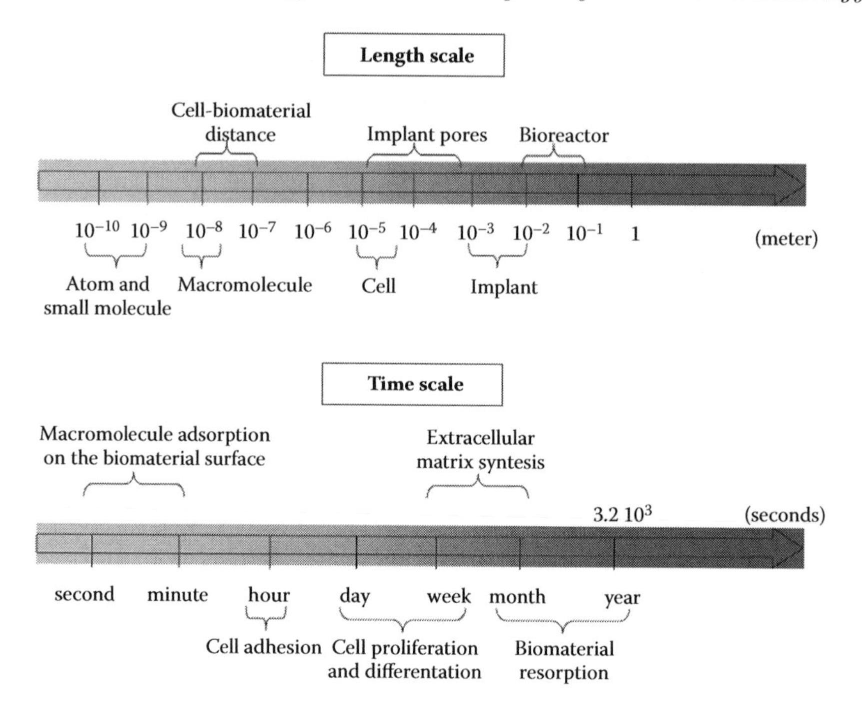

FIGURE 3.6
Time and length scales associated with biological phenomena occurring during the implant development.

osteoarticular tissue-engineered constructs. Examples of our methodological approaches aimed at enabling an integrated study of the solute transport and mechanical phenomena taking place in perfusion bioreactors are reviewed here. In a first part, the study is focused on oxygen transport. In a second part, this approach is extended to ionic species under electrochemical coupled effects.

At the pore length scale a, the variables and parameters are written in a nondimensional form (the "prime" notation is used to signal the nondimensional form of the considered quantity). Thus the oxygen concentration C_{O_2}, the flow direction coordinate X and the transverse coordinate $Y$and Z, the perfusion velocity $\vec{u}$, and the fluid pressure p are reduced, respectively, owing to a reference concentration c_0 (the inlet oxygen concentration for instance), the pore size a, a reference velocity u_0 (the inlet velocity for instance), and the dynamic viscosity of the perfusing fluid η:

$$C_{O_2} = C' \times c_0;\ \ X = X' \times a;\ \ Y = Y' \times a; Z = Z' \times a;\ \ \vec{u} = \vec{U}' \times u_0$$
$$p = P' \times \eta u_0 / a \tag{3.10}$$

Moreover, nondimensional numbers comparing different physical phenomena occurring in the transport process are also introduced:

$$Pe = (u_0 \times a)/D_{O_2}; \quad \mathrm{Re} = (u_0 \times a)/r$$
$$Da = (\sigma_{cel} \times V_{max} \times a)/(c_0 \times D_{O_2}); \quad \lambda_M = K_M/c_0 \tag{3.11}$$

As already mentioned, the Péclet number Pe represents the ratio of convective and diffusive effects for problems in mass transport. It is the equivalent to the Reynolds number Re for the momentum transfer in fluid dynamics phenomena, with v standing for the kinematic viscosity of the fluid in place of the solute diffusion coefficient D_{O_2}. Furthermore, the Damkölher number Da represents the ratio of reactive (for instance, O_2 consumption of the cell layer with surface density σ_{cel}) and diffusive effects. And finally the dimensionless Michaelis constant λ_M is reduced using the reference concentration.

In this framework, the distribution of the oxygen concentration within the channel is governed by the following stationary diffusion–convection equation, written in its dimensionless form:

$$\vec{\nabla}'.\left[-\vec{\nabla}'(C') + Pe \times \vec{U'} \times C'\right] = 0 \tag{3.12}$$

Since cells are only attached on the pore wall, their consumption is represented by a flux (noted $\vec{N'}_{O_2}$) boundary condition following the Michaelis–Menten kinetics:

$$-\vec{N}'_{O_2} \cdot \vec{n} = -Da \times f(C') \tag{3.13}$$

where $f(C') = C'/(C' + \lambda_M)$ (approximately, $f(C') = C'$ for $C' \leq \lambda_M$ and $f(C') = 1$ for $C' \geq \lambda_M$) and $\vec{n}$ is the unit vector normal to the element of wall surface.

Other boundary conditions associated with this transport equations are as follows:

– At the inlet, the given reference oxygen concentration:

$$C'_{\text{inlet}} = C'_0 = 1 \tag{3.14}$$

– A convective flux at the outlet of the pore:

$$\vec{n}_{\text{outlet}}.\left(-\vec{\nabla}'C'\right) = 0 \tag{3.15}$$

– Symmetric boundary conditions everywhere else on connecting pore surfaces:

$$\vec{n}_{\text{sym}}.\left(-\vec{\nabla}'C' + Pe \times \vec{U'} \times C'\right) \tag{3.16}$$

The velocity field $\vec{U'}$ appearing in the convective term of equation (3.12) is obtained by solving the dimensionless Navier–Stokes equations, considering the steady flow of an incompressible Newtonian fluid:

$$Re\left(\vec{U'}.\vec{\nabla'}\right)\vec{U'} = -\vec{\nabla}'(P') + \Delta'\left(\vec{U'}\right) \quad \text{and} \quad \vec{\nabla}'.\vec{U'} = 0 \tag{3.17}$$

Boundary conditions associated with Navier–Stokes equations are as follows:

– Blunt profile at the inlet:

$$\vec{U'} = U'_0 \vec{X'} \text{ with } U'_0 = 1 \tag{3.18}$$

– Arbitrary pressure at the outlet:

$$P' = 0 \tag{3.19}$$

– No-slip boundary condition at the cell layer:

$$\vec{U'} = \vec{0} \tag{3.20}$$

– Symmetric boundary conditions everywhere else on connecting pore surfaces:

$$\vec{n}.\vec{U'} = 0 \quad \text{and} \quad \vec{t} \cdot \left(-P'\overline{\overline{I}} + \left(\overline{\overline{\nabla' U'}} + \overline{\overline{\nabla' U'}}^{\mathrm{T}} \right) \right) \cdot \vec{n} = 0 \tag{3.21}$$

where $\vec{t}$ is a unit vector tangent to the element of surface.

Moreover, a fair approach in mean oxygen concentration repartition along the principal X' perfusion direction of the pore channel can be obtained in one-dimension formulation (Pierre and Oddou 2007) by solving the following stationary diffusion–convection–reaction equation:

$$d^2C'/dX'^2 - Pe \times dC'/dX' - Da \times f(C') = 0 \tag{3.22}$$

In this equation, the main transport mechanisms are diffusion and hydraulic convection (quantified by the Péclet number), associated with reaction due to the cell oxygen consumption (quantified by the Damköhler number).

Similar equation can be obtained for ionic nutrients when considering porous media presenting large pores. Nevertheless, for small hydraulic permeability, such as in the case of the mature cartilage for instance, the other physicochemical phenomena described in Section 3.2.4 have to be taken into account. Indeed, in addition to the Fickian diffusion involving the gradient of the ionic force C_i, the electromigration of charged species in response to the gradient of the reduced streaming potential $\overline{\psi} = F\psi/RT$ has to be considered. Thus, the ionic flux becomes (Lemaire et al. 2008):

$$\overrightarrow{J_\pm} = -D_\pm \exp(\mp\overline{\varphi}) \left(\nabla C_i \pm C_i \nabla \overline{\psi} \right) \tag{3.23}$$

Owing to the double-layer effects, the cationic or anionic diffusion coefficient $D_\pm$ is weighted by a Boltzmann-like term involving the double-layer reduced potential $\overline{\varphi}$. Moreover the velocity used in the definition of the Péclet number is the sum of three contributions in response to hydraulic, osmotic,

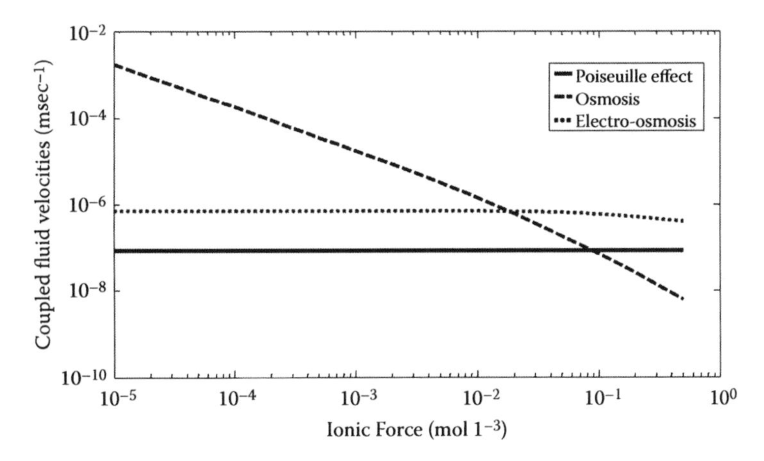

FIGURE 3.7
Coupled fluid velocities versus the ionic force considering a model of plane channel for the water-saturated electrolyte pore with 10 nm thickness and negative surface charge of 0.2 C/m^2 density: (i) hydraulic (Poiseuille) gradient of 10 MPa/m; (ii) osmotic gradient of 10^4 mol/m^4; (iii) electroosmotic gradient of 10 V/m.

and electroosmotic driving gradients as exhibited by the generalized Poiseuille law (Lemaire et al. 2008):

$$\overrightarrow{u} = -\kappa_{\mathbf{P}} \nabla p - \kappa_{\mathbf{C}} \nabla C_i - \kappa_{\mathbf{E}} \nabla \overline{\psi} \tag{3.24}$$

The coupled permeability tensors κ_i are calculated using homogenization procedures adapted from Moyne and Murad (2002). In Figure 3.7, using the obtained expressions of these permeability parameters, we present the resulting fluid velocity as a function of the ionic force. Here, we consider a model of plane channel for a representative pore. This channel is filled with water-saturated electrolyte, has 10 nm thickness and negative surface charge of 0.2 C/m^2 density. Flow is generated by driving gradients corresponding to a physiological situation (Figure 3.7). If the velocity contribution due to the hydraulic effect remains constant since it only depends on the geometry of the pore, the electrochemical velocities do change with the ionic force variations. In particular, the lower the ionic force is, the higher the double-layer thickness is, and the more efficient the osmotic effect is. Moreover, in this model, since the surface charge density, which governs electroosmosis efficiency, is not modified by the ionic force changes, electroosmotic effect is only slightly affected for the largest ionic force values. Thus electroosmosis seems to be the main fluid transport mechanism through very thin pores for physiological biochemical conditions.

Using similar nondimensional scaling as mentioned before, the stationary chemical transport equation becomes the well-known Nernst–Planck equation:

$$P_e \nabla' . \left(c' \exp\left(\mp\overline{\varphi}\right) \overrightarrow{U}' \right) + J_d \nabla' . \overrightarrow{J_{\pm}}' = Da \times f\left(c'\right) \qquad (3.25)$$

In this equation, a new nondimensional number $J_d = I_0 L / D_0 F c_0$ appears. It involves a reference electric current I_0, a macroscopic reference length L, the Faraday constant F, the reference solute diffusion coefficient D_0, and a reference concentration c_0. This nondimensional number compares electric and diffusive effects.

For usual model-driven experiments involving tissue engineering at the microfluidic scale, these electrokinetic phenomena are unimportant in a first approach and the classical convection–diffusion–reaction equations can be used.

In the following paragraphs, two model-driven studies will be presented to illustrate the complementarity between experimental approach and theoretical analysis. In a first example, bone tissue implants are questioned. The design of the porous biomaterials is particularly investigated. In a second example, we present a numerical investigation attempting to complement previously observed experimental responses to flow by simulating oxygen transport within cartilage constructs of different geometries and as a function of flow rate.

3.5.3 Computational Models: Two Examples of Model-Driven Experimental Approaches

3.5.3.1 Modeling of Transport Processes in Bone Tissue-Engineered Implants

For *in vivo* osseous tissue regeneration in large defects, the interest for the use of large porous implants of biocompatible and biodegradable coral have been the object of particularly promising studies realized by the biologists in our laboratory (Petite et al. 2000). However, problems of *in vitro* culture in implants of big dimensions (i.e., a few cube centimeters) remain the major concern, in particular the design of the porous biomaterial's microarchitecture and the control of hydrodynamic flow and mass transport during the cell and tissue culture process. In this purpose, the mass and momentum transport phenomena occurring in such highly porous and interconnected active media have to be modeled. To achieve that, mass and momentum transport equations previously presented were solved using a three-dimensional model at the pore scale taking into account realistic implant geometry. The structure of the local flow field within implant pores, the related shear-stress distributions, and the nutrient transport effects have been analyzed.

Among different types of bioresorbable and biocompatible implants (ceramic composite, natural, or artificial polymers), polymeric implants made of PCL by fused deposition modeling (a rapid prototyping technique [Zein et al. 2002]) present a periodic fibrous architecture with sufficient simple geometry well adapted to modeling. They constitute a bioresorbable and porous structure made of a simple network displaying fully interconnected pores (porosity of the order of 60%, pore and fiber diameters about $a = h = 250\,\mu\text{m}$). A sufficiently simple three-dimensional geometry of the pore channel was then defined taking into account the symmetries of the implant architecture: we considered here that each constitutive fibrous layer has a fiber direction perpendicular to the previous one (as schematically shown in Figure 3.8). More precisely, an elementary representative test section containing 10 cylindrical fibers ($l_{Ch} = 2.5\,\text{mm}$) was studied. Liquid flows from left to right perpendicularly to the inlet edge as seen on Figure 3.9, and the cell layer was considered as a homogeneous thin monolayer attached around the fibers.

The reference values of entry data model's parameters, coming from experimental conditions of our biologist colleagues or from the literature, are given in Table 3.2.

As expected for such a flow dominated by viscous effects (Reynolds number of the order of 10^{-2}), the structure of the velocity field reproduces the waviness, periodicity, and symmetry of the substrate structure, as shown by the flow field and spiraling streamlines given in Figure 3.9.

Despite this smallness of the Reynolds number, a secondary flow due to the cross crenellation of the channel is generated by a tiny amount of inertial effects as shown in the helicity of the streamlines and the vortical structure of this secondary flow field.

The analysis of such a flow field is relevant of the fluid dynamics inside slowly curved pipe, where the response depends upon the Dean number Dn defined as (Dean 1927)

$$Dn = Re \times \left({}^{a}\!/_{r_c}\right)^{\frac{1}{2}} \tag{3.26}$$

where r_c is the radius of curvature of the streamlines.

The resulting velocity, u_t, characterizing the transverse secondary flow takes then the following approximate value:

$$u_t \approx u_0 \times Dn \times (a/r_c)^{\frac{1}{2}} \tag{3.27}$$

Such a velocity is to compare with the apparent velocity u_D for oxygen diffusion as defined by

$$u_D \approx D_{O_2}/a \tag{3.28}$$

The result is

$$u_t/u_D \approx Pe \times Dn \times (a/r_c)^{\frac{1}{2}} \tag{3.29}$$

The field of this secondary transverse flow is illustrated by arrows in Figure 3.10.

For the reference data concerning this model, the obtained magnitude (around $10^{-6}\,\text{m} \times \text{sec}^{-1}$) of such a recirculation flow velocity is slightly lower

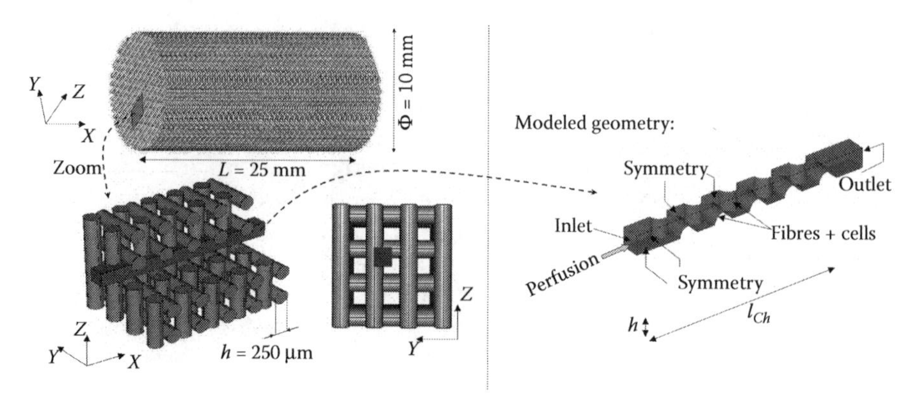

FIGURE 3.8
Three-dimensional geometry of a PCL fibrous implant at the different length scales. The modeled geometry of the representative elementary volume of one pore is defined by taking into account the symmetrical properties of the structure and by considering 10 fibrous layers localized within the implant.

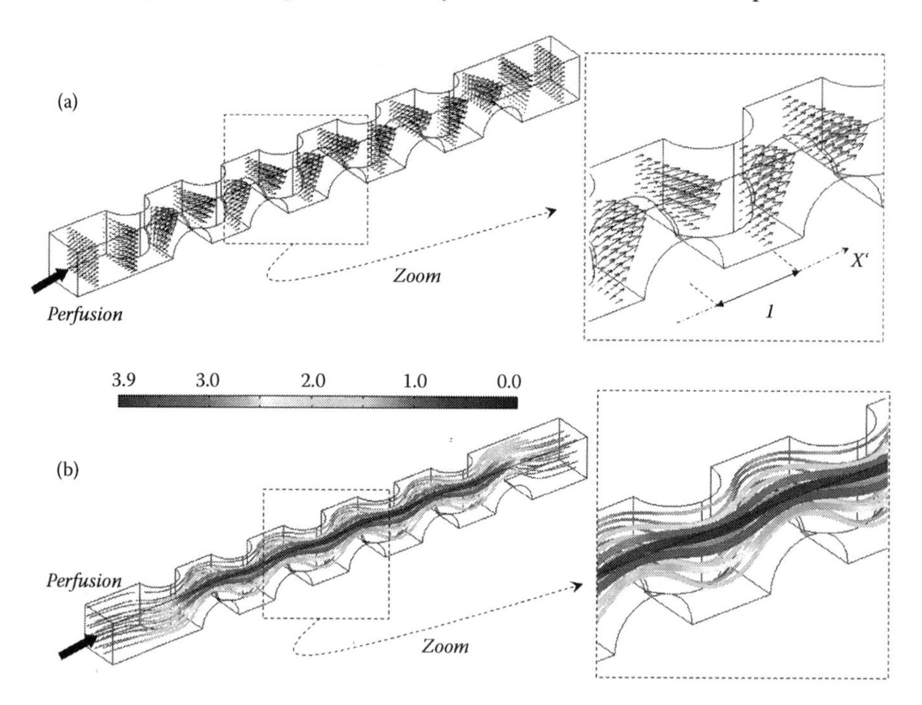

FIGURE 3.9
Flow velocity field (a) and streamlines (b) within a representative elementary volume constituted by one quarter section of a pore. Notice the obvious waviness of the flow field with a reduced velocity U'maximum in the vicinity of the axis, as shown on the gray scale. This maximum magnitude is of the order of 4, as expected by the mass conservation and the section area obstruction due to the presence of fibers.

TABLE 3.2
Parameters Values (Named Reference Values in the Text) Introduced in the Simulations

Parameter	**Value**	**References**
Diameter of polymeric fibers h	0.25×10^{-3} m	–
Implant length L	25×10^{-3} m	–
Length of the representative elementary volume l_{Ch}	25×10^{-4} m	–
Mean velocity at the inlet u_0	10^{-4} m $\times$ sec^{-1}	–
Liquid density ρ	10^3 kg $\times$ m^{-3}	Peng and Palson 1996
Liquid dynamic viscosity η	0.7×10^{-3} Pa $\times$ sec	Peng and Palson 1996
Liquid kinematic viscosity ν	0.7×10^{-6} m^2 $\times$ sec^{-1}	Peng and Palson 1996
Oxygen concentration at inlet c_0	0.2 mol $\times$ m^{-3}	–
Oxygen binary diffusivity in water D_{O_2}	3×10^{-9} m^2 $\times$ sec^{-1}	Williams et al. 2002
Michaelis constant k_M	6×10^{-3} mol $\times$ m^{-3}	Obradovic et al. 2000
Cell density on fiber surface σ_{Cell}	2×10^9 cell $\times$ m^{-2}	–
Maximal oxygen consumption rate V_{Max}	4×10^{-17} mol $\times$ sec^{-1} $\times$ cell^{-1}	Komarova et al. 2000
Damköhler number Da	3×10^{-2}	*
Péclet number Pe	8	*
Reynolds number Re	3.5×10^{-2}	*
Dimensionless Michaelis constant λ_M	3×10^{-2}	*

than the characteristic diffusion velocity of oxygen (about 10^{-5} m $\times$ sec^{-1}) at the length scale a of the pore. Nevertheless, such a secondary flow is capable of significantly contributing to the increase in the nutrient transport processes of oxygen molecules from the center of the channel toward its periphery, where the consumers of the cell's oxygen lie.

Thus, the mixing effects generated by this flow are contributing to the homogenization of the oxygen distribution in the vicinity of the fiber surface as shown in Figure 3.11, where a relative variation of less than 10% around the mean value of the oxygen concentration is noted.

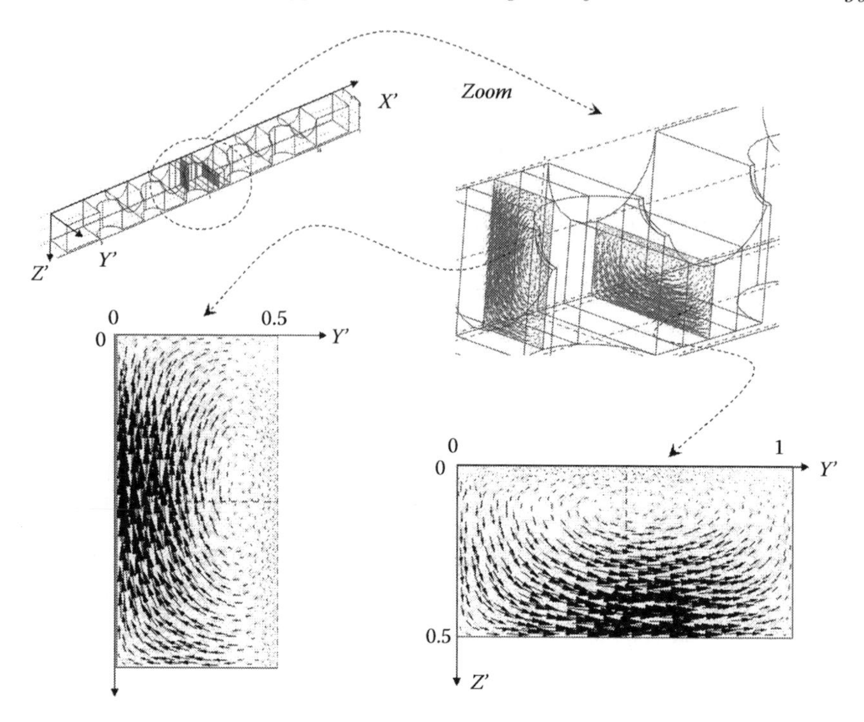

FIGURE 3.10
Vortex structure of the secondary flow generated by the waviness of the pore and helicity of the perfusion flow. The higher the Dean number is, the stronger the vortex is. As the fluid spins in the curved channel, a control mass of fluid travels in the transverse direction, eventually reaching the outer wall, where it must change direction toward a return path. Such a transverse mass transfer produced by this secondary flow, from the axis of the pore toward the wall where the cells are localized, is likely to highly supplement diffusion effects.

These results, calculated for the same reference values of the parameters, are significantly different from those obtained with a one-dimensional analytical approach (equation 3.22) where higher amplitudes and more uniform distribution in downstream direction of the oxygen concentration were found. Indeed, the presence of stagnation zones in the flow field within such a crenellated duct leads to a drastic decrease in local concentration in the stream wise direction, as revealed by the waviness of the profile given in Figure 3.12.

Moreover, such a flow generates in the vicinity of the fiber surface a nonunidirectional and nonhomogeneous repartition of viscous stresses, the magnitude of which is of the order of 10^{-3}–10^{-2} Pa on the major part of the

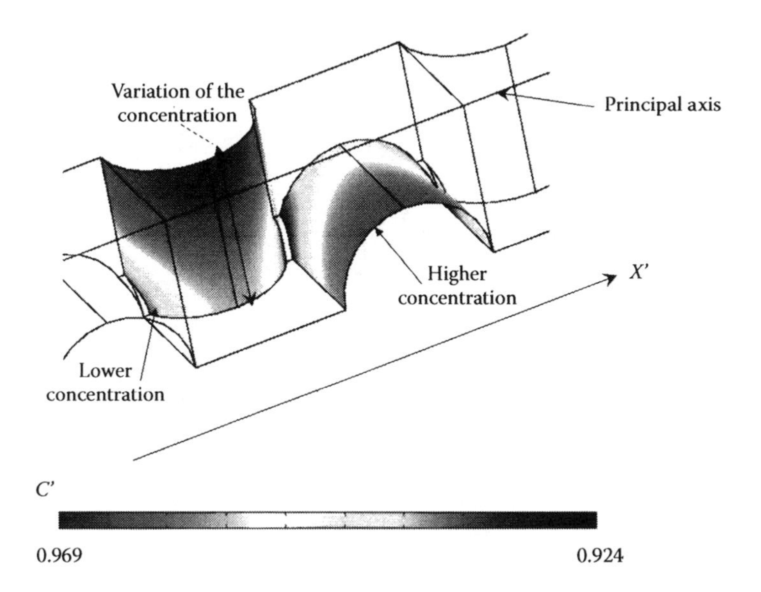

FIGURE 3.11
Spatial distribution of the dimensionless oxygen concentration in the vicinity of two adjacent fiber surfaces where the cells are lying. The mixing effects owing to the swirling flow with transverse velocity components are contributing to the homogenization of the oxygen distribution. In this tested part involving an overall relative variation in oxygen concentration of about 10% around the mean value is noted.

fibers. These results are significantly different (one or two orders of magnitude lower) from those obtained with a two-dimensional model, where higher amplitudes and more uniform distribution of shear stresses were found (Pierre and Oddou 2007). Moreover, they are also significantly different and lower from those experimentally obtained in *in vitro* mechanotransduction assays using parallel-plate flow chambers. Such results can bring additional insight, leading to a detailed knowledge of the cell mechanotransduction phenomena taking place in bone tissue engineering. Indeed, as recently shown (Pedersen and Swartz 2005), they point out "the relevance and importance of dimensionality in mediating cellular responses to the biophysical environment."

3.5.3.2 Microfluidic Bioreactor: A Numerical Driven Experiment for Cartilage Culture

In this section, the behavior of a microfluidic bioreactor is investigated using a modeling approach. Four different cases corresponding to experimental conditions were defined regarding the tissue thickness (250 μm or 1 mm, depending upon the initial cell seeding conditions) and the intensity of the applied

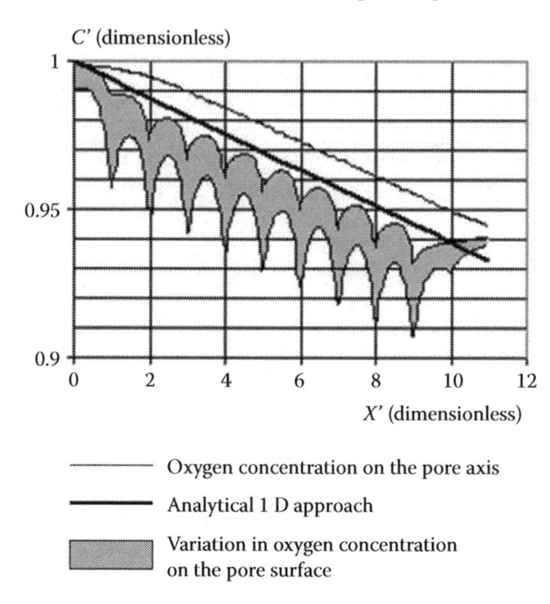

FIGURE 3.12
Longitudinal distribution profile of the dimensionless oxygen concentration in the representative elementary volume of the pore channel. Comparison between one- and three-dimensional modeling approaches shows that, under the assumed experimental conditions and at the scale of the overall pore length, the transport phenomena are convection dominated. The variation in local surface distribution of the feeding oxygen concentration (gray shaded area) is mainly due to the presence of stagnation zones in the vicinity of the close contact between two adjacent fibers.

mechanical stimulation (0, 0.1, and 1.0 dyne cm^{-2} viscous shear stress depending on the flow rate). In this framework, a mathematical model of oxygen transport within the dual chamber parallel-plate bioreactor presented in Figure 3.13 was defined and numerically solved (Pierre et al. 2008).

The parallel-plate bioreactor (Gemmiti and Guldberg 2006) is constituted of three distinct domains: a tissue-engineered rectangular slab of cartilage, which is enveloped by an upper chamber in which the media flows, and a lower chamber which acts as a reservoir of nutrient. Via the control of nutrient media flow rate in the upper chamber, different mechanical stimulation of cells were applied to obtain a tissue construct with mechanical properties (Young's modulus, ultimate strength) and matrix composition approaching those of natural cartilage. The question is then to know if varying the flow rate in the bioreactor consequently leads to a significant and quantifiable change in the oxygen feeding of the cells embedded in the engineered tissue.

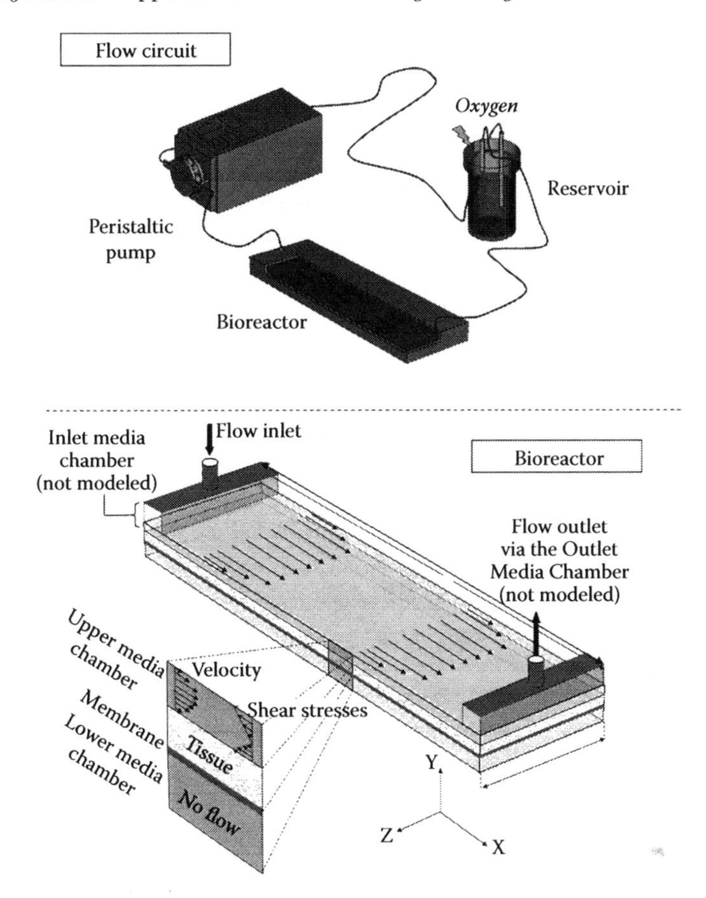

FIGURE 3.13
Sketch of the parallel-plate bioreactor designed to apply a given level of fluid-flow-induced shear stress to tissue-engineered articular cartilage. Noticeable improvement in matrix composition and mechanical properties of the native tissue seeded with chondrocytes was reported (Gemmiti and Guldberg 2006). Mechanical stimulation of the cells as well as enhanced nutrient mass transport associated with the hydrodynamic environment of this system design may reveal to be an effective functional strategy in tissue engineering.

Four different cases, corresponding to the stimulation phases of the thinner and thicker tissues as well as the controls (kept in the preculture conditions), were investigated (see Figure 3.14).

The basic convection–diffusion–reaction equations of the nutrient transport model introduced in Section 3.4.2 have been adapted to this bioreactor schematically designed in two dimension. It is thus possible to simulate the time evolution of the space variations in oxygen concentration within the engineered cartilage tissue. In the upper chamber flow channel, neglecting the

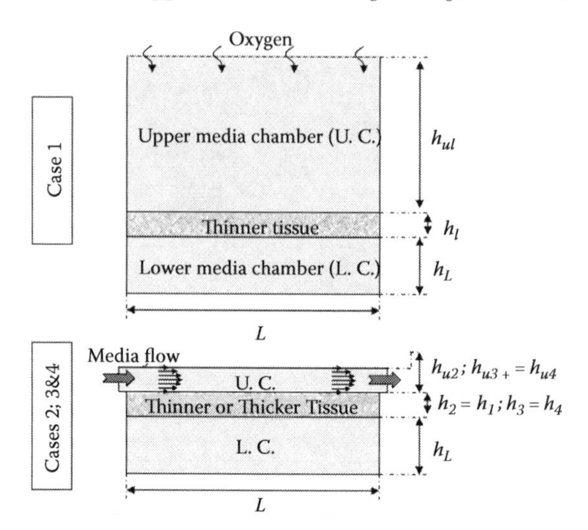

FIGURE 3.14
Models of the four cases considered in the culture process of the cartilage tissue inside the microfluidic bioreactor. Case 1: Control culture of the thinner tissue (static culture, no shear stress); Case 2: Stimulation culture of the thinner tissue (dynamic culture under the higher shear stress); Case 3: Control culture of the thicker tissue (dynamic culture under the lower shear stress); Case 4: Stimulation culture of the thicker tissue (dynamic culture under the higher shear stress).

entry effects, a plane Poiseuille flow field was assumed to be representative of the profiles (Reynolds number, $Re \approx 5$). In the tissue part of this chamber, no intratissue convective flow was considered, because of the very low tissue permeability ($K \approx 3 \cdot 5 \times 10^{-17}$ m^2). At the interface between the channel and the tissue, the continuity in the oxygen flux across the boundary was imposed. The oxygen transport phenomena are governed by the magnitude of the two already mentioned parameters: the Damköhler number defined at the scale of the overall tissue thickness ($Da \approx 10^{-1}$), and the Péclet number ($Pe \approx 10^3$) defined at the scale of the overall channel length, which represents the ratio between the diffusive time along the transverse direction and the convective transit time along the downstream direction. For the high values of the Péclet number, it is thus expected that only the oxygen molecules contained within a thin boundary layer have enough time to diffuse into the tissue.

Figure 3.15 gives the model results providing the range of computed oxygen concentrations within the tissue at $t \approx 36$ hours. When compared with the physiologic values (normoxia) reported by Obradovic et al. (1999), these results show that for the control and stimulation cultures of the thinner tissue (cases 1 and 2), the oxygenation states are very homogeneous, leading either to hypoxia or hyperoxia situations. Indeed, the mean oxygen concentration of the

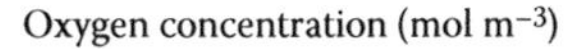

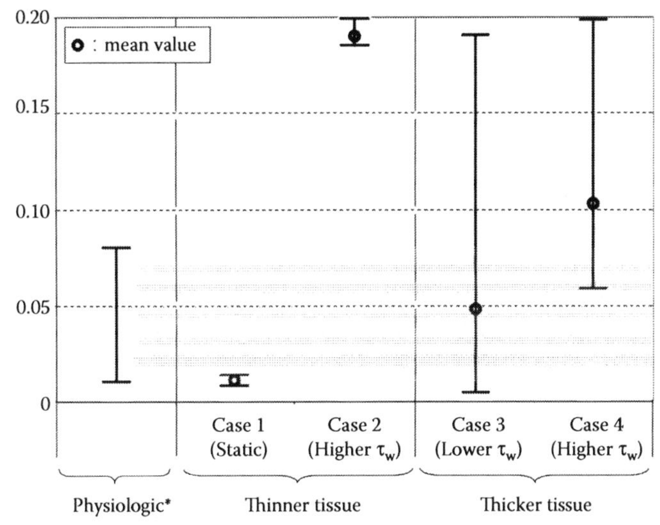

FIGURE 3.15
Range (illustrated by bars) of computed oxygen concentrations in the tissue at $t \approx 36$ h, for each culture case. The open circles on these bars represents the mean value (surface averaged value) on the overall tissue. The term "Physiologic*" corresponds to normoxia values reported by Obradovic et al. (1999).

control (case 1) is close to the lower boundary of the *in vivo* range, whereas during the stimulation phase (case 2) the thin tissue is under a hyperoxygenation state. For the thick tissue, on the contrary, the low shear conditions (case 3) present more similitude with the physiologic state as compared to the high shear conditions (case 4): roughly two-third versus only one-fourth of the tissues are under normoxic conditions, respectively. Finally, the thick tissues (cases 3 and 4) present a wider range of oxygen concentrations than those present in the thin tissues (cases 1 and 2). This is owing to the extent of the domain of oxygen consumption.

3.6 Conclusion

As recalled here, a variety of perfusion systems have been recently proposed in an effort to enhance the development of three-dimensional tissue constructs *in vitro*. Nevertheless, tissue engineering remains actually limited by our lack of knowledge and technological means of quantification in transport phenomena within biological tissues as well as cell mechanotransduction response and angiogenesis processes. Thus, culturing of cells within a large porous implant placed in a perfusion bioreactor remains a challenge for tissue engineering. This challenge can be surmounted only by a multidisciplinary approach where the modeling analysis, a typical research activity of the engineering sciences, has to fully play its role.

Nevertheless, it is often useless and indeed not practical to attempt to reproduce exactly all aspects of cell behavior through a single model. In fact, the model has to include assumptions neglecting the inappropriate level of detail for the particular research questions to be answered. As a consequence, for use in establishing tissue-engineering protocols, a balance between the predictive efficiency of the model, its complexity, and the range of physico-biological parameters involved has to be reached. Therefore, the model may either provide a phenomenological description, with its range of applicability consequently limited to a well-defined experimental system, or include more mechanistic aspects to actually explain phenomena in tissue regeneration. In parallel, the biological knowledge of the cell response to physical solicitations is an avenue worth exploring since this domain remains to be one key limiting factor of transport phenomena simulation. This is why we adopted a multi-scale and multiphysics viewpoint with emphasis on the fluid dynamics and transport phenomena through porous substrates at the mesoscopic scale of the pore.

Numerical models have been widely used for problems involving fluid flow through porous media outside the area of biomechanics. Nevertheless, three-dimensional simulations of cell culture media flowing through a perfused three-dimensional construct that estimates local nutrient cell uptake and shear stresses at the pore scale are not extensively studied. Indeed, transport phenomena within porous substrate and the resulting threshold in concentration of nutrient molecules such as oxygen above which cells can survive seems to be one of the key factors. Thus, an improved understanding of the local shear stress and nutrient feeding experienced by cells under flow conditions in three-dimensional scaffolds as a function of flow rate and microarchitecture is required for identifying culture conditions that would impart appropriate properties for enhanced cell proliferation and activity.

Given the complexity of cell behavior and the numerous interactions with the evolving cell and tissue environments, computational approaches, such as presented here, contribute to a better understanding of the different contributing phenomena and mechanisms involved in the tissue-engineering culture: effective role of the substrate geometry, culture fluid transport, mechanical stresses induced by the perfusing fluid flow, and possibly in the next future cell attachment, cell–cell interaction, and population dynamics. Thus, modeling parametric studies are necessary to determine the effects of diffusion and convection processes on the penetration of cells within the porous substrate. The models here designed and developed have then brought significant unpredictable insight, leading to a better knowledge of the cell mechanotransduction phenomena taking place in osteoarticular tissue engineering. It can then be shown that a compromise between a sufficient oxygen supply and an adapted mechanical load has to be found in the regulation of the perfusion flow.

These results suggest that scaffold architectural properties such as porosity and tortuosity as well as physicochemical properties of the materials are important in determining optimal scaffold parameters allowing adequate environmental conditions for cell activity. Our results provide a basis for the completion of more exhaustive quantitative studies to further assess the relationship between perfusion, at known microfluid dynamic conditions, and tissue growth *in vitro.*

Particularly, they have shown that the conception of a performing perfusion bioreactor has to take into account the porous active properties of the involved substrate-cells medium. The reorganization of the porous matrix due to the coupling between perfusion fluid flow and biological consolidation processes would have to be approached by such a model in the next future. Under certain conditions that have been emphasized here, electrical properties of the substrate have also to be taken into account, leading to further integrated experimental-computational way of research. Such an approach has to be conducted in a multidisciplinary scientific environment where the biomechanics field has to play a major role by its characteristic multiscale and multiphysics methodology.

3.7 References

Abousleiman, R. I. and Sikavitsas, V. I. (2006). Bioreactors for tissues of the musculoskeletal system. *Advances in Experimental Medicine and Biology,* **585**:243–259.

Akhyari, P., Fedak, P. W. M., Weisel, R. D., Lee, T.-Y. J., Verma, S., Mickle, D. A. G., and Li, R. K. (2002). Mechanical stretch regimen enhances the formation of bioengineered autologous cardiac muscle grafts. *Circulation,* **106:**I137–I142.

Altman, G., Horan, R., Martin, I., Farhadi, J., Stark, P., Volloch, V., Vunjak-Novakovic, G., Richmond, J., and Kaplan, D. L. (2002). Cell differentiation by mechanical stress. *FASEB Journal,* **16**:270–272.

Arnett, T., Gibbons, D., Utting, J., Orriss, I., Hoebertz, A., Rosendaal, M., and Meghji, S. (2003). Hypoxia is a major stimulator of osteoclast formation and bone resorption. *Journal of Cellular Physiology,* **196**:2–8.

Atlan, G., Delattre, O., Berland, S., LeFaou, A., Nabias, G., Cot, D., and Lopez, E. (1999). Interface between bone and nacre implants in sheep. *Biomaterials,* **20**:1017–1022.

Bancroft, G. N., Sikavitsas, V. I., van den Dolder, J., Sheffield, T. L., Ambrose, C. G., Jansen, J. A., and Mikos, A. G. (2002). Fluid flow increases mineralized matrix deposition in 3D perfusion culture of marrow stromal osteoblasts in a dose-dependent manner. *Proceedings of the National Academy of Sciences, USA,* **99**:12600–12605.

Bancroft, G., Sikavitsas, V., and Mikos, A. (2003). Design of a flow perfusion bioreactor system for bone tissue-engineering applications. *Tissue Engineering,* **9**:549–554.

Bilgen, B., Chang-Mateu, I. M., and Barabino, G. A. (2005). Characterization of mixing in a novel wavy-walled bioreactor for tissue engineering. *Biotechnology and Bioengineering,* **92**:907–919.

Bilgen, B., Sucosky, P. G., Neitzel, P., and Barabino, G. A. (2006). Flow characterization of a wavy-walled bioreactor for cartilage tissue engineering. *Biotechnology and Bioengineering,* **95**:1009–1022.

Botchwey, E. A., Pollack, S. R., Levine, E. M., and Laurencin, C. T. (2001). Bone tissue engineering in a rotating bioreactor using a microcarrier matrix system. *Journal of Biomedical Materials Research,* **55**:242–253.

Botchwey, E. A., Dupree, M. A., Pollack, S. R., Levine, E. M., and Laurencin, C. T. (2003). Tissue engineered bone: measurement of nutrient transport in three-dimensional matrices. *Journal of Biomedical Materials Research,* **67A**:357–367.

Botchwey, E. A., Pollack, S. R., Levine, E. M., Johnston, E. D., and Laurencin, C. T. (2004). Quantitative analysis of three-dimensional fluid flow in rotating bioreactors for tissue engineering. *Journal of Biomedical Materials Research,* **69A**:205–215.

Brekke, J. H. and Toth, J. M. (1998). Principles of tissue engineering applied to programmable osteogenesis. *Journal of Biomedical Materials Research,* **43**:380–398.

Butler, D. L., Goldstein, S. A., and Guilak, F. (2000). Functional tissue engineering: the role of biomechanics. *Journal of Biomechanical Engineering,* **122**:570–575.

Cartmell, S. H., Porter, B. D., Garcia, A. J., and Guldberg, R. E. (2003). Effects of medium perfusion rate on cell-seeded three-dimensional bone constructs in vitro. *Tissue Engineering,* **9**:1197–1203.

Chen, C. S. (2008). Mechanotransduction—a field pulling together? *Journal of Cell Science,* **121**:3285–3292.

Cimetta, E., Flaibani, M., Mella, M., Serena, E., Boldrin, L., De Coppi, P., and Elvassore, N. (2007). Enhancement of viability of muscle precursor cells on

3D scaffold in a perfusion bioreactor. *The International Journal of Artificial Organs,* **30**:415–428.

David, B. (2002). Mise en place et validation d'un modèle in vitro pour l'étude des propriétés mécaniques, diffusives et métaboliques d'un foie bioartificiel à lit fluidisé. PhD Thesis, University of Technology of Compiègne-France.

Davisson, T., Kunig, S., Chen, A., Sah, R., and Ratcliffe, A. (2002). Static and dynamic compression modulate matrix metabolism in tissue engineered cartilage. *Journal of Orthopaedic Research,* **20**:842–848.

Dean, W. R. (1927). Note on the motion of fluid in a curved pipe. *Physician's Management,* **20**:208–223.

Démarteau, O., Jakob, M., Schäfer, D., Heberer, M., and Martin, I. 2003a. Development and validation of a bioreactor for physical stimulation of engineered cartilage. *Biorheology,* **40**:331–336.

Démarteau, O., Wendt, D., Braccini, A., Jakob, M., Schäfer, D., Heberer, M., and Martin, I. 2003b. Dynamic compression of cartilage constructs engineered from expanded human articular chondrocytes. *Biochemical and Biophysical Research Communications,* **310**:580–588.

Discher, D. E., Janmey, P., and Wang, Y. L. (2005). Tissue cells feel and respond to the stiffness of their substrate. *Science,* **310**:1139–1143.

Donnan, F. G. (1924). The theory of membrane equilibrium. *Chemical Reviews,* **1**:73–90.

Drury, J. L. and Mooney, D. J. (2003). Hydrogels for tissue engineering: scaffold design variables and applications. *Biomaterials Synthesis Biomimetic Polymers,* **24**:4337–4351.

Dulong, J. L. and Legallais, C. (2005). What are the relevant parameters for the geometrical optimization of an implantable bioartificial pancreas? *Journal of Biomechanical Engineering,* **127**:1054–1061.

El-Ghannam, A., Ducheyne, P., and Shapiro, I. M. (1997). Porous bioactive glass and hydroxyapatite ceramic affect bone cell function in vitro along different time lines. *Journal of Biomedical Materials Research,* **36**:167–180.

Funk, R. H. W., Monsees, T., and Ozkucur, N. (2009). Electromagnetic effects—from cell biology to medicine. *Progress in Histochemistry and Cytochemistry,* **43**:177–264.

Gemmiti, C. V. and Guldberg, R. E. (2006). Fluid flow increases type II collagen deposition and tensile mechanical properties in bioreactor-grown tissue-engineered cartilage. *Tissue Engineering,* **12**:469–479.

Glowacki, J., Mizuno, S., and Greenberger, J. S. (1998). Perfusion enhances functions of bone marrow stromal cells in three-dimensional culture. *Cell Transplantation,* **7**:319–326.

Goldstein, A. S., Juarez, T. M., Helmke, C. D., Gustin, M. C., and Mikos, A. G. (2001). Effect of convection on osteoblastic cell growth and function in biodegradable polymer foam scaffolds. *Biomaterials*, **22**:1279–1288.

Gu, W. Y., Lai, W. M., and Mow, V. C. (1998). A mixture theory for charged-hydrated soft tissues containing multi-electrolytes: passive transport and swelling behaviour. *Journal of Biomechanical Engineering,* **120**:169–180.

Guillemin, G., Meunier, A., Dallant, P., Christel, P., Pouliquen, J. C., and Sedel, L. (1989). Comparison of coral resorption and bone apposition with two natural corals of different porosities. *Journal of Biomedical Materials Research,* **23**:765–779.

Guyon, E., Hulin J. P., and Petit, L. (2001). *Hydrodynamique Physique.* EDP Sciences CNRS Eds, Paris.

Healy, Z. R., Lee, N. H., Gao, X., Goldring, M. B., Talalay, P., Kensler, T. W., and Konstantopoulos, K. (2005). Divergent responses of chondrocytes and endothelial cells to shear stress: cross-talk among COX-2, the phase 2 response, and apoptosis. *Proceedings of the National Academy of Sciences, USA,* **102**:14010–14015.

Hunter, R. J. (1981). *Zeta Potential in Colloid Science: Principles and Applications.* Academic Press, New York.

Hunter, R. J. (2001). *Foundations of Colloid Science.* Oxford University Press, New York.

Hutmacher, D. W., Schantz, T., Zein, I., Ng, K. W., Teoh, S. H., and Tan, K. C. (2001). Mechanical properties and cell cultural response of polycaprolactone scaffolds designed and fabricated via fused deposition modeling. *Journal of Biomedical Materials Research,* **55**:203–216.

Hutmacher, D. W., Sittinger, M., and Risbud, M. V. (2004). Scaffold-based tissue engineering: rationale for computer-aided design and solid free-form fabrication systems. *Trends in Biotechnology,* **22**:354–362.

Huyghe, J. M. and Janssen, J. D. (1997). Quadriphasic mechanics of swelling incompressible porous media. *International Journal of Engineering Science,* **35**:793–802.

Ishaug, S. L., Crane, G. M., Miller, M. J., Yasko, A. W., Yaszemski, M. J., and Mikos, A. G. (1997). Bone formation by three-dimensional stromal osteoblast culture in biodegradable polymer scaffolds. *Journal of Biomedical Materials Research,* **36**:17–28.

Israelachvili, J. (1991). *Intermolecular and Surface Forces.* Academic Press, New York.

Jagur-Grodzinski, J. (2006). Polymers for tissue engineering, medical devices, and regenerative medicine. Concise general review of recent studies. *Polymers for Advanced Technologies,* **17**:395–418.

Janssen, F. W., Oostra, J., Oorschot, v. A., van Blitterswijk, C. A. (2006). A perfusion bioreactor system capable of producing clinically relevant volumes of tissue-engineered bone: in vivo bone formation showing proof of concept. *Biomaterials*, **27**:315–323.

Kelly, D. J. and Prendergast, P. J. (2005). Mechano-regulation of stem cell differentiation and tissue regeneration in osteochondral defects. *Journal of Biomechanics,* **38**:1413–1422.

Kim, B. S., Nikolovski, J., Bonadio, J., and Mooney, D. J. (1999). Cyclic mechanical strain regulates the development of engineered smooth muscle tissue. *Nature Biotechnology*, **17**:979–983.

Kim, S. S., Penkala, R., and Abrahimi, P. (2007). A perfusion bioreactor for intestinal tissue engineering. *The Journal of Surgical Research,* **142**: 327–331.

Komarova, S. V., Ataullakhanov, F. I., and Globus, R. K. (2000). Bioenergetics and mitochondrial transmembrane potential during differentiation of cultured osteoblasts. *AJP-Cell Physio*, **279**:C1220–C1229.

Langmuir, I. (1938). The role of attractive and repulsive forces in the formation of tactoids, thixotropic gels, protein crystal, and coacervates. *The Journal of Chemical Physics,* **6**:873–896.

Lanza, R., Langer, R., and Vacanti, J. P. (2002). *Principles of Tissue Engineering,* second edition. Academic Press, London, UK.

Leclerc, E., David, B., Griscom, L., Lepioufle, B., Fujii, T., Layrolle, P., and Legallais, C. (2006). Study of osteoblastic cells in a microfluidic environment. *Biomaterials*, **27**:586–595.

Lemaire, T., Naïli, S., and Rémond, A. (2006). Multiscale analysis of the coupled effects governing the movement of interstitial fluid in cortical bone. *Biomechanics and Modeling in Mechanobiology,* **5**:39–52.

Lemaire, T., Naïli, S., and Rémond, A. (2008). Study of the role of fibrous pericellular matrix in the cortical interstitial fluid movement with hydro-electro-chemical effects. *Journal of Biomechanical Engineering,* **130**: 1–11.

Lewis, M. C., MacArthur, B. D., Malda, J., Pettet, G., and Please, C. P. (2005). Heterogeneous proliferation within engineered cartilaginous tissue: the role of oxygen tension. *Biotechnology and Bioengineering,* **91**: 607–615.

Liu, G., Zhao, L., Cui, L., Liu, W., and Cao, Y. (2007). Tissue-engineered bone formation using human bone marrow stromal cells and novel beta-tricalcium phosphate. *Biomedical Materials,* **2**:78–86.

Liu, G., Zhao, L., Zhang, W., Cui, L., Liu, W., and Cao, Y. (2008). Repair of goat tibial defects with bone marrow stromal cells and beta-tricalcium phosphate. *Journal of Materials Science. Materials in Medicine,* **19**: 2367–2376.

Malda, J., Blitterswijk, C. A. van, Geffen, M. van, Martens, D. E., Tramper, J., and Riesle, J. (2004). Low oxygen tension stimulates redifferentiation of dedifferentiated adult human nasal chondrocytes. *Osteoarthritis Cartilage,* **12**:306–313.

Martin I., Obradovic B., Freed L. E., and Vunjak-Novakovic, G. (1999). Method for quantitative analysis of glycosaminoglycan distribution in cultured natural and engineered cartilage. *Annals of Biomedical Engineering,* **27:**656–662.

Martin, I., Wendt, D., and Heberer, M. (2004). The role of bioreactors in tissue engineering. *Trends in Biotechnology,* **22**:82–86.

Martin, Y. and Vermette, P. (2005). Bioreactors for tissue mass culture: Design, characterization, and recent advances. *Biomaterials,* **26**:7481–7503.

Maton, A., Hopkins, J., Johnson, S., LaHart, D., Warner, M. Q., and Wright, J. D. (1997). *Cells Building Blocks of Life.* Prentice Hall, Upper Saddle River, NJ.

Mauck, R. L., Soltz, M. A., Wang, C. C. B., Wong, D. D., Chao, P-H. G., Valhmu. W. B., Hung, C. T., and Ateshian, G. A. (2000). Functional tissue engineering of articular cartilage through dynamic loading of chondrocyte-seeded agarose gels. *Journal of Biomechanical Engineering,* **122**:252–260.

Maurin, A. C., Fromental, R., Cantaloube, D., and Caterini, R. (2005). Etude de la colonisation par des ostéoblastes humains de métaux poreux à base de nitinol ou de tantale dans un modèle de culture cellulaire tridimensionnelle. *Implantodontie,* **14**:44–50.

Moyne, C. and Murad, M. A. (2002). Electro-chemo-mechanical couplings in swelling clays derived from a micro/macro-homogenization procedure. *International Journal of Solids and Structures,* **39**:6159–6190.

Nauman, E. A., Satcher, R. L., Keaveny, T. M., Halloran, B. P., and Bikle, D. D. (2001). Osteoblasts respond to pulsatile fluid flow with short-term increases in PGE(2) but no change in mineralization. *Journal of Applied Physics*, **90**:1849–1854.

Niklason, L. E., Gao, J., Abbott, W. M., Hirschi, K. K., Houser, S., Marini, R., Langer, R. (1999). Functional arteries grown in vitro. *Science*, **284**: 489–493.

Obradovic, B., Carrier, R. L., Vunjak-Novakovic, G., and Freed, L. E. (1999). Gas exchange is essential for bioreactor cultivation of tissue engineered cartilage. *Biotechnology and Bioengineering,* **63**:197–205.

Obradovic, B., Meldon, J. H., Freed, L. E., and Vunjak-Novakovic, G. (2000). Glycosaminoglycan deposition in engineered cartilage: experiments and mathematical model. *AIChE Journal,* **46**:1860–1871.

Pedersen, J. A. and Swartz M. A. (2005). Mechanobiology in the third dimension. *L'Annee endocrinologique,* **33**:1469–1490

Peng, C. A. and Palsson, B. O. (1996). Cell growth and differentiation on feeder layers is predicted to be influenced by bioreactor geometry. *Biotechnology and Bioengineering,* **50**:479–492.

Petite, H., Viateau, V., Bensaïd, W., Meunier, A., de Pollak, C., Bourguignon, M., Oudina, K., Sedel, L., and Guillemin, G. (2000). Tissue-engineered bone regeneration. *Nature Biotechnology,* **18**:959–963.

Pierre, J. and Oddou C. (2007). Engineered bone culture in a perfusion bioreactor: a 2D computational study of stationary mass and momentum transport. *Computer Methods in Biomechanics and Biomedical Engineering,* **10–6**:429–438

Pierre, J., Gemmiti, C. V., Kolambkar, Y. M., Oddou, C., and Guldberg, R. E. (2008). Theoretical analysis of engineered cartilage oxygenation: influence of construct thickness and media flow rate. *Biomechanics and Modeling in Mechanobiology,* **7**:497–510.

Pothuaud, L., Fricain, J.-C., Pallu, S., Bareille, R., Renard, M., Durrieu, M.-C., Dard, M., Vernizeau, M., and Amedee, J. (2005). Mathematical modelling of the distribution of newly formed bone in bone tissue engineering. *Biomaterials,* **26**:6788–6797.

Powell, C. A., Smiley, B. L., Mills, J., and Vandenburgh, H. H. (2002). Mechanical stimulation improves tissue-engineered human skeletal muscle. *American Journal of Physiology. Cell Physiology,* **283**:C1557–C1565.

Raimondi, M. T., Boschetti, F., Falcone, L., Fiore, G. B., Remuzzi, A., Marinoni, E., Marazzi, M., and Pietrabissa, R. (2002). Mechanobiology of engineered cartilage cultured under a quantified fluid-dynamic environment. *Biomechanics and Modeling in Mechanobiology,* **1**:69–82.

Rezwan, K., Chen, Q. Z., Blaker, J. J., and Boccaccini, A. R. (2006). Biodegradable and bioactive porous polymer/inorganic composite scaffolds for bone tissue engineering. *Biomaterials,* **27**:3413–3431.

Ringeisen, B. R., Othon, C. M., Barron, J. A., Young, D., Spargo, B. J. (2006). Jet-based methods to print living cells. *Biotechnology Journal,* **1**:930–948.

Rose, F. R., Cyster, L. A., Grant, D. M., Scotchford, C. A., Howdle, S. M., and Shakesheff, K. M. (2004). In vitro assessment of cell penetration into porous hydroxyapatite scaffolds with a central aligned channel. *Biomaterials,* **25**:5507–5514.

Saini, S. and Wick, T. M. (2003). Concentric cylinder bioreactor for production of tissue engineered cartilage: effect of seeding density and hydrodynamic loading on construct development. *Biotechnology Progress,* **19**:510–521.

Sengers, B. G., Taylor, M., Please, C. P., and Oreffo, R. O. C. (2007). Computational modelling of cell spreading and tissue regeneration in porous scaffolds. *Biomaterials,* **28**:1928–1940.

Service, R. F. (2005). Tissue engineering. Technique uses body as 'bioreactor' to grow new bone. *Science,* **309**:683.

Shors, E. C. (1999). Coralline bone graft substitutes. *The Orthopedic Clinics of North America,* **30**:599–613.

Sikavitsas, V. I., Bancroft, G. N., Holtorf, H. L., Jansen, J. A., and Mikos, A. G. (2003). Mineralized matrix deposition by marrow stromal osteoblasts in 3D perfusion culture increases with increasing fluid shear forces. *Proceedings of the National Academy of Sciences, USA,* **100**:14683–14688.

Singh, H., Teoh, S. H., Low, H. T., and Hutmacher, D. W. (2005). Flow modelling within a scaffold under the influence of uni-axial and bi-axial bioreactor rotation. *Journal of Biotechnology,* **119**:181–196.

Sucosky, P., Osorio, D. F., Brown, J. B., and Neitzel, G. P. (2004). Fluid mechanics of a spinner-flask bioreactor. *Biotechnology and Bioengineering,* **85**:34–46.

Sutherland, R. M., Sordat, B., Bamat, J., Gabbert, H., Bourrat, B., and Mueller-Klieser, W. (1986). Oxygenation and differentiation in multicellular spheroids of human colon carcinoma. *Cancer Research,* **46**:5320–5329.

Swartz, M. A. and Fleury, M. E. (2007). Interstitial flow and its effects in soft tissues. *Annual Review of Biomedical Engineering,* **9**:229–256.

Tierney, C. M., Haugh, M. G., Liedl, J., Mulcahy, F., Hayes, B., and O'Brien, F. J. (2009). The effects of collagen concentration and crosslink density on the biological, structural and mechanical properties of collagen-GAG scaffolds for bone tissue engineering. *Journal of the Mechanical Behavior of Biomedical Materials,* **2**:202–209.

Tuncay, O. C., Ho, D., and Barker, M. K. (1994). Oxygen tension regulates osteoblast function. *American Journal of Orthodontics and Dentofacial Orthopedics,* **105**:457–463.

van den Dolder, J., Bancroft, G. N., Sikavitsas, V. I., Spauwen, P. H. M., Jansen, J. A., and Mikos, A. G. (2003). Flow perfusion culture of marrow stromal osteoblasts in titanium fiber mesh. *Journal of Biomedical Materials Research. Part A*, **64A**:235–241.

Vunjak-Novakovic, G., Martin, I., Obradovic, B., Treppo, S., Grodzinsky, A. J., Langer, R., and Freed, L. E. (1999). Bioreactor cultivation conditions modulate the composition and mechanical properties of tissue-engineered cartilage. *Journal of Orthopaedic Research,* **17**:130–138.

Waldman, S. D, Spiteri, C. G., Grynpas, M. D., Pilliar, R. M., Hong, J., Kandel, R. A. (2003). Effect of biomechanical conditioning on cartilaginous tissue formation in vitro. *The Journal of Bone and Joint surgery (American),* **85**:101–105.

Williams, K. A., Saini, S., and Wick, T. M. (2002). Computational fluid dynamics modelling of steady-state momentum and mass transport in a bioreactor for cartilage tissue engineering. *Biotechnology Progress,* **18**: 951–963.

Wilson, C. G., Bonassar, L. J., and Kohles, S. S. (2002). Modelling the dynamic composition of engineered cartilage. *Archives of Biochemistry and Biophysics,* **408**:246–254.

Zein, I., Hutmacher, D., Cheng Tan, K., and Hin Teoh, S. (2002). Fused deposition modelling of novel scaffold architectures for tissue engineering applications. *Biomaterials,* **23**:1169–1185.

Zenati, R., Fantozzi, G., Chevalier, J., and Arioua, M. (2006). Porous bioactive glass and preparation method thereof. *Patent. Pub. No.: WO/2006/018531. International Application* No.:PCT/FR2005/001921. Filed Jul 25, 2005, published Feb. 23, 2006, Institut National Des Sciences Appliquees.

Zhao, F., Chella, R., and Ma, T. (2007). Effects of shear stress on 3-D human mesenchymal stem cell construct development in a perfusion bioreactor system: experiments and hydrodynamic modelling. *Biotechnology and Bioengineering,* **96**:584–595.

4

Biomedical Implications of the Porosity of Microbial Biofilms

H. Ben-Yoav
Department of Physical Electronics, School of Electrical Engineering, Faculty of Engineering, Tel Aviv University, Tel Aviv, Israel

N. Cohen-Hadar
Department of Molecular Microbiology and Biotechnology, Faculty of Life Sciences, Tel Aviv University, Tel Aviv, Israel

Amihay Freeman
Department of Molecular Microbiology and Biotechnology, Faculty of Life Sciences, Tel Aviv University, Tel Aviv, Israel

CONTENTS

4.1 Introduction

Biofilms are mostly considered as a slimy layer of microorganisms adhering to solid surface (de Beer and Stoodley 2006). The earliest phase of studying biofilms was focused on the physical properties of the solid surface, for example, roughness, hydrophobicity, and hydrophilicity. As electron microscopy developed, a more detailed picture of the structure of microbial biofilms emerged. The subsequent development of confocal scanning laser microscopy, coupled with fluorescent markers, allowed visualization of live hydrated biofilms as three-dimensional architecture (Costerton et al. 1995).

4.1.1 What Is a Biofilm?

Biofilms may be broadly defined as communities of microbial cells associated with a surface, typically encased in an extracellular matrix (Ghannoum and O'Toole 2004). This definition is applicable to biofilms formed on solid–liquid interfaces as well as on semisolid–air interfaces and relates to a wide range of surfaces, including steel pipes, soils, medical implants, biomaterials, tissues, and epithelial cells. The most abundant biofilms (>90% of microorganisms) are associated with environmental biofilms found in nature or throughout biotechnological applications (Lens et al. 2003; Tandoi et al. 2006). Studies of these biofilms were dominated by phenotypic behavior, for example, biofilm life cycle and resistance to antibiotics (O'Toole et al. 2000b; Chandra et al. 2001; Reysenbach and Shock 2002). Ongoing research in the field aims at understanding the community physiology, metabolism, ecology, structure/function relationships, and the role of gene exchange, with the overall objective of providing a database for the development of efficient strategies to control biofilm development and methods to eradicate them. The number of studies directed toward better understanding of the physical characteristics of biofilm, for example, porosity and three-dimensional structures of biofilms is substantially low.

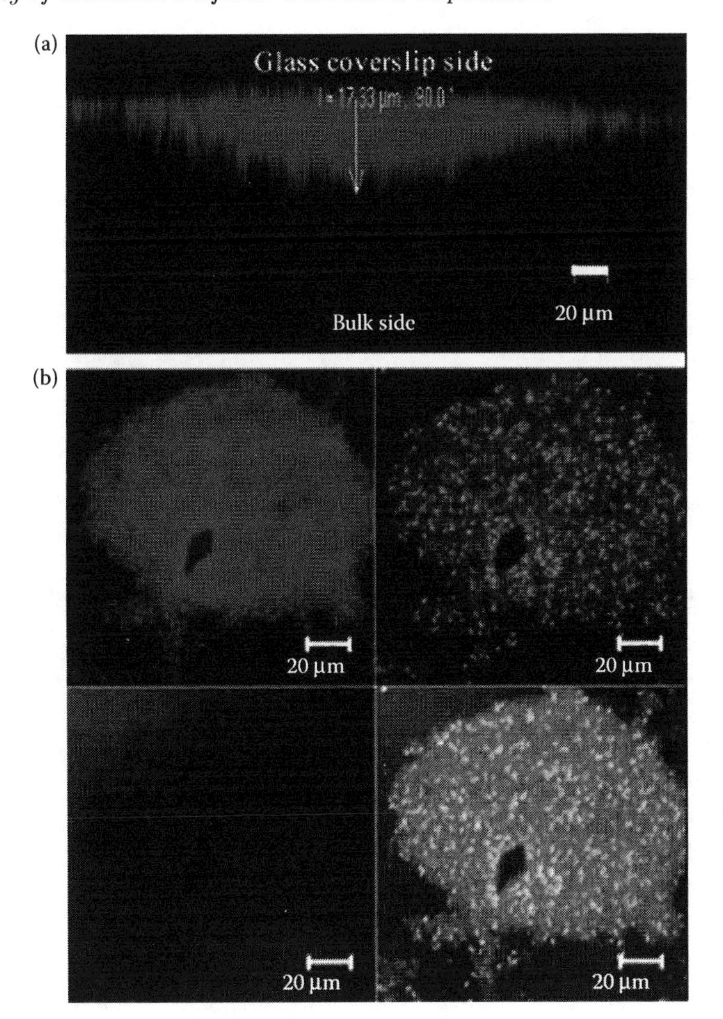

FIGURE 4.1
Images of a mature *Escherichia coli* biofilm on glass substrate from a scanning confocal laser microscopy (SCLM). (a) A two-dimensional X–Z plane view and (b) a two-dimensional X–Y plane view (upper left box—bacterial cells and extracellular matrix; upper right box—dead bacterial cells; lower left box—transmitted light image; and lower right box—all three boxes together). Bar: 20 μm. (Ben-Yoav, H. and Freeman, A., *J. Drug Deliv. Sci. Tech.*, 18, 1, 1998. With permission.)

The dominating natural appearance of bacterial cells as colonies and films (Figure 4.1) makes "multicellularity" the most common mode of interrelationship in bacterial communities (O'Toole et al. 2000b). Broadly defined, multicellularity is a physical state composed of many cells communicating and coordinating their activities as a group. Thus, multicellularity is distinguishable from the unicellular behavior by cell–cell cooperation resulting, in some cases, in individual cells "sacrificing" their capability to reproduce for communal benefit (Velicer 2003).

4.1.2 Biofilms in Medicine

Biofilm-based infections can occur on abiotic or biotic surfaces (Pace et al. 2006). Infections involving biofilm formation are most frequently associated with microbial colonization of the abiotic surfaces of indwelling medical devices (IMDs) and are commonly referred to as foreign body infections (FBIs) (Lynch and Robertson 2008; Romeo 2008; Shirtliff and Leid 2009). Although clinical biofilms are typically of lower cell densities when compared to biofilms grown *in vitro*, their overall architecture, as revealed by electron microscopy methods, is similar. In the United States, device infections associated with low-attributable mortality have been estimated to have initial rates of infection that vary from 1% to 3% for mammary implants, penile implants, and joint prosthesis and 10% to 30% for urinary catheters (Donlan and Costerton 2002; Thomas et al. 2006). Recent studies have also provided evidence on interactions of infectious biofilms and their host tissues: although most biofilm-related infections are associated with colonization of abiotic surfaces, there is a growing number of reports on biofilm colonization of natural surfaces, for example, urinary tract infections caused by uropathogenic *Escherichia coli* (Anderson et al. 2003).

Biofilm infections affect a wide spectrum of tissues and structures, including ear, nose, throat, mouth, eye, lung, heart, kidney, gall bladder, pancreas, nervous system, skin, bone, in addition to contamination of the surfaces of implanted medical devices (Donlan and Costerton 2002). The Centers for Disease Control and Prevention estimate that more than 80% of infections in humans are caused by bacteria growing as biofilms (National Institutes of Health 1999). Should biofilm infections be considered as a single disease category, the prevalence of this disease and the mortality associated with it is substantial.

A typical biofilm infection is mostly treated first with antibiotics. As this medication is ineffective in many of these cases a broader spectrum of antibiotics, sometimes combined with steroids, is used. Should the combined treatment fail, a decision often is made to remove the infected tissue or biomaterial by surgical intervention.

An early-stage diagnostic method that would allow for detection of the early stages of tissue (Mansson et al. 2007; Dowd et al. 2008; Nett and Andes 2008; Trampuz and Zimmerli 2008) or biomedical implant infection (Bauer et al. 2006; Selan et al. 2008) is now emerging. Xiong and colleagues (Xiong et al. 2005) report a rapid, continuous method for real-time monitoring of biofilm development, both *in vitro* and in a mouse infection model, through noninvasive imaging of bioluminescent bacteria. An alternative to this bioluminescence method is the noninvasive detection of fluorescent bacteria using multiphoton laser scanning microscopy. As bacterial expression of fluorescent protein, for example, green fluorescent protein (GFP) and its variants, does not require any substrate for endogenous production, it is the method of choice for prokaryotic labeling. Unfortunately, use of GFP is restricted to only those bacterial species that can produce the fluorescent protein, which eliminates the tracking of some key pathogens, such as *Staphylococcus epidermidis* or obligate anaerobes.

4.2 The Life Cycle of Biofilms

Microbial cells embedded within a biofilm ("sessile cells") differ from their freely suspended counterparts ("planktonic cells") by structural organization assigned as "biofilm architecture." Characklis (Characklis 1990a) identified up to eight different processes describing microbial attachment to surfaces resulting in biofilm growth and development. These can be reduced to three main stages: attachment, growth, and detachment from biofilm surface (Figure 4.2).

4.2.1 Microbial Attachment

The solid–liquid interface between a semisolid surface and an aqueous medium, for example, blood vessel, provides an ideal environment for the attachment and growth of microorganisms. Understanding the attachment mechanism

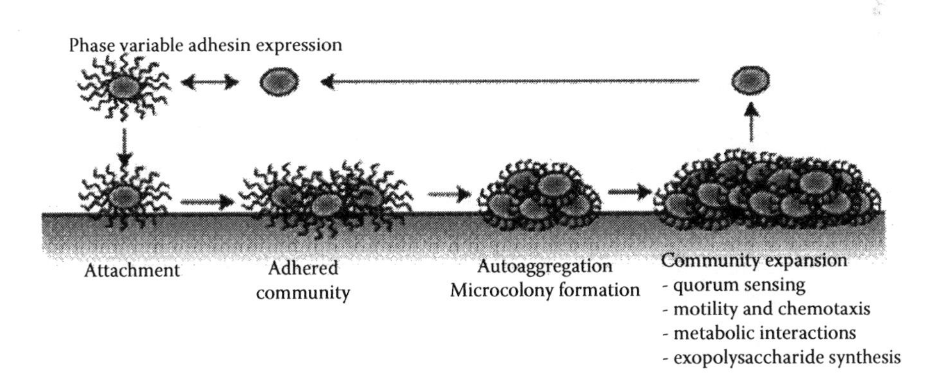

FIGURE 4.2
Model describing the specific stages in the development of microbial biofilms. (From Schembri, M.A., Givskov, M., and Klemm, P., *Science's STKE*, 2002. Reprinted with permisison from AAAS.)

of microbial cells to surfaces is a crucial factor in the development of prevention and therapy methods against persistent biofilms. The initial attachment of microbial cells to a given substrate is affected by various variables (Anderson and O'Toole 2008). A suitable host substratum, which can satisfy the nutritional requirements of the microorganisms, is essential for microbial growth and reproduction (Wilson 2005). Capability to move using flagella or pili is often a prerequisite for efficient cell-to-surface attachment (Geesey 2001). Furthermore, capability of bacterial cells to "sense" contact with substrate surface and to respond by expressing adhesins was also suggested as an important factor (Otto and Silhavy 2002). Studies have demonstrated that microorganisms typically attach more rapidly to hydrophobic surfaces such as plastics than to hydrophilic glass or metals (Donlan 2002). An exception to this is *Listeria monocytogenes,* which forms biofilms more rapidly on hydrophilic, for example, stainless steel medical devices, than on hydrophobic surfaces (Chavant et al. 2002). Attachment cannot be obtained without the favorable effects of the substratum surface, formation of conditioning films, medium characteristics, and complementary properties of the adhering cell surface.

4.2.1.1 Substratum Effects

The physical properties of the solid support may have a strong impact on the cell attachment. Characklis et al. (1990) noted that microbial colonization appears to increase with surface roughness. This seems to result from reduction in shear forces and higher surface area. The physicochemical properties of the surface may also exert a strong impact on the rate and extent of the attachment. Increasing rate of attachment was also observed for hydrophobic, nonpolar surfaces, such as Teflon as well as glass or metallic surfaces (Bendinger et al. 1993).

4.2.1.2 Conditioning Films

Biomaterial surface exposed to an aqueous medium, for example, blood, will be inevitably and almost immediately coated by proteins from the contacting medium. The resulting surface modification will affect the rate and extent of microbial attachment. The nature of conditioning films may be quite different for the surfaces exposed to different mediums. A prime example may be the proteinaceous conditioning film called "acquired pellicle," which develops on tooth enamel surfaces in oral cavities. Pellicle is composed of albumin, lysozyme, glycoproteins, phosphoproteins, lipids, and gingival crevice fluid (Marsh 1995); bacteria from the oral cavity colonize the pellicle-conditioned surfaces within hours from exposure to these surfaces. Mittelman noted that a number of host-produced conditioning films such as blood, tears, urine, saliva, intervascular fluid, and respiratory secretions strongly affect the attachment of bacteria to biomaterials (Mittelman 1996). Ofek and Doyle (1994) also noted that the surface energy of the suspending medium may affect hydrodynamic interactions of microbial cells with surfaces by altering the substratum characteristics.

4.2.1.3 Hydrodynamics

Flow velocity of fluids in direct contact with substratum interface is mostly negligible. This zone is defined as the hydrodynamic boundary layer. Its thickness depends on the linear velocity of the outer medium; the higher the velocity, the thinner the boundary layer. The flow in the region outside the boundary layer is dominated by mixing or turbulence. Slow flow is characterized by laminar or minimal turbulence. The hydrodynamic boundary may enhance cell–substratum interactions. Suspended microbial cells behave as particles in a liquid, and the rate of settling and association with a submerged surface will largely depend on their carrier velocity. Under very low linear velocities, the cells must cross the hydrodynamic boundary layer. As the velocity increases, the thickness of the boundary layer decreases and cells will be subjected to vigorous turbulence and mixing. Low linear velocities would enable rapid association with the surface while high velocities will exert substantial shear forces, resulting in cell detachment (Characklis 1990b). These effects were confirmed by studies carried out by Rijnaarts et al. (1993) and Zheng et al. (1994). Mass transport of substrates and antimicrobial agents into the biofilm is also affected by turbulence and boundary layer (Wanner et al. 2006).

4.2.1.4 Characteristics of the Contacting Aqueous Medium

Other characteristics of the aqueous medium, such as pH, nutrient levels, ionic strength, and temperature, may also play a role in the rate of microbial attachment to a substratum. Several studies have shown a seasonal effect on bacterial attachment and biofilm formation in different aqueous systems (Donlan et al. 1994). This effect may be due to water temperature and other seasonally affected parameters. Fletcher (1988) found that an increase in the concentration of several cations affected the attachment of *Pseudomonas fluorescens* to glass surfaces, presumably by reducing the repulsive forces between the negatively charged bacterial cell's surface and the glass surfaces. Cowan et al. (1991) showed correlation of increase in nutrient concentration with an increase in the number of attached bacterial cells.

4.2.1.5 Cell Properties

Cell-surface hydrophobicity, presence of fimbriae and flagella, and production of exopolysaccharide (EPS) all affect the rate and extent of microbial cell attachment. The hydrophobicity of many bacterial cell surfaces plays an important role in their adhesion along with their surface negative charge, as noted by Rosenberg and Kjelleberg (1986). Fimbriae, that is, nonflagellar appendages contribute to cell-surface hydrophobicity and play a role in cell attachment by overcoming the initial electrostatic repulsion barrier, which exists between segments of cell surface and the substratum (Rosenberg and Kjelleberg 1986). Other cell-surface components may also facilitate attachment: Several studies provided evidence for the impact of proteins on attachment (Bashan and Levanony 1988). The O antigen component of lipopolysaccharide (LPS) was also shown to confer hydrophilic properties in gram-negative bacteria (Williams and Fletcher 1996). Electrostatic force was also shown to participate in microbial attachment. Cations were shown to cross-link with the anionic groups of polysaccharides, resulting in contraction (Fletcher et al. 1991). Beech and Gaylarde (1989) found that lectins inhibited but did not prevent attachment. Binding of lectins with the cells would minimize the attachment sites and interfere with cell attachment mediated by polysaccharides (Zottola 1991).

In view of these findings, cell-surface components such as fimbriae, various proteins, LPS, EPS, and flagella, all play an important role in the attachment process. Cell-surface polymers with nonpolar sites such as fimbriae, proteins, and components of certain gram-positive bacteria (mycolic acids) appear to dominate attachment to hydrophobic substrata, while EPS and LPSs are more important in attachment to hydrophilic materials. Flagella are also important in attachment, though their role may be to overcome repulsive forces rather than to act as adhesive (Korber et al. 1989).

4.2.2 Biofilm Growth

Cells irreversibly attached to the supporting surface will undergo cell division, form microcolonies, and produce the extracellular polymers (EPS) acting as the intercellular adhesive matrix of the biofilm. These polysaccharides are highly hydrated (>95% water) and tenaciously bound to the underlying surface.

The structure of the biofilm is not merely a homogeneous slimy monolayer but a heterogeneous one, changing in space and over time. Biofilms have a composite structure comprising "water channels" that allow diffusion-controlled transport of essential nutrients and oxygen to the cells embedded within (Figure 4.3). Biofilms may also act as filters to entrap suspended particles of various kinds, including minerals and host components such as fibrin, red blood cells (RBCs), and platelets.

4.2.2.1 Quorum Sensing

Quorum sensing is a term used to describe intercellular signaling in bacteria. Although several quorum-sensing systems are known, the most thoroughly investigated systems are the acyl-homoserine lactone (acyl-HSL) system of many Gram-negative species and the peptide-based signaling system of many Gram-positive species (Sturme et al. 2002; Parsek and Greenberg 2005). For

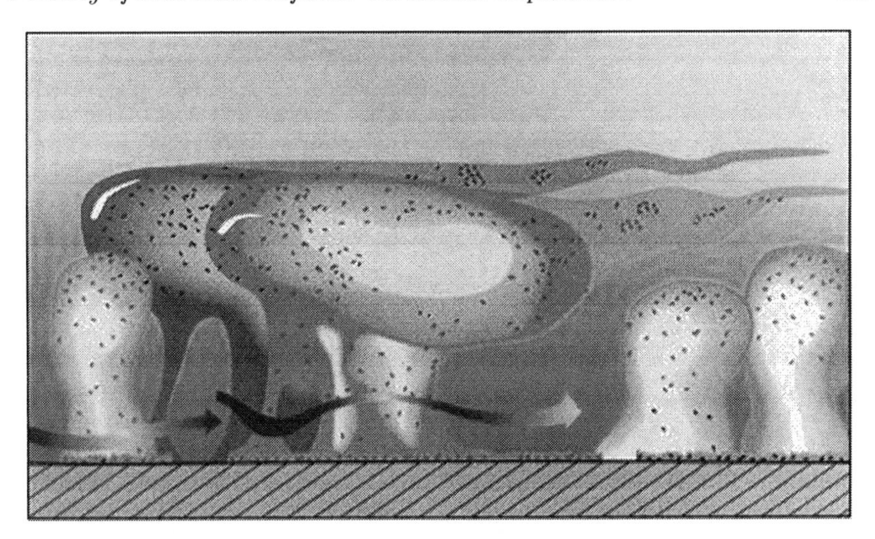

FIGURE 4.3
Biofilm structure layout. (Donlan, R.M. and Costerton, J.W., *Clin. Microbiol. Rev.*, 15, 2, 2002. Reproduced with permission from American Society for Microbiology.)

acyl-HSL quorum sensing, a single enzyme is required for the synthesis of the signal from cellular metabolites (Parsek et al. 1999). Peptide-based signaling usually involves the production of small linear or cyclic peptides that are translated as a larger pro-peptide inside the cell; these are then further processed during secretion (Dunny and Leonard 1997). In contrast to acyl-HSL-based signaling, peptide signals are not detected inside the cell. In some cases, a membrane-bound sensor protein belonging to the two-component signal transduction family interacts with the peptide. Peptide-bound sensor then activates an associated response regulator, which modulates expression of quorum-sensing-regulated genes.

4.2.3 Detachment

Cell detachment from biofilm surface may be affected by either the shedding of daughter cells from dividing cells, the detachment as a result of changes in nutrient levels, the quorum sensing, or the shearing of small segments of the biofilm as a result of shear forces.

In case of infectious biofilms, such detached cells may affect systemic infection, depending on a number of factors, including the response of the host immune system. Gilbert et al. (1993) showed that detached microbial cell's surface hydrophobicity was low for newly dispersed cells but steadily increased upon cell growth. Boyd and Chakrabarty (1994) suggested that the enzyme alginate lyase may play a role in the release of cells from the solid surfaces of biofilms.

Detachment affected by physical shear forces was investigated in greater detail. Brading et al. (1995) emphasized the importance of physical forces in detachment, stating that the three main processes for detachment are erosion/shearing (continuous removal of small portions of the biofilm), sloughing (rapid and massive removal), and abrasion (detachment due to collision of particles from the bulk fluid with the biofilm). The mode of cell dispersal apparently affects the phenotypic characteristics of the organisms. Eroded or sloughed aggregates from the biofilm are likely to retain certain potential biofilm characteristics, for example, antimicrobial resistance, whereas cells that have been shed as a result of growth may quickly revert to the planktonic phenotype.

4.3 Infectious Microbial Biofilms—Structural and Biological Characteristics

4.3.1 Bacterial Biofilms

Biofilms may be composed of either single or multiple microbial species and are formed on a range of biotic and abiotic surfaces. Though mixed-species biofilms mostly dominate, single-species biofilms do exist in some infections as well as on the surface of medical implants (Adal and Farr 1996). *Pseudomonas aeruginosa* was particularly investigated as single-species, biofilm-forming Gram-negative bacterial cells. Other Gram-negative bacteria, for example, *Pseudomonas fluorescens*, *Escherichia coli*, and *Vibrio cholera* have also been studied in detail. Gram-positive biofilm-forming bacteria studied included *S. epidermidis*, *Staphylococcus aureus*, and the *Enterococci.*

4.3.1.1 Biofilms Composed of Gram-Negative Bacteria

The first stage in the formation of Gram-negative biofilms is primarily regulated by the initial attachment. Biofilm formation is assumed to begin when suspended bacterial cells sense environmental conditions, triggering a shift to life on a surface (Pratt and Kolter 1998; Stoodley et al. 1999; O'Toole et al. 2000a). These environmental signals may be different for different organisms, for example, *P. aeruginosa* and *P. fluorescens* will form biofilms under almost any condition enabling growth (O'Toole and Kolter 1998) while *E. coli* K-12 and *V. cholera* will not form biofilms in minimal medium unless supplemented with amino acids (Pratt and Kolter 1998; Watnick et al. 1999). In addition to the nutritional impact, other environmental factors that may affect biofilm formation include temperature, osmolarity, pH, iron, and oxygen (O'Toole and Kolter 1998; Stoodley et al. 1999).

The complexity of signals triggering biofilm development is evident from the fact that organisms have multiple genetic pathways that control this behavior. In *P. fluorescens*, multiple pathways control biofilm formation and function under different growth conditions (Hinsa and O'Toole 2006). *V. cholerae* may have at least three different means for adhering to surfaces depending on whether this organism is within its human host or in an aqueous environment (Zampini et al. 2005), while *E. coli* K-12 can form biofilms on abiotic surfaces under a wide range of environmental conditions (Beloin et al. 2008).

Following attachment to a surface, bacteria undergo further adaptation to life as biofilm. Two properties are often associated with this phenomenon: enhanced biosynthesis of EPS (Beloin et al. 2008) and the development of antibiotic resistance (Mah and O'Toole 2001). These features appear to create a protective environment, converting the biofilms to a tenacious clinical problem. Bacterial biofilms may also develop other properties, including increased resistance to UV light, increased rates of genetic exchange, altered biodegradative capabilities, and increased secondary metabolite production (Goodman et al. 1994; Brazil et al. 1995; Moller et al. 1998).

Detachment and return to the planktonic growth mode is an important portion of the biofilm development pathway. Boyd and Chakrabarty (1994) reported that the enzyme alginate lyase may play a role in the detachment phase in *P. aeruginosa.* Allison and colleagues (1998) showed that a biofilm of *P. fluorescens* decreased in size following an extended incubation, attributed, at least in part, to loss of EPS.

4.3.1.2 Biofilms Composed of Gram-Positive Bacteria

A number of Gram-positive bacterial infections, caused by *S. epidermidis*, *S. aureus,* and the *Enterococci*, have proved to be particularly difficult to eradicate with currently available antibiotic therapies, in part due to their high-level natural resistance to antimicrobial compounds. Furthermore, these organisms become resistant to the highest doses allowed for antibiotics when growing in a biofilm (Raad et al. 1995).

The initiation of the attachment event has proven to be a critical stage in biofilm development. Mutants of *S. epidermidis*, which could be isolated with a microtiter dish-based assay, were not able to form biofilms on abiotic surfaces (Heilmann and Gotz 1998), a capability that is strongly correlated to the initiation of diseases under clinical settings (Deighton and Balkau 1990). Genetic studies have also led to the conclusion that biofilm development by this organism occurs initially via cell–surface interaction (Heilmann and Gotz 1998). These interactions may be mediated through a number of factors, including uncharacterized surface proteins (Hussain et al. 1997), extracellular proteins (Schumacher-Perdreau et al. 1994), capsular polysaccharide/adhesion (PS/A) (McKenney et al. 1998), and the cell surface-localized autolysin (Heilmann et al. 1997). Cramton and colleagues showed that *S. aureus*, like

S. epidermidis, has the ability to produce intracellular adhesion mediated by the *ica* locus. This data suggests that the early stages in biofilm formation may be similar for the two organisms (Cramton et al. 1999). *Enterococci* are important pathogens in device-related infections and, like other Gram-positive organisms, the formation of enterococcal biofilms on medical implants is increasingly becoming an important clinical issue (Shay et al. 1995).

In similarity to Gram-negative organisms, Gram-positive bacterial cells also produce extracellular polysaccharides (often referred to as "slime") when growing on a surface. Deighton and Borland showed that *S. epidermidis* increased slime production in iron-limited medium and later in the growth phase when nutrients were exhausted. They suggested that this observation may be an important signal *in vivo*, when iron nutrient levels may be limiting (Deighton and Borland 1993). *S. aureus* also produces EPS or slime in a phase-variable manner. The increased EPS correlates to its capability to form a biofilm in an *in vitro* system (Baselga et al. 1993).

There is no much data on the potential detachment of *Staphylococci* or *Enterococci* from their biofilms as well as on the control of this process. Expression of the *ica* locus may be phase variable, as Ziebuhr and colleagues (1999) proposed that the "switch-off" of *ica* by the IS256 insertion element may be the mechanism by which individual *S. epidermidis* (and possibly also *S. aureus*) cells can leave the biofilm and colonize on new surfaces.

4.3.2 Fungal Biofilms

Fungal infections are abundant especially in immunocompromised patients and in patients treated with high doses of antibiotics, and are a growing concern. It has been recognized that, like bacteria, fungi and yeast, for example, *Candida albicans,* are also capable of biofilm formation on medical implants (Kojic and Darouiche 2004; Hawser and Islam 2006). Hawser and Douglas reported the use of disks made of catheter material as a simple assay for biofilm development *in vitro* (Hawser and Douglas 1994; Hawser and Islam 2006). Using this system, they showed that *C. albicans* could form biofilms on a wide range of abiotic surfaces and that biofilm formation occurred best on latex, poly(vinylchloride), or silicone elastomer, but less on polyurethane or pure latex (Hawser and Douglas 1994). Baillie and Douglas (1999) went on to use this model system in a series of elegant experiments to show that switching between yeast form and hyphal growth form plays an important role in fungal biofilm development. Using scanning electron microscopy as their research tool, they observed that the biofilm layer closest to the surface was primarily composed of yeast cells, while the upper portion of the biofilm was composed of a layer of hyphae. Baillie and Douglas (1999) also showed that the surface to which the yeast cells attached could influence the structure of the biofilm formed and that the structure of the fungal biofilms formed resembled their bacterial counterparts.

The fungal biofilm developed, however, new physiological properties that were different from their planktonic counterparts. Douglas and colleagues used classical assays to address this question. Using their standard assay for biofilm formation in conjunction with electron microscopy, they showed that *C. albicans* biofilms became resistant to five antifungal compounds routinely used in clinical settings (Hawser and Douglas 1995). This drug resistance phenotype is often associated with bacterial biofilms. Furthermore, the observed increase in antibiotic resistance could not be simply attributed to the decrease in growth rate observed for fungi growing on surfaces (Baillie and Douglas 1998). Another species of *Candida*, *C. dubliniensis* has the ability to adhere and form biofilms with structural heterogeneity, typical microcolony, and water channel architecture similar to that described for bacterial and *C. albicans* biofilms (O'Toole et al. 2000b; Ramage et al. 2001). In addition, resistance of *C. dubliniensis* to fluconazole, as well as its increased resistance to amphotericin B treatment (Moran et al. 1998; Quindos et al. 2000; Martinez et al. 2002), were also demonstrated.

Other opportunistic biofilm-forming *Candida* species were also associated with catheter-related bloodstream infections (CRBSIs) including *C. parapsilosis, C. tropicalis, C. lusitaniae, C. krusei*, and *C. glabrata* (Hawser and Islam 2006). Biofilm colonization of ventricular shunt catheters, peritoneal dialysis fistulas, and cardiac valves by *Crytococcus neoformans* have also been reported and are of particular concern in view of the growing use of ventriculo-peritoneal shunts to manage intracranial hypertension associated with cryptococcal meningoencephalitis in immunocompromised patients. Recurrent meningitis in patients with ventriculo-peritoneal shunts has also been associated with biofilm colonization by *Coccidioides immitis*. Endocarditis associated with infections of prosthetic valves and other cardiac devices by *Aspergillus* species is a growing concern in immunocompromised patients, although a definitive role of biofilms has yet to be established. Although *Candida* and *Aspergillus* species are the etiological agents in only about 8% of implant infections, they are emerging as dangerous pathogens affecting patient survival rate in some settings of 50% (Anderson and Marchant 2000).

4.3.3 Microbial Interactions in Mixed-Species Biofilms

Data accumulating in the past few years on biofilms suggests that there seems to be a specific "biofilm life mode" distinctive from the "planktonic life mode" (Stoodley et al. 2002). It appears that biofilm development creates structured communities in which several different subpopulations may be present, in spite of the fact that the cell population is isogenic (Tolker-Nielsen and Molin 2000). These features are in accord with the assumption that biofilms are bacterial analogues or a reflection of tissues. The observed heterogeneities point toward putative phenotypic interactions between the different subpopulations, in analogy with what is known from ecological studies on microbial subpopulations in the environment.

In microbial biofilms most of the cells are present in the form of various types of aggregates, with extremely small intercellular distances, creating ideal conditions for cell to cell interactions, as well as aggregate to aggregate interaction, including aggregates formed from different microbial species. Several bacterial pathogens were shown to be associated with this phenomenon, including *Legionella pneumophila* (Murga et al. 2001), *S. aureus* (Raad et al. 1992), *Listeria monocytogenes* (Wirtanen et al. 1996), *Campylobacter* spp. (Buswell et al. 1998), *E. coli* O157:H7 (Camper et al. 1998), *Salmonella typhimurium* (Hood and Zottola 1997), *V. cholera* (Watnick and Kolter 1999), and *Helicobacter pylori* (Stark et al. 1999). Although all these organisms are capable of attaching themselves to surfaces to form biofilms, most, if not all, appear incapable of forming single-species biofilms of their own. This may be due to their fastidious growth requirements or due to their inability to compete with indigenous organisms. The mechanism of interaction and growth apparently varies with the pathogens involved, and at least for *L. pneumophila*, which appears to require the presence of free-living protozoa to grow in biofilm form (Murga et al. 2001). Survival and growth of pathogenic organisms within biofilms might also be enhanced by the association and metabolic interactions with indigenous organisms. Camper and colleagues (1998) showed that *Salmonella typhimurium* persisted in a model distribution system containing undefined heterotrophic bacteria from an unfiltered reverse osmosis water system for more than 50 days, which suggests that the normal biofilm flora of this water system provided niche conditions supporting the growth of this microorganism.

The isolation of multiple discrete species from biofilm-colonized implants derived from patients provides the most compelling evidence of the medical importance of polymicrobial (or heterotypic) biofilms (Costerton et al. 1999). Studies of polymicrobial biofilms formed by *C. albicans* and *S. epidermidis* indicate that the exopolymeric matrix produced by the fungal species may protect the bacteria against antibiotics, while the bacterial matrix protects the fungi from antifungal agents. Other studies of polymicrobial biofilms have provided evidence of enhanced interspecies transfer of antimicrobial resistance traits, symbiotic interactions, and sequential colonization patterns. These observations raise the intriguing possibility that virulence traits associated with biofilm commensalism may be coselected through enhanced persistence and antimicrobial tolerance of such polymicrobial biofilms (Wargo and Hogan 2006; Hansen et al. 2007).

4.3.4 Antimicrobial Resistance in Infectious Bacterial Biofilms

Biofilms of pathogenic bacteria and fungi have substantial impact on public health, as they are difficult to eradicate and affect chronic or recurrent infections. Biofilm infections constitute a number of clinical challenges, including diseases involving uncultivable species, chronic inflammation, impaired wound healing, and rapidly acquired antibiotic resistance.

Biofilm-based antimicrobial resistance is considered as the main cause of the increasing abundance of antibiotic-resistant infections, previously associated with intensive-care units and now recovered with increasing frequency in extended-care facilities, outpatients, and home-care settings. The bacterium *S. epidermidis* has thus evolved into one of the leading causes of nosocomial sepsis; this opportunistic pathogen is now the most common organism isolated from nosocomial infections (Bryers 2008).

Biofilm-based antimicrobial resistance differs from other antimicrobial resistances by a variety of mechanisms (Anderl et al. 2000). Extracellular matrix associated with biofilms was commonly assumed to restrict penetration of antimicrobial agents. This may not be, however, the main factor in some cases according to several recent studies. Diffusional limitation of antimicrobial agents into biofilm does not always seem to play a significant role in biofilm resistance. Fluoroquinolones rapidly diffused deeply into the biofilms of *P. aeruginosa* (Vrany et al. 1997) and *Klebsiella pneumonia* (Anderl et al. 2000). Tetracycline-penetrated *E. coli* biofilms (Stone et al. 2002) and vancomycin accessed the inner layers of *S. epidermidis* biofilms (Darouiche et al. 1994). A notable exception (del Pozo and Patel 2007) were the aminoglycosides, possibly due to the interaction that these positively charged antimicrobial agent had with the negatively charged extracellular matrix (Shigeta et al. 1997).

The increased bacterial density within biofilm microcolonies results in the accumulation of waste and the alteration of the microenvironment (e.g., low pH, low pO_2, high pCO_2, low divalent cation and pyrimidine concentration, low hydration level), which may compromise antimicrobial action deep within the biofilm. Absence of oxygen reduces the antimicrobial activity of aminoglycosides (Tack and Sabath 1985). It was shown that although ciprofloxacin is capable of penetrating *P. aeruginosa* biofilms, it is only active against cells located in zones with high pO_2 and high metabolic activity (Walters et al. 2003). One reason may be the spatial physiological heterogeneity of the microbial cells within the biofilm because of oxygen availability (Xu et al. 1998).

Antimicrobial agents may be trapped within the biofilm matrix wherein they may be chelated by inactivating enzymes. For example, it was observed that ampicillin is rapidly destroyed by β-lactamases inside *K. pneumoniae* biofilms (Anderl et al. 2000). Furthermore, biofilms may facilitate the spread of conventional antimicrobial resistance by promoting horizontal gene transfer. Although planktonic and biofilm antimicrobial resistance mechanisms differ, bacteria resistant to a particular antimicrobial agent in the planktonic state would not be susceptible to the agent at the biofilm state; accordingly, horizontal gene transfer within biofilms can affect antimicrobial susceptibility in the biofilm state. For example, Mah and colleagues (2003) reported a gene locus (*ndvB*), which is required for the synthesis of periplasmic glucans, which interact physically with tobramycin. They suggested that these glucose polymers may prevent antibiotics efficiency in *P. aeruginosa* biofilms. Therefore, this locus is a potential candidate for horizontal gene transfer within the biofilm resulting in an increased antimicrobial resistance.

Quorum-sensing signaling systems synchronize target gene expression and coordinate biological activity within biofilms (Dong and Zhang 2005). As an example, *N*-acylhomoserine lactones are secreted by *P. aeruginosa* within the deep regions of the biofilm. These quorum sensors interact with transcriptional activators to direct expression of several factors that facilitate bacterial persistence, for example, enabling *P. aeruginosa* to overcome the effects of antimicrobial agents (Prince 2002).

Adaptation to survival in the biofilm state requires changes in metabolic and catabolic pathways that can alter the intrinsic activity of antimicrobial agents in view of their mode of action (Fux et al. 2005). For example, experimental biofilms formed by coagulase negative *Staphylococcus* (CoNS) are highly resistant to antibiotics that target cell wall biosynthesis while remaining susceptible to antibiotics that target ribonucleic acid (RNA) and protein synthesis (Cerca et al. 2005). This response is consistent with a diminished role for cell wall biosynthesis in the biofilm population and reflects an ongoing role for transcription and translation in biofilm establishment, maturation, and propagation. It is interesting that so-called small colony variants (SCVs) of bacteria, characterized by a reduced *in vitro* growth rate due to genotypic changes in metabolic pathways (Proctor et al. 2006), have been isolated from both experimental biofilms and patients with biofilm-associated persistent infections. These include *P. aeruginosa* SCVs isolated from biofilms grown *in vitro* and from cystic fibrosis patients (von Gotz et al. 2004), an *E. coli* SCV isolated from a chronic prosthetic hip infection (Roggenkamp et al. 1998), and *S. aureus* SCVs isolated from patients with cystic fibrosis, osteomyelitis, and device-related infections (Chatterjee et al. 2007). Genotypic adaptations that alter metabolic capacity and decrease growth rate may therefore contribute to antibiotic resistance of some biofilm-related infections.

As most antimicrobial agents target rapidly growing cells, growth rate may play an important role in mediating biofilm-associated antimicrobial resistance. Deep within the biofilm, a small fraction of bacteria may differentiate into a protected phenotypic state, that is, slow or nongrowing bacteria (Lewis 2001). These cells have been referred to as "persisters" and may be present in relatively high numbers in the deeper section of a given biofilm. Antimicrobial treatment of bacterial biofilms may lead to eradication of most of the susceptible population while the small fraction of persister cell variants may survive and affect biofilm reconstitution following discontinuation of antimicrobial therapy. This, however, may depend on the specific antimicrobial agent. Penicillin, for example, does not kill nongrowing cells. Some of the newer β-lactam agents, for example, cephalosporins, carbapenems, aminoglycosides, and fluoroquinolones can kill nongrowing bacteria, although they are more effective at killing growing cells (Brooun et al. 2000).

Bacteria in biofilms can upregulate stress-response genes and can switch to more resistant phenotypes upon exposure to environmental stresses. As an example, biofilm bacteria may increase their ability to express chromosomal β-lactamases following prolonged exposure to β-lactam drugs (Bagge et al. 2004). Multidrug resistance pumps play an important role in the resistance of planktonic *P. aeruginosa* to antimicrobial agents, probably by biofilm mode of growth affecting an increase in the expression of efflux pumps (Brooun et al. 2000; Gillis et al. 2005). Genes encoding antibiotic transporters (or their regulators) has also been observed in studies of biofilm formed by uropathogenic *E. coli* (Hancock and Klemm 2007) and *Candida albicans* (Andes et al. 2004). However, genes associated with four well-characterized multidrug-resistant efflux pumps were not overexpressed inside a biofilm (De Kievit et al. 2001).

4.3.5 Porosity and Diffusional Limitations in Biofilms

Data accumulating from biofilm studies clearly indicate that biofilm structure and, in particular, transport of nutrients and antimicrobial agents into the deeper layers of the biofilm affect the major impact on its characteristics. The physiological responses of the microbial cells comprising the biofilm are not homogeneous throughout a biofilm, and the cells are strongly affected by their local environment. The metabolic activities of the cells, together with diffusional limitations, result in concentration gradients of nutrients, signaling compounds, and bacterial waste within the biofilm. As the cells respond to these gradients, they become adopted to the local chemical conditions, which can change over time as the biofilm develops. As a result, biofilms exhibit considerable structural, chemical, and biological heterogeneity. Cells growing in biofilms are therefore not only physiologically different from planktonic cells, but also vary from each other, in place, and over time as the biofilm develops (Stewart and Franklin 2008).

Solutes are transported within microbial biofilms by a combination of convection and diffusion. The heterogeneous structure of many biofilms permits convective transport within the larger voids and water channels penetrating the biofilm (Lewandowski et al. 1995). Within local cell aggregates or clusters, molecular diffusion is the dominant mode of mass transport (de Beer and Stoodley 1995). The rate of the diffusion process within biofilm cell clusters is characterized as diffusion coefficient (Stewart 2003).

The common starting point in the evaluation of a biofilm diffusion coefficient is an estimate of its value for a given solute in water. The presence of microbial cells, extracellular polymeric substances (EPS), and inorganic materials all affect diffusional limitations within the biofilm and reduce the diffusion coefficient from its value in pure water. This reduction is characterized by calculating the ratio of the effective diffusion coefficient in the biofilm to the diffusion coefficient recorded in the medium in contact with the biofilm.

Stewart (1998) distinguishes between two parameters that are both referred to as effective diffusion coefficients. Following the terminology of Libicki and colleagues (Libicki et al. 1988), D_e^d denotes the effective diffusion coefficient and D_e^p denotes the effective diffusive permeability. Both parameters are defined by Fick's first law, but with different definitions of the solute concentration in the concentration gradient.

The mass transport in biofilms is affected by the biofilm structure and its composition. A quantitative understanding on how biofilm structure is linked to mass transport is essential for understanding the biofilm behavior. Two main approaches can be used to correlate biofilm structure to mass transport. One approach is to explicitly describe the complex three-dimensional structure of the different biofilm components where the three-dimensional structure can be obtained from direct imaging of biofilms (Staudt et al. 2003) or from mathematical modeling (van Loosdrecht et al. 2002). These approaches require detailed information of the three-dimensional structure and the specific diffusion coefficients throughout the various parts of the biofilm system, that is, cell clusters and local different types of extracellular polymeric substances (EPS). Such specific microscopic diffusion coefficients are, however, difficult to measure experimentally. Therefore, in most three-dimensional mathematical models usually a diffusion coefficient uniform throughout the entire biofilm was assumed. Another approach was to correlate the overall biofilm diffusion to the biofilm structure based on macroscale parameters, such as overall biofilm density and porosity. A disadvantage of the latter approach is that the spatial resolution of the three-dimensional biofilm structure is lost. It carries, however, the advantage that established methods for measuring parameters describing the overall biofilm structure and the overall diffusion coefficients become available.

Biofilms are mainly composed of water, and the macroscale diffusion coefficient for the biofilm (D_F) is often related to the diffusion coefficient in pure water (D_W) as

$$D_F = f_D D_W \tag{4.1}$$

where f_D is the relative diffusivity (Hinson and Kocher 1996). Values for f_D reported in the literature range from 0.1 to 1.0 depending on the characteristics of the biofilm and of the solute (Hinson and Kocher 1996; Stewart 1998). Three main approaches have been used to quantify f_D experimentally: (1) The two-chamber method, where a biofilm on a membrane is placed between two chambers and the rate of diffusion through it is quantified by bulk-phase measurements in the two chambers (Matson and Characklis 1976); (2) microelectrode measurements, where f_D is determined from the change in the concentration gradient measured above and inside the biofilm matrix (Cronenberg and van den Heuvel 1991); and (3) quantification of the overall substrate removal and assuming a substrate conversion rate inside the biofilm (Yano et al. 1961). These three methods have been applied to a large variety of biofilms ranging from biofilms grown directly on membrane surfaces (Siegrist and Gujer 1985), on detached biofilm, or on activated sludge filtered by a membrane (Matson and Characklis 1976) as well as to model systems such as yeast cells entrapped in alginate (Cronenberg and van den Heuvel 1991). Beuling and colleagues (1998) applied nuclear magnetic resonance (NMR) technique to determine the self-diffusion of water in artificial and natural biofilms. f_D values for water within the range of 0.59–0.89 were measured depending on biofilm density.

Several studies determined diffusion coefficients for biofilms (Cronenberg and van den Heuvel 1991; Lewandowski et al. 1991; Bryers and Drummond 1998). For example, Yano and colleagues (Yano et al. 1961) evaluated biopellets from *Aspergillus niger,* which were used for citric acid production, where f_D was calculated on the basis of overall conversion rates resulting in values ranging from 0.11 to 0.9 for aggregate densities of 170 and 30 kg/m^3, respectively. The variability of these results was, however, high.

Detailed studies and accumulating data from traditional light microscopy or from laser confocal scanning microscopy confirmed that biofilm structure determines the mass transport mechanism within the biofilm and mass transport rates at the surface of the biofilm (de Beer et al. 1994b; Bishop and Rittmann 1995; Yang and Lewandowski 1995; Bishop 1997). As biofilms display different structures, it was assumed that the structural differences reflect biofilm formation. Furthermore, environmental conditions, for example, hydrodynamics, seem to have an impact on these structures as well as on the chemical composition of the contacting medium and the chemical and physical properties of the surface supporting the biofilm. The fundamental processes, for example, attachment/detachment and growth occurring within the biofilms are also indirectly affected. For example, it is known that biofilms grown at high shear stress develop elongated microcolonies (Lewandowski and Stoodley 1995). It is also known that dense biofilms develop either as a result of high shear stress or as a result of starvation (Beyenal and Lewandowski 2000).

Quantitative structural parameters of biofilms are currently solely available from the data worked out from microscope micrographs. The best of these are derived from confocal laser scanning microscopy as it allows the acquisition of images of fully hydrated biofilms at high spatial three-dimensional resolution. Many researchers use image analysis to calculate values of parameters characterizing biofilm structures by available technologies such as Bioquant Meg-IV software (R&M Biometrics, USA), MOCHA Image Analysis Software (Jandel Scientific, San Rafael, California), and IDL Interactive Data Language (Research System Incorporated, Boulder, Colorado). Some researchers developed their own software, for example, COMSTAT (Heydorn et al. 2000) and image structure analyzer (ISA) (Yang et al. 2000). These parameters include fractal dimension characterization of the variability of biofilm structures as well as porosity/voids array. Some researchers calculated more exotic parameters such as energy or textural entropy (Beyenal et al. 2004b).

Recognizing the importance of quantifying biofilm structure, ISA was used to calculate a series of parameters, for example, textural entropy, homogeneity, energy, areal porosity, average horizontal and vertical run lengths, diffusion distance, and fractal dimension obtained from digital biofilm images (Yang et al. 2000). To make the results of the image analysis less dependent on the operator, an automatic thresholding algorithm was developed and integrated with ISA (Yang et al. 2001). The performance of ISA was tested by several researchers. Lewandowski and colleagues (Lewandowski et al. 1999) used ISA to monitor temporal variations in the structure of biofilms composed of *P. aeruginosa*, *P. fluorescens*, and *K. pneumoniae* and concluded that some of the structural parameters may reach stable values. Lewandowski (Lewandowski 2000) used ISA to calculate areal porosities of layered biofilms and developed a method for calculating volumetric biofilm porosity. Beyenal and Lewandowski (Beyenal and Lewandowski 2001) used ISA to quantify the areal porosity and fractal dimension of mixed population bacterial biofilms comprising *Desulfovibrio desulfuricans* and *P. fluorescens* and concluded that the extent of biofilm heterogeneity was directly correlated with the flux of H_2S from the cell clusters. Purevdorj and colleagues (Purevdorj et al. 2002) used ISA to quantify the areal porosity, fractal dimension, average horizontal run length, average diffusion distance, textural entropy and energy of biofilms of wild-type *P. aeruginosa* PAO1, and cell–cell signaling using *lasI* mutant PAO1-JP1 under laminar and turbulent flows. They concluded that both cell signaling and hydrodynamics influenced biofilm structure. Recently, Beyenal and colleagues (Beyenal et al. 2004a) presented a three-dimensional version of the ISA software, which calculated three-dimensional parameters characterizing biofilm structure in terms of heterogeneity, size, and morphology of biomass.

Over the past few years, modeling of heterogeneous biofilms gained much attention. The biofilm's inner space is referred to as porous media: porosity, diffusion, and permeability are hence the parameters of choice for attempts to quantify the extent of biofilm structural heterogeneity. For example, cells that are located near the biofilm–bulk-fluid interface will be considered to be provided with both substrate and oxygen. Deeper in the biofilm, there could be a region in which both oxygen and substrate have been depleted; in this zone, the cells might become resting cells or dead cells (Stewart and Franklin 2008). Older biofilm models, like AQUASIM, which accept heterogeneous structure of biofilms, required porosity as an input parameter (Wanner and Reichert 1996; Horn and Hempel 1997). A newer approach to biofilm modeling, cellular automaton (Wimpenny and Colasanti 1997; Picioreanu et al. 1998) allows prediction of biofilm porosity from first principles.

In general, there are two different approaches used in modeling of biofilms. The more prevalent approach is to consider biofilm as a continuous layer. Another approach is based on considering it as patchy aggregates that accumulate in pore throats. For cases with sufficient nutrients, continuous layer assumption matches better with reality. For the low-load cases, it is important to distinguish between continuous and discontinuous biofilms to have a more accurate prediction for spatial distribution and permeability reduction (Rittmann 1993). It should be assumed that a biofilm is composed of two different parts, that is, base and surface films. The base film components are packed and continuous, while the surface film is discontinuous (Gujer and Wanner 1990). One of the very common discrete-stochastic methods used in this field is cellular automata (CA), in which nutrient and biomass are simulated as individual particles (Wanner et al. 2006). In this approach, the biofilm is represented as a continuous layer and its properties primarily change in vertical direction.

A number of studies focused on the estimation of the steady state thickness of biofilms (Fouad and Bhargava 2005; Qi and Morgenroth 2005; Gapes et al. 2006). In some of these (Rittmann and McCarty 1980; Rittmann and Manem 1992), it was assumed that there is a minimum substrate concentration that can support biofilm development and that below that level biofilm will not develop. Wanner and Gujer (1986) developed a multicomponent but homogeneous model to predict biofilm growth. Later, they expanded their model (Gujer and Wanner 1990) to consider liquid and solid phases within the biofilm. The model was modified to include the solid (particulate) phase diffusive flux and the liquid (dissolved) phase advective flux (Wanner et al. 1995; Alpkvist et al. 2006; Lee and Park 2007).

Detailed metabolic information on the microscale is critical to understanding and exploiting beneficial biofilms and combat antibiotic-resistant, disease-associated biofilm-forming species. These temporally and spatially variable metabolic gradients, which occur at the microscale, are extremely difficult to measure. Microelectrodes offer a mean to map a single chosen parameter, for example, pH or pO_2 (Revsbech 2005); it should be mentioned, however, that these probes physically perforate the sample, thereby changing its permeability and metabolism. Confocal laser scanning microscopy (CLSM) is widely used for biofilm investigations because of its noninvasive nature and its three-dimensional resolution capability. In spite of some technical limitations, for example, limited optical penetration depth due to optical absorption and scattering, it is widely used (Vroom et al. 1999). Two-photon CLSM methods overcome and improve depth penetration. Both two-photon and CLSM require, however, the addition of fluorescent tracers when detecting metabolic activity that may have undesirable effects on cellular functions (Ullrich et al. 1996). Fluorescence *in situ* hybridization and microautoradiography provide species and substrate-uptake information at the single-cell level, but are destructive and detect just one substrate per sample. Thus, very few techniques can continuously detect biofilm metabolite profiles in a truly noninvasive and nondestructive manner with adequate time and spatial resolution.

Nuclear magnetic resonance is a noninvasive method that provides a noninvasive, subatomic view of molecular, chemical, physical, and transport processes. NMR is nondestructive, nonsample consuming, and is insensitive to sample opacity. NMR spectroscopy techniques provide detailed metabolic information by analysis of cell extracts, supernatants, biological fluids, and live biological samples. ^{1}H and ^{1}H-detected ^{13}C NMR provide direct, time-resolved monitoring of metabolite concentrations. The disadvantage of NMR is its inherent low sensitivity due to the low energies involved, requiring careful optimization to reduce measurement times and lower concentration detection thresholds. NMR has been used to measure flow and diffusion in biofilm systems (Van As and Lens 2001; Seymour et al. 2004; Hornemann et al. 2008).

Recently, the capability of a combined CLSM and NMR microscope combination was demonstrated to measure time and depth-resolved metabolite concentrations under flow in viable biofilms (Majors et al. 2005). Unlike the traditional flow-tube geometry used for biofilm imaging and diffusion studies, this application employs a growth media perfusable sample chamber of planar geometry that supports a sample grown on a thin surface. Using this system, one can apply a series of combined techniques to perform magnetic resonance imaging (MRI), magnetic resonance spectroscopy (MRS), and also diffusion in a novel way on a single biofilm sample. McLean and colleagues (McLean et al. 2007) studied, with the help of this system, biomass volume and distributions as well as time- and depth-resolved metabolite concentrations in active *Shewanella oneidensis* and *S. mutans* biofilms.

4.4 Infectious Microbial Biofilms—Treatment Modalities and Resistance

Commonly employed antimicrobial agents were not specifically designed to target bacterial biofilms. As described earlier, bacterial biofilms may have more than one antimicrobial resistance mechanism, and that these diverse mechanisms may act concurrently, and in some cases, synergistically (Drenkard 2003). Antibiofilm strategies must address and target a range of potential resistance mechanisms and take into account the substantial heterogeneity inside biofilms. It was shown, for example, that the dynamics and spatial distribution of β-lactamase induction in *P. aeruginosa* cells growing in biofilms is highly heterogeneous (De Kievit et al. 2001). In addition, as bacterial cells growth rate within biofilms is not synchronized, a nonuniform susceptibility pattern to antimicrobial agents should be anticipated.

4.4.1 Antibacterial and Antifungal Treatment Modalities of Infectious Biofilms

Prophylactic use of antibiotics and biocides can reduce the incidence of biofilm-associated infections of medical devices. Strategies for prophylaxis include device coatings, device immersion, surgical site irrigation, antibiotic-loaded cements, and antibiotic lock therapy. Although the use of antibiotic prophylaxis is controversial because of its potential to increase antimicrobial resistance, it is increasingly common in high-risk patient groups. With regard to device coatings, a recent and comprehensive meta-analysis of randomized-controlled trials of rifampin-impregnated central venous catheters (CVCs) suggest that they are both safe and effective in reducing the rate of catheter colonization and CRBSIs (Falagas et al. 2007). Similarly, on six independent studies of the efficacy of antibiotic lock therapy in the prevention of CRBSIs in hemodialysis patients, an overall reduction of 64%–100% in CRBSIs was observed (Manierski and Besarab 2006).

Replacement or removal of an infected medical device, combined with systemic antibiotic and/or antifungal therapy, is the most effective treatment in most settings. Standard practice involves either a one-stage or a two-stage procedure (Trampuz and Zimmerli 2006, 2008). For managing medical-device infections without surgical intervention, long-term antimicrobial suppressive therapy remains the only option, and current salvage rates are highest with early diagnosis (Stein et al. 2000). Recommendations of antibiotic therapies for the management of biofilm-associated infections have been driven largely by empiric observations and typically involve the use of combinations over extended periods.

The biofilm intercellular matrix is essential for biofilm structure. Substances capable of depolymerizing, dissolving, or inhibiting the synthesis of this matrix may thus convert the status of biofilm cells to planktonic, fully exposed to the impact of environmental conditions and access of antibiotic. *In vitro* treatment of clarithromycin-resistant *S. epidermidis* and *P. aeruginosa* biofilms with a relatively low concentration of clarithromycin resulted in the decrease of the quantity of biofilm matrix and in an increased penetration of other antimicrobial agents (Yasuda et al. 1994). Furthermore, an equal combination of streptokinase and ofloxacin had an additive effect on *S. aureus* biofilms *in vitro* (Nemoto et al. 2000).

A major component of the extracellular matrix of several bacteria, for example, *S. aureus*, *S. epidermidis*, *E. coli*, is a linear polymer composed of *N*-acetylglucosamine residues in β-(1,6)-linkage called *polysaccharide intercellular adhesin*. Kaplan and colleagues (2003) reported that *Actinobacillus actinomycetemcomitans* produces a soluble glycoside hydrolase called *dispersin B*, which degrades polysaccharide intercellular adhesin. It seems possible that dispersin B may be used to detach polysaccharide intercellular adhesin-containing biofilms from the surface of colonized biomaterials.

New approaches to antimicrobial activity against persistent biofilms are crucial to effective control and eradication of infections. New compounds or effective new modes need to be developed and tested for their penetration into biofilms. Genes responsible for persistence may be considered as targets for new drugs. Development of drugs that disable the persistent phenotype is likely to provide an effective therapy for biofilm-associated infections. Ideally, an inhibitor of persistence development may be combined with a conventional antimicrobial agent to achieve biofilm eradication (Lewis 2001).

Another option is to use a new type of delivery of known antimicrobial activity. Ionic silver and silver nanoparticles are antimicrobial agents, which are commonly used in public health therapy (Silver et al. 2006; Chopra 2007). Although extensive topical use of silver may cause argyria (Payne et al. 1992; Tomi et al. 2004), it is considered safe. The authors of this chapter used novel enzyme-silver hybrids to affect enzymatically attenuated *in situ* release of silver ions to eradicate *E. coli* biofilms *in vitro* (Ben-Yoav and Freeman 2008). The hybrids were prepared by a new approach directing the process of electroless deposition of silver to the surface of single, soluble enzyme glucose oxidase molecules (Dagan-Moscovich et al. 2007). In the presence of oxygen, glucose oxidase will oxidize glucose to gluconolactone, a process accompanied by the formation of hydrogen peroxide as by-product. The hydrogen peroxide thus obtained will oxidize, in turn, some of the metallic silver coating of the enzyme, affecting local controlled release of silver ions (Ben-Yoav and Freeman 2008). The solubility and the nanometric size of the silver–enzyme hybrid allowed its diffusion and penetration into the targeted biofilm followed by scavenging residual glucose molecules present within the biofilm's pores, resulting in effective antibacterial activity. It should be mentioned within this context that increasing numbers of silver-resistant bacteria are accumulating (Pirnay et al. 2003; Silver 2003).

Recently, a number of bacterial enzymes that are capable of enzymatic inactivation of *N*-acylhomoserine lactones were identified. These enzymes show considerable promise as "quenchers" of quorum sensing taking place within biofilms (Roche et al. 2004). Over the past few years, a range of quorum-quenching enzymes and inhibitors were identified from different sources. Several potential quorum-sensing targets for the inhibition of *P. aeruginosa* virulence were described (Dong and Zhang 2005). RNAIII-inhibiting peptide is a heptapeptide reported to inhibit *S. aureus* biofilm formation by obstructing quorum-sensing mechanisms. It was shown that RNAIII-inhibiting peptide on its own reduces bacterial load and enhances the effect of coadded vancomycin, ciprofloxacin, and imipenem in the treatment of an experimental model of *S. aureus* catheter-related infection (Cirioni et al. 2006). Identification of compounds capable of interacting with quorum-sensing genes or their products, which could be used as adjuvants with conventional antimicrobials, is a promising approach.

Another interesting possibility for biofilm elimination arises from the observation of biofilm self-destruction. *Pseudomonas fluorescens* readily forms a biofilm in a well-oxygenated environment. As oxygen is depleted by the growing biofilm mass, a specific EPS lyase is induced and digests the biofilm matrix, liberating the cells (Allison et al. 1998). It is yet unclear whether a biofilm will self-destruct in response to any type or specific types of nutrient limitation or in the presence of chlorhexidine at the site of infection (Lewis 2001).

Physical forces can also affect the susceptibility of mature biofilm to antimicrobial treatments. Several physical approaches to biofilm eradication are being evaluated, including the use of an electric current (Jass et al. 1995), radiofrequency electric current (Caubet et al. 2004), electromagnetic fields (McLeod et al. 1999), and ultrasound (Rediske et al. 2000), in combination with antimicrobial therapy.

Guidelines for the clinical evaluation of antibiotic treatments are now available for the CRBSI setting only; comparative studies have rarely been undertaken in any medical-device infections (Lynch and Robertson 2008). Out of all agents investigated, a rifampin-antibiotic combination for staphylococcal medical-device infections has perhaps been most thoroughly evaluated in clinical trials. In these studies, rifampin was combined with quinolones (Isiklar et al. 1999; Schrenzel et al. 2004), β-lactams (Widmer et al. 1992) or fusidic acid (Drancourt et al. 1997), and rifampin-containing regimens are now established as standard therapies for a range of device-associated infections (Karchmer 2000; Steckelberg and Osmon 2000; Yogev and Bisno 2000; Zimmerli et al. 2004; Trampuz and Widmer 2006).

Fungal biofilm-associated infections are notoriously difficult to treat systemically with antifungal agents. This poor *in vivo* efficacy is not surprising, given the *in vitro* demonstrated resistance of biofilms formed by *C. albicans* and related species to various classes of antifungal agents (Chandra and Ghannoum 2004; Kojic and Darouiche 2004; Hawser and Islam 2006). It appears, however, that agents of the echinocandin class exhibited *in vitro* fungicidal activity on established biofilms (Ramage et al. 2005) and in suppressing biofilm colonization on biomaterials (Soustre et al. 2004; Shuford et al. 2006). Lipid-based formulations of amphotericin B have also proven effective in *in vitro* assays of biofilm activity. The management of fungal biofilm-associated medical-device infections involved exchange of the infected device, whenever possible, combined with systemic antifungal therapy (Chandra and Ghannoum 2004). Data accumulating from *in vitro* and preclinical animal studies suggest that antifungals of the echinocandin class (caspofungin, micafungin, anildafungin) and amphotericin B lipid formulations represent the currently known, best available options for the management of infections caused by fungal biofilms.

4.4.2 The Impact of Porosity and Diffusional Limitations on Treatment Efficacy

Though the details of development of antimicrobial biofilm resistance is only partially understood, recent studies have already used a variety of model systems to determine how and why biofilms are so resistant to commonly used antimicrobial agents (see above). As the importance of biofilms in nosocomial infections has increased, efforts are being directed toward the study of the impact of antimicrobial agents on these surface-attached communities. Four leading hypotheses explain the reduced susceptibility of biofilms: poor antimicrobial penetration, deployment of adaptive stress responses, physiological heterogeneity in the biofilm population; and the presence of phenotypic variants or persister cells (Stewart 2002; Davies 2003). It seems likely that combination(s) of these factors determine the overall resistance of a given biofilm (Stewart and Costerton 2001; Stewart 2002).

Biofilms are enclosed within an exopolymer matrix that can restrict the diffusion of substances as well as bind certain antimicrobials. This will provide effective resistance for biofilm cells against large molecules such as antimicrobial proteins, lysozyme, and complement. The diffusion barrier is also probably effective against smaller antimicrobial peptides and their analogs. The negatively charged EPS is very effective in protecting cells from positively charged aminoglycoside antibiotics by restricting their permeation, possibly through electrostatic interactions (Kumon et al. 1994; Shigeta et al. 1997; Meers et al. 2008). An example is the work of Gordon and colleagues (Gordon et al. 1988) that examined the diffusion of several antimicrobial agents (ceftazidime, cefsulodin, piperacillin, gentamicin, and tobramycin) through synthetic and naturally produced alginate gels and found that β-lactam antibiotics diffused into the matrix more rapidly than did aminoglycosides. Aminoglycosides were found to initially bind to the alginates, but diffusion increased after an 80- to 100-min lag period. To make matters even more complicated, subinhibitory levels of aminoglycosides were reported to induce biofilm formation (Drenkard and Ausubel 2002; Hoffman et al. 2005).

In most cases involving small antimicrobial molecules, the barrier of the polysaccharide matrix only postpones the death of cells rather than provide full protection. An illustration of such a case was provided by a study on fluoroquinolone antibiotics, which readily equilibrated across the biofilm (Shigeta et al. 1997; Vrany et al. 1997; Ishida et al. 1998; Anderl et al. 2000). Fluoroquinolones are indeed very effective in arresting the growth of a biofilm (Brooun et al. 2000). At the same time, restricted diffusion can protect the biofilm from a degradable antimicrobial. Retarded diffusion will decrease the concentration of the antibiotic within the biofilm, helping an enzyme like β-lactamase to destroy the incoming antibiotic. This synergy between retarded diffusion and degradation provides effective resistance to *P. aeruginosa* biofilms expressing β-lactamase (Giwercman et al. 1991). The synergistic relationship between diffusion retardation and degradation has been convincingly analyzed in a mathematical model based on these experimental observations (Stewart 1996).

Another interesting case of a diffusion barrier that helps protect the cells was described for hydrogen peroxide. Unlike planktonic cells of *P. aeruginosa,* which were very sensitive to 50 mM H_2O_2, the same cells in biofilm were found to be protected: these cells had lower levels of catalase (KatA), which effectively degraded the invading hydrogen peroxide (Elkins et al. 1999; Hassett et al. 1999). A restricted penetration of this small molecule coupled with its destruction by the microbial cells was apparently responsible for the resistance observed. It may be expected that any mechanism of antibiotic destruction or modification (like acetylation of aminoglycosides) will be especially effective when coupled with diffusion barriers.

Other studies examined antimicrobial agent penetration and interaction with the extracellular polymeric substance material of the biofilms. Hatch and Schiller (Hatch and Schiller 1998) showed that a 2% suspension of alginate isolated from *P. aeruginosa* inhibited diffusion of gentamicin and tobramycin, and that this effect was reversed by using alginate lyase. Souli and Giamarellou (Souli and Giamarellou 1998) demonstrated the ability of *S. epidermidis* slime to hinder the antimicrobial susceptibility of *Bacillus subtilis* to a large number of agents. Not all antimicrobial agents were equally effective: glycopeptides such as vancomycin and teicoplanin were more effective, whereas agents such as rifampin, clindamycin, and the macrolides were either ineffective or mildly effective.

One of the factors that is generally conceded to have a role in antibiotic resistance by biofilms is the limited access of the antibiotic to all areas of the biofilm. Several studies in which antibiotic penetration has been assessed by detecting the concentration of the antibiotic at the base of the biofilm were reported. In one such series of experiments, the penetration of the antibiotic ciprofloxicin was investigated for its ability to cross biofilms of *P. aeruginosa* to reach the surface of a germanium crystal substratum placed in an infrared (IR) field. Germanium crystal is transparent to IR radiation, which passes through the crystal to create an evanescent field extending 0.2 μm above the surface. The IR signature of a material (such as an antibiotic) that is located within the evanescent field is, therefore, detectable and can be monitored *in situ* in real time. Results from these experiments demonstrated that the biofilm affected diffusional limitations reducing—but not blocking—antibiotic penetration (Suci et al. 1994). These and subsequent results also showed that penetration rates through biofilms depended on the specific antibiotic used and were not directly correlated with the efficacy of the antibiotic against the tested biofilm (Vrany et al. 1997). In another work, wild-type *K. pneumoniae* grown on filter discs were shown to have reduced-antibiotic penetration for ampicillin compared to ciprofloxicin. However, β-lactamase-deficient *K. pneumoniae* biofilms—in which ampicillin was shown to reach full penetration—were still resistant to treatment, with a log reduction of 0.18 for the mutant strain, compared to 0.06 for the wild type in biofilm and >4 for the wild type in planktonic culture (Anderl et al. 2000). These results indicated that reduced antibiotic penetration might be important in the protection of microbial biofilms from certain antibiotics, but that this reduction could not account for the overall resistance of biofilms to antibiotic treatment.

In an effort to address the question of antibiotic penetration into dense cell aggregates located within biofilms, Matin and colleagues used direct microscopic observation of tetracycline penetration into *E. coli* biofilms. This study demonstrated that biofilms formed over 2 days on a polystyrene surface were less susceptible to the antibiotic than were planktonic cells; however, the biofilms showed tetracycline-mediated fluorescence distributed throughout the entire biofilm following exposure to the antibiotic for 7.5–10 min (Stone et al. 2002). Although this study did not provide quantitative data on the concentration of tetracycline within the biofilm, it nonetheless demonstrated that the antibiotic was able to penetrate to all observable parts of the biofilm. Furthermore, Huang and colleagues (1995) demonstrated gradients in specific respiratory activity within biofilms in response to disinfection with monochloramine.

Most published studies of diffusion in biofilms focus on an endpoint after a number of hours and fail to address the rate after which antibiotics are transported. The rate of transport in biofilms is important, as mixing of an antibiotic with a suspension of planktonic bacteria rapidly exposes all cells to the full antibiotic dose. If, however, the rate of antibiotic penetration through a biofilm is decreased with respect to the rate of transport through a liquid, then the bacteria may be exposed to gradually increasing doses of the antibiotic and may require time to overcome the biofilm's defensive response. de Beer and colleagues (1994a) demonstrated a limited penetration of chlorine into the biofilm matrix reducing the efficacy of this biocide as compared to its biocidal action on planktonic cells. In support of this idea, bacteria have been shown to increase transcription of stress-associated genes, such as heat shock protein homologues and cell wall-synthesis genes, within an hour of exposure to low doses of cell wall-active antibiotics (Utaida et al. 2003). An additional problem with many previous studies is that they demonstrated that antibiotics move from one side of an intact biofilm to the other but do not prove that they actually reach their cellular target (Zheng and Stewart 2002). Thus, antibiotics could traverse the biofilm through the exopolymeric matrix and the larger water channels without interacting with bacterial cells within bacterial clusters.

The kinetics of the penetration of the antimicrobial agents is a key factor in elucidating the antimicrobial efficiency. Suci and colleagues (Suci et al. 1994) demonstrated a delayed penetration of ciprofloxacin into *P. aeruginosa* biofilms: what normally required 40 sec for a sterile surface required 21 minutes for a biofilm-containing surface. Hoyle and colleagues (Hoyle et al. 1992) found that dispersed bacterial cells were 15 times more susceptible to tobramycin than were same cells in intact biofilms. DuGuid and colleagues (Duguid et al. 1992) examined *S. epidermidis's* susceptibility to tobramycin and concluded that the organization of these cells within biofilms could, in part, explain the resistance of this organism to this antimicrobial agent.

A variety of mathematical models and computer simulations have been proposed to investigate biofilm resistance to antimicrobial disinfection (Kissel et al. 1984; Stewart 1994; Stewart and Raquepas 1995; Stewart et al. 1996; Dodds et al. 2000; Roberts and Stewart 2004; Cogan et al. 2005; Szomolay et al. 2005; Wanner et al. 2006). Since biofilm processes are highly complex, all these models included simplifications, for example, spatially homogeneous and flat biofilm layer, fixed biofilm morphology, steady state assumptions for antimicrobial agents, and hydrostatic environment (Demaret et al. 2008). Stewart (1996) provided quantitative framework for the discussion and analysis of processes affecting antibiotic penetration into microbial biofilms. Later,

Demaret and colleagues (Demaret et al. 2008) demonstrated comprehensive modeling and simulation of the diffusion limitations dominating antimicrobial agent penetration into bacterial biofilms.

One promising solution to the problem of antibiotic penetration has emerged from the manipulation of electrical fields that surround bacteria in a biofilm. This "Bioelectric Effect"—a term coined by J. W. Costerton and colleagues—has been postulated to electrically alter the configuration of the EPS matrix, as well as to enhance the penetration of antimicrobial agents across the bacterial-cell envelope (Stoodley et al. 1997). Using alternating-current densities of less than $100\,\mu A/cm^2$, it was found that the antibiotic concentrations required to kill biofilm cells were significantly reduced compared to untreated bacterial biofilm (Stoodley et al. 1997; Caubet et al. 2004). These concentrations, however, were still higher than those required to kill planktonic bacteria of the same species (Costerton et al. 1994; McLeod et al. 1999). It was proposed that this phenomenon is due to the electrostatic interactions between negatively charged groups in the structure of the biofilm and the charged electrode (Stoodley et al. 1997), which will increase "fluidity" of the matrix, allowing a better penetration of the antibiotics (Caubet et al. 2004).

4.5 Concluding Remarks

Accumulating data on structural characteristics of medically important biofilms provides a complicated and challenging description of nonhomogeneous, multicomponent, and continuously changing complexes of mixed population of microbial cells. The porosity created throughout the buildup of a biofilm by attachment followed by cell division and secretion of EPS matrix is inherently irregular, presenting larger channels in between condensed cell clusters, which contain much smaller channels. Both channel categories are partially or fully packed with viscous secreted polysaccharide. Gradients of nutrients, oxygen, pH, and secretions are created, affecting, in turn, the metabolic state of the film forming mixed microbial cell populations embedded within clusters of various sizes. Moreover, biological response to the continuously changing environmental conditions affects genetic, biochemical, and structural local changes at different paces. Detailed structural information on some of the parameters affecting the structure and porosity of biofilms is yet very limited.

It clearly appears that medically important infectious biofilms should not be considered as readily characterized physical three-dimensional structures, handled with tools developed for such systems: thorough understanding and awareness of the complex nature of systems composed of viable complex cell communities continuously responding to environmental conditions is crucial—along with physical and chemical tools—for the completion of our understanding of these complicated systems and development of more effective means for their eradication.

4.6 References

Adal, K. A. and Farr, B. M. (1996). Central venous catheter-related infections: a review. *Nutrition,* **12**(3):208–213.

Allison, D. G., Ruiz, B. SanJose, C., Jaspe, A., and Gilbert, P. (1998). Extracellular products as mediators of the formation and detachment of *Pseudomonas fluorescens* biofilms. *FEMS Microbiology Letters,* **167**(2): 179–184.

Alpkvist, E., Picioreanu, C., M. van Loosdrecht, M. C., and Heyden, A. (2006). Three-dimensional biofilm model with individual cells and continuum EPS matrix. *Biotechnology and Bioengineering,* **94**(5):961–979.

Anderl, J. N., Franklin, M. J., and Stewart, P. S. (2000). Role of antibiotic penetration limitation in *Klebsiella pneumoniae* biofilm resistance to ampicillin and ciprofloxacin. *Antimicrobial Agents and Chemotherapy,* **44**(7): 1818–1824.

Anderson, G. G. and O'Toole, G. A. (2008). Innate and induced resistance mechanisms of bacterial biofilms. In *Bacterial Biofilms*, ed. T. Romeo, Springer, Berlin Heidelberg, pp. 85–106.

Anderson, G. G., Palermo, J. J., Schilling, J. D., Roth, R., Heuser, J. and Hultgren, S. J. (2003). Intracellular bacterial biofilm-like pods in urinary tract infections. *Science,* **301**(5629):105–107.

Anderson, J. M. and Marchant, R. E. (2000). Biomaterials: factors favoring colonization and infection. In *Infections Associated with Indwelling Medical Devices*, ed. F. A. Waldvogel and A. L. Bisno, ASM Press, Washington, DC, pp. 89–109.

Andes, D., Nett, J., Oschel, P., Albrecht, R., Marchillo, K., and Pitula, A. (2004). Development and characterization of an in vivo central venous catheter *Candida albicans* biofilm model. *Infection and Immunity,* **72**(10):6023–6031.

Bagge, N., Hentzer, M., Andersen, J. B., Ciofu, O., Givskov, M., and Hoiby, N. (2004). Dynamics and spatial fistribution of b-lactamase expression in *Pseudomonas aeruginosa* biofilms. *Antimicrobial Agents and Chemotherapy,* **48**(4):1168–1174.

Baillie, G. S. and Douglas, L. J. (1998). Effect of growth rate on resistance of *Candida albicans* biofilms to antifungal agents. *Antimicrobial Agents and Chemotherapy,* **42**(8):1900–1905.

Baillie, G. S. and Douglas, L. J. (1999). Role of dimorphism in the development of *Candida albicans* biofilms. *Journal of Medical Microbiology,* **48**(7): 671–679.

Baselga, R., Albizu, I., De La Cruz, M., Del Cacho, E., Barberan, M., and Amorena, B. (1993). Phase variation of slime production in *Staphylococcus aureus*: implications in colonization and virulence. *Infection and Immunity,* **61**(11):4857–4862.

Bashan, Y. and Levanony, H. (1988). Active attachment of *Azospirillum brasilense* Cd to quartz sand and to a light-textured soil by protein bridging. *The Journal of General Microbiology,* **134**:2269–2279.

Bauer, T. W., Parvizi, J., Kobayashi, N., and Krebs, V. (2006). Diagnosis of periprosthetic infection. *Journal of Bone and Joint Surgery,* **88**(4):869–882.

Beech, I. B. and Gaylarde, C. C. (1989). Adhesion of *Desulfovibrio desulfuricans* and *Pseudomonas fluorescens* to mild steel surfaces. *The Journal of Applied Bacteriology,* **67**:2017.

Beloin, C., Roux, A., and Ghigo, J. M. (2008). *Escherichia coli* biofilms. *Current Topics in Microbiology and Immunology,* **322**:249–289.

Ben-Yoav, H. and Freeman, A. (2008). Enzymatically attenuated *in situ* release of silver ions to combat bacterial biofilms: a feasibility study. *Journal of Drug Delivery Science and Technology,* **18**(1):25–29.

Bendinger, B., Rijnaarts, H. H. M., Altendorf, K., and Zehnder, A. J. B. (1993). Physicochemical cell surface and adhesive properties of *Coryneform bacteria* related to the presence and chain length of mycolic acids. *Applied and Environmental Microbiology,* **59**(11):3973–3977.

Beuling, E. E., van Dusschoten, D., Lens, P., van den Heuvel, J. C., Van As, H., and Ottengraf, S. P. P. (1998). Characterization of the diffusive properties of biofilms using pulsed field gradient-nuclear magnetic resonance. *Biotechnology and Bioengineering,* **60**(3):283–291.

Beyenal, H. and Lewandowski, Z. (2000). Combined effect of substrate concentration and flow velocity on effective diffusivity in biofilms. *Water Research,* **34**(2):528–538.

Beyenal, H. and Lewandowski, Z. (2001). Mass-transport dynamics, activity, and structure of sulfate-reducing biofilms. *AIChE Journal,* **47**(7): 1689–1697.

Beyenal, H., Donovan, C., Lewandowski, Z., and Harkin, G. (2004a). Three-dimensional biofilm structure quantification. *Journal of Microbiological Methods,* **59**(3):395–413.

Beyenal, H., Lewandowski, Z., and Harkin, G. (2004b). Quantifying biofilm structure: facts and fiction. *Biofouling,* **20**:1–23.

Bishop, P. L. (1997). Biofilm structure and kinetics. *Water Science and Technology,* **36**(1):287–294.

Bishop, P. L. and Rittmann, B. E. (1995). Modelling heterogeneity in biofilms: report of the discussion session. *Water Science and Technology,* **32**(8): 263–265.

Boyd, A. and Chakrabarty, A. M. (1994). Role of alginate lyase in cell detachment of *Pseudomonas aeruginosa. Applied and Environmental Microbiology,* **60**:2355–2359.

Brading, M. G., Jass, J., and Lappin-Scott, H. M. (1995). Dynamics of bacterial biofilm formation. In *Microbial Biofilms*, ed. H. M. Lappin-Scott and J. W. Costerton, Cambridge University Press, Cambridge, pp. 46–63.

Brazil, G., Kenefick, L., Callanan, M., Haro, A., de Lorenzo, V., Dowling, D., and O'Gara, F. (1995). Construction of a rhizosphere pseudomonad with potential to degrade polychlorinated biphenyls and detection of bph gene expression in the rhizosphere. *Applied and Environmental Microbiology,* **61**(5):1946–1952.

Brooun, A., Liu, S., and Lewis, K. (2000). A dose-response study of antibiotic resistance in *Pseudomonas aeruginosa* biofilms. *Antimicrobial agents and chemotherapy,* **44**:640–646.

Bryers, J. D. (2008). Medical biofilms. *Biotechnology and Bioengineering,* **100**(1):1–18.

Bryers, J. D. and Drummond, F. (1998). Local macromolecule diffusion coefficients in structurally non-uniform bacterial biofilms using fluorescence recovery after photobleaching (FRAP). *Biotechnology and Bioengineering,* **60**(4):462–473.

Buswell, C. M., Herlihy, Y. M., Lawrence, L. M., McGuiggan, J. T. M., Marsh, P. D., Keevil, C. W., and Leach, S. A. (1998). Extended survival and persistence of *Campylobacter* spp. in water and aquatic biofilms and their detection by immunofluorescent-antibody and -rRNA staining. *Applied and Environmental Microbiology,* **64**(2):733–741.

Camper, A. K., Warnecke, M., Jones, W. L., and McFeters, G. A. (1998). *Pathogens in Model Distribution System Biofilms.* American Water Works Association Research Foundation, Denver.

Caubet, R., Pedarros-Caubet, F., Chu, M., Freye, E., de Belem Rodrigues, M., Moreau, J. M., and Ellison, W. J. (2004). A radio frequency electric current enhances antibiotic efficacy against bacterial biofilms. *Antimicrobial Agents and Chemotherapy,* **48**(12):4662–4664.

Cerca, N., Martins, S., Cerca, F., Jefferson, K. K., Pier, G. B., Oliveira, R., and Azeredo, J. (2005). Comparative assessment of antibiotic susceptibility of coagulase-negative *Staphylococci* in biofilm versus planktonic culture as assessed by bacterial enumeration or rapid XTT colorimetry. *The Journal of Antimicrobial Chemotherapy,* **56**(2):331–336.

Chandra, J. and Ghannoum, A. (2004). Fungal biofilms. In *Microbial Biofilms,* eds. M. Ghannoum and G. A. O'Toole. ASM Press, Washington, DC, pp. 30–42.

Chandra, J., Kuhn, D. M., Mukherjee, P. K., Hoyer, L. L., McCormick, T., and Ghannoum, M. A. (2001). Biofilm formation by the fungal pathogen Candida albicans: development, architecture, and drug resistance. *Journal of Bacteriology,* **183**(18):5385–5394.

Characklis, W. G. (1990a). Biofilm processes. In *Biofilms,* ed. W. G. Characklis and K. C. Marshall. John Wiley & Sons, New York, pp. 195–231.

Characklis, W. G. (1990b). Microbial fouling. In *Biofilms,* ed. W. G. Characklis, and K. C. Marshall. John Wiley & Sons, New York, pp. 523–584.

Characklis, W. G., McFeters, G. A., Marshall, K. C., Characklis, W. G., and Marshall, K. C. (1990). Physiological ecology in biofilm systems. In *Biofilms,* ed. John Wiley & Sons, New York, pp. 341–394.

Chatterjee, I., Herrmann, M., Proctor, R. A., Peters, G., and Kahl, B. C. (2007). Enhanced post-stationary-phase survival of a clinical thymidine-dependent small-colony variant of *Staphylococcus aureus* results from lack of a functional tricarboxylic acid cycle. *Journal of Bacteriology,* **189**(7):2936–2940.

Chavant, P., Martinie, B., Meylheuc, T., Bellon-Fontaine, M.-N., and Hebraud, M. (2002). Listeria monocytogenes LO28: surface physicochemical properties and ability to form biofilms at different temperatures and growth phases. *Applied and Environmental Microbiology,* **68**(2):728–737.

Chopra, I. (2007). The increasing use of silver-based products as antimicrobial agents: a useful development or a cause for concern? *The Journal of Antimicrobial Chemotherapy,* **59**(4):587–590.

Cirioni, O., Giacometti, A., Ghiselli, R., Dell'Acqua, G., Orlando, F., Mocchegiani, F., Silvestri, C., Licci, A., Saba, V., Scalise, G., and Balaban, N. (2006). RNAIII-inhibiting peptide significantly reduces bacterial load and enhances the effect of antibiotics in the treatment of central venous catheter-associated *Staphylococcus aureus* infections. *The Journal of Infectious Diseases,* **193**(2): 180–186.

Cogan, N., Cortez, R., and Fauci, L. (2005). Modeling physiological resistance in bacterial biofilms. *Bulletin of Mathematical Biology,* **67**(4):831–853.

Costerton, J. W., Ellis, B., Lam, K., Johnson, F., and Khoury, A. E. (1994). Mechanism of electrical enhancement of efficacy of antibiotics in killing biofilm bacteria. *Antimicrobial Agents and Chemotherapy,* **38**(12): 2803–2809.

Costerton, J. W., Lewandowski, Z., Caldwell, D. E., Korber, D. R., and Lappin-Scott, H. M. (1995). Microbial biofilms. *Annual Review of Microbiology,* **49**:711–736.

Costerton, J. W., Stewart, P. S., and Greenberg, E. P. (1999). Bacterial biofilms: a common cause of persistent infections. *Science,* **284**(5418): 1318–1322.

Cowan, M. M., Warren, T. M., and Fletcher, M. (1991). Mixed species colonization of solid surfaces in laboratory biofilms. *Biofouling,* **3**:23–34.

Cramton, S. E., Gerke, C., Schnell, N. F., Nichols, W. W., and Gotz, F. (1999). The intercellular adhesion (ica) locus is present in *Staphylococcus aureus* and is required for biofilm formation. *Infection and Immunity,* **67**(10): 5427–5433.

Cronenberg, C. C. H. and van den Heuvel, J. C. (1991). Determination of glucose diffusion coefficients in biofilms with micro-electrodes. *Biosensors and Bioelectronics,* **6**(3):255–262.

Dagan-Moscovich, H., Cohen-Hadar, N., Porat, C., Rishpon, J., Shacham-Diamand, Y., and Freeman, A. (2007). Nanowiring of the catalytic site of novel molecular enzyme-metal hybrids to electrodes. *Journal of Physical Chemistry,* ***C*** **111**(15):5766–5769.

Darouiche, R. O., Dhir, A., Miller, A. J., Landon, G. C., Raad, II, and Musher, D. M. (1994). Vancomycin penetration into biofilm covering infected prostheses and effect on bacteria. *The Journal of Infectious Diseases,* **170**(3):720–723.

Davies, D. (2003). Understanding biofilm resistance to antibacterial agents. *Nature Reviews Drug Discovery,* **2**(2):114–122.

de Beer, D. and Stoodley, P. (1995). Relation between the structure of an aerobic biofilm and transport phenomena. *Water Science and Technology,* **32**(8):11–18.

de Beer, D. and Stoodley, P. (2006). Microbial biofilms. In *The Prokaryotes*, ed. M. Dworkin, S. Falkow, E. Rosenberg, K.-H. Schleifer, and E. Stackebrandt. Springer, New York, pp. 904–937.

de Beer, D., Srinivasan, R., and Stewart, P. S. (1994a). Direct measurement of chlorine penetration into biofilms during disinfection. *Applied and Environmental Microbiology,* **60**(12):4339–4344.

de Beer, D., Stoodley, P., Roe, F., and Lewandowski, Z. (1994b). Effects of biofilm structures on oxygen distribution and mass transport. *Biotechnology and Bioengineering,* **43**(11):1131–1138.

De Kievit, T. R., Parkins, M. D., Gillis, R. J., Srikumar, R., Ceri, H., Poole, K., Iglewski, B. H., and Storey, D. G. (2001). Multidrug efflux pumps: expression patterns and contribution to antibiotic resistance in *Pseudomonas aeruginosa* biofilms. *Antimicrobial Agents and Chemotherapy,* **45**(6): 1761–1770.

Deighton, M. A. and Balkau, B. (1990). Adherence measured by microtiter assay as a virulence marker for *Staphylococcus epidermidis* infections. *Journal of Clinical Microbiology,* **28**(11):2442–2447.

Deighton, M. and Borland, R. (1993). Regulation of slime production in *Staphylococcus epidermidis* by iron limitation. *Infection and Immunity,* **61**(10):4473–4479.

del Pozo, J. L. and Patel, R. (2007). The challenge of treating biofilm-associated bacterial infections. *Clinical Pharmacology and Therapeutics,* **82**(2):204–209.

Demaret, L., Eberl, H., Efendiev, M., and Lasser, R. (2008). Analysis and simulation of a meso-scale model of diffusive resistance of bacterial biofilms to penetration of antibiotics. *Advances in Mathematical Sciences and Applications,* **18**(1):269–304.

Dodds, M. G., Grobe, K. J., and Stewart, P. S. (2000). Modeling biofilm antimicrobial resistance. *Biotechnology and Bioengineering,* **68**(4):456–465.

Dong, Y. H. and Zhang, L. H. (2005). Quorum sensing and quorum-quenching enzymes. *Journal of Microbiology,* **43**, Spec No:101–109.

Donlan, R. M. (2002). Biofilms: microbial life on surfaces. *Emerging Infectious Diseases,* **8**(9):881–890.

Donlan, R. M. and Costerton, J. W. (2002). Biofilms: survival mechanisms of clinically relevant microorganisms. *Clinical Microbiology Reviews,* **15**(2):167–193.

Donlan, R. M., Pipes, W. O., and Yohe, T. L. (1994). Biofilm formation on cast iron substrata in water distribution systems. *Water Research,* **28**(6): 1497–1503.

Dowd, S., Sun, Y., Secor, P., Rhoads, D., Wolcott, B., James, G., and Wolcott, R. (2008). Survey of bacterial diversity in chronic wounds using Pyrosequencing, DGGE, and full ribosome shotgun sequencing. *BMC Microbiology,* **8**(1):43.

Drancourt, M., Stein, A., Argenson, J., Roiron, R., Groulier, P., and Raoult, D. (1997). Oral treatment of *Staphylococcus* spp. infected orthopaedic implants with fusidic acid or ofloxacin in combination with rifampicin. *The Journal of Antimicrobial Chemotherapy,* **39**(2):235–240.

Drenkard, E. (2003). Antimicrobial resistance of *Pseudomonas aeruginosa* biofilms. *Microbes and Infection,* **5**(13):1213–1219.

Drenkard, E. and Ausubel, F. M. (2002). Pseudomonas biofilm formation and antibiotic resistance are linked to phenotypic variation. *Nature,* **416**(6882):740–743.

Duguid, I. G., Evans, E., Brown, M. R. W., and Gilbert, P. (1992). Effect of biofilm culture upon the susceptibility of *Staphylococcus epidermidis* to tobramycin. *The Journal of Antimicrobial Chemotherapy,* **30**:803–810.

Dunny, G. M. and Leonard, B. A. B. (1997). Cell-cell communication in Gram-positive bacteria. *Annual Review of Microbiology,* **51**(1):527–564.

Elkins, J. G., Hassett, D. J., Stewart, P. S., Schweizer, H. P., and McDermott, T. R. (1999). Protective role of catalase in *Pseudomonas aeruginosa* biofilm resistance to hydrogen peroxide. *Applied and Environmental Microbiology,* **65**(10):4594–4600.

Falagas, M. E., Fragoulis, K., Bliziotis, I. A., and Chatzinikolaou, I. (2007). Rifampicin-impregnated central venous catheters: a meta-analysis of randomized controlled trials. *Journal of Antimicrobial Chemotherapy,* **59**(3):359–369.

Fletcher, M. (1988). Attachment of Pseudomonas fluorescens to glass and influence of electrolytes on bacterium-substratum separation distance. *Journal of Bacteriology,* **170**:2027–2030.

Fletcher, M., Lessman, J. M., and Loeb, G. I. (1991). Bacterial surface adhesives and biofilm matrix polymers of marine and freshwater bacteria. *Biofouling,* **4**:129–140.

Fouad, M. and Bhargava, R. (2005). A simplified model for the steady-state biofilm-activated sludge reactor. *Journal of Environmental Management,* **74**(3):245–253.

Fux, C. A., Costerton, J. W., Stewart, P. S., and Stoodley, P. (2005). Survival strategies of infectious biofilms. *Trends in Microbiology,* **13**(1):34–40.

Gapes, D., Prez, J., Picioreanu, C., and van Loosdrecht, M. (2006). Corrigendum to "Modeling biofilm and floc diffusion processes based on analytical solution of reaction-diffusion equations" [*Water Res*earch, **39**(2005): 1311–1323]. *Water Research,* **40**(16):3144–3145.

Geesey, G. G. (2001). Bacterial behavior at surfaces. *Current Opinion in Microbiology,* **4**(3):296–300.

Ghannoum, M. and O'Toole, G. A. (2004). *Microbial Biofilms.* ASM Press, Washington, DC.

Gilbert, P., Evans, D. J., and Brown, M. R. W. (1993). Formation and dispersal of bacterial biofilms in vivo and in situ. *Journal of Applied Microbiology,* **74**(S22):67S–78S.

Gillis, R. J., White, K. G., Choi, K.-H., Wagner, V. E., Schweizer, H. P., and Iglewski, B. H. (2005). Molecular basis of azithromycin-resistant *Pseudomonas aeruginosa* biofilms. *Antimicrobial Agents and Chemotherapy,* **49**(9):3858–3867.

Giwercman, B., Jensen, E. T., Hoiby, N., Kharazmi, A., and Costerton, J. W. (1991). Induction of beta-lactamase production in *Pseudomonas aeruginosa* biofilm. *Antimicrobial Agents and Chemotherapy,* **35**(5):1008–1010.

Goodman, A. E., Marshall, K. C., and Hermansson, M. (1994). Gene transfer among bacteria under conditions of nutrient depletion in simulated and natural aquatic environments. *FEMS Microbiology Ecology,* **15**(1–2): 55–60.

Gordon, C. A., Hodges, N. A., and Marriott, C. (1988). Antibiotic interaction and diffusion through alginate and exopolysaccharide of cystic fibrosis-derived *Pseudomonas aeruginosa. Journal of Antimicrobial Chemotherapy,* **22**(5):667–674.

Gujer, W. and Wanner, O. (1990). Modeling mixed population biofilms. In *Biofilms*, ed. W. G. Characklis and K. C. Marshall. John Wiley & Sons, New York, pp. 397–443.

Hancock, V. and Klemm, P. (2007). Global gene expression profiling of asymptomatic bacteriuria *Escherichia coli* during biofilm growth in human urine. *Infection and Immunity,* **75**(2):966–976.

Hansen, S. K., Rainey, P. B., Haagensen, J. A. J., and Molin, S. (2007). Evolution of species interactions in a biofilm community. *Nature,* **445**(7127): 533–536.

Hassett, D. J., Ma, J.-F., Elkins, J. G., McDermott, T. R., Ochsner, U. A., West, S. E. H., Huang, C.-T., Fredericks, J., Burnett, S., Stewart, P. S., McFeters, G., Passador, L., and Iglewski, B. H. (1999). Quorum sensing in *Pseudomonas aeruginosa* controls expression of catalase and superoxide dismutase genes and mediates biofilm susceptibility to hydrogen peroxide. *Molecular Microbiology,* **34**(5):1082–1093.

Hatch, R. A. and Schiller, N. L. (1998). Alginate lyase promotes diffusion of aminoglycosides through the extracellular polysaccharide of mucoid *Pseudomonas aeruginosa. Antimicrobial Agents and Chemotherapy,* **42**(4): 974–977.

Hawser, S. and Islam, K. (2006). Candida. In *Biofilms, Infection, and Antimicrobial Therapy,* ed. J. L. Pace, M. E. Rupp, and R. G. Finch. CRC Press, Boca Raton, FL, pp. 171–184.

Hawser, S. P. and Douglas, L. J. (1994). Biofilm formation by Candida species on the surface of catheter materials in vitro. *Infection and Immunity,* **62**(3):915–921.

Hawser, S. P. and Douglas, L. J. (1995). Resistance of Candida albicans biofilms to antifungal agents in vitro. *Antimicrobial Agents and Chemotherapy,* **39**(9):2128–2131.

Heilmann, C. and Gotz, F. (1998). Further characterization of *Staphylococcus epidermidis* transposon mutants deficient in primary attachment or intercellular adhesion. *Zentralblatt für Bakteriologie: International Journal of Medical Microbiology,* **287**(1–2):69–83.

Heilmann, C., Hussain, M., Peters, G., and Gotz, F. (1997). Evidence for autolysin-mediated primary attachment of *Staphylococcus epidermidis* to a polystyrene surface. *Molecular Microbiology,* **24**(5):1013–1024.

Heydorn, A., Nielsen, A. T., Hentzer, M., Sternberg, C., Givskov, M., Ersboll, B. K., and Molin, S. (2000). Quantification of biofilm structures by the novel computer program COMSTAT. *Microbiology,* **146**(10):2395–2407.

Hinsa, S. M. and O'Toole, G. A. (2006). Biofilm formation by Pseudomonas fluorescens WCS365: a role for LapD. *Microbiology,* **152**(5):1375–1383.

Hinson, R. K. and Kocher, W. M. (1996). Model for effective diffusivities in aerobic biofilms. *Journal of Environmental Engineering,* **122**(11): 1023–1030.

Hoffman, L. R., D'Argenio, D. A., MacCoss, M. J., Zhang, Z., Jones, R. A., and Miller, S. I. (2005). Aminoglycoside antibiotics induce bacterial biofilm formation. *Nature,* **436**(7054):1171–1175.

Hood, S. K. and Zottola, E. A. (1997). Adherence to stainless steel by food-borne microorganisms during growth in model food systems. *International Journal of Food Microbiology,* **37**(2–3):145–153.

Horn, H. and Hempel, D. C. (1997). Substrate utilization and mass transfer in an autotrophic biofilm system: experimental results and numerical simulation. *Biotechnology and Bioengineering,* **53**(4):363–371.

Hornemann, J. A., Lysova, A. A., Codd, S. L., Seymour, J. D., Busse, S. C., Stewart, P. S., and Brown, J. R. (2008). Biopolymer and water dynamics in microbial biofilm extracellular polymeric substance. *Biomacromolecules,* **9**(9):2322–2328.

Hoyle, B. D., Wong, C. K., and Costerton, J. W. (1992). Disparate efficacy of tobramycin on Ca(2+)-, Mg(2+)-, and HEPES-treated *Pseudomonas aeruginosa* biofilms. *Canadian Journal of Microbiology,* **38**(11):1214–1218.

Huang, C., Yu, F., McFeters, G., and Stewart, P. (1995). Nonuniform spatial patterns of respiratory activity within biofilms during disinfection. *Applied and Environmental Microbiology,* **61**(6):2252–2256.

Hussain, M., Herrmann, M., von Eiff, C., Perdreau-Remington, F., and Peters, G. (1997). A 140-kilodalton extracellular protein is essential for the accumulation of *Staphylococcus epidermidis* strains on surfaces. *Infection and Immunity,* **65**(2):519–524.

Ishida, H., Ishida, Y., Kurosaka, Y., Otani, T., Sato, K., and Kobayashi, H. (1998). In vitro and in vivo activities of levofloxacin against biofilm-producing *Pseudomonas aeruginosa. Antimicrobial Agents and Chemotherapy,* **42**(7):1641–1645.

Isiklar, Z. U., Demirors, H., Akpinar, S., Tandogan, R. N., and Alparslan, M. (1999). Two-stage treatment of chronic Staphylococcal orthopaedic implant-related infections using vancomycin impregnated PMMA spacer and rifampin containing antibiotic protocol. *Bulletin (Hospital for Joint Diseases (New York, N.Y.)),* **58**(2):79–85.

Jass, J., Costerton, J. W., and Lappin-Scott, H. M. (1995). The effect of electrical currents and tobramycin on *Pseudomonas aeruginosa* biofilms. *Journal of Industrial Microbiology and Biotechnology,* **15**(3): 234–242.

Kaplan, J. B., Ragunath, C., Ramasubbu, N., and Fine, D. H. (2003). Detachment of Actinobacillus actinomycetemcomitans biofilm cells by an endogenous {beta}-hexosaminidase activity. *Journal of Bacteriology,* **185**(16):4693–4698.

Karchmer, A. W. (2000). Infections of prosthetic heart valves. In *Infections Associated with Indwelling Medical Devices*, ed. F. A. Waldvogel and A. L. Bisno. ASM Press, Washington, DC, pp. 145–172.

Kissel, J. C., McCarty, P. L., and Street, R. L. (1984). Numerical simulation of mixed-culture biofilm. *Journal of Environmental Engineering,* **110**(2): 393–411.

Kojic, E. M. and Darouiche, R. O. (2004). Candida infections of medical devices. *Clinical Microbiology Reviews,* **17**(2):255–267.

Korber, D. R., Lawrence, J. R., Sutton, B., and Caldwell, D. E. (1989). Effect of laminar flow velocity on the kinetics of surface recolonization by Mot^+ and Mot^- *Pseudomonas fluorescens. Microbial Ecology,* **18**: 1–19.

Kumon, H., Tomochika, K., Matunaga, T., Ogawa, M., and Ohmori, H. (1994). A sandwich cup method for the penetration assay of antimicrobial agents through Pseudomonas exopolysaccharides. *Microbiology and Immunology,* **38**(8):615–619.

Lee, M. W. and Park, J. M. (2007). One-dimensional mixed-culture biofilm model considering different space occupancies of particulate components. *Water Research,* **41**(19):4317–4328.

Lens, P. N. L., Moran, A. P., Mahony, T., O'flaherty, V., and Stoodley, P. (2003). *Biofilms in Medicine, Industry and Environmental Biotechnology: Characteristics, Analysis and Control.* IWA Publishing, London.

Lewandowski, Z. (2000). Notes on biofilm porosity. *Water Research,* **34**(9): 2620–2624.

Lewandowski, Z. and Stoodley, P. (1995). Flow induced vibrations, drag force, and pressure drop in conduits covered with biofilm. *Water Science and Technology,* **32**(8):19–26.

Lewandowski, Z., Stoodley, P., and Altobelli, S. (1995). Experimental and conceptual studies on mass transport in biofilms. *Water Science and Technology,* **31**(1):153–162.

Lewandowski, Z., Walser, G., and Characklis, W. G. (1991). Reaction kinetics in biofilms. *Biotechnology and Bioengineering,* **38**(8):877–882.

Lewandowski, Z., Webb, D., Hamilton, M., and Harkin, G. (1999). Quantifying biofilm structure. *Water Science and Technology,* **39**(7):71–76.

Lewis, K. (2001). Riddle of biofilm resistance. *Antimicrobial Agents and Chemotherapy,* **45**(4):999–1007.

Libicki, S. B., Salmon, P. M., and Robertson, C. R. (1988). The effective diffusive permeability of a nonreacting solute in microbial cell aggregates. *Biotechnology and Bioengineering,* **32**(1):68–85.

Lynch, A. S. and Robertson, G. T. (2008). Bacterial and fungal biofilm infections. *Annual Review of Medicine,* **59**(1):415–428.

Mah, T.-F. C. and O'Toole, G. A. (2001). Mechanisms of biofilm resistance to antimicrobial agents. *Trends in Microbiology,* **9**(1):34–39.

Mah, T.-F., Pitts, B., Pellock, B., Walker, G. C., Stewart, P. S., and O'Toole, G. A. (2003). A genetic basis for *Pseudomonas aeruginosa* biofilm antibiotic resistance. *Nature,* **426**(6964):306–310.

Majors, P. D., McLean, J. S., Fredrickson, J. K., and Wind, R. A. (2005). NMR methods for in-situ biofilm metabolism studies: spatial and temporal resolved measurements. *Water Science and Technology,* **52**:7–12.

Manierski, C. and Besarab, A. (2006). Antimicrobial locks: putting the lock on catheter infections. *Advances in Chronic Kidney Disease,* **13**(3):245–258.

Mansson, L. E., Melican, K., Boekel, J., Sandoval, R. M., Hautefort, I., Tanner, G. A., Molitoris, B. A., and Richter-Dahlfors, A. (2007). Real-time studies of the progression of bacterial infections and immediate tissue responses in live animals. *Cellular Microbiology,* **9**(2):413–424.

Marsh, P. D. (1995). Dental plaque. In *Microbial biofilms*, ed. H. M. Lappin-Scott and J. W. Costerton. Cambridge University Press, Cambridge, pp. 282–300.

Martinez, M., Lopez-Ribot, J. L., Kirkpatrick, W. R., Coco, B. J., Bachmann, S. P., and Patterson, T. F. (2002). Replacement of *Candida albicans* with *C. dubliniensis* in human immunodeficiency virus-infected patients with oropharyngeal candidiasis treated with fuconazole. *Journal of Clinical Microbiology,* **40**(9):3135–3139.

Matson, J. V. and Characklis, W. G. (1976). Diffusion into microbial aggregates. *Water Research,* **10**(10):877–885.

McKenney, D., Hubner, J., Muller, E., Wang, Y., Goldmann, D. A., and Pier, G. B. (1998). The ica locus of *Staphylococcus epidermidis* encodes production of the capsular polysaccharide/adhesin. *Infection and Immunity,* **66**(10):4711–4720.

McLean, J. S., Ona, O. N., and Majors, P. D. (2007). Correlated biofilm imaging, transport and metabolism measurements via combined nuclear magnetic resonance and confocal microscopy. *The ISME Journal,* **2**(2): 121–131.

McLeod, B. R., Fortun, S., Costerton, J. W., and Stewart, P. S. (1999). Enhanced bacterial biofilm control using electromagnetic fields in combination with antibiotics. *Methods in Enzymology,* **310**:656–670.

Meers, P., Neville, M., Malinin, V., Scotto, A. W., Sardaryan, G., Kurumunda, R., Mackinson, C., James, G., Fisher, S., and Perkins, W. R. (2008). Biofilm penetration, triggered release and in vivo activity of inhaled liposomal amikacin in chronic *Pseudomonas aeruginosa* lung infections. *Journal of Antimicrobial Chemotherapy,* **61**(4):859–868.

Mittelman, M. W. (1996). Adhesion to biomaterials. In *Bacterial Adhesion: Molecular and Ecological Diversity,* ed. M. Fletcher. Wiley-Liss, New York, Inc., pp. 89–127.

Moller, S., Sternberg, C., Andersen, J. B., Christensen, B. B., Ramos, J. L., Givskov, M., and Molin, S. (1998). In situ gene expression in mixed-culture biofilms: evidence of metabolic interactions between community members. *Applied and Environmental Microbiology,* **64**(2):721–732.

Moran, G. P., Sanglard, D., Donnelly, S. M., Shanley, D. B., Sullivan, D. J., and Coleman, D. C. (1998). Identification and expression of multidrug transporters responsible for fluconazole resistance in *Candida dubliniensis. Antimicrobial Agents and Chemotherapy,* **42**(7):1819–1830.

Murga, R., Forster, T. S., Brown, E., Pruckler, J. M., Fields, B. S., and Donlan, R. M. (2001). The role of biofilms in the survival of *Legionella pneumophila* in a model potable-water system. *Microbiology,* **147**(11):3121–3126.

National Institutes of Health. April 21, 1999. SBIR/STTR Study and Control of Microbial Biofilms. Access Date: March 27, 2009. http://grants.nih.gov/grants/guide/pa-files/PA-99–084.html.

Nemoto, K., Hirota, K., Ono, T., Murakami, K., Nagao, D., and Miyake, Y. (2000). Effect of varidase (streptokinase) on biofilm formed by *Staphylococcus aureus. Chemotherapy,* **46**(2):111–115.

Nett, J. and Andes, D. (2008). Review of techniques for diagnosis of catheter-related *Candida* biofilm infections. *Current Fungal Infection Reports,* **2**(4):237–243.

O'Toole, G. A. and Kolter, R. (1998). Flagellar and twitching motility are necessary for *Pseudomonas aeruginosa* biofilm development. *Molecular Microbiology,* **30**(2):295–304.

O'Toole, G. A., Gibbs, K. A., Hager, P. W., Phibbs, P. V. Jr., and Kolter, R. (2000a). The global carbon metabolism regulator Crc is a component of a signal transduction pathway required for biofilm development by *Pseudomonas aeruginosa. Journal of Bacteriology,* **182**(2):425–431.

O'Toole, G. A., Kaplan, H. B., and Kolter, R. (2000b). Biofilm formation as microbial development. *Annual Review of Microbiology,* **54**:49–79.

Ofek, I. and Doyle, R. J. (1994). *Bacterial Adhesion to Cells and Tissues.* Chapman & Hall, New York.

Otto, K. and Silhavy, T. J. (2002). Surface sensing and adhesion of Escherichia coli controlled by the Cpx-signaling pathway. *Proceedings of the National Academy of Sciences, USA*, **99**(4):2287–2292.

Pace, J. L., Rupp, M. E., and Finch, R. G. (2006). *Biofilm, Infection, and Antimicrobial Therapy.* CRC Press, Boca Raton, FL.

Parsek, M. R. and Greenberg, E. P. (2005). Sociomicrobiology: the connections between quorum sensing and biofilms. *Trends in Microbiology,* **13**(1): 27–33.

Parsek, M. R., Val, D. L., Hanzelka, B. L., Cronan, J. E., and Greenberg, E. P. (1999). Acyl homoserine-lactone quorum-sensing signal generation. *Proceedings of the National Academy of Sciences, USA*, **96**(8):4360–4365.

Payne, C., Bladin, C., Colchester, A. C. F., Bland, J., Lapworth, R., and Lane, D. (1992). Argyria from excessive use of topical silver sulfadiazine. *Lancet,* **340**(8811):126–126.

Picioreanu, C., van Loosdrecht, M. C. M., and Heijnen, J. J. (1998). Mathematical modeling of biofilm structure with a hybrid differential-discrete cellular automaton approach. *Biotechnology and Bioengineering,* **58**(1): 101–116.

Pirnay, J.-P., De Vos, D., Cochez, C., Bilocq, F., Pirson, J., Struelens, M., Duinslaeger, L., Cornelis, P., Zizi, M., and Vanderkelen, A. (2003). Molecular epidemiology of *Pseudomonas aeruginosa* colonization in a burn unit: persistence of a multidrug-resistant clone and a silver sulfadiazine-resistant clone. *Journal of Clinical Microbiology,* **41**(3):1192–1202.

Pratt, L. A. and Kolter, R. (1998). Genetic analysis of *Escherichia coli* biofilm formation: roles of flagella, motility, chemotaxis and type I pili. *Molecular Microbiology,* **30**(2):285–293.

Prince, A. S. (2002). Biofilms, antimicrobial resistance, and airway infection. *The New England Journal of Medicine,* **347**(14):1110–1111.

Proctor, R. A., von Eiff, C., Kahl, B. C., Becker, K., McNamara, P., Herrmann, M., and Peters, G. (2006). Small colony variants: a pathogenic form of bacteria that facilitates persistent and recurrent infections. *Nature Reviews. Microbiology,* **4**(4):295–305.

Purevdorj, B., Costerton, J. W., and Stoodley, P. (2002). Influence of hydrodynamics and cell signaling on the structure and behavior of *Pseudomonas aeruginosa* biofilms. *Applied and Environmental Microbiology,* **68**(9): 4457–4464.

Qi, S. and Morgenroth, E. (2005). Modeling steady-state biofilms with dual-substrate limitations. *Journal of Environmental Engineering,* **131**(2): 320–326.

Quindos, G., Carrillo-Munoz, A. J., Arevalo, M. P., Salgado, J., Alonso-Vargas, R., Rodrigo, J. M., Ruesga, M. T., Valverde, A., Peman, J., Canton, E., Martin-Mazuelos, E., and Ponton, J. (2000). In vitro susceptibility of *Candida dubliniensis* to current and new antifungal agents. *Chemotherapy,* **46**(6):395–401.

Raad, I. I., Sabbagh, M. F., Rand, K. H., and Sherertz, R. J. (1992). Quantitative tip culture methods and the diagnosis of central venous catheter-related infections. *Diagnostic Microbiology and Infectious Disease,* **15**(1):13–20.

Raad, I., R. Darouiche, R., Hachem, R., Sacilowski, M., and Bodey, G. (1995). Antibiotics and prevention of microbial colonization of catheters. *Antimicrobial Agents and Chemotherapy,* **39**(11):2397–2400.

Ramage, G., Saville, S. P., Thomas, D. P., and Lopez-Ribot, J. L. (2005). Candida biofilms: an update. *Eukaryotic Cell,* **4**(4):633–638.

Ramage, G., Vande Walle, K., Wickes, B. L., and Lopez-Ribot, J. L. (2001). Biofilm formation by *Candida dubliniensis. Journal of Clinical Microbiology,* **39**(9):3234–3240.

Rediske, A. M., Roeder, B. L., Nelson, J. L., Robison, R. L., Schaalje, G. B., Robison, R. A., and Pitt, W. G. (2000). Pulsed ultrasound enhances the killing of *Escherichia coli* biofilms by aminoglycoside antibiotics in vivo. *Antimicrobial Agents and Chemotherapy,* **44**(3):771–772.

Revsbech, N. P. (2005). Analysis of microbial communities with electrochemical microsensors and microscale biosensors. *Methods in Enzymology,* **397**:147–166.

Reysenbach, A.-L. and Shock, E. (2002). Merging genomes with geochemistry in hydrothermal ecosystems. *Science,* **296**(5570):1077–1082.

Rijnaarts, H. H. M., Norde, W., Bouwer, E. J., Lyklema, J., and Zehnder, A. J. B. (1993). Bacterial adhesion under static and dynamic conditions. *Applied and Environmental Microbiology,* **59**(10):3255–3265.

Rittmann, B. E. (1993). The significance of biofilms in porous media. *Water Resources Research,* **29**(7):2195–2202.

Rittmann, B. E. and Manem, J. A. (1992). Development and experimental evaluation of a steady-state, multispecies biofilm model. *Biotechnology and Bioengineering,* **39**(9):914–922.

Rittmann, B. E. and McCarty, P. L. (1980). Evaluation of steady-state-biofilm kinetics. *Biotechnology and Bioengineering,* **22**(11):2359–2373.

Roberts, M. E. and Stewart, P. S. (2004). Modeling antibiotic tolerance in biofilms by accounting for nutrient limitation. *Antimicrobial Agents and Chemotherapy,* **48**(1):48–52.

Roche, D. M., Byers, J. T., Smith, D. S., Glansdorp, F. G., Spring, D. R., and Welch, M. (2004). Communications blackout? Do N-acylhomoserine-lactone-degrading enzymes have any role in quorum sensing? *Microbiology,* **150**(7):2023–2028.

Roggenkamp, A., Sing, A., Hornef, M., Brunner, U., Autenrieth, I. B., and Heesemann, J. (1998). Chronic prosthetic hip infection caused by a small-colony variant of *Escherichia coli. Journal of Clinical Microbiology,* **36**(9):2530–2534.

Romeo, T. (2008). *Bacterial Biofilms*: Springer, Berlin Heidelberg.

Rosenberg, M. and Kjelleberg, S. (1986). Hydrophobic interactions in bacterial adhesion. *Advances in Microbial Ecology,* **9**:353–393.

Schembri, M. A., Givskov, M., and Klemm, P. (2002). An attractive surface: gram-negative bacterial biofilms. *Science's STKE,* **132**:re6.

Schrenzel, J., Harbarth, S., Schockmel, G., Genne, D., Bregenzer, T., Flueckiger, U., Petignat, C., Jacobs, F., Francioli, P., Zimmerli, W., and Lew, D. P. (2004). A randomized clinical trial to compare fleroxacin-rifampicin with flucloxacillin or vancomycin for the treatment of staphylococcal infection. *Clinical Infectious Diseases,* **39**(9):1285–1292.

Schumacher-Perdreau, F., Heilmann, C., Peters, G., Gotz, F., and Pulverer, G. (1994). Comparative analysis of a biofilm-forming *Staphylococcus epidermidis* strain and its adhesion-positive, accumulation-negative mutant M7. *FEMS Microbiology Letters,* **117**(1):71–78.

Selan, L., Kofonow, J., Scoarughi, G. L., Vail, T., Leid, J. G., and Artini, M. (2008). Use of immunodiagnostics for the early detection of biofilm infections. In *The Role of Biofilms in Device-Related Infections*, ed. M. Shirtliff and J. G. Leid, Springer, Berlin Heidelberg, pp. 219–238.

Seymour, J. D., Codd, S. L., Gjersing, E. L., and Stewart, P. S. (2004). Magnetic resonance microscopy of biofilm structure and impact on transport in a capillary bioreactor. *Journal of Magnetic Resonance,* **167**(2):322–327.

Shay, D. K., Goldmann, D. A., and Jarvis, W. R. (1995). Reducing the spread of antimicrobial-resistant microorganisms. Control of vancomycin-resistant enterococci. *Pediatric Clinics of North America,* **42**(3):703–716.

Shigeta, M., Tanaka, G., Komatsuzawa, H., Sugai, M., Suginaka, H., and Usui, T. (1997). Permeation of antimicrobial agents through Pseudomonas aeruginosa biofilms: a simple method. *Chemotherapy,* **43**(5):340–5.

Shirtliff, M. and Leid, J. G. (2009). *The Role of Biofilms in Device-Related Infections.* Springer, Berlin Heidelberg.

Shuford, J. A., Rouse, M. S., Piper, K. E., Steckelberg, J. M., and Patel, R. (2006). Evaluation of caspofungin and amphotericin B deoxycholate against *Candida albicans* biofilms in an experimental intravascular catheter infection model. *The Journal of Infectious Diseases,* **194**(5): 710–713.

Siegrist, H. and Gujer, W. 1985. Mass transfer mechanisms in a heterotrophic biofilm. *Water Research,* **19**(11):S. 1369–1378.

Silver, S. (2003). Bacterial silver resistance: molecular biology and uses and misuses of silver compounds. *FEMS Microbiology Reviews,* **27**(2–3): 341–353.

Silver, S., Phung, L., and Silver, G. (2006). Silver as biocides in burn and wound dressings and bacterial resistance to silver compounds. *Journal of Industrial Microbiology and Biotechnology,* **33**(7):627–634.

Souli, M. and Giamarellou, H. (1998). Effects of slime produced by clinical isolates of coagulase-negative Staphylococci on activities of various antimicrobial agents. *Antimicrobial Agents and Chemotherapy,* **42**(4):939–941.

Soustre, J., Rodier, M.-H., Imbert-Bouyer, S., Daniault, G., and Imbert, C. (2004). Caspofungin modulates in vitro adherence of *Candida albicans* to plastic coated with extracellular matrix proteins. *The Journal of Antimicrobial Chemotherapy,* **53**(3):522–525.

Stark, R. M., Gerwig, G. J., Pitman, R. S., Potts, L. F., Williams, N. A., Greenman, J., Weinzweig, I. P., Hirst, T. R., and Millar, M. R. (1999). Biofilm formation by *Helicobacter pylori. Letters in Applied Microbiology,* **28**(2):121–126.

Staudt, C., Horn, H., Hempel, D. C., and Neu, T. R. (2003). Screening of lectins for staining lectin-specific glycoconjugates in the EPS of biofilms. In *Biofilms in Industry, Medicine & Environmental Biotechnology,* ed. V. O'Flaherty, P. Moran, P. Lens, and P. Stoodley. IWA Publishing, London, pp. 308–327.

Steckelberg, J. M. and Osmon, D. R. (2000). Prosthetic joint infections. In *Infections Associated with Indwelling Medical Devices*, ed. F. A. Waldvogel and A. L. Bisno. ASM Press, Washington, DC, pp. 173–209.

Stein, A., Drancourt, M., and Raoult, D. (2000). Ambulatory management of infected orthopedic implants. In *Infections Associated with Indwelling Medical Devices*, ed. F. A. Waldvogel and A. L. Bisno. ASM Press, Washington, DC, pp. 211–230.

Stewart, P. S. (1994). Biofilm accumulation model that predicts antibiotic resistance of *Pseudomonas aeruginosa* biofilms. *Antimicrobial Agents and Chemotherapy,* **38**(5):1052–1058.

Stewart, P. S. (1996). Theoretical aspects of antibiotic diffusion into microbial biofilms. *Antimicrobial Agents and Chemotherapy,* **40**(11): 2517–2522.

Stewart, P. S. (1998). A review of experimental measurements of effective diffusive permeabilities and effective diffusion coefficients in biofilms. *Biotechnology and Bioengineering,* **59**(3):261–272.

Stewart, P. S. (2002). Mechanisms of antibiotic resistance in bacterial biofilms. *International Journal of Medical Microbiology,* **292**(2):107–113.

Stewart, P. S. (2003). Diffusion in biofilms. *Journal of Bacteriology,* **185**(5): 1485–1491.

Stewart, P. S. and Costerton, J. W. (2001). Antibiotic resistance of bacteria in biofilms. *The Lancet,* **358**(9276):135–138.

Stewart, P. S. and Franklin, M. J. (2008). Physiological heterogeneity in biofilms. *Nature Reviews Microbiology,* **6**(3):199–210.

Stewart, P. S. and Raquepas, J. B. (1995). Implications of reaction-diffusion theory for the disinfection of microbial biofilms by reactive antimicrobial agents. *Chemical Engineering Science,* **50**(19):3099–3104.

Stewart, P. S., Hamilton, M. A., Goldstein, B. R., and Schneider, B. T. (1996). Modeling biocide action against biofilms. *Biotechnology and Bioengineering,* **49**(4):445–455.

Stone, G., Wood, P., Dixon, L., Keyhan, M., and Matin, A. (2002). Tetracycline rapidly reaches all the constituent cells of uropathogenic *Escherichia coli* biofilms. *Antimicrobial Agents and Chemotherapy,* **46**(8):2458–2461.

Stoodley, P., deBeer, D., and Lappin-Scott, H. M. (1997). Influence of electric fields and pH on biofilm structure as related to the bioelectric effect. *Antimicrobial Agents and Chemotherapy,* **41**(9):1876–1879.

Stoodley, P., Dodds, I., Boyle, J. D., and Lappin-Scott, H. M. (1999). Influence of hydrodynamics and nutrients on biofilm structure. *Journal of Applied Microbiology,* **85**(1):19S-28S.

Stoodley, P., Sauer, K., Davies, D. G., and Costerton, J. W. (2002). Biofilms as complex differentiated communities. *Annual Review of Microbiology,* **56**(1):187–209.

Sturme, M., Kleerebezem, M., Nakayama, J., Akkermans, A., Vaughan, E., and de Vos, W. (2002). Cell to cell communication by autoinducing peptides in gram-positive bacteria. *Antonie van Leeuwenhoek,* **81**(1):233–243.

Suci, P. A., Mittelman, M. W., Yu, F. P., and Geesey, G. G. (1994). Investigation of ciprofloxacin penetration into *Pseudomonas aeruginosa* biofilms. *Antimicrobial Agents and Chemotherapy,* **38**(9):2125–2133.

Szomolay, B., Klapper, I., Dockery, J., and Stewart, P. S. (2005). Adaptive responses to antimicrobial agents in biofilms. *Environmental Microbiology,* **7**(8):1186–1191.

Tack, K. J. and SabathL. D. (1985). Increased minimum inhibitory concentrations with anaerobiasis for tobramycin, gentamicin, and amikacin, compared to latamoxef, piperacillin, chloramphenicol, and clindamycin. *Chemotherapy,* **31**(3):204–210.

Tandoi, V., Jenkins, D., and Wanner, J. (2006). *Activated Sludge Separation Problems: Theory, Control Measures, Practical Experiences.* IWA Publishing, London.

Thomas, J. G., Litton, I., and Rinde, H. (2006). Economic impact of biofilms on treatment costs. In *Biofilms, Infection, and Antimicrobial Therapy,* ed. J. L. Pace, M. E. Rupp, and R. G. Finch. CRC Press, Boca Raton, FL, pp. 21–38.

Tolker-Nielsen, T. and Molin, S. (2000). Spatial organization of microbial biofilm communities. *Microbial Ecology,* **40**(2):75–84.

Tomi, N. S., Kranke, B., and Aberer, W. (2004). A silver man. *Lancet,* **363**(9408):532.

Trampuz, A. and Widmer, A. F. (2006). Infections associated with orthopedic implants. *Current Opinion in Infectious Diseases,* **19**(4):349–356.

Trampuz, A. and Zimmerli, W. (2006). Antimicrobial agents in orthopaedic surgery—Prophylaxis and treatment. *Drugs,* **66**(8):1089–1105.

Trampuz, A. and Zimmerli, W. (2008). Diagnosis and treatment of implant-associated septic arthritis and osteomyelitis. *Current Infectious Disease Reports,* **10**(5): 394–403.

Ullrich, S., Karrasch, B., Hoppe, H., Jeskulke, K., and Mehrens, M. (1996). Toxic effects on bacterial metabolism of the redox dye 5-cyano-2,3-ditolyl tetrazolium chloride. *Applied and Environmental Microbiology,* **62**(12):4587–4593.

Utaida, S., Dunman, P. M., Macapagal, D., Murphy, E., Projan, S. J., Singh, V. K., Jayaswal, R. K., and Wilkinson, B. J. (2003). Genome-wide transcriptional profiling of the response of *Staphylococcus aureus* to cell-wall-active antibiotics reveals a cell-wall-stress stimulon. *Microbiology,* **149**(10): 2719–2732.

Van As, H. and Lens, P. (2001). Use of 1H NMR to study transport processes in porous biosystems. *Journal of Industrial Microbiology and Biotechnology,* **26**(1):43–52.

van Loosdrecht, M. C. M., Heijnen, J. J., Eberl, H., Kreft, J., and Picioreanu, C. (2002). Mathematical modelling of biofilm structures. *Antonie van Leeuwenhoek,* **81**(1):245–256.

Velicer, G. J. (2003). Social strife in the microbial world. *Trends in Microbiology,* **11**(7):330–337.

von Gotz, F., Haussler, S., Jordan, D., Saravanamuthu, S. S., Wehmhoner, D., Strussmann, A., Lauber, J., Attree, I., Buer, J., Tummler, B., and Steinmetz, I. (2004). Expression analysis of a highly adherent and cytotoxic small colony variant of *Pseudomonas aeruginosa* isolated from a lung of a patient with cystic fibrosis. *Journal of Bacteriology,* **186**(12):3837–3847.

Vrany, J., Stewart, P., and Suci, P. (1997). Comparison of recalcitrance to ciprofloxacin and levofloxacin exhibited by *Pseudomonas aeruginosa* bofilms displaying rapid-transport characteristics. *Antimicrobial Agents and Chemotherapy,* **41**(6):1352–1358.

Vroom, J. M., De Grauw, K. J., Gerritsen, H. C., Bradshaw, D. J., Marsh, P. D., Watson, J. J. Birmingham, G. K., and Allison, C. (1999). Depth penetration and detection of pH gradients in biofilms by two-photon excitation microscopy. *Applied and Environmental Microbiology,* **65**(8):3502–3511.

Walters, M. C. III, Roe, F., Bugnicourt, A., Franklin, M. J., and Stewart, P. S. (2003). Contributions of antibiotic penetration, oxygen limitation, and low metabolic activity to tolerance of Pseudomonas aeruginosa biofilms to ciprofloxacin and tobramycin. *Antimicrobial Agents and Chemotherapy,* **47**(1):317–323.

Wanner, O. and Gujer, W. (1986). A multispecies biofilm model. *Biotechnology and Bioengineering,* **28**(3):314–328.

Wanner, O. and Reichert, P. (1996). Mathematical modeling of mixed-culture biofilms. *Biotechnology and Bioengineering,* **49**(2):172–84.

Wanner, O., Cunningham, A. B., and Lundman, R. (1995). Modeling biofilm accumulation and mass transport in a porous medium under high substrate loading. *Biotechnology and Bioengineering,* **47**(6):703–712.

Wanner, O., Eberl, H. J., Morgenroth, E., Noguera, D. R., Picioreanu, C., Rittmann, B. E., and van Loosdrecht, M. C. M. (2006). *Mathematical Modeling of Biofilms.* IWA Publishing, London.

Wargo, M. J. and Hogan, D. A. (2006). Fungal–bacterial interactions: a mixed bag of mingling microbes. *Current Opinion in Microbiology,* **9**(4): 359–364.

Watnick, P. I. and Kolter, R. (1999). Steps in the development of a Vibrio cholerae El Tor biofilm. *Molecular Microbiology,* **34**(3):586–595.

Watnick, P. I., Fullner, K. J., and Kolter, R. (1999). A role for the mannose-sensitive hemagglutinin in biofilm formation by Vibrio cholerae El Tor. *Journal of Bacteriology,* **181**(11):3606–3609.

Widmer, A. F., Gaechter, A., Ochsner, P. E., and Zimmerli, W. (1992). Antimicrobial treatment of orthopedic implant-related infections with rifampin combinations. *Clinical Infectious Diseases,* **14**(6):1251–1253.

Williams, V. and Fletcher, M. (1996). Pseudomonas fluorescens adhesion and transport through porous media are affected by lipopolysaccharide composition. *Applied and Environmental Microbiology,* **62**:1004.

Wilson, M. (2005). *Microbial Inhabitants of Humans: Their Ecology and Role in Health and Disease.* Cambridge University Press, Cambridge, UK.

Wimpenny, J. W. T. and Colasanti, R. (1997). A unifying hypothesis for the structure of microbial biofilms based on cellular automaton models. *FEMS Microbiology Ecology,* **22**(1):1–16.

Wirtanen, G., Alanko, T., and Mattila-Sandholm, T. (1996). Evaluation of epifluorescence image analysis of biofilm growth on stainless steel surfaces. *Colloids and Surfaces B: Biointerfaces,* **5**(6):319–326.

Xiong, Y. Q., Willard, J., Kadurugamuwa, J. L., Yu, J., Francis, K. P., and Bayer, A. S. (2005). Real-time in vivo bioluminescent imaging for evaluating the efficacy of antibiotics in a rat *Staphylococcus aureus* endocarditis model. *Antimicrobial Agents and Chemotherapy,* **49**(1):380–387.

Xu, K. D., Stewart, P. S., Xia, F., Huang, C.-T., and McFeters, G. A. (1998). Spatial physiological heterogeneity in *Pseudomonas aeruginosa* biofilm is determined by oxygen availability. *Applied and Environmental Microbiology,* **64**(10):4035–4039.

Yang, S. and Lewandowski, Z. (1995). Measurement of local mass transfer coefficient in biofilms. *Biotechnology and Bioengineering,* **48**(6):737–744.

Yang, X., Beyenal, H., Harkin, G., and Lewandowski, Z. (2000). Quantifying biofilm structure using image analysis. *Journal of Microbiological Methods,* **39**(2):109–119.

Yang, X., Beyenal, H., Harkin, G., and Lewandowski, Z. (2001). Evaluation of biofilm image thresholding methods. *Water Research,* **35**(5):1149–1158.

Yano, T., Kodama, T., and Yamada, K. (1961). Fundamental studies on the aerobic fermentation Part VIII. oxygen transfer within a mold pellet. *Agriculture Biological Chemistry,* **25**(7):580–584.

Yasuda, H., Ajiki, Y., Koga, T., and Yokota, T. (1994). Interaction between clarithromycin and biofilms formed by *Staphylococcus epidermidis. Antimicrobial Agents and Chemotherapy,* **38**(1):138–141.

Yogev, R. and Bisno, A. L. (2000). Infections of central nervous system shunts. In *Infections Associated with Indwelling Medical Devices*, ed. F. A. Waldvogel and A. L. Bisno. ASM Press, Washington, DC, pp. 231–246.

Zampini, M., Pruzzo, C., Bondre, V. P., Tarsi, R., Cosmo, M., Bacciaglia, A., Chhabra, A., Srivastava, R., and Srivastava, B. S. (2005). Vibrio cholerae persistence in aquatic environments and colonization of intestinal cells: involvement of a common adhesion mechanism. *FEMS Microbiology Letters,* **244**(2):267–273.

Zheng, D., Taylor, G. T., and Gyananath, G. (1994). Influence of laminar flow velocity and nutrient concentration on attachment of marine bacterioplankton. *Biofouling,* **8**(2):107–120.

Zheng, Z. and Stewart, P. S. (2002). Penetration of rifampin through *Staphylococcus epidermidis* biofilms. *Antimicrobial Agents and Chemotherapy,* **46**(3):900–903.

Ziebuhr, W., Krimmer, V., Rachid, S., LoBner, I., Gotz, F., and Hacker, J. (1999). A novel mechanism of phase variation of virulence in Staphylococcus epidermidis: evidence for control of the polysaccharide intercellular adhesin synthesis by alternating insertion and excision of the insertion sequence element IS256. *Molecular Microbiology,* **32**(2):345–356.

Zimmerli, W., Trampuz, A., and Ochsner, P. E. (2004). Prosthetic-joint infections. *The New England Journal of Medicine*, **351**(16):1645–1654.

Zottola, E. A. (1991). Characterization of the attachment matrix of Pseudomonas fragi attached to non-porous surfaces. *Biofouling,* **5**(1–2):37–55.

5

Influence of Biofilms on Porous Media Hydrodynamics

Robin Gerlach
Department of Chemical and Biological Engineering, Center for Biofilm Engineering, Montana State University, Bozeman, MT

Alfred B. Cunningham
Department of Civil Engineering, Center for Biofilm Engineering, Montana State University, Bozeman, MT

CONTENTS

Acknowledgments
Financial support was provided by the Office of Science (BER), U.S. Department of Energy, Grant No. DE-FG-02-09ER64758 and by the National Science Foundation (NSF) Award No.: DMS-0934696. Logan Schultz is acknowledged for his various contributions. Assistance in the preparation of figures by Peg Dirckx and Chantel Naylor is gratefully acknowledged.

5.1 Introduction and Overview

Microbial biofilms form in natural and engineered systems and can significantly affect the hydrodynamics in porous media. Microbial biofilms develop through the attachment and growth of microorganisms, which encase themselves in self-produced extracellular polymeric substances (EPS). Microbial biofilms are, in general, more resistant to environmental stresses, such as mechanical stress, temperature, pH, and water potential fluctuations, than planktonic cells. Biofilm growth in porous media influences porosity, permeability, dispersion, diffusion, and mass transport of reactive and nonreactive solutes. Understanding and controlling biofilm formation in porous media will maximize the potential benefit and will minimize the detrimental effects of porous media biofilms. Subsurface remediation, enhanced oil recovery, and carbon sequestration are only a few examples of beneficial porous media biofilm applications.

5.2 An Introduction to Biofilms

Microbial biofilms have probably been known to exist for as long as we have known about microorganisms. When Anthony van Leeuwenhoek described the "scuff" (plaque) from his teeth in 1683 (see http://www.ucmp.berkeley.edu/history/leeuwenhoek.html [accessed Jan. 09, 2009] or Dixon 2009), the discovery of these "many very little living animalcules, very prettily a-moving," i.e., the mere existence of microorganisms, overshadowed the fact that they were associated with a surface, that is, the teeth. It might be due to the fact that microorganisms associated with surfaces are very difficult to study that the scientific community almost exclusively focused on the study of free-floating ("suspended" or "planktonic") microorganisms well into the twentieth century. The importance of attached microorganisms in nature and engineered systems was really not described until the end of the 1970s when Characklis and Costerton et al. clearly described the abundance of biofilms in many environments (Characklis 1973a,b; Costerton et al. 1978). Bill Characklis' legacy to the biofilm field was recently acknowledged with the republication of a "vintage article" along with a commentary in biotechnology and bioengineering (Characklis and Bryers 2008).

It took until 1990 for the first book on biofilms to appear. Two of the pioneers in biofilm research, Kevin Marshall and Bill Characklis coedited *Biofilms*, a book still very worthwhile for the beginning biofilm engineer or microbiologist (Characklis and Marshall 1990). Over the past 20 years, numerous books on biofilms have been published, and the interested reader is encouraged to consult those for more detail (Characklis and Marshall 1990; Lappin-Scott and Costerton 1995; Bryers 2000; Evans 2000; Ghannoum and O'Toole 2004; Costerton 2007; Lewandowski and Beyenal 2007). In addition, the Center for Biofilm Engineering at Montana State University is spearheading an effort to develop the "Biofilm Hypertextbook," which can currently be accessed at http://www.biofilmbook.com/.

Biofilms are complex three-dimensional microbial communities attached to a surface (Figure 5.1). There is no one commonly accepted definition for the term "biofilm," although it is generally agreed upon that a biofilm is an aggregate of microorganisms, such as bacteria, algae, fungi, or protozoa, attached to a surface and embedded in a self-produced matrix of EPS. Biofilms can be found on various surfaces and in various industrially, environmentally, and medically relevant systems. They form in completely saturated as well as unsaturated environments such as pipelines, soils, medical implants, blood vessels, biomaterials, tissues, biofilters, cooling towers, ship hulls, river rocks, and a variety of other environments. One of the major differences between suspended cells and biofilms is the commonly large amount of EPS present in biofilms, which, among other potential roles, provides biofilms with structural support. Biofilm communities have been argued to be the predominant form of microbial life in many environments (Costerton et al. 1995; Stoodley et al. 2002b; Costerton 2007), and especially in systems with high surface area to volume ratios, such as porous media, biofilm communities can be expected to dominate (VanLoosdrecht et al. 1990; Bouwer et al. 2000).

The biofilm mode of growth can have significant competitive advantages. Immobilized organisms (e.g., a biofilm on soil particles) in a continuous-flow system (e.g., flowing groundwater containing growth substrates) will have a continuous supply of substrates and nutrients from the flowing fluid or the porous medium itself.

In general, three stages of biofilm development are to be considered: (1) Microbial transport and attachment, (2) biofilm growth, and (3) microbial detachment and propagation. Two animations (Movies 7 and 8) were published as supplementary materials to "The Biofilm Primer" at http://www.springer.com/life+sciences/microbiology?SGWID=0-10037-12-322199-0 (Costerton 2007) and are recommended to be viewed by the novice in this field.

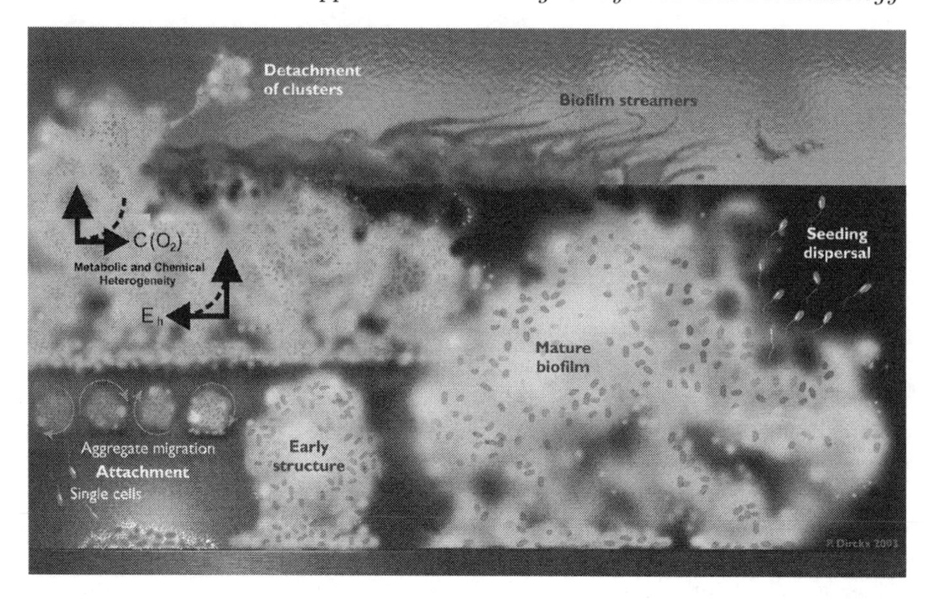

FIGURE 5.1
Schematic of the life cycle of a biofilm. Starting on the bottom left, attachment; growth and development of biofilm structure (bottom center); and fully developed biofilm with detaching cells (bottom right, in this case through seeding dispersal). Top: developed heterogeneous biofilm. Different shades within the biofilm indicate differences in physicochemical conditions (e.g., pH, concentration of nutrient, availability of electron acceptors, etc.). Throughout, mechanically induced stress and strain (e.g., through fluid flow) can cause detachment of clusters, aggregate migration, streamer formation, as well as single cell detachment. Detached cells can colonize surfaces elsewhere to form more biofilm.

5.2.1 Microbial Transport and Attachment

In most situations, if a surface is present, microorganisms tend to attach and make the transition from the planktonic to the attached (sessile) state. The mechanisms and kinetics of attachment in porous media are most frequently described using the colloid filtration theory (Yao et al. 1971). The removal of microorganisms from the flowing fluid has been found to be governed by a large number of parameters, including the properties of the microbial cells, solution chemistry, porous media characteristics, and hydrodynamics. This chapter provides only a brief overview of the parameters and conditions influencing microbial transport in porous media. More detailed reviews are provided, for example, by Bouwer et al. (2000) and Ginn et al. (2002), and the reader is referred to these publications as well as the references therein.

Cell surface properties such as the presence of cell surface molecules (e.g., proteins or carbohydrates), pili, flagella, as well as cell surface hydrophobicity and charge have been shown to play a role in microbial attachment. Motility, chemotaxis, cell buoyant density, size, and shape are cell properties that can influence the transition of cells from the planktonic to the attached state (Bouwer et al. 2000). Many of these parameters can change with the physiological state of the microbes, and recent studies have demonstrated that bacterial starvation can enhance bacterial transport in porous media (Cunningham et al. 2007).

Solute characteristics such as solution ionic strength, pH, temperature, and the presence of organic compounds, such as nutrients or surfactants, can influence the tendency of microbes to remain in suspension or to attach to a surface in porous media (Bouwer et al. 2000). In natural systems, the ability to control these parameters is somewhat limited because of the large amount of (ground) water that would have to be manipulated.

The same is true for porous media characteristics. Pore size and pore size distribution; grain and grain size distribution; mineralogy; roughness; the presence of sorbed, dissolved, or suspended organic matter; and porous media hydrophobicity have been shown to influence microbial attachment and transport in porous media, but these parameters might be difficult to control or manipulate in natural systems (Bouwer et al. 2000).

Many of the models describing microbial transport in porous media are based on the advection dispersion equation and attempt to correlate one or multiple of the parameters listed above to the collision efficiency factor to create a link between the fairly well-understood hydrodynamics and the less well-understood microbe–surface interactions. However, apparent scale dependencies of these parameters, possibly due to the heterogeneity of porous media and microbial populations, have made it difficult to reliably couple our knowledge of the micro- and nano-scale interactions between microbial cells and the porous media with models describing the advective transport of microorganisms (see discussions and references in Ginn et al. [2002] and Thullner and Baveye [2008] for more information).

5.2.2 Biofilm Growth

Once attached to a surface, the development of a biofilm structure will depend on a number of parameters, including availability of growth-limiting nutrients, presence of inhibitors, as well as the prevailing hydrodynamics.

As will be discussed later in detail, direct observations of biofilm formation and growth in porous media is a formidable challenge due to the opaque nature of most porous media. Hence, most of the existing knowledge is based on very simple flow cell experiments with plane (nonporous) surfaces to facilitate microscopic investigations.

The type and extent of biofilm formation depends on the ability of the attached microorganisms to grow and reproduce. Major factors contributing toward the ability of biofilms to form are as follows: (1) the availability of nutrients and energy (e.g., electron donors and acceptors; macro- and micronutrients), (2) appropriate geochemical conditions (pH, temperature, osmotic pressure, etc.), (3) absence of inhibitors (toxins, antimicrobial agents, waste products), (4) tolerable level of biofilm consuming (e.g., grazing) organisms, and (5) hydrodynamics, which influence mass transport of solutes and can result in mechanical stress acting upon biofilms.

Biofilms are distinctly different from the long-studied planktonic cells (Stoodley et al. 2002b), and research over the past decade has clearly revealed intricate spatial organization in biofilms. A recent review by Stewart and Franklin (2008) nicely summarizes the chemical, physical, and biological (genetic) heterogeneity of microbial biofilms and strategies on how to assess and describe these heterogeneities.

Biofilm communities appear to organize themselves spatially to form continuous films and distinct colonies (also often referred to as mushrooms, towers, streamers, etc.), which can vary in density and spatial organization depending on the culture conditions. Much effort is being expended into understanding organizational structures and processes within biofilms as evidenced in several review papers regarding biofilms as well as recent experimental work investigating differential gene expression in single and multispecies biofilms (Costerton et al. 1995; O'Toole et al. 2000; Tolker-Nielsen and Molin 2000; Hall-Stoodley et al. 2004; Stewart and Franklin 2008; Lenz et al. 2008). Biofilms have been suggested to have tissue-like characteristics (Costerton et al. 1995; Neu et al. 2002), and it has even been postulated that biofilms behave more like a multicellular organism than single cells or even a community of unicellular organisms (Velicer 2003; Crespi and Springer 2003).

Cell–cell communication, the process by which microorganisms can influence each other's behavior via small molecular weight chemical molecules has received significant attention in this context, and the reader is referred to the highly cited works in this area as well as book chapters (Singh et al. 2000; Miller and Bassler 2001; Chen et al. 2002; Sauer et al. 2002; Hentzer et al. 2004; Hall-Stoodley et al. 2004).

It is clear that the biofilm mode of growth offers a number of competitive advantages to microorganisms, including, but not limited to, protection from chemical and physical environmental stress factors, the trapping of nutrients in systems with low nutrient concentrations, symbiotic or mutualistic community interactions, and enhanced exchange of genetic material (Tolker-Nielsen and Molin 2000; Tolker-Nielsen et al. 2000; Hentzer et al. 2004; Cvitkovitch 2004; Molin et al. 2004).

The presence of EPS, which stabilizes the spatial organization of biofilm communities, appears to be crucial in this context. EPS largely immobilize or at least drastically reduce movement of microbial cells, resulting in a stable, yet not rigid, three-dimensional community, which can provide competitive advantages owing to mutualistic or symbiotic relationships among organisms.

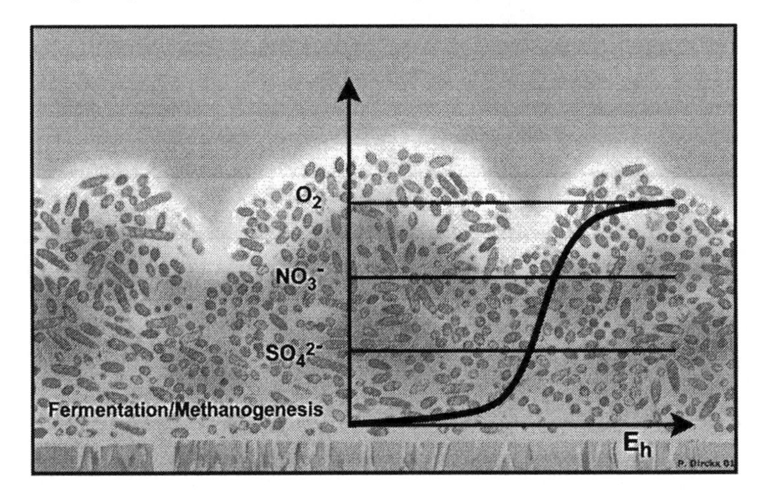

FIGURE 5.2
Schematic of possible spatial organization in a natural biofilm community. Depending on the availability of terminal electron acceptors (e.g., oxygen, nitrate, sulfate), aerobic organisms might establish themselves in the top layer of the biofilm, followed by nitrate-reducing organisms whose activity will increase once oxygen availability decreases, followed by sulfate-reducing organisms, which require a much lower redox potential (indicated by E_h). In the absence of these electron acceptors, anaerobic hydrolysis of complex carbon sources, fermentation, as well as methanogenesis might occur, which can produce sugars, organic acids, alcohols, and other small organic compounds, which can serve as electron donors and carbon sources for the respiratory organisms (aerobes, nitrate- and sulfate-reducing organisms) in the top layers of the biofilm. Carbon cycling not shown in this figure to reduce complexity.

It can be imagined, for instance, that in a natural environment a number of aerobic organisms establish themselves in the top layers of a biofilm consuming the available oxygen and provide anaerobic conditions for nitrate-reducing, sulfate-reducing, and fermentative organisms. The fermentative organisms in turn might produce short-chain organic acids and alcohols, which could become inhibitory in a solely fermentative community but are being consumed by the nitrate-, sulfate-, and oxygen-reducing organisms. Experimental evidence of such stratification has been published (Kuhl and Jorgensen 1992; Ramsing et al. 1993; Okabe et al. 1999) and is outlined in Figure 5.2.

Even without the protection from other community members, the EPS can present a diffusion barrier for solutes allowing organisms deeper inside a biofilm community potentially more time for the initiation of a protective stress response. For reactive solutes, such as oxidative antimicrobials (e.g., chlorine), EPS can also present a reactive barrier. Owing to the high-reaction rate of chlorine with extracellular matrix components, its penetration can be highly limited (Stewart 2003). Growth-dependent antimicrobials, such as certain antibiotics, might also be less effective against biofilms owing to the presence of slow-growing organisms even in pure culture biofilms (Brown et al. 1988; Gilbert et al. 1990). Overviews of these and other possible mechanisms of biofilm resistance are summarized in a number of articles and book chapters that have been published over the past years (Stewart et al. 2000; Stewart 2003; Fux et al. 2005; Costerton 2007).

The morphology of biofilms in porous media can be highly variable and can range from patchy, colony-like biofilms to continuous films with varying thickness. The significance of such different biofilm morphologies and thicknesses to mass transport of solutes in biofilms and biofilm-affected porous media will be discussed in more detail later (Biofilms in Porous Media and Their Effect on Hydrodynamics).

5.2.3 Microbial Detachment and Propagation

The detachment of microorganisms from microbial biofilms plays an important role in the propagation of biofilms and might ultimately determine the ability of biofilm organisms to survive. Detachment is probably the least-understood process in the life cycle of a biofilm. As shown in nonporous model systems, detachment of biomass can occur in the form of erosion, the loss of single cells, or small clusters of cells from the surface of the biofilm, through hollowing of microcolonies or in the form of small to large sloughing events (e.g., Tolker-Nielsen et al. 2000; Sauer et al. 2002; Stoodley et al. 2002a).

While erosion-like detachment appears to be most highly influenced by hydrodynamics (e.g., changes in shear stress), substrate availability, and the amount of EPS present (Peyton et al. 1995; Paulsen et al. 1997; Kim and Fogler 2000; van Loosdrecht et al. 2002; Ramasamy and Zhang 2005; Ross et al. 2007), massive sloughing events seem to be controlled more by the physiology of the cells. Chemical signaling as well as bacteriophage-induced detachment events have been described in the literature (Stoodley et al. 2001; Wilson et al. 2004; Purevdorj-Gage et al. 2005).

There are reports that detaching cells are metabolically more active than those that remain (Rice et al. 2003) but also reports that bacterial starvation, which should result in decreased activity, increases detachment (Ross et al. 2007).

Detachment of cells is likely to be followed by attachment of cells with subsequent biofilm development as long as the environmental conditions permit. The sequence of attachment, growth and biofilm development, detachment, followed by (re-)attachment is often referred to as the life cycle of a biofilm (outlined in Figure 5.1).

As will be discussed below, the ability to directly observe attachment, biofilm growth, and detachment processes is limited, especially in porous media. However, a thorough understanding of the behavior of biofilms in porous media offers significant opportunities for the development of environmental and industrial processes as well as the control of detrimental biofilms in medicine and industry. Hence, a number of experimental systems and approaches have been developed to study biofilms in porous media.

5.3 Experimental Systems and Techniques for the Investigation of Biofilms in Porous Media

The direct observation of biofilm processes in porous media is challenging due to the irregular shape and opaque nature of most porous media. However, for the development of effective porous media biofilm technologies it is often necessary to observe the spatial and temporal distribution *and* activity of biofilm cells as well as of the EPS.

A number of reactor designs can be imagined for the investigation of biofilm processes in porous media, and the exact design of laboratory (and industrial) scale reactors will depend on the goal of each study or application. In addition, the mere existence of biofilm cells does not necessarily correlate with metabolic activity since it has been shown that a large portion of biofilm cells can be basically metabolically inactive (Mclean et al. 1999; Hunt et al. 2004; Werner et al. 2004; Sharp et al. 2005; Rani et al. 2007; Kim et al. 2009). The activity of biofilm organisms is governed by a number of mass transfer processes, such as the transport of solvents and solutes into reactors, the possible mass transfer from the gaseous to the liquid phase (e.g., for oxygen as an electron acceptor), the mass transfer of solutes from the liquid phase to the biofilm surface, and simultaneous reaction, diffusion, as well as the sorption of solutes within the biofilm and the supporting porous medium (Mclean et al. 1999).

The remainder of this chapter will mostly focus on approaches for better understanding the influence of biofilms on porous media porosity, permeability, and hydrodynamics. It should be kept in mind that the activity of microorganisms in porous media will ultimately determine the success of advanced biofilm technologies, and, activity measurements should be utilized as much as possible. However, spatially and temporally resolved measurements of biofilm activity are a challenge even in the absence of porous media.

Approaches for measuring biofilm activity include, but are not limited to, traditional microbiological culturing techniques (e.g., plate counts and most probable number techniques), substrate consumption measurements, enzyme assays to assess specific activities, and molecular techniques, such as mRNA (messenger ribonucleic acid) measurements, gene-specific quantitative PCR (polymerase chain reaction), and reporter gene constructs. The interested reader is referred to past (e.g., Fletcher 1979; Poulsen et al. 1993; Lazarova and Manem 1995; Dorn et al. 2004; Teal et al. 2006; Stewart and Franklin 2008; Lenz et al. 2008) and future literature evaluating techniques suitable for activity assessments in porous media environments. A recent review article by

Geesey and Mitchell reemphasizes the need for experimental systems in which hydrodynamics and biological activity can be measured directly (Geesey and Mitchell 2008).

5.3.1 The Challenge of Imaging Biofilms in Porous Media

Destructive (e.g., end-point) and nondestructive measurements have been used to assess the presence, structure, and distribution of biofilms in porous media. Such techniques include visual imaging using high-resolution photography and microscopy, scanning or transmission electron microscopy (SEM or TEM), as well as more recently developed methods such as x-ray tomography, nuclear magnetic resonance (NMR) spectroscopy-, or ultrasound-based imaging techniques.

Noninvasive low-energy techniques such as photography, brightfield or reflective microscopy, NMR imaging, or ultrasound have the advantage of having no or negligible effects on biofilms but often suffer a lack of resolution, depth penetration, or selectivity.

Currently available microscopes and image analysis programs allow determining the thickness of stained or unstained biofilms. However, the applicability of optical techniques to observe biofilms in porous media is limited as most porous media particles are not flat (e.g., sand grains, glass beads). Porous media surfaces are usually rather irregularly shaped, resulting in increased background signal and image blurriness. Confocal scanning laser microscopy (CSLM) combined with fluorescent labeling techniques, and three-dimensional image analysis can reduce the amount of background fluorescence and thus potentially allow for continuous and spatially resolved observation of biofilms in porous media. However, since porous media are opaque and the working distance of high-resolution microscopy objectives is limited, the depth of field of observation is usually very limited.

There have been a few studies in which the refractive index of the fluid and porous media used were chosen in a way to maximize the ability to observe biofilm formation *in situ* over time (e.g., Leis et al. 2005). However, such studies have remained rare and the choice of porous media materials is limited if aqueous solutions are to be used to grow biofilms.

Paulsen et al. (1997) described a model system in which noninvasive microscopic observation combined with measurements of local-flow velocity allowed for estimating the influence of biofilm morphology on convective mass transport. However, such detailed measurements are usually limited to specifically designed experimental systems and one- or pseudo–two-dimensional geometries.

Electron microscopy techniques such as SEM and TEM have been used to estimate the thickness of biofilms on surfaces (Vandevivere and Baveye 1992b; Rinck-Pfeiffer et al. 2000; Hand et al. 2008), however, they cannot be applied directly to porous media systems and can therefore suffer artifacts due to destructive sampling, sample preparation, and the vacuum conditions necessary for the analysis (Nam et al. 2000). For instance, in one study, the mean biofilm thickness estimates based on SEM images were about 60%–82% less than those obtained through optical microscopy (Jean et al. 2004). Environmental scanning electron microscopy (ESEM) can avoid some of the artifacts possibly introduced during sample preparation and imaging compared to SEM. ESEM does not necessarily require coating of the sample with a conductive material and much less stringent vacuum conditions; however, it can still suffer artifacts associated with the destructive sampling of the porous media.

X-ray tomography techniques for imaging porous media are available, but microscale imaging of biofilms in porous media has not yet been established completely. Some promising work has been published in the recent years, which shows the principal feasibility of performing microtomography on porous media to image the transport and deposition of fluids and colloids in porous media (e.g., Wildenschild et al. 2005; Li et al. 2006; Gaillard et al. 2007). However, the lack of x-ray–detectable absorption properties that are specific to the presence of biofilms or suitable biofilm-labeling techniques have slowed the progress in this field. Furthermore, the applicability of high-energy techniques to biological systems is limited because of the potential damage that can occur during exposure.

Nonoptical techniques such as NMR (e.g., Hoskins et al. 1999; Paterson-Beedle et al. 2001; Seymour et al. 2004a,b, 2007; Metzger et al. 2006; McLean et al. 2007; Hornemann et al. 2009), ultrasound (e.g., Shemesh et al. 2007), or complex conductivity (Davis et al. 2006) imaging of biofilms have immense potential for imaging biofilms in opaque media. However, their application is currently restricted because of their limited availability, resolution, and because of the interference that natural substrates, such as natural soils, sand, or stone cores, have on their signal.

5.3.2 Porous Media Biofilm Reactors

A large variety of bench- and pilot-scale porous media biofilm reactors have been used to evaluate the effect of biofilm accumulation on porous media hydrodynamics and mass transport.

Columns, ranging from a few millimeters to several meters in length and millimeter to approximately 1 m diameter, are probably the most commonly used reactor types. Unfortunately, natural porous media themselves as well as most column materials are opaque and do not lend themselves to direct optical interrogation. Hence, high-optical quality microscopic flowcells and flat-plate reactors, which can represent natural porous media or fractures more or less closely, have been used frequently to investigate fundamental processes of biofilm development in porous media (e.g., Cunningham et al. 1991, 1995; Sharp et al. 1999a, 2005; Yarwood et al. 2002, 2006; Nambi et al. 2003; Knutson et al. 2005; Ross et al. 2007; Willingham et al. 2008). These reactors can consist of materials that contain certain patterns or can be filled with thin layers of porous media (sand, glass beads, or similar). Depending on their size and optical quality, these reactors allow for direct optical or microscopic observations of biofilm formation and mass transport (see Figure 5.3 for examples). Capillary flowcells also allow for direct microscopic interrogations and are sometimes used to represent single pores.

While there are no review articles available, which summarize use and operation of reactors for the investigation of biofilm formation in porous media, there are reviews and special issues available related to the possibilities of visualizing colloid transport in porous media, such as a recent special section in *Water Resources Research* (2006, Vol. 42, Issue 12) and, in specific, an article by Ochiai et al. (2006).

The challenges in both fields—colloid transport and biofilm formation in porous media—are similar: (1) Aqueous phase measurements are simply not sufficient to completely understand the fundamental microscale processes affecting colloid transport or biofilm development and (2) The direct observation of microscale processes is complicated because of the opaque nature of porous media.

Nevertheless, a fairly large body of literature on the influence of biofilm formation on porous media hydrodynamics is available and will be described in the next section. Such studies generally rely on measurements of differences in hydraulic head, flow rate, and (particulate or dissolved, reactive or nonreactive) tracer through characteristics, sometimes combined with direct, real-time observation of biofilm distribution in columns, capillaries, network models, or larger lysimeter-like (two-dimensional/quasi three-dimensional) reactors (Paulsen et al. 1997; Sharp et al. 1999a; Rinck-Pfeiffer et al. 2000; Yarwood et al. 2002; Thullner et al. 2002a; VanGulck and Rowe 2004; Sharp et al. 2005; Arnon et al. 2005a; Yarwood et al. 2006; Castegnier et al. 2006; Seki et al. 2006; Ross et al. 2007; Rees et al. 2007).

More recently, with the broader availability of computer-controlled microscale fabrication opportunities, microscale reactors have been developed and employed in our and other laboratories, which allow for significantly improved possibilities for the noninvasive observation of reactive transport processes in porous media model reactors (Wan et al. 1994; Dupin and McCarty 1999; Kim and Fogler 2000; Nambi et al. 2003; Knutson et al. 2005; Willingham et al. 2008; and Figure 5.4).

Larger scale laboratory studies have also been conducted, mostly as precursors to field-scale demonstrations (Figure 5.5). In general, it has been observed that the increased complexity of two- or pseudo–three-dimensional systems can result in lesser permeability reductions, presumably due to the possibility of flow around biofilm- or mineral-clogged areas (Cunningham et al. 1997, 2003; Kildsgaard and Engesgaard 2001; Thullner et al. 2002a; Seki et al. 2006). Direct observations and simulations clearly show diverted flow around biofilm-clogged areas (Thullner et al. 2002a; Seki et al. 2006).

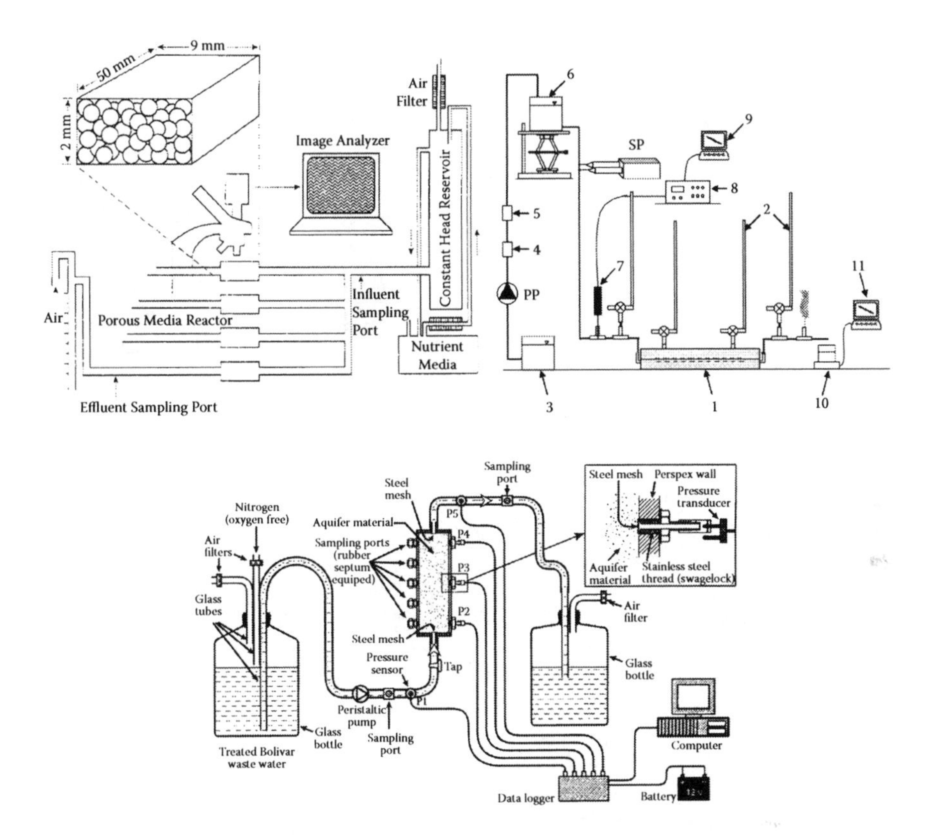

FIGURE 5.3
Schematics of experimental setups to study biofilm formation in porous media. Top left: Capillary flowcells used by Cunningham et al. (1991); Top right: artificial fracture setup used by Castegnier et al. (2006); Bottom: column setup utilized by Rinck-Pfeiffer et al. (2000). The two flowcells systems (top) both allow for direct optical interrogation of the system. The column setup (bottom) allows for more realistic porous media conditions but does not permit direct optical interrogation. The fracture and column flow systems allow for limited spatially resolved assessments of hydraulic conductivity due to multiple piezometers or pressure transducers. See cited sources for more detailed explanations of the experimental setups. (Top Left: Reprinted with permission from Cunningham, A. B., Characklis, W. G., Abedeen, F., and Crawford, D., *Env. Sci. Tech.*, 25, 1991. ©1991, American Chemical Society. Top Right: Reprinted from, Castegnier, F., Ross, N., Chapuis, R.P., Deschenes, L., and Samson, R., *Water Res.*, 40, 2006. ©2006—with permission from Elsevier. Bottom: Reprinted from, Rinck-Pfeiffer, S., Ragusa, S., Sztajnbok, P., and Vandevelde, T., *Water Res.*, 34, 2000. ©2006—with permission from Elsevier.)

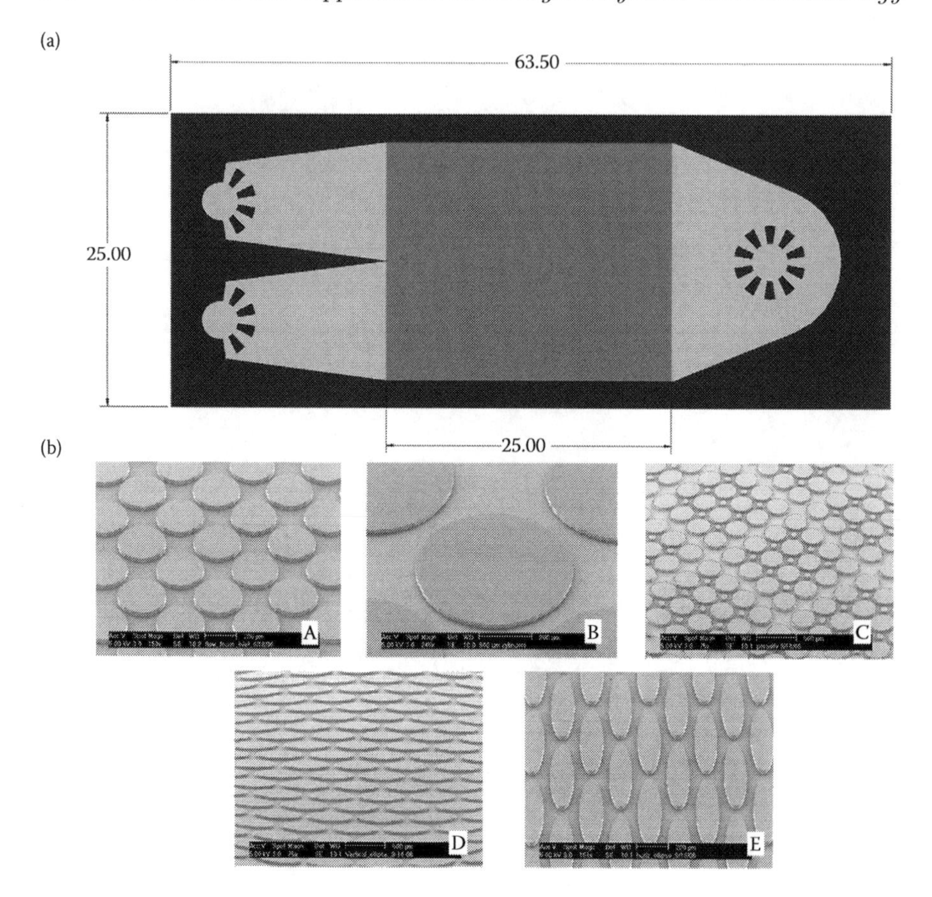

FIGURE 5.4
Technical drawing and pictures of selected microscale reactors for the study of biofilm formation in porous media. (a) Microscale porous media reactor developed at Montana State University; (b) SEM images of porous media structures employed by Willingham et al. (2008). (Reprinted with permission, Willingham, T.W., Werth, C.J., and Valocchi, A.J., *Env. Sci. Tech.*, 42, 2008. ©2008, American Chemical Society.)

5.4 Biofilms in Porous Media and Their Effect on Hydrodynamics

5.4.1 The Relationship of Porous Media Hydrodynamics and Biofilm Structure

This section will discuss the influence of biofilm growth on porous media hydrodynamics including porosity, permeability, and dispersivity. During this

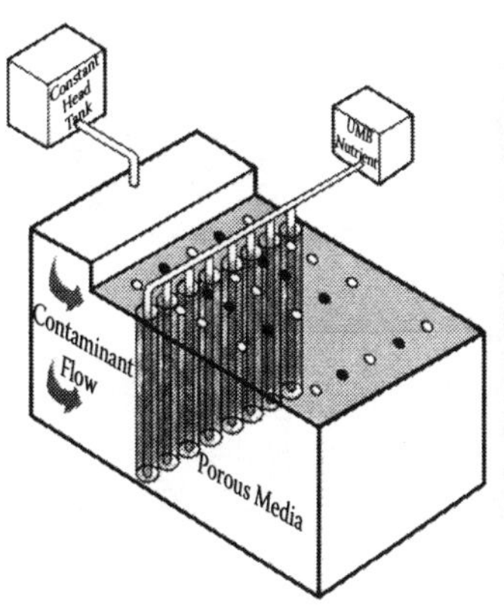

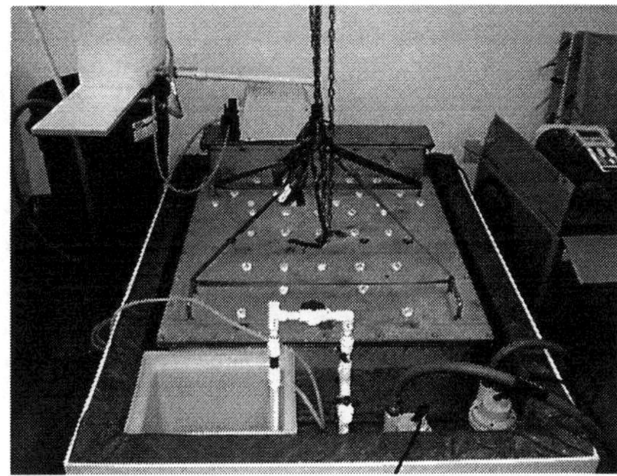

FIGURE 5.5
Schematic and picture of $3 \times 4 \times 1$ ft^3 rectangular lysimeter utilized by Cunningham et al. (1997).

discussion, the reader should keep in mind that not only biofilm growth influences the hydrodynamics of porous media but that the prevailing hydrodynamic conditions also influence biofilm growth characteristics; that is, it is likely that, over time, biofilm structure responds to changes in mass transport and adapts by adjusting its structure to ultimately optimize the mass transport of substrates and products as well as the stability of the biofilm. This interplay of hydrodynamics and biofilm growth will be discussed in more detail later.

It is commonly accepted that EPS and the microbial cells themselves, can affect the hydrodynamics in porous media. Moreover, there is significant experimental evidence that biofilms do not always accumulate homogeneously in porous media but that there can be significant spatial and temporal variability.

It has been observed repeatedly that discrete clusters are formed at least during the initial phase of biofilm accumulation and that the distribution of biofilms in porous media is influenced by the availability of nutrients, the effective porous media particle size, and local hydrodynamics (Taylor and Jaffe 1990b; Baveye et al. 1992; Vandevivere and Baveye 1992b; Rittmann 1993; Vandevivere 1995; Vandevivere et al. 1995; Sharp et al. 1999a; Rowe et al. 2000; Nam et al. 2000; Hill and Sleep 2002; Thullner et al. 2002a; Thullner et al. 2002b; Sharp et al. 2005; Seifert and Engesgaard 2007; Thullner and Baveye 2008).

Continuous biofilms occasionally form in porous media, but it is often observed that flow channels and low-permeability zones dominate (Sharp et al. 1999a, 2005; Stewart and Fogler 2001; Hill and Sleep 2002; Seymour et al. 2004a, 2007; Viamajala et al. 2008).

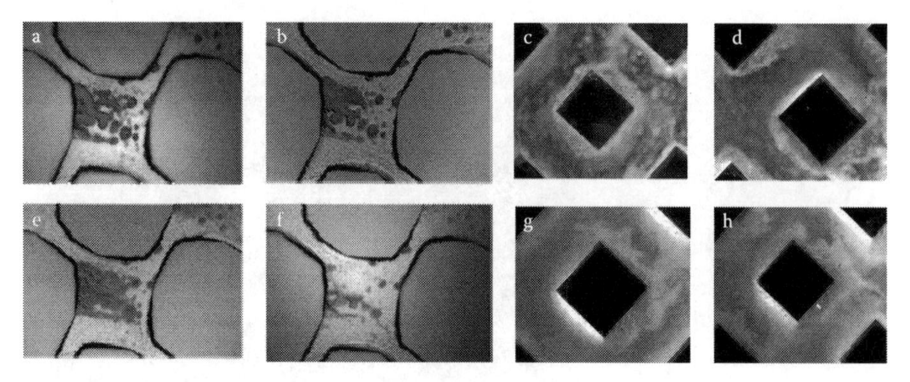

FIGURE 5.6
Microscopy images of biofilm forming in continuous-flow porous media environments. Flow from left to right in all images. Images (a), (b), (e), (f) are from Nambi et al. (2003), biofilms appear to be forming preferentially on the downstream edge of the porous media elements, which are approximately 300 μm in diameter; pore throats are approximately 35 μm in size. (a) 25 days, (b) 32 days, (e) 39 days, and (f) 44 days after inoculation. Images (c), (d), (g), (h) are depicting thick biofilm development in a porous media reactor over time. Porous media elements (black) are 1 mm^3; cross-section of flow channels is 1 mm^2. The reactor design is explained in more detail in Section 5.4.5. Images taken at effluent, (c and d) and influent region (g and h) of reactor. (Reprinted with permission Nambi, I.M., Werth, C.J., Sanford, R.A., and Valocchi, A.J. *Env. Sci. Tech.*, 37, 2003. ©2003, American Chemical Society.)

On the pore scale, it appears that biofilms accumulate initially in regions that are somewhat protected (Figure 5.6), that is, areas low in shear stress, and biofilm growth subsequently expands on the leeward side of porous media particles (Nambi et al. 2003; Stoodley et al. 2005; Knutson et al. 2007). The relatively high-shear forces in the pore throats of porous media are generally not favorable for initial attachment and thick biofilm development. High-shear environments generally result in relatively thin and smooth biofilms while lower shear environments usually produce thicker and rougher biofilms (Nam et al. 2000).

Oscillations in pressure drop across reactors have been related to changes in biomass distribution, the development of flow channels within the porous media filled with biomass, and creation of plugged regions where advective flow had occurred previously (Stewart and Fogler 2001). This behavior is probably more pronounced under constant flow conditions, which are employed more commonly in the laboratory than under constant head conditions, which are more commonly observed in the subsurface.

Under both types of flow conditions (constant head and constant flow), the most significant effects are usually observed in the influent regions of biofilm-affected porous media, where electron acceptor and donor supply is the greatest (see case study below as well as Hosokawa et al. 1992; Chen et al. 1994; Jennings et al. 1995; Ince et al. 2000). Decreasing availability of electron donors, acceptors, or other important nutrients (e.g., phosphorous or nitrogen) can be as responsible for a decrease in biofilm formation as the presence of inhibitory compounds. There are definite reports that the availability of carbon can influence the formation of biofilms in porous media (Hand et al. 2008). However, there are also reports that indicate increased biofilm growth in the effluent regions as well as where carbon might be limited (Kim et al. 2006; Wheeler 2009). This might be partially explained by the increased availability of another limiting growth factor, for example, the availability of oxygen. In column systems, which are repeatedly opened or which are connected to gas-permeable tubing, such as silicon tubing, the entry of oxygen into the systems can result in increased microbial growth in those regions.

Changes in hydrodynamics in biofilm-affected porous media are not always solely due to biofilm growth itself but also often due to the biofilm induced formation of minerals, such as di- or tri-valent (e.g., Fe, Ca, Mg) carbonate, sulfate, sulfide, and phosphate minerals (Mclean et al. 1997; Benner et al. 1999; Rinck-Pfeiffer et al. 2000; VanGulck and Rowe 2004). In at least one case of investigating the effect of biofilm growth in porous media, the majority of the "clog material" was identified as calcium carbonate (Rowe et al. 2000).

5.4.2 Porosity

When describing the porosity of biofilm-affected porous media one has to consider at least three types of porosity: (1) the overall porosity of the porous medium, (2) the effective porosity of the porous medium, and (3) the internal porosity of the biofilm itself.

The internal porosity of biofilms was initially believed to be negligible. Biofilms were treated as hydrogels with little relevant internal structure until the mid-1990s when it was demonstrated that advective flow can occur within biofilms through channels formed during biofilm growth (Stoodley et al. 1994; Okabe et al. 1998). Such channels can be tens of micrometers in size, allowing for significant advective mass transport.

However, in most cases the influence of biofilm growth on porosity has been investigated on scales at which the internal porosity of biofilms becomes insignificant. Hence, in most experimental and modeling work, the overall (or bulk) porosity and effective (i.e., available for advective fluid flow) porosity have been assessed. However, one recent study suggests the importance of including the internal porosity (and thus permeability) of biofilms to more accurately describe permeability changes in biofilm-affected porous media (Thullner and Baveye 2008).

Differences in effective and overall porosity can be drastic, and their effect on localized and bulk permeability depends on the location of the formed biofilm plugs. As biofilms grow, they initially decrease the free pore space; however, the overall porosity might not be significantly affected. If biofilms form on the inside of large pores or fractures, even a large change in porosity due to the presence of relatively thick biofilms can have a negligible effect on overall permeability. In contrast, if biofilms form in regions where their potential to affect fluid flow is great, such as pore throats, fracture entrances, and so on, a small change in overall porosity can have a significant effect on localized or overall permeability. In addition, as discussed by Sharp et al. (2005), Paulsen et al. (1997) and in Section 5.4.5, areas basically excluded from flow through thick biofilm growth can become accessible to flowing fluid owing to sudden detachment events, such as sloughing of large biofilm clusters that previously had blocked certain pores. Hence, methods, which allow spatially resolved measurements of porosity and differentiation between bulk and effective porosity are necessary to assess spatial and temporal changes.

Changes in porosity have mostly been assessed through direct microscopic observation combined with image analysis or are based on tracer breakthrough curves. Changes in media porosity can vary widely depending on the pore size distribution of the media as well as the method of measurement. Seifert and Engesgaard (2007) discuss problems with tracer breakthrough curve-based porosity estimates of biofilm-affected porous media in detail and discuss the validity of dual-porosity models for the description of hydrodynamics in biofilm-affected porous media (Seifert and Engesgaard 2007). Along with many other authors, Seifert and Engesgaard and Bielefeldt et al. point out that traditional porosity, permeability relationships, such as the Kozeny-Carman equation, generally underpredict the change in hydraulic conductivity (Bielefeldt et al. 2002b; Seifert and Engesgaard 2007).

5.4.3 Permeability

By far the most frequently used parameter to describe the influence of biofilm formation on porous media properties is permeability (k), expressed in dimensions of L^2 (length squared, e.g., m^2 or meter squared). In many published research papers permeability is expressed in the form of hydraulic conductivity (K, L/T, e.g., m/d or meters per day). The conversion is simple and uses the fluid density (ρ, M/L^3, e.g., kg/m^3), gravitational constant (g, L/T^2, e.g., m/sec^2), and viscosity (μ, (M/L)/T, e.g., (kg/m)/sec) according to the following equation:

$$k = K\frac{\mu}{\rho \bullet g}$$

For the purpose of comparing different studies (Table 5.1), 15°C was utilized as a reference with water as the fluid of interest so that $\rho = 999.099$ kg/m^3; $\mu = 1.14*10^{-3}$ (kg/m)/sec; $g = 9.807$ m/sec^2. Table 5.1 summarizes

TABLE 5.1
Initial and Final (or Lowest) Permeability Reported for Experiments in which Biofilm Growth in Porous Media or Fractures Was Promoted

Initial Permeability (cm^2)	Final or Lowest Permeability (cm^2)	Log (Reduction)	Porous Medium	Inoculum	Source
1.75E–04	1.53E–04	0.1	Sand	*Klebsiella oxytoca* and *Burkholderia cepacia*	Komlos et al. 2004
6.42E–08	5.71E–08	0.1	Sintered glass bead model cores	*Klebsiella pneumoniae* (starved)	MacLeod et al. 1988
3.49E–06	2.33E–06	0.2	Berea sandstone	*Pseudomonas aeruginosa* (killed)	Kalish et al. 1964
1.75E–04	1.01E–04	0.2	Sand	*Klebsiella oxytoca* and *Burkholderia cepacia*	Komlos et al. 2004
6.82E–08	4.17E–08	0.2	Sintered glass bead model cores	*Pseudomonas* sp. (killed)	Shaw et al. 1985
3.49E–07	1.75E–07	0.3	Berea sandstone	*Pseudomonas aeruginosa* (killed)	Kalish et al. 1964
3.49E–06	1.16E–06	0.5	Berea sandstone	*Micrococcus roseus* (killed)	Kalish et al. 1964
3.49E–06	1.05E–06	0.5	Berea sandstone	*Proteus vulgaris* (killed)	Kalish et al. 1964
3.49E–06	1.05E–06	0.5	Sandstone cores	*Klebsiella pneumonia*	Lappin-Scott et al. 1988b
6.42E–08	1.86E–08	0.5	Sintered glass bead model cores	*Klebsiella pneumoniae* (starved)	MacLeod et al. 1988

(*Continued*)

TABLE 5.1
(Continued)

Initial Permeability (cm^2)	Final or Lowest Permeability (cm^2)	Log (Reduction)	Porous Medium	Inoculum	Source
2.33E–07	5.82E–08	0.6	Berea sandstone	*Bacillus cereus* (killed)	Kalish et al. 1964
1.75E–04	3.68E–05	0.7	Sand	*Klebsiella oxytoca* and *Burkholderia cepacia*	Komlos et al. 2004
3.49E–06	4.65E–07	0.9	Berea sandstone	*Bacillus* sp.	Raiders et al. 1986
2.64E–01	2.88E–02	1.0	Artificial fracture	Groundwater community	Castegnier et al. 2006
2.33E–07	2.33E–08	1.0	Berea sandstone	*Micrococcus flavus* (killed)	Kalish et al. 1964
1.14E–05	1.16E–06	1.0	Sand	Mixed community, previously exposed to nitrate	Kildsgaard and Engesgaard 2001
4.65E–05	3.49E–06	1.1	Sand	*Pseudomonas* sp.	Cusack et al. 1992
2.33E–09	1.75E–10	1.1	Sand plus kaolinite	*Beijerinckia indica*	Dennis and Turner 1998
1.05E–05	8.35E–07	1.1	Sandy limestone aquifer material	Indigenous community	Rinck-Pfeiffer et al. 2000
6.98E–06	4.65E–07	1.2	Hanford loam	Indigenous	Allison 1947
4.89E–04	3.42E–05	1.2	Screened soil	*Pseudomonas fluorescens*	Cunningham et al. 2003
5.82E–06	3.49E–07	1.2	Berea sandstone, aerobic	Indigenous	Raiders et al. 1986
4.65E–05	2.33E–06	1.3	Sandy loam	Indigenous	Allison 1947

9.70E–07	5.00E–08	1.3	0.12 mm sand	*Pseudomonas aeruginosa*	Cunningham et al. 1991
4.65E–05	2.33E–06	1.3	Sand	*Pseudomonas* sp.	Cusack et al. 1992
3.49E–06	1.75E–07	1.3	Berea sandstone	*Micrococcus flavus* (killed)	Kalish et al. 1964
5.70E–05	2.09E–06	1.4	#70 quartz sand	Airport soil community	Bielefeldt et al. 2002b
1.51E–06	5.82E–08	1.4	Berea sandstone	*Bacillus subtilis* (killed)	Hart et al. 1960
1.97E–09	7.90E–11	1.4	Sandstone cores	*Klebsiella pneumonia*	Lappin-Scott et al. 1988b
3.95E–09	1.54E–10	1.4	Sandstone cores	*Klebsiella pneumonia*	Lappin-Scott et al. 1988b
2.10E–05	7.35E–07	1.5	1 mm glass beads	*Pseudomonas aeruginosa*	Cunningham et al. 1991
3.49E–04	1.16E–05	1.5	Ottawa 50–70 sand	Indigenous	Gupta and Swartzendruber 1962
2.09E–01	6.98E–03	1.5	artificial fracture	Top soil community	Hill and Sleep 2002
6.98E–06	2.33E–07	1.5	Berea sandstone, anaerobic	Indigenous	Raiders et al. 1986
3.49E–05	8.14E–07	1.6	Exeter sandy loam	Indigenous	Allison 1947
2.17E–06	5.00E–08	1.6	0.54 mm sand	*Pseudomonas aeruginosa*	Cunningham et al. 1991
4.77E–04	1.12E–05	1.6	Field scale	*Pseudomonas* sp.	Cunningham et al. 2003
3.95E–09	8.69E–11	1.7	Sandstone cores	*Klebsiella pneumonia*	Lappin-Scott et al. 1988b

(*Continued*)

TABLE 5.1
(Continued)

Initial Permeability (cm^2)	Final or Lowest Permeability (cm^2)	Log (Reduction)	Porous Medium	Inoculum	Source
3.19E–06	5.00E–08	1.8	0.7 mm sand	*Pseudomonas aeruginosa*	Cunningham et al. 1991
1.86E–06	2.68E–08	1.8	Sandy soil	*Azotobacter chroococcum*	Kim et al. 2006
3.49E–06	4.65E–08	1.9	Sandstone cores	*Klebsiella pneumonia*	Lappin-Scott et al. 1988b
9.87E–08	1.18E–09	1.9	300 um glass beads	*Leuconostoc mesenteroides*	Stewart and Fogler 2001
5.82E–04	6.40E–06	2.0	#30 quartz sand	Airport soil community	Bielefeldt et al. 2002b
1.97E–09	1.97E–11	2.0	Sandstone cores	Produced water	Lappin-Scott et al. 1988b
3.95E–09	3.95E–11	2.0	Sandstone cores	*Klebsiella pneumonia*	Lappin-Scott et al. 1988b
6.42E–08	6.42E–10	2.0	Sintered glass bead model cores	*Klebsiella pneumonia*	MacLeod et al. 1988
1.05E–05	7.37E–07	2.0	Quartz sand	Subsurface strains	Vandevivere and Baveye 1992a
1.28E–06	1.06E–08	2.1	Sandy soil	*Azotobacter chroococcum*	Kim et al. 2006
1.16E–06	9.08E–09	2.1	Sandy soil	*Azotobacter chroococcum*	Kim et al. 2006
6.59E–02	5.27E–04	2.1	Limestone fracture	groundwater community	Ross et al. 2001
6.43E–08	4.50E–10	2.2	Sintered glass bead model cores	*Pseudomonas* sp.	Shaw et al. 1985
6.46E–08	3.88E–10	2.2	Sintered glass bead model cores	*Pseudomonas* sp.	Shaw et al. 1985

1.16E–08	5.82E–11	2.3	Sand plus kaolinite	*Beijerinckia indica*	Dennis and Turner 1998
3.95E–09	1.97E–11	2.3	Sandstone cores	*Klebsiella pneumonia*	Lappin-Scott et al. 1988b
4.30E–04	1.75E–06	2.4	Quartz sand	Airport soil community	Bielefeldt et al. 2002a
4.07E–04	1.28E–06	2.5	#50 quartz sand	Airport soil community	Bielefeldt et al. 2002b
4.02E–01	1.21E–03	2.5	Artificial fracture	Groundwater community	Castegnier et al. 2006
5.84E–08	1.75E–10	2.5	Sintered glass bead model cores	*Pseudomonas* sp.	Shaw et al. 1985
5.82E–08	1.40E–10	2.6	Sand plus kaolinite	*Beijerinckia indica*	Dennis and Turner 1998
3.95E–09	7.90E–12	2.7	Sandstone cores	*Klebsiella pneumonia*	Lappin-Scott et al. 1988b
6.42E–08	1.28E–10	2.7	Sintered glass bead model cores	*Pseudomonas* sp. (killed)	Shaw et al. 1985
1.05E–07	1.86E–10	2.8	Sand plus kaolinite	*Beijerinckia indica*	Dennis and Turner 1998
6.98E–08	1.05E–10	2.8	Sand plus kaolinite	*Beijerinckia indica*	Dennis and Turner 1998
1.16E–07	1.75E–10	2.8	Sand plus kaolinite	*Beijerinckia indica*	Dennis and Turner 1998
4.32E–01	4.32E–04	3.0	Artificial fracture	Groundwater community	Castegnier et al. 2006
1.16E–07	1.16E–10	3.0	Compacted, silty sand	*Beijerinckia indica*	Dennis and Turner 1998

(Continued)

TABLE 5.1
(Continued)

Initial Permeability (cm^2)	Final or Lowest Permeability (cm^2)	Log (Reduction)	Porous Medium	Inoculum	Source
1.16E–07	1.16E–10	3.0	Sand plus kaolinite	*Beijerinckia indica*	Dennis and Turner 1998
3.95E–09	3.95E–12	3.0	Sandstone cores	produced water	Lappin-Scott et al. 1988b
3.95E–09	3.95E–12	3.0	Sandstone cores	*Klebsiella pneumonia*	Lappin-Scott et al. 1988b
8.38E–05	8.38E–08	3.0	Sand	*Arthrobacter* sp.	Vandevivere and Baveye 1992b
2.33E–07	1.63E–10	3.2	Sand plus kaolinite	*Beijerinckia indica*	Dennis and Turner 1998
2.95E–03	1.47E–06	3.3	Porous medium	wastewater treatment plant	Taylor and Jaffe 1990b
3.07E–04	6.46E–08	3.7	Loamy soil	Indigenous	Frankenberger et al. 1979
8.14E–04	8.14E–08	4.0	Two-dimensional lysimeter	*Klebsiella oxytoca*	Cunningham et al. 1997
1.05E–05	7.37E–09	4.0	Quartz sand	Subsurface strains	Vandevivere and Baveye 1992a
3.03E–03	1.16E–07	4.4	Sand	*Klebsiella pneumonia*	Cunningham et al. 1997
3.46E–02	1.16E–07	5.5	Sand	*Klebsiella pneumonia*	Cunningham et al. 1997
3.87E–01	1.16E–06	5.5	Glass beads	Landfill leachate, evidence of $CaCO_3$	VanGulck and Rowe 2004

the changes in permeability observed during studies in which biological growth reduced the permeability of porous media or fractures.

The reduction of porous media permeability is generally believed to occur through the accumulation of biomass and polysaccharides. Permeability reductions of three orders of magnitude or less are generally reported, but greater reductions have been reported. The largest reductions reported have been clearly associated with the coprecipitation of carbonate minerals (VanGulck and Rowe 2004; Castegnier et al. 2006).

Like biofilm accumulation, permeability reduction is usually not homogeneous but rather spatially and temporally varied with higher production of biomass, and thus greater permeability reduction, at the influent (Rinck-Pfeiffer et al. 2000; VanGulck and Rowe 2004; Castegnier et al. 2006). Permeability reduction also appears to be more pronounced in fine-textured materials than in coarse-textured ones (Vandevivere 1995; Vandevivere et al. 1995).

Once achieved, decreased porous media permeability can often be maintained even during periods of environmental stress for the biofilm organisms, such as starvation, toxic upset or similar, as demonstrated in one- and two-dimensional flow fields in the laboratory (Cunningham et al. 1997; Kim and Fogler 2000; Kim et al. 2006) as well as in a field-scale demonstration (Cunningham et al. 2003). This persistence is usually attributed to the presence of large amounts of EPS that do not degrade readily. These observations are supported by Kim and Fogler who observed no EPS degradation in batch experiments for a period of about 2 years (Kim and Fogler 1999).

5.4.4 Dispersion and Diffusion

Just like permeability and porosity, dispersivity (which relates the dispersion coefficient to velocity) is influenced by biofilm growth in porous media. Most of the experiments reveal a two- to eight-fold increase in dispersivity (Sharp et al. 1999b, 2005; Hill and Sleep 2002; Bielefeldt et al. 2002b; Arnon et al. 2005b), although order of magnitude changes in dispersivity have also been observed (Taylor and Jaffe 1990a; Bielefeldt et al. 2002a; Seifert and Engesgaard 2007). In general, dispersivity increases over time in biofilm-affected porous media but often reaches semistable values once the biofilm has reached a pseudosteady state (Sharp et al. 1999a, 2005; Bielefeldt et al. 2002b). Hill and Sleep stated that the dispersion coefficient increased logarithmically with hydraulic conductivity reduction in biofilm-affected fractures (Hill and Sleep 2002).

Increases in dispersivity have mostly been attributed to increases in tortuosity of the porous medium, the development of no flow zones, or increased influence of diffusive transport into and out of the developed biofilms (see for example, Sharp et al. 1999a, 2005; Seymour et al. 2004a,b). Seymour et al. demonstrated that diffusion out of no flow zones can clearly influence the macroscale dispersion in biofilm-affected porous media (Seymour et al. 2004b).

In most cases, the dispersivity in biofilm-affected porous media has been estimated on the basis of tracer tests and subsequent parameter fitting in mathematical models (e.g., Taylor and Jaffe 1990a; Sharp et al. 1999b; Kildsgaard and Engesgaard 2001; Hill and Sleep 2002; Bielefeldt et al. 2002a,b; Sharp et al. 2005; Arnon et al. 2005b). However, recently more advanced techniques have been developed.

Magnetic resonance microscopy (MRM) techniques are ideally suited to provide nondestructive, spatially, and temporally resolved measurements of hydrodynamic dispersion and velocity in porous media while allowing limited biofilm imaging (10–100 sec of μm resolution). The recent work by Seymour et al. clearly demonstrates the strength of this technique and already indicates the development of anomalous hydrodynamic dispersion in biofilm-affected porous media (Seymour et al. 2004b; Seymour et al. 2007).

In the future, this technique will allow for the correlation of local and bulk dispersivity with the distribution of biofilms in porous media environments. This capability will ultimately allow for the development of improved models for the description of reactive transport in biofilm-affected porous media. MRM techniques as well as other techniques capable of spatially resolving dispersion and diffusion phenomena will also allow for the evaluation of the importance of diffusive transport processes in biofilms.

The importance of diffusion on biofilm processes in general has been reported widely (Williamson and McCarty 1976a,b; Debeer et al. 1994; Stewart 1996, 1998; Xu et al. 1998, 2003; Stewart and Costerton 2001) and its influence on reactive transport in porous media affected by biofilms should be considered to explain micro- and macroscale processes (Seymour et al. 2004a,b, 2007).

5.4.5 Constant Head versus Constant Flow

Most experimental systems for the investigation of the influence of biofilm on porous media hydrodynamics have been operated under constant flow conditions by using pumps. Although there are situations in industry and the environment where constant flow conditions are encountered, constant head conditions are more commonly encountered in subsurface environments. For instance, most aquifers are subject to constant head conditions. This section will give an example that clearly demonstrates differences in biofilm development and its influence on porous media hydrodynamics under constant flow and constant head conditions.

Significant biofilm growth in porous media will reduce the pore space available for advective flow. Figures 5.7 and 5.8 provide a comparison of tracer studies and images, which compare solute transport through two-dimensional mesoscale (3.8 cm × 8.5 cm) porous media reactors. Similar studies demonstrating the utility of these reactors for monitoring reactive and nonreactive transport in porous media were conducted with *Vibrio fischeri* (Sharp et al. 2005). It became apparent in the studies by Sharp et al. (2005) that electron

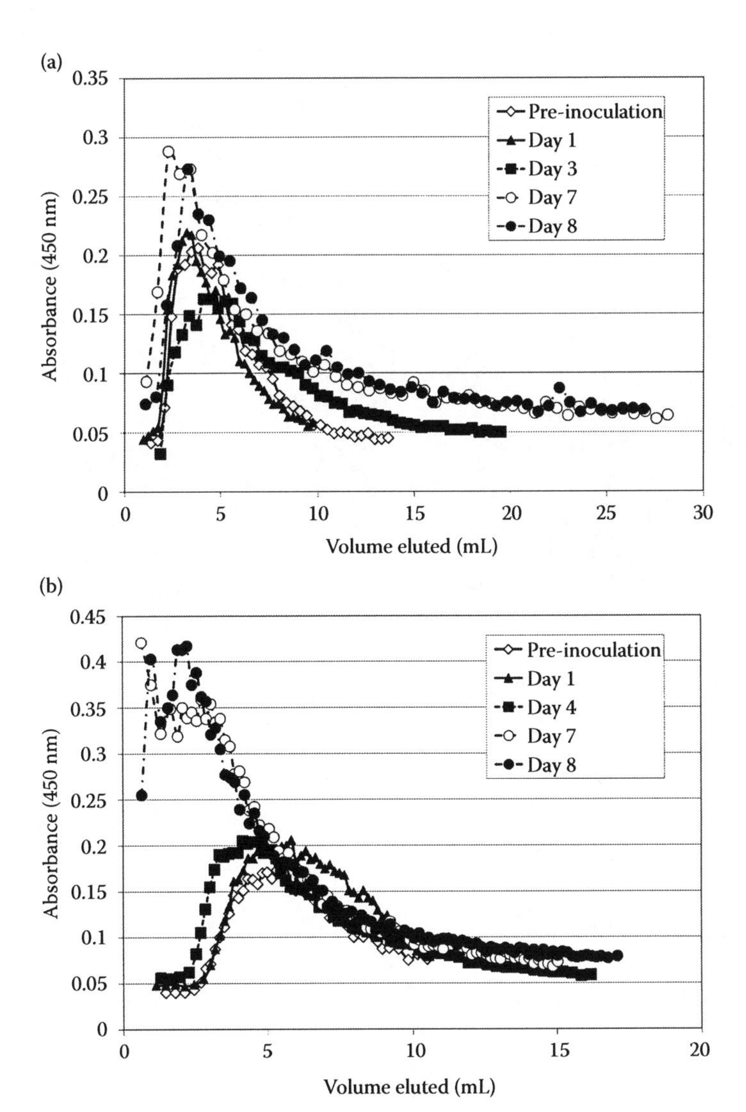

FIGURE 5.7
Tracer breakthrough curves acquired over time for two different flat-plate porous media reactor systems: a constant flow system (a) and a constant head system (b). The constant flow system clearly shows an accelerated breakthrough after 7 days, while the constant head system shows increased retention and dispersion of the dye tracer beginning on day 3. Figure 5.8 shows pictures for the 7 day tracer study for each system (as well as for a clean ["pre-inoculation"] reactor).

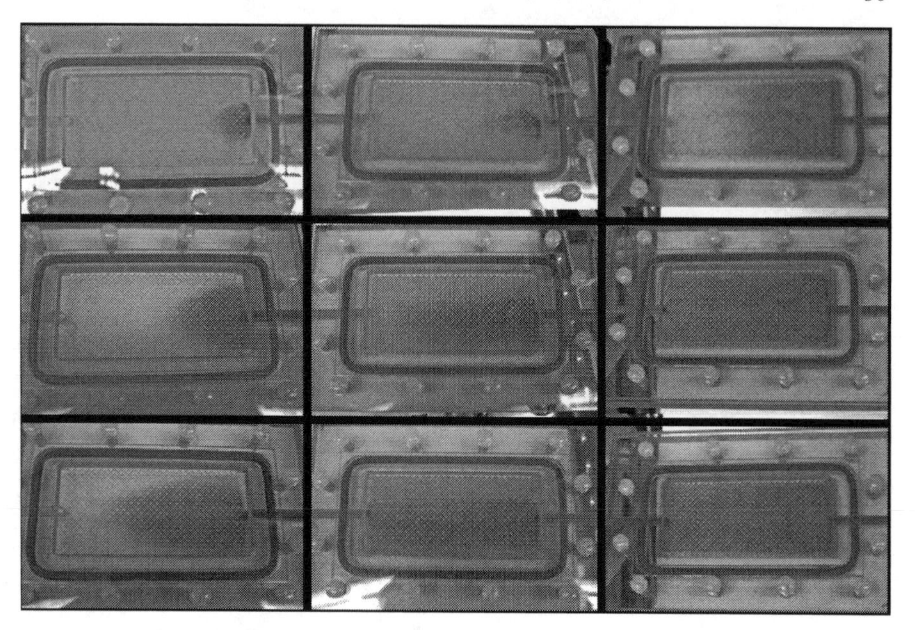

FIGURE 5.8
Images of flat-plate porous media reactor systems during representative times of tracer studies. Left column: clean reactor before inoculation; center column: constant flow system after 7 days; right column: constant head system. Duration from time of tracer injection increases from top to bottom of each column. The pictures correspond to the day 7 tracer curves in Figure 5.7.

acceptor (i.e., oxygen) availability limited biofilm growth and activity. To circumvent electron acceptor availability-related limitation of biofilm growth, a fermentative microorganism, *Cellulomonas* sp. strain ES6 (Sani et al. 2002; Smith et al. 2002; Viamajala et al. 2008) was utilized in the studies summarized in Figures 5.7 and 5.8.

A comparison of biofilm development and its influence on hydrodynamics between constant head and constant flow conditions becomes immediately obvious. The tracer breakthrough characteristics in the clean (not biofilm-affected) reactors are similar for the constant head and constant flow conditions (Figure 5.7). However, differences in the characteristics of tracer breakthrough become obvious very quickly (compare images in Figure 5.8, taken after 7 days in both reactors). Under constant flow conditions significant changes in tracer breakthrough characteristics become obvious after 7 days, when the initial breakthrough of tracer occurs very quickly, indicating a significant reduction in effective porosity. In addition, some of the dye injected tends to remain in the biofilm-affected constant flow reactor for an extended period of time as indicated by elevated absorbance readings at large eluted volumes after 7 and 9 days. This prolonged retention of dye is likely due to no flow regions of the biofilm-affected reactor and diffusion-limited transport of dye back into the main flow path after the initial front of dye has passed through the reactor. Prolonged retention of dye is even more evident in the biofilm reactor operated under constant head. While the initial breakthrough of dye seems to be relatively unaffected under constant head conditions throughout the experiment, dye retention due to slow (likely diffusion-limited) transport of dye out of the biofilm toward the end of each tracer study becomes significant after 3 days already.

It is clear from the research in our laboratories that there are significant differences between biofilm growth patterns under the different flow regimes and that there is a need for a more thorough understanding of the reasons for these differences. Under constant flow conditions a decrease in effective porosity will increase the fluid velocity and thus in most cases the influent pressure as well as shear stress within the porous medium (data not shown). Biofilms grown under continuous-flow conditions in a two-dimensional flat-plate reactor appear to reach a pseudosteady state in which the average hydraulic residence time changes only slightly although the location of the primary flow path seems to be changing with time (Sharp et al. 1999a, 2005; Arnon et al. 2005b). The time until a pseudosteady state is reached and the extent of porosity reduction depends on the microorganism(s) present, growth conditions (e.g., temperature, electron acceptor availability, pH, etc.), flowrate, and similar parameters. Equivalent observations for constant flow conditions were obtained in our laboratories using MRM in pseudo one-dimensional porous media columns (Seymour et al. 2004b, 2007).

The existence of a "critical shear stress," that is, a shear stress above which significant detachment occurs, has been proposed (Kim and Fogler 2000) and makes intuitive sense under constant flow conditions. Once biofilm growth begins to restrict pore spaces the localized flowrate and associated shear stress increase with decreasing effective porosity and can result in increased biofilm detachment.

In constant head systems, such as many shallow aquifers, a decrease in pore space results in a decrease in overall porosity and, likely, permeability. Since the influent head remains constant, such a decrease in overall permeability will result in a decrease in flowrate according to Darcy's Law.

$$Q = -KA\frac{dh}{dl}$$

where Q is the flow rate (L^3/T, e.g., m^3/sec), K is the hydraulic conductivity (L/T, e.g., m/d), A is the cross-sectional area of flow (L^2, e.g., m^2), dh is the difference in hydraulic head across the reactor (L, e.g., m), and dl the length of the reactor (L, e.g., m).

Overall, the prediction and control of biofilm formation in porous media remains difficult since the ability to observe biofilm development spatially and temporally is severely limited even in laboratory systems. Mathematical modeling can aid in the development of biofilm-based technologies, but the development of conceptual and mathematical models is not only limited by a lack of highly resolved experimental data but also by computational challenges.

5.5 A Few Notes on Modeling

The experimental observations summarized above clearly show that biomass is often distributed heterogeneously in porous media. To properly model biofilm processes in porous media, models should take into account the influence of microscale heterogeneities and distributions. Unfortunately, it can become computationally burdensome to model a large (meter to hundreds of meter) scale system on the microscale. Hence, compromises have to be made with respect to the desired accuracy at the microscale and the computational feasibility for larger-scale systems.

5.5.1 Macroscopic versus Microscopic Models

Bulk-scale models using analytical or fast numerical solutions are computationally efficient and work well for cases where large volume averaging is appropriate, that is, where microscale processes are negligible in relation to the overall behavior of the system (Clement et al. 1996). Such models are capable of modeling bulk changes for parameters, such as porosity, specific surface area, permeability, and dispersivity, in dependence of biofilm formation in porous media but in general they do not assume any specific pattern for microbial growth but instead use macroscopic estimates of the average biomass concentration.

Microscale models, which, in contrast, treat biofilm-affected porous media as multidimensional on the pore scale, can potentially predict localized clogging, which can have a significant effect on the overall hydrodynamics of a system, but become computationally demanding if they are to predict the effect of localized (i.e., pore scale) biofilm growth on the bulk properties of the porous medium.

Over the past years, a number of papers have been published that address the issues associated with bulk-scale modeling, while attempting to obtain computationally efficient, appropriate descriptions of bioclogging processes in porous media. The pore-network model approach allows for the simulation of porosity and hydraulic conductivity changes without having to describe the process of biofilm development on the microscale for every single pore (Dupin and McCarty 2000; Thullner et al. 2002b, 2004). In this context, it is rightfully pointed out by Thullner et al. that there are no published, theoretically-derived hydraulic conductivity versus porosity relationships that account for interpore connections in more than one dimension and heterogeneous biofilm distributions (Thullner et al. 2002b).

In addition, it has been proposed to consider the importance of advective flow within biofilms themselves. There is experimental evidence that advective flow occurs within biofilms (Stoodley et al. 1994; Debeer et al. 1994) but until relatively recently these findings had not been included into mathematical modeling approaches. It is now being suggested to model the biofilm phase itself as a porous medium (Zhang and Bishop 1994; Nguyen et al. 2005; Zacarias et al. 2005; Kapellos et al. 2007a,b). The inclusion of fluid flow through the EPS matrix and biofilm microchannels has been demonstrated to result in improved description of biofilm processes (Seifert and Engesgaard 2007; Thullner and Baveye 2008).

5.5.2 Mixed Domain (Hybrid) Models

More recently, the use of multiscale models has been described, which solve the Navier–Stokes and Brinkman equation numerically and combine the approach with a cellular automaton approach or Lagrangian-type simulations of detached fragment trajectories (Kapellos et al. 2007a,b).

A hybrid Lagrangian particle dynamics model capable of describing biofilm formation in porous media is described in detail in Chapter 7 of this book. Such a model might be better suited to incorporate the heterogeneity of porous media and microbial populations, and to ultimately improve our ability to describe hydrodynamics and mass transport in biofilm-affected porous media.

5.6 Porous Media Biofilms in Nature and Technology

Biofilms are recognized to be present in many industrial, environmental, and medical systems. Initial work mostly focused on the eradication of biofilms by treatment with antimicrobials, but more recently many of the "typical" biofilm properties have been recognized to be potentially advantageous for engineered applications (Petrozzi et al. 1993; Bouwer et al. 2000; Sauer et al. 2002; Stoodley et al. 2002b; Cvitkovitch 2004; Molin et al. 2004; Massoudieh et al. 2007; Costerton 2007; Stewart and Franklin 2008).

Despite the fact that biofilm growth and organization are not yet completely understood (as discussed earlier), it is clear that the establishment of organized biofilm communities can be utilized for benefit. The close proximity of organisms to each other, allowing for the possibility of cell–cell communication, exchange of genetic elements (DNA, RNA), colocation of physiologically different organisms that can facilitate the exchange of metabolites, and the increased resistance to environmental stresses, make biofilms very attractive for technology development.

Owing to their increased tolerance to environmental stress and toxic compounds, biofilm reactors are frequently proposed to be used for the treatment of recalcitrant compounds. The range of such compounds spans from surfactants to herbicides and organic solvents to dyes (Mol et al. 1993; Petrozzi et al. 1993; Jerabkova et al. 1999; Mondragon-Parada et al. 2008). In addition, biofilms can influence colloid transport (Kim and Corapcioglu 1997; Stevik et al. 2004; Muris et al. 2005; Morales et al. 2007). The transport of pathogens (biological colloids) and colloid-mediated contaminant transport through porous media are both areas of importance for public health.

Biofilm technologies are available, and there are ongoing research and development efforts in the areas of subsurface biofilm barriers, hazardous waste treatment, biofilters, enhanced oil recovery, acid mine drainage treatment, filters, and infiltration systems in water and wastewater treatment, biofilms as biosorbents, air pollution control, and municipal solid waste leachate control.

Changes in porosity, permeability, dispersion, and diffusion can be desirable or detrimental for a given application. For instance, in the case of subsurface biofilm barriers for the control of groundwater flow, maximum porosity and permeability reduction is desired. In contrast, in biotrickling filters for wastewater treatment an optimal biofilm thickness is desired, which results in maximum removal of solutes while maintaining fairly high permeability.

To limit the discussion of biofilm technologies in porous media in this chapter, it should be pointed out that moving bed reactors, such as suspended (fluidized) bed reactors, will be excluded from consideration. The extent of mixing and abrasion of biofilm by the carrier material is significantly different from, for instance, filters, soils, and packed bed reactors, and the topic would become too complex to be discussed in sufficient detail here. Membrane systems such as reverse osmosis filtration cartridges are also excluded from consideration, although these networks of flow channels could be considered a special form of porous medium, and biofilm growth (biofouling) in these systems can significantly affect flow as indicated by increasing backpressure in (bio-) fouled membrane systems.

A major factor in the performance of biofilm reactors is the limited mass transport of solutes into the biofilm where the active biomass is located. Transport limitations of electron acceptors and donors can significantly affect the performance of porous media biofilm reactors, especially if the reaction depends on oxygen, which has a very limited solubility in water (Kirchner et al. 1992; Joannis-Cassan et al. 2007). Hence, it is extremely important to understand flow dynamics and solute reactive transport in biofilm-affected porous media (Iliuta and Larachi 2004).

It should also be kept in mind that, in the environment and industrial systems, biofilms are likely to accumulate solutes and possibly significant amounts of minerals. Iron oxides, calcium carbonate, and sulfur-containing minerals are the most frequently described inorganic constituents of biofilms in porous media (Mclean et al. 1997; Cooke et al. 2005). The precipitation of minerals can result in semipermanent to permanent encrustation of biofilm cells or clogging of certain areas in a porous medium. These effects might or might not be desirable depending on the goal of a given application.

The following sections will summarize a few biofilm technologies in porous media in detail and discuss research and development needs associated with the further development of these technologies.

5.6.1 Subsurface Biofilm Barriers for the Control and Remediation of Contaminated Groundwater

Subsurface biofilm barriers are engineered structures developed through the growth and activity of microorganisms in soils. The establishment of permeable, impermeable, and semipermeable biofilm barriers has been proposed for the control and remediation of contaminated soil and groundwater (Cunningham et al. 1997; Waybrant et al. 1998; Benner et al. 1999; Hiebert et al. 2001; Nyman et al. 2002; Ludwig et al. 2002; Cunningham et al. 2003; Komlos et al. 2004).

The establishment of low-permeability biofilm barriers (Figure 5.9) has been the focus of a number of research and development efforts. These barriers are designed to provide maximum reduction of permeability by promoting thick biofilm growth to either reduce the flow of groundwater through certain areas of the subsurface or direct groundwater into a certain direction (e.g., an area where treatment occurs). Maximum permeability reduction is usually obtained by promoting the production of copious amounts of EPS by indigenous microorganisms or through bioaugmentation with organisms known to produce copious amounts of EPS.

These barriers have been shown to reduce porous media permeability by several orders of magnitude (Cunningham et al. 1997, 2003; Ross et al. 1998, 2007; Hiebert et al. 2001; Komlos et al. 2004). More recently, the biofilm-promoted precipitation of carbonate minerals has been proposed to be advantageous for long-term biofilm barrier stability (Gerlach et al. 2009).

In addition, these subsurface biofilm barriers have been shown to effectively remove solutes. The consumption of soluble electron acceptors such as oxygen as well as the removal of contaminants of concern such as nitrate (Hiebert et al. 2001; Cunningham et al. 2003) has been shown. Cunningham et al. (2003) described the establishment of a nitrate-remediating biofilm barrier in a 130 ft wide, 180 ft long, 21 ft deep test cell. Starved cells of *Pseudomonas fluorescens* strain CPC211a were injected and biofilm growth was stimulated through the injection of a growth nutrient mixture composed of molasses, nitrate, and other additives. The biofilm barrier reduced the soil hydraulic conductivity by 99% from an initial value of 0.042 cm/sec and reliably reduced nitrate concentrations by 93% or more from an initial value

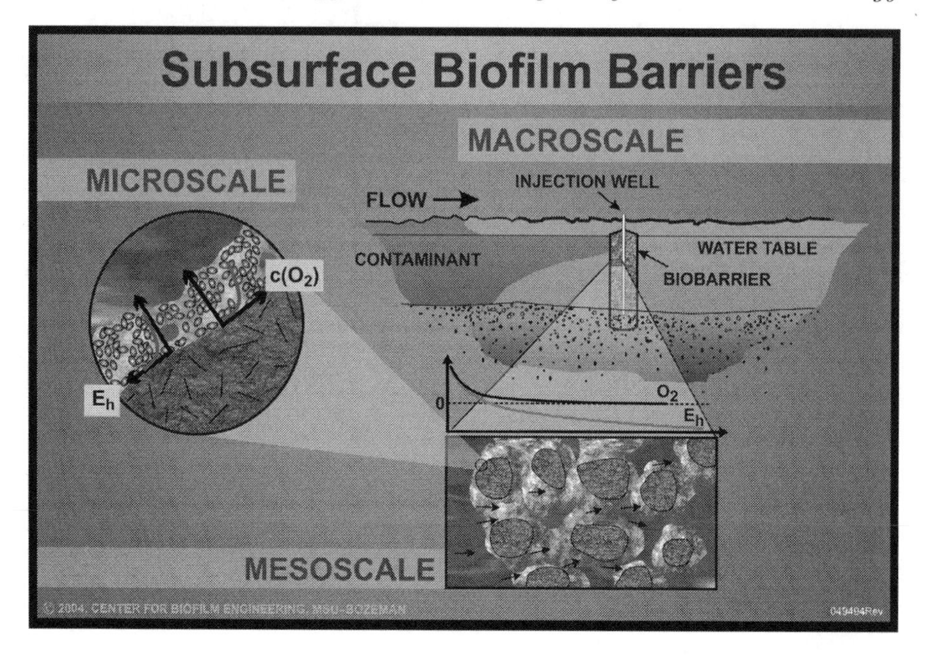

FIGURE 5.9
Schematic of a low-permeability subsurface biofilm barrier. Such barriers have been established at the field scale through the injection of bacteria and/or nutrients. Thick biofilm growth is promoted in the pore space, which reduces soil permeability. Nutrient and electron acceptor consumption create gradients (as indicated by decreasing oxygen concentrations and redox potential [E_h]) along the direction of groundwater flow (see mesoscale schematic) and transversally into the biofilm (see microscale schematic).

of approximately 100 mg/L (Cunningham et al. 2003). This biofilm barrier was designed for maximum reduction of permeability; however, the observation of nearly complete nitrate removal has led to additional research efforts designed to optimize permeability and contaminant transformation rates in reactive subsurface biofilm barriers.

Low permeability and reactive as well as permeable and reactive subsurface biofilm barriers have been proposed (Figure 5.10). Komlos et al. (2004, 2006) evaluated the possibility of establishing biofilm barriers capable of simultaneously providing a reduction in soil permeability and trichloroethylene (TCE) degradation. These studies led to the development of readily deployable strategies for controlling the relative abundance of different species in defined mixed-culture biofilms in porous media (Komlos et al. 2004, 2005, 2006).

Permeable reactive subsurface barriers for the treatment of chlorinated solvents, heavy metals, and radionuclides have been proposed and employed in

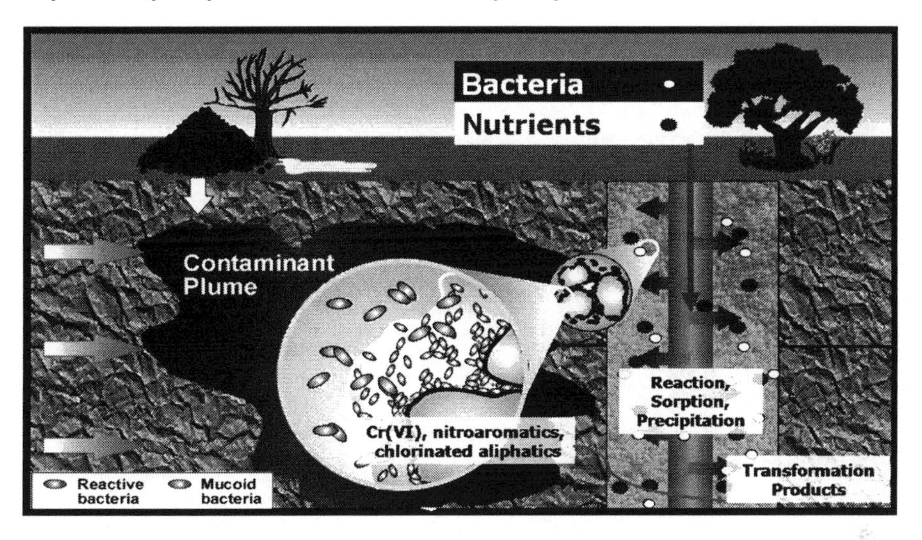

FIGURE 5.10
Schematic of a reactive biofilm barrier. Depending on the goal of the barrier (reduction in permeability [groundwater flow] *and* degradation of contaminants or only degradation of contaminants but no hydraulic manipulation) a different degree of permeability reduction, and thus biofilm growth, is desired. The barrier can be established through the injection of nutrients (carbon source, electron donor, electron acceptor, etc.) and bacteria (if necessary). Contaminant transformation, sorption, and precipitation will all play a role in the ultimate fate of the contaminants.

field situations. Zero-valent iron-based barriers are probably the most common type of permeable subsurface barrier employed (Gould 1982; Cantrell et al. 1995; O'Hannesin and Gillham 1998; Fiedor et al. 1998; Scherer et al. 2000; Wilkin et al. 2003), and the influence of attached microorganisms on their performance has been noted (Gerlach et al. 2000; Gu et al. 2002; Shin et al. 2007). Furthermore, biologically active permeable reactive barriers for the control and treatment of acidic mine wastewater have been proposed and demonstrated in the field (Waybrant et al. 1998; Benner et al. 1999; Ludwig et al. 2002; Golab et al. 2006; Davis et al. 2007). Chemical *in situ* redox manipulation has also been demonstrated in situations where trench and fill methods are impractical owing to depth limitations (Istok et al. 1999; Seaman et al. 1999).

On the basis of the success and relative ease of implementation of these technologies, combined with the possibility of establishing biologically active zones in the deep subsurface through the careful control of distribution and activity of microorganisms, the development of permeable subsurface biofilm barriers has immense economical potential. Especially the possibility of employing such treatment strategies in the deep subsurface through the use of injection wells has resulted in a flurry of research and development efforts that contribute toward the continued development of such permeable reactive subsurface biofilm barriers.

5.6.2 Deep Subsurface Biofilms for Enhanced Oil Recovery and Carbon Sequestration

The initial recovery of oil from deep subsurface reservoirs is typically estimated to reach between 10% and 35% of a reservoir's oil. Secondary recovery, which most often involves waterflooding, can increase recovery by 20% or more. Tertiary recovery, often also called secondary enhanced oil recovery, further increases the recovery of oil from these reservoirs. Thermal recovery, chemical flooding, miscible displacement (such as gas injection), and microbially enhanced oil recovery (MEOR) have been explored as tertiary techniques.

Microbially enhanced oil recovery is proposed to be applicable in a number of different ways, such as production of biosurfactants or gases to enhance oil mobility, selective plugging of high-permeability channels in reservoir rock to increase sweep efficiency, and *in situ* biocracking, during which microbes break down long alkane chains to produce higher solubility shorter alkane chains.

A flurry of research and development efforts in the 1980s and 1990s investigated the use of microorganisms to enhance oil recovery from deep subsurface oil-bearing formations, and detailed information can be found in a number of books and reviews (Zajic and Donaldson 1985; Kosaric et al. 1987; Donaldson et al. 1989; Yen 1990).

Biofilms are thought to be effective in selectively plugging water-filled high-permeability areas, also called thief zones. Selectively reducing the permeability of such high-permeability areas would allow fluids used for enhanced secondary or tertiary oil recovery, such as water or supercritical CO_2 ($scCO_2$), to more effectively mobilize residual oil in low-permeability areas of the formation. Significant research and development efforts have been conducted to develop this technology (e.g., Shaw et al. 1985; MacLeod et al. 1988; Lappin-Scott et al. 1988a,b; Cusack et al. 1992; Lappin-Scott and Costerton 1992) yet widespread reports of successful implementation of this technology are lacking. This may be due to difficulties in evaluating the economics of using biofilm-mediated plugging of high-permeability areas in the deep subsurface. Determining the location of biofilms in the deep subsurface is even more challenging than the imaging of biofilms in laboratory-scale reactors as described earlier. Hence, MEOR is continuing to be proposed for application in the oilfield but has yet to gain widespread acceptance, and thus application.

The potential of using biofilms to enhance the deep subsurface sequestration of carbon dioxide has been proposed as well (Davis et al. 2006; Mitchell et al. 2008a; Mitchell et al. 2009). Recent work by Mitchell et al. (2008, 2009) demonstrated the utility of biofilms in sandstone cores to reduce the permeability by more than 95% at elevated pressures (8.9 MPa) and moderate temperatures (32°C). Biofilm organisms under these conditions were also demonstrated to survive exposure to $scCO_2$ and maintain the decreased permeability of the core even under starvation conditions.

5.6.3 Porous Media Biofilm Reactors in Industry and Waste Treatment

Other industrial and environmental systems and applications in which porous media biofilms play a significant role are mostly related to water and waste treatment.

Attached microorganisms in water filtration, which can form within, at the influent, or effluent of filtration devices, can have detrimental or beneficial effects on water treatment processes. Obviously, excessive microbial growth can reduce the hydraulic conductivity of filtration devices and result in increased maintenance requirements, reduced run times, increased frequency of backflushing cycles, or increased need to exchange filter materials. However, porous media biofilms have also been observed to facilitate the sorption or degradation of organic particulates and solutes as well as the oxidative precipitation of problem metals, such as reduced iron and manganese, during drinking water treatment.

The use of trickling filters and infiltration systems in wastewater treatment has been practiced for a long time and microorganisms, more or less immobilized in porous media, have been used for the removal of wastewater constituents. Both, large-scale applications of these technologies as well as fundamental research assessing the role of biofilms in these technologies are ongoing (Ileri and Muslu 1996; Iliuta and Larachi 2004; Wanko et al. 2005, 2006; Mauclaire et al. 2006).

Some of the characteristics discussed earlier, such as enhanced resistance to adverse conditions and enhanced-metabolic capabilities of biofilm communities, have also led to an increased use of immobilized microbial communities in hazardous waste treatment. The treatment of synthetic dyes or tannery wastewater in fixed-bed biofilm reactors are two examples of such processes (Song et al. 2003; Tse and Yu 2003).

Gas-phase treatment using biofilters has also been practiced for an extended period of time, especially for the control of odors, such as hydrogen sulfide, and volatile organic compounds, such as methanol or acetone. While such technologies are already being applied on the industrial scale (Shareefdeen and Baltzis 1994; Kennes and Thalasso 1998; Shareefdeen et al. 2002; Elmrini et al. 2004), fundamental research targeted at understanding the physico- and biochemical processes controlling the efficiency of such vapor-filtration devices is ongoing (Hwang et al. 1997; Tang et al. 1997; Wani et al. 1997; Ramirez et al. 2008). Overall, biofilm processes in unsaturated porous media have been studied much less extensively than in saturated porous media and much work has yet to be done to better understand and consequently capitalize on the existing potential.

5.7 Conclusions and Outlook

Biofilms can have a significant influence on hydrodynamics and thus on mass transport in porous media. Biofilm growth quickly induces heterogeneity into even highly uniform porous media and the dynamics of biofilm development are—at this point—somewhat unpredictable owing to the influence biofilm growth and hydrodynamics have on each other. Hydrodynamics, nutrient availability, physicochemical conditions (e.g., temperature, pH), presence (or production) of inhibitory or stimulating compounds, community composition and structure are only some of the factors that have been shown to influence the development of biofilms in porous media. The interplay of hydrodynamics and biofilm growth leads to an oscillatory behavior, which appears to reach a pseudosteady state at certain scales in time and space. The ability to predict the size of these relevant scales remains a challenge.

In general, porous media permeability and porosity decrease during biofilm formation while dispersivity generally increases. Although rarely observed, opposite effects might also occur.

The immense beneficial (and detrimental) potential of porous media biofilms has led to the development of large-scale technologies, such as subsurface biofilm barriers and biofilm reactors, as well as strategies on how to control biofilms in porous media environments (e.g., around injection wells where excessive biofilm growth is undesirable).

Owing to the challenges of studying porous media biofilms in detail, a number of these developments have been based more or less on science-based understanding. The (mostly microscale) research summarized in this chapter has been designed to develop the fundamental knowledge necessary to develop science-based engineering strategies for porous media biofilm technologies.

Even without a complete understanding of biofilm processes in porous media, the potential of promoting subsurface biofilm growth for benefit has been widely acknowledged and technology development in the areas of oil production, contaminant remediation, carbon sequestration, soil stabilization, and waste treatment proves its economic potential.

The widely accepted inherent resistance of biofilm organisms to environmental stresses has made biofilm barriers an attractive strategy. Even exposure for extended periods of time to toxins; absence of nutrientsl; low water potential; highly saline, acidic, or basic conditions; and supercritical carbon dioxide have been shown to affect biofilm barriers in porous media only slightly (Sturman et al. 1995; Warwood et al. 1995; Cunningham et al. 1997; Dennis and Turner 1998; Bouwer et al. 2000; Hiebert et al. 2001; Cunningham et al. 2003; Komlos et al. 2004; Mitchell et al. 2008a,b, 2009).

5.8 References

Allison, L. E. (1947). Effect of microorganisms on permeability of soil under prolonged submergence. *Soil Science*, **63**:439–450.

Arnon, S., Adar, E., Ronen, Z., Yakirevich, A., and Nativ, R. (2005a). Impact of microbial activity on the hydraulic properties of fractured chalk. *Journal of Contaminant Hydrology*, **76**:315–336.

Arnon, S., Ronen, Z., Adar, E., Yakirevich, A., and Nativ, R. (2005b). Two-dimensional distribution of microbial activity and flow patterns within naturally fractured chalk. *Journal of Contaminant Hydrology*, **79**:165–186.

Baveye, P., Vandevivere, P., and Delozada, D. (1992). Biofilm growth and the related changes in the physical-properties of a porous-medium. 1. Experimental investigation—comment. *Water Resources Research*, **28**:1481–1482.

Benner, S. G., Blowes, D. W., Gould, W. D., Herbert, R. B., and Ptacek, C. J. (1999). Geochemistry of a permeable reactive barrier for metals and acid mine drainage. *Environmental Science & Technology*, **33**:2793–2799.

Bielefeldt, A. R., Illangasekare, T., Uttecht, M., and LaPlante, R. (2002a). Biodegradation of propylene glycol and associated hydrodynamic effects in sand. *Water Research*, **36**:1707–1714.

Bielefeldt, A. R., McEachern, C., and Illangasekare, T. (2002b). Hydrodynamic changes in sand due to biogrowth on naphthalene and decane. *Journal of Environmental Engineering-Asce*, **128**:51–59.

Bouwer, E. J., Rijnaarts, H. H. M., Cunningham, A. B., and Gerlach, R. (2000). *Biofilms in Porous Media, Biofilms II: Process Analysis and Applications*, ed. J. D. Bryers. Wiley-Liss, Inc., New York, 123–158.

Brown, M. R. W., Allison, D. G., and Gilbert, P. (1988). Resistance of bacterial biofilms to antibiotics—a growth-rate related effect. *Journal of Antimicrobial Chemotherapy*, **22**:777–780.

Bryers, J. D. (2000). *Biofilms II: Process Analysis and Applications.* Wiley, New York.

Cantrell, K. J., Kaplan, D. I., and Wietsma, T. W. (1995). Zero-valent iron for the in-situ remediation of selected metals in groundwater. *Journal of Hazardous Materials*, **42**:201–212.

Castegnier, F., Ross, N., Chapuis, R. P., Deschenes, L., and Samson, R. (2006). Long-term persistence of a nutrient-starved biofilm in a limestone fracture. *Water Research*, **40**:925–934.

Characklis, W. G. (1973a). Attached microbial growths. 2. Frictional resistance due to microbial slimes. *Water Research*, **7**:1249–1258.

Characklis, W. G. (1973b). Attached microbial growths. 1. Attachment and growth. *Water Research*, **7**:1113–1127.

Characklis, W. G. and Bryers, J. D. (2008). Bioengineering report: fouling biofilm development: a process analysis. *Biotechnology and Bioengineering*, **102**:309–347.

Characklis, W. G. and Marshall, K. C. (1990). *Biofilms*. Wiley, New York.

Chen, C. I., Mueller, R. F., and Griebe, T. (1994). Kinetic-analysis of microbial sulfate reduction by desulfovibrio-desulfuricans in an anaerobic upflow porous-media biofilm reactor. *Biotechnology and Bioengineering*, **43**: 267–274.

Chen, X., Schauder, S., Potier, N., Van Dorsselaer, A., Pelczer, I., Bassler, B. L., and Hughson, F. M. (2002). Structural identification of a bacterial quorum-sensing signal containing boron. *Nature*, **415**:545–549.

Clement, T. P., Hooker, B. S., and Skeen, R. S. (1996). Macroscopic models for predicting changes in saturated porous media properties caused by microbial growth. *Ground Water*, **34**:934–942.

Cooke, A. J., Rowe, R. K., VanGulck, J., and Rittmann, B. E. (2005). Application of the BioClog model for landfill leachate clogging of gravel-packed columns. *Canadian Geotechnical Journal*, **42**:1600–1614.

Costerton, J. W. (2007). *The Biofilm Primer*. Springer, Berlin Heidelberg, New York.

Costerton, J. W., Geesey, G. G., and Cheng, K. J. (1978). How bacteria stick. *Scientific American*, **238**:86–95.

Costerton, J. W., Lewandowski, Z., Caldwell, D. E., Korber, D. R., and Lappin-Scott, H. M. (1995). Microbial biofilms. *Annual Review of Microbiology*, **49**:711–745.

Crespi, B. and Springer, S. (2003). Social slime molds meet their match. *Science*, **299**:56–57.

Cunningham, A. B., Characklis, W. G., Abedeen, F., and Crawford, D. (1991). Influence of biofilm accumulation on porous-media hydrodynamics. *Environmental Science & Technology*, **25**:1305–1311.

Cunningham, A. B., Sharp, R. R., Caccavo, F., and Gerlach, R. (2007). Effects of starvation on bacterial transport through porous media. *Advances in Water Resources*, **30**:1583–1592.

Cunningham, A. B., Sharp, R. R., Hiebert, R., and James, G. (2003). Subsurface biofilm barriers for the containment and remediation of contaminated groundwater. *Bioremediation Journal*, **7**:151–164.

Cunningham, A. B., Visser, E., Lewandowski, Z., and Abrahamson, M. (1995). Evaluation of a coupled mass transport-biofilm process model using dissolved oxygen microsensors. *Water Science and Technology*, **32**: 107–114.

Cunningham, A. B., Warwood, B., Sturman, P., Horrigan, K., James, G., Costerton, J. W., and Hiebert, R. (1997). Biofilm processes in porous media—practical applications. In *The Microbiology of the Terrestrial Deep Subsurface*, ed. P. S. Amy and D. L. Haldeman. CRC Press, Boca Raton, FL, pp. 325–344.

Cusack, F., Singh, S., McCarthy, C., Grecco, J., De Rocco, M., Nguyen, D., Lappin-Scott, H. M., and Costerton, J. W. (1992). Enhanced oil recovery—three dimensional sandpack simulation of ultramicrobacteria resuscitation in reservoir formations. *Journal of General Microbiology*, **138**: 647–655.

Cvitkovitch, D. G. (2004). Genetic exchange in biofilms. In *Microbial Biofilms*, ed. G. O'Toole and M. Ghannoum. ASM Press, Washington, DC, pp. 192–205.

Davis, A. C., Patterson, B. M., Grassi, M. E., Robertson, B. S., Prommer, H., and Mckinley, A. J. (2007). Effects of increasing acidity on metal(loid) bioprecipitation in groundwater: column studies. *Environmental Science & Technology*, **41**:7131–7137.

Davis, C. A., Atekwana, E., Atekwana, E., Slater, L. D., Rossbach, S., and Mormile, M. R. (2006). Microbial growth and biofilm formation in geologic media is detected with complex conductivity measurements. *Geophysical Research Letters*, **33**. L18403. doi: 10.1029/2006GL027312.

Debeer, D., Stoodley, P., Roe, F., and Lewandowski, Z. (1994). Effects of biofilm structures on oxygen distribution and mass-transport. *Biotechnology and Bioengineering*, **43**:1131–1138.

Dennis, M. L. and Turner, J. P. (1998). Hydraulic conductivity of compacted soil treated with biofilm. *Journal of Geotechnical and Geoenvironmental Engineering*, **124**:120–127.

Dixon, B. (2009). *Animalcules: The Activities, Impacts, and Investigators of Microbes.* ASM Press, Washington, DC.

Donaldson, E. C., Chillingarian, G. V., and Yen, T. F. (1989). *Microbial Enhanced Oil Recovery: Fundamentals and Analysis. Development in Petroleum Sciences.* Elsevier, Amsterdam.

Dorn, J. G., Mahal, M. K., Brusseau, M. L., and Maier, R. M. (2004). Employing a novel fiber optic detection system to monitor the dynamics of in situ lux bioreporter activity in porous media: system performance update. *Analytica Chimica Acta*, **525**:63–74.

Dupin, H. J. and McCarty, P. L. (1999). Mesoscale and microscale observations of biological growth in a silicon pore imaging element. *Environmental Science & Technology*, **33**:1230–1236.

Dupin, H. J. and McCarty, P. L. (2000). Impact of colony morphologies and disinfection on biological clogging in porous media. *Environmental Science & Technology*, **34**:1513–1520.

Elmrini, H., Bredin, N., Shareefdeen, Z., and Heitz, M. (2004). Biofiltration of xylene emissions: bioreactor response to variations in the pollutant inlet concentration and gas flow rate. *Chemical Engineering Journal*, **100**: 149–158.

Evans, L. V. (2000). *Biofilms: Recent Advances in their Study and Control.* Harwood Academic Publishers, Amsterdam.

Fiedor, J. N., Bostick, W. D., Jarabek, R. J., and Farrell, J. (1998). Understanding the mechanism of uranium removal from groundwater by zero-valent iron using X-ray photoelectron spectroscopy. *Environmental Science & Technology*, **32**:1466–1473.

Fletcher, M. (1979). Micro-auto-radiographic study of the activity of attached and free-living bacteria. *Archives of Microbiology*, **122**:271–274.

Frankenberger, W. T., Troeh, F. R., and Dumenil, L. C. (1979). Bacterial effects on hydraulic conductivity of soils. *Soil Science Society of America Journal*, **43**:333–338.

Fux, C. A., Costerton, J. W., Stewart, P. S., and Stoodley, P. (2005). Survival strategies of infectious biofilms. *Trends in Microbiology*, **13**:34–40.

Gaillard, J. F., Chen, C., Stonedahl, S. H., Lau, B. L. T., Keane, D. T., and Packman, A. I. (2007). Imaging of colloidal deposits in granular porous media by X-ray difference micro-tomography. *Geophysical Research Letters*, **34**. L18404. doi: 10.1029/2006GL027312.

Geesey, G. G. and Mitchell, A. C. (2008). Need or direct measurements of coupled microbiological and hydrological processes at different scales in porous media systems. *Journal of Hydrologic Engineering*, **13**:28–36.

Gerlach, R., Cunningham, A. B., and Caccavo, F. (2000). Dissimilatory iron-reducing bacteria can influence the reduction of carbon tetrachloride by iron metal. *Environmental Science & Technology*, **34**:2461–2464.

Gerlach, R., Cunningham, A. B., Ferris, F. G., and Mitchell, A. C. (2009). *Metal and Carbon Dioxide Sequestration through Biologically Induced Mineral Precipitation: Influence of Hydrodynamics.* Platform presentation. 237th ACS National Meeting. Division of Geochemistry. Coprecipitation of Metals during Chemically and Biologically Induced Mineral Precipitation. Salt Lake City, Utah. March 22–26.

Ghannoum, M. and O'Toole, G. (2004). *Microbial Biofilms.* ASM Press, Washington, DC.

Gilbert, P., Collier, P. J., and Brown, M. R. W. (1990). Influence of growth-rate on susceptibility to antimicrobial agents—biofilms, cell-cycle, dormancy, and stringent response. *Antimicrobial Agents and Chemotherapy,* **34**:1865–1868.

Ginn, T. R., Wood, B. D., Nelson, K. E., Scheibe, T. D., Murphy, E. M., and Clement, T. P. (2002). Processes in microbial transport in the natural subsurface. *Advances in Water Resources*, **25**:1017–1042.

Golab, A. N., Peterson, M. A., and Indraratna, B. (2006). Selection of potential reactive materials for a permeable reactive barrier for remediating acidic groundwater in acid sulphate soil terrains. *Quarterly Journal of Engineering Geology and Hydrogeology*, **39**:209–223.

Gould, J. P. (1982). The kinetics of hexavalent chromium reduction by metallic iron. *Water Research*, **16**:871–877.

Gu, B. H., Watson, D. B., Wu, L. Y., Phillips, D. H., White, D. C., and Zhou, J. Z. (2002). Microbiological characteristics in a zero-valent iron reactive barrier. *Environmental Monitoring and Assessment*, **77**: 293–309.

Gupta, R. P. and Swartzendruber, D. (1962). Flow-associated reduction in the hydraulic conductivity of quartz sand. *Soil Science Society of America Proceedings*, **26**:6–10.

Hall-Stoodley, L., Costerton, J. W., and Stoodley, P. (2004). Bacterial biofilms: from the natural environment to infectious diseases. *Nature Reviews Microbiology*, **2**:95–108.

Hand, V. L., Lloyd, J. R., Vaughan, D. J., Wilkins, M. J., and Boult, S. (2008). Experimental studies of the influence of grain size, oxygen availability and organic carbon availability on bioclogging in porous media. *Environmental Science & Technology*, **42**:1485–1491.

Hart, R. T., Fekete, T., and Flock, D. L. (1960). The plugging effect of bacteria in sandstone systems. *Canadian Mining Metallurgical Bulletin*, **20**:495–501.

Hentzer, M., Givskov, M., and Ebihara, T. (2004). Quorum sensing in biofilms: gossip in slime city. In *Microbial Biofilms*, ed. G. O'Toole and M. Ghannoum. ASM Press, Washington, DC, pp. 118–140.

Hiebert, R., Sharp, R. R., Cunningham, A. B., and James, G. (2001). Development and demonstration of subsurface biofilm barriers using starved bacterial cultures. *Contaminated Soil Sediment & Water*, August 45–47.

Hill, D. D. and Sleep, B. E. (2002). Effects of biofilm growth on flow and transport through a glass parallel plate fracture. *Journal of Contaminant Hydrology*, **56**:227–246.

Hornemann, J. A., Codd, S. L., Romanenko, K. V., and Seymour, J. D. (2009). T2-T2 exchange in biofouled porous media. *Diffusion Fundamentals*, **10**:1.1–1.3.

Hoskins, B. C., Fevang, L., Majors, P. D., Sharma, M. M., and Georgiou, G. (1999). Selective imaging of biofilms in porous media by NMR relaxation. *Journal of Magnetic Resonance*, **139**:67–73.

Hosokawa, Y., Ootsuki, T., and Niwa, C. (1992). Channel experiments on coastal water-purification by porous bed using crushed stones. *Water Science and Technology*, **26**:2007–2010.

Hunt, S. M., Werner, E. M., Huang, B. C., Hamilton, M. A., and Stewart, P. S. (2004). Hypothesis for the role of nutrient starvation in biofilm detachment. *Applied and Environmental Microbiology,* **70**:7418–7425.

Hwang, S. J., Tang, H. M., and Wang, W. C. (1997). Modeling of acetone biofiltration process. *Environmental Progress*, **16**:187–192.

Ileri, R. and Muslu, Y. (1996). Unsaturated flow phenomena in porous medium as in a trickling filter. *Journal of Chemical Technology and Biotechnology*, **66**:25–34.

Iliuta, I. and Larachi, F. (2004). Biomass accumulation and clogging in trickle-bed bioreactors. *Aiche Journal*, **50**:2541–2551.

Ince, O., Ince, B. K., and Donnelly, T. (2000). Attachment, strength and performance of a porous media in an upflow anaerobic filter treating dairy wastewater. *Water Science and Technology*, **41**:261–270.

Istok, J. D., Amonette, J. E., Cole, C. R., Fruchter, J. S., Humphrey, M. D., Szecsody, J. E., Teel, S. S., Vermeul, V. R., Williams, M. D., and Yabusaki, S. B. (1999). In situ redox manipulation by dithionite injection: intermediate-scale laboratory experiments. *Ground Water*, **37**:884–889.

Jean, J. S., Tsao, C. W., and Chung, M. C. (2004). Comparative endoscopic and SEM analyses and imaging for biofilm, growth on porous quartz sand. *Biogeochemistry*, **70**:427–445.

Jennings, D. A., Petersen, J. N., Skeen, R. S., Hooker, B. S., Peyton, B. M., Johnstone, D. L., and Yonge, D. R. (1995). Effects of slight variations in nutrient loadings on pore plugging in soil columns. *Applied Biochemistry and Biotechnology*, **51**–**52**:727–734.

Jerabkova, H., Kralova, B., and Nahlik, J. (1999). Biofilm of Pseudomonas C12B on glass support as catalytic agent for continuous SDS removal. *International Biodeterioration & Biodegradation*, **44**:233–241.

Joannis-Cassan, C., Delia, M., and Riba, J. P. (2007). Biofilm growth kinetics on hydrocarbon in a porous medium under biostimulation conditions. *Environmental Progress*, **26**:140–148.

Kalish, P. J., Stewart, J. A., Bennett, E. O., and Rogers, W. F. (1964). Effect of bacteria on sandstone permeability. *Journal of Petroleum Technology*, **16**:805–814.

Kapellos, G. E., Alexiou, T. S., and Payatakes, A. C. (2007a). A multiscale theoretical model for diffusive mass transfer in cellular biological media. *Mathematical Biosciences*, **210**:177–237.

Kapellos, G. E., Alexiou, T. S., and Payatakes, A. C. (2007b). Hierarchical simulator of biofilm growth and dynamics in granular porous materials. *Advances in Water Resources*, **30**:1648–1667.

Kennes, C. and Thalasso, F. (1998). Waste gas biotreatment technology. *Journal of Chemical Technology and Biotechnology*, **72**:303–319.

Kildsgaard, J. and Engesgaard, P. (2001). Numerical analysis of biological clogging in two-dimensional sand box experiments. *Journal of Contaminant Hydrology*, **50**:261–285.

Kim, D. S. and Fogler, H. S. (1999). The effects of exopolymers on cell morphology and culturability of Leuconostoc mesenteroides during starvation. *Applied Microbiology and Biotechnology*, **52**:839–844.

Kim, D. S. and Fogler, H. S. (2000). Biomass evolution in porous media and its effects on permeability under starvation conditions. *Biotechnology and Bioengineering*, **69**:47–56.

Kim, G., Lee, S., and Kim, Y. (2006). Subsurface biobarrier formation by microorganism injection for contaminant plume control. *Journal of Bioscience and Bioengineering*, **101**:142–148.

Kim, J., Hahn, J. S., Franklin, M. J., Stewart, P. S., and Yoon, J. (2009). Tolerance of dormant and active cells in *Pseudomonas aeruginosa* PA01 biofilm to antimicrobial agents. *Journal of Antimicrobial Chemotherapy*, **63**:129–135.

Kim, S. and Corapcioglu, M. Y. (1997). The role of biofilm growth in bacteria-facilitated contaminant transport in porous media. *Transport in Porous Media*, **26**:161–181.

Kirchner, K., Wagner, S., and Rehm, H. J. (1992). Exhaust-gas purification using biocatalysts (fixed bacteria monocultures)—the influence of biofilm diffusion rate (O_2) on the overall reaction-rate. *Applied Microbiology and Biotechnology*, **37**:277–279.

Knutson, C., Valocchi, A., and Werth, C. (2007). Comparison of continuum and pore-scale models of nutrient biodegradation under transverse mixing conditions. *Advances in Water Resources*, **30**:1421–1431.

Knutson, C. E., Werth, C. J., and Valocchi, A. J. (2005). Pore-scale simulation of biomass growth along the transverse mixing zone of a model two-dimensional porous medium. *Water Resources Research,* **41**. WO7007. doi: 10.1029/2004 WR003459.

Komlos, J., Cunningham, A. B., Camper, A. K., and Sharp, R. R. (2004). Biofilm barriers to contain and degrade dissolved trichloroethylene. *Environmental Progress*, **23**:69–77.

Komlos, J., Cunningham, A. B., Camper, A. K., and Sharp, R. R. (2005). Interaction of *Klebsiella oxytoca* and *Burkholderia cepacia* in dual-species batch cultures and biofilms as a function of growth rate and substrate concentration. *Microbial Ecology*, **49**:114–125.

Komlos, J., Cunningham, A. B., Camper, A. K., and Sharp, R. R. (2006). Effect of substrate concentration on dual-species biofilm population densities of *Klebsielia oxytoca* and *Burkholderia cepacia* in porous media. *Biotechnology and Bioengineering*, **93**:434–442.

Kosaric, N., Cairns, W. L., and Gray, N. C. C. (1987). *Biosurfactants and Biotechnology.* Marcel Dekker, New York.

Kuhl, M. and Jorgensen, B. B. (1992). Microsensor measurements of sulfate reduction and sulfide oxidation in compact microbial communities of aerobic biofilms. *Applied and Environmental Microbiology.* **58**:1164–1174.

Lappin-Scott, H. M. and Costerton, J. W. (1992). Ultramicrobacteria and their biotechnological applications. *Current Opinion in Biotechnology,* **3**:283–285.

Lappin-Scott, H. M. and Costerton, J. W. (1995). *Microbial Biofilms.* Cambridge University Press, Cambridge.

Lappin-Scott, H. M., Cusack, F., MacLeod, F. A., and Costerton, J. W. (1988a). Starvation and nutrient resuscitation of *Klebsiella pneumoniae* isolated from oil well waters. *The Journal of Applied Bacteriology,* **64**:541–549.

Lappin-Scott, H. M., Cusack, F., MacLeod, F. A., and Costerton, J. W. (1988b). Nutrient resuscitation and growth of starved cells in sandstone cores: a novel approach to enhanced oil recovery. *Applied and Environmental Microbiology,* **54**:1373–1382.

Lazarova, V. and Manem, J. (1995). Biofilm characterization and activity analysis in water and waste-water treatment. *Water Research,* **29**:2227–2245.

Leis, A. P., Schlicher, S., Franke, H., and Strathmann, M. (2005). Optically transparent porous medium for nondestructive studies of microbial biofilm architecture and transport dynamics. *Applied and Environmental Microbiology,* **71**:4801–4808.

Lenz, A. P., Williamson, K. S., Pitts, B., Stewart, P. S., and Franklin, M. J. (2008). Localized gene expression in *Pseudomonas aeruginosa* biofilms. *Applied and Environmental Microbiology,* **74**:4463–4471.

Lewandowski, Z. and Beyenal, H. (2007). *Fundamentals of Biofilm Research.* CRC Press, Boca Raton.

Li, X. Q., Lin, C. L., Miller, J. D., and Johnson, W. P. (2006). Pore-scale observation of microsphere deposition at grain-to-grain contacts over assemblage-scale porous media domains using X-ray microtomography. *Environmental Science & Technology,* **40**:3762–3768.

Ludwig, R. D., McGregor, R. G., Blowes, D. W., Benner, S. G., and Mountjoy, K. (2002). A permeable reactive barrier for treatment of heavy metals. *Ground Water,* **40**: 59–66.

MacLeod, F. A., Lappin-Scott, H. M., and Costerton, J. W. (1988). Plugging of a model rock system by using starved bacteria. *Applied and Environmental Microbiology,* **54**:1365–1372.

Massoudieh, A., Mathew, A., Lambertini, E., Nelson, K. E., and Ginn, T. R. (2007). Horizontal gene transfer on surfaces in natural porous media: conjugation and kinetics. *Vadose Zone Journal,* **6**:306–315.

Mauclaire, L., Schurmann, A., and Mermillod-Blondin, F. (2006). Influence of hydraulic conductivity on communities of microorganisms and invertebrates in porous media: a case study in drinking water slow sand filters. *Aquatic Sciences,* **68**:100–108.

McLean, J. S., Ona, O. N., and Majors, P. D. (2007). Correlated biofilm imaging, transport and metabolism measurements via combined nuclear magnetic resonance and confocal microscopy. *ISME J,* **2**:121–131.

Mclean, R. J. C., Jamieson, H. E., and Cullimore, D. R. (1997). Formation of nesquehonite and other minerals as a consequence of biofilm dehydration. *World Journal of Microbiology & Biotechnology*, **13**:25–28.

Mclean, R. J. C., Whiteley, M., Hoskins, B. C., Majors, P. D., and Sharma, M. M. (1999). Laboratory techniques for studying biofilm growth, physiology, and gene expression in flowing systems and porous media. *Methods in Enzymology*, **310**:248–264.

Metzger, U., Lankes, U., Hardy, E. H., Gordalla, B. C., and Frimmel, F. H. (2006). Monitoring the formation of an *Aureobasidium pullulans* biofilm in a bead-packed reactor via flow-weighted magnetic resonance imaging. *Biotechnology Letters,* **28**:1305–1311.

Miller, M. B. and Bassler, B. L. (2001). Quorum sensing in bacteria. *Annual Review of Microbiology*, **55**:165–199.

Mitchell, A. C., Phillips, A. J., Hamilton, M. A., Gerlach, R., Hollis, W. K., Kaszuba, J. P., and Cunningham, A. B. (2008a). Resilience of planktonic and biofilm cultures to supercritical CO_2. *Journal of Supercritical Fluids*, **47**:318–325.

Mitchell, A. C., Phillips, A. J., Kaszuba, J. P., Hollis, W. K., Cunningham, A. L. B., and Gerlach, R. (2008b). Microbially enhanced carbonate mineralization and the geologic containment of CO_2. *Geochimica et Cosmochimica Acta*, **72**:A636.

Mitchell, A. C., Phillips, A. J., Hiebert, R., Gerlach, R., Spangler, L. H., and Cunningham, A. B. (2009). Biofilm enhanced geologic sequestration of supercritical CO_2. *International Journal of Greenhouse Gas Control,* **3**: 90–99.

Mol, N., Kut, O. M., and Dunn, I. J. (1993). Adsorption of toxic shocks on carriers in anaerobic biofilm fluidized-bed reactors. *Water Science and Technology*, **28**:55–65.

Molin, S., Tolker-Nielsen, T., and Hansen, S. K. (2004). Microbial interactions in mixed-species biofilms, In *Microbial Biofilms*, ed. G. O'Toole and M. Ghannoum. ASM Press, Washington, DC, pp. 206–221.

Mondragon-Parada, M. E., Ruiz-Ordaz, N., Tafoya-Garnica, A., Juarez-Ramirez, C., Curiel-Quesada, E., and Galindez-Mayer, J. (2008). Chemostat selection of a bacterial community able to degrade s-triazinic compounds: continuous simazine biodegradation in a multi-stage packed bed biofilm reactor. *Journal of Industrial Microbiology & Biotechnology*, **35**: 767–776.

Morales, C. F. L., Strathmann, M., and Flemming, H. C. (2007). Influence of biofilms on the movement of colloids in porous media. Implications for colloid facilitated transport in subsurface environments. *Water Research*, **41**:2059–2068.

Muris, M., Delolme, C., Gaudet, J. P., and Spadini, L. (2005). Assessment of biofilm destabilisation and consequent facilitated zinc transport. *Water Science and Technology*, **51**:21–28.

Nam, T. K., Timmons, M. B., Montemagno, C. D., and Tsukuda, S. M. (2000). Biofilm characteristics as affected by sand size and location in fluidized bed vessels. *Aquacultural Engineering*, **22**:213–224.

Nambi, I. M., Werth, C. J., Sanford, R. A., and Valocchi, A. J. (2003). Pore-scale analysis of anaerobic halorespiring bacterial growth along the transverse mixing zone of an etched silicon pore network. *Environmental Science & Technology*, **37**:5617–5624.

Neu, T. R., Kuhlicke, U., and Lawrence, J. R. (2002). Assessment of fluorochromes for two-photon laser scanning microscopy of biofilms. *Applied and Environmental Microbiology*, **68**:901–909.

Nguyen, V. T., Morgenroth, E., and Eberl, H. J. (2005). A mesoscale model for hydrodynamics in biofilms that takes microscopic flow effects into account. *Water Science and Technology*, **52**:167–172.

Nyman, J. L., Caccavo Jr., F., Cunningham, A. B., and Gerlach, R. (2002). Biogeochemical elimination of chromium (VI) from contaminated water. *Bioremediation Journal*, **6**:39–55.

O'Hannesin, S. F. and Gillham, R. W. (1998). Long-term performance of an in situ "iron wall" for remediation of VOCs. *Ground Water*, **36**: 164–170.

O'Toole, G., Kaplan, H. B., and Kolter, R. (2000). Biofilm formation as microbial development. *Annual Review of Microbiology*, **54**:49–79.

Ochiai, N., Kraft, E. L., and Selker, J. S. (2006). Methods for colloid transport visualization in pore networks. *Water Resources Research*, **42**. W12S06. doi: 10.1029/2006WR004961.

Okabe, S., Itoh, T., Satoh, H., and Watanabe, Y. (1999). Analyses of spatial distributions of sulfate-reducing bacteria and their activity in aerobic wastewater biofilms. *Applied and Environmental Microbiology,* **65**: 5107–5116.

Okabe, S., Kuroda, H., and Watanabe, Y. (1998). Significance of biofilm structure on transport of inert particulates into biofilms. *Water Science and Technology*, **38**:163–170.

Paterson-Beedle, M., Nott, K. P., Macaskie, L. E., and Hall, L. D. (2001). Study of biofilm within a packed-bed reactor by three-dimensional magnetic resonance imaging. *Microbial Growth in Biofilms*, pp. 285–305. See http://www.ncbi.nlm.nih.gov/pubmed/11398437 for more information. Accessed April 6, 2010.

Paulsen, J. E., Oppen, E., and Bakke, R. (1997). Biofilm morphology in porous media, a study with microscopic and image techniques. *Water Science and Technology*, **36**:1–9.

Petrozzi, S., Kut, O. M., and Dunn, I. J. (1993). Protection of biofilms against toxic shocks by the adsorption and desorption capacity of carriers in anaerobic fluidized-bed reactors. *Bioprocess Engineering*, **9**:47–59.

Peyton, B. M., Skeen, R. S., Hooker, B. S., Lundman, R. W., and Cunningham, A. B. (1995). Evaluation of bacterial detachment rates in porous-media. *Applied Biochemistry and Biotechnology*, **51–52:**785–797.

Poulsen, L. K., Ballard, G., and Stahl, D. A. (1993). Use of ribosomal-RNA fluorescence in situ hybridization for measuring the activity of single cells in young and established biofilms. *Applied and Environmental Microbiology,* **59**:1354–1360.

Purevdorj-Gage, B., Costerton, W. J., and Stoodley, P. (2005). Phenotypic differentiation and seeding dispersal in non-mucoid and mucoid *Pseudomonas aeruginosa* biofilms. *Microbiology-Sgm*, **151**:1569–1576.

Raiders, R. A., McInerney, M. J., Revus, D. E., Torbati, H. M., Knapp, R. M., and Jenneman, G. E. (1986). Selectivity and depth of microbial plugging in berea sandstone cores. *Journal of Industrial Microbiology*, **1**:195–203.

Ramasamy, P. and Zhang, X. (2005). Effects of shear stress on the secretion of extracellular polymeric substances in biofilms. *Water Science and Technology*, **52**:217–223.

Ramirez, A. A., Benard, S., Giroir-Fendler, A., Jones, J. P., and Heitz, M. (2008). Treatment of methanol vapours in biofilters packed with inert materials. *Journal of Chemical Technology and Biotechnology*, **83**:1288–1297.

Ramsing, N. B., Kuhl, M., and Jorgensen, B. B. (1993). Distribution of sulfate-reducing bacteria, O_2, and H_2S in photosynthetic biofilms determined by oligonucleotide probes and microelectrodes. *Applied and Environmental Microbiology,* **59**:3840–3849.

Rani, S. A., Pitts, B., Beyenal, H., Veluchamy, R. A., Lewandowski, Z., Davison, W. M., Buckingham-Meyer, K., and Stewart, P. S. (2007). Spatial patterns of DNA replication, protein synthesis, and oxygen concentration within bacterial biofilms reveal diverse physiological states. *Journal of Bacteriology*, **189**:4223–4233.

Rees, H. C., Oswald, S. E., Banwart, S. A., Pickup, R. W., and Lerner, D. N. (2007). Biodegradation processes in a laboratory-scale groundwater contaminant plume assessed by fluorescence imaging and microbial analysis. *Applied and Environmental Microbiology*, **73**:3865–3876.

Rice, A. R., Hamilton, M. A., and Camper, A. K. (2003). Movement, replication, and emigration rates of individual bacteria in a biofilm. *Microbial Ecology*, **45**:163–172.

Rinck-Pfeiffer, S., Ragusa, S., Sztajnbok, P., and Vandevelde, T. (2000). Interrelationships between biological, chemical, and physical processes as an analog to clogging in aquifer storage and recovery (ASR) wells. *Water Research*, **34**:2110–2118.

Rittmann, B. E. (1993). The significance of biofilms in porous-media. *Water Resources Research*, **29**:2195–2202.

Ross, N., Deschenes, L., Bureau, J., Clement, B., Comeau, Y., and Samson, R. (1998). Ecotoxicological assessment and effects of physicochemical factors on biofilm development in groundwater conditions. *Environmental Science & Technology*, **32**:1105–1111.

Ross, N., Novakowski, K. S., Lesage, S., Deschenes, L., and Samson, R. (2007). Development and resistance of a biofilm in a planar fracture during biostimulation, starvation, and varying flow conditions. *Journal of Environmental Engineering and Science*, **6**:377–388.

Ross, N., Villemur, R., Deschenes, L., and Samson, R. (2001). Clogging of a limestone fracture by stimulating groundwater microbes. *Water Research*, **35**:2029–2037.

Rowe, R. K., Armstrong, M. D., and Cullimore, D. R. (2000). Particle size and clogging of granular media permeated with leachate. *Journal of Geotechnical and Geoenvironmental Engineering*, **126**:775–786.

Sani, R. K., Peyton, B. M., Smith, W. A., Apel, W. A., and Petersen, J. N. (2002). Dissimilatory reduction of Cr(VI), Fe(III), and U(VI) by *Cellulomonas* isolates. *Applied Microbiology and Biotechnology*, **60**: 192–199.

Sauer, K., Camper, A. K., Ehrlich, G. D., Costerton, J. W., and Davies, D. G. (2002). *Pseudomonas aeruginosa* displays multiple phenotypes during development as a biofilm. *Journal of Bacteriology*, **184**:1140–1154.

Scherer, M. M., Richter, S., Valentine, R. L., and Alvarez, P. J. J. (2000). Chemistry and microbiology of permeable reactive barriers for *in situ* groundwater clean up. *Critical Reviews in Microbiology*, **26**:221–264.

Seaman, J. C., Bertsch, P. M., and Schwallie, L. (1999). In situ Cr(VI) reduction within coarse-textured, oxide-coated soil and aquifer systems using Fe(II) solutions. *Environmental Science & Technology*, **33**:938–944.

Seifert, D. and Engesgaard, P. (2007). Use of tracer tests to investigate changes in flow and transport properties due to bioclogging of porous media. *Journal of Contaminant Hydrology*, **93**:58–71.

Seki, K., Thullner, M., Hanada, J., and Miyazaki, T. (2006). Moderate bioclogging leading to preferential flow paths in biobarriers. *Ground Water Monitoring and Remediation*, **26**:68–76.

Seymour, J. D., Codd, S. L., Gjersing, E. L., and Stewart, P. S. (2004a). Magnetic resonance microscopy of biofilm structure and impact on transport in a capillary bioreactor. *Journal of Magnetic Resonance*, **167**:322–327.

Seymour, J. D., Gage, J. P., Codd, S. L., and Gerlach, R. (2004b). Anomalous fluid transport in porous media induced by biofilm growth. *Physical Review Letters* 93.

Seymour, J. D., Gage, J. P., Codd, S. L., and Gerlach, R. (2007). Magnetic resonance microscopy of biofouling induced scale dependent transport in porous media. *Advances in Water Resources*, **30**:1408–1420.

Shareefdeen, Z. and Baltzis, B. C. (1994). Biofiltration of toluene vapor under steady-state and transient conditions—theory and experimental results. *Chemical Engineering Science*, **49**:4347–4360.

Shareefdeen, Z., Herner, B., Webb, D., Polenek, S., and Wilson, S. (2002). Removing volatile organic compound (VOC) emissions from a printed circuit board manufacturing facility using pilot- and commercial-scale biofilters. *Environmental Progress*, **21**:196–201.

Sharp, R. R., Cunningham, A. B., Komlos, J., and Billmayer, J. (1999a). Observation of thick biofilm accumulation and structure in porous media and corresponding hydrodynamic and mass transfer effects. *Water Science and Technology*, **39**:195–201.

Sharp, R. R., Gerlach, R., and Cunningham, A. B. Bacterial transport issues related to subsurface biobarriers. In *Engineered Approaches for in situ Bioremediation of Chlorinated Solvent Contamination,* ed. A. L. Leeson and B. C. Alleman. Battelle Press, Columbus, Richland, pp. 211–216.

Sharp, R. R., Stoodley, P., Adgie, M., Gerlach, R., and Cunningham, A. (2005). Visualization and characterization of dynamic patterns of flow, growth and activity of biofilms growing in porous media. *Water Science and Technology*, **52**:85–90.

Shaw, J. C., Bramhill, B., Wardlaw, N. C., and Costerton, J. W. (1985). Bacterial fouling of a model core system. *Applied and Environmental Microbiology,* **49**:693–701.

Shemesh, H., Goertz, D. E., van der Sluis, L. W. M., de Jong, N., Wu, M. K., and Wesselink, P. R. (2007). High frequency ultrasound imaging of a single-species biofilm. *Journal of Dentistry,* **35**:673–678.

Shin, H. Y., Singhal, N., and Park, J. W. (2007). Regeneration of iron for trichloroe–thylene reduction by *Shewanella alga* BrY. *Chemosphere,* **68:**1129–1134.

Singh, P. K., Schaefer, A. L., Parsek, M. R., Moninger, T. O., Welsh, M. J., and Greenberg, E. P. (2000). Quorum-sensing signals indicate that cystic fibrosis lungs are infected with bacterial biofilms. *Nature,* **407**:762–764.

Smith, W. A., Apel, W. A., Petersen, J. N., and Peyton, B. M. (2002). Effect of carbon and energy source on bacterial chromate reduction. *Bioremediation Journal,* **6**:205–215.

Song, Z., Williams, C. J., and Edyvean, R. G. J. (2003). Tannery wastewater treatment using an upflow anaerobic fixed biofilm reactor (UAFBR). *Environmental Engineering Science,* **20**:587–599.

Stevik, T. K., Aa, K., Ausland, G., and Hanssen, J. F. (2004). Retention and removal of pathogenic bacteria in wastewater percolating through porous media: a review. *Water Research,* **38**:1355–1367.

Stewart, P. S. (1996). Theoretical aspects of antibiotic diffusion into microbial biofilms. *Antimicrobial Agents Chemotherapy,* **40**:2517–2522.

Stewart, P. S. (1998). A review of experimental measurements of effective diffusive permeabilities and effective diffusion coefficients in biofilms. *Biotechnology and Bioengineering,* **59**:261–272.

Stewart, P. S. (2003). Diffusion in biofilms. *Journal of Bacteriology,* **185**: 1485–1491.

Stewart, P. S. and Costerton, J. W. (2001). Antibiotic resistance of bacteria in biofilms. *Lancet,* **358**:135–138.

Stewart, P. S. and Franklin, M. J. (2008). Physiological heterogeneity in biofilms. *Nature Reviews Microbiology,* **6**:199–210.

Stewart, P. S., McFeters, G. A., and Huang, C.-T. (2000). Biofilm control by antimicrobial agents. In *Biofilms II—Process Analysis and Applications*, ed. J. D. Bryers. John Wiley & Sons, New York, pp. 373–405.

Stewart, T. L. and Fogler, H. S. (2001). Biomass plug development and propagation in porous media. *Biotechnology and Bioengineering*, **72**:353–363.

Stoodley, P., Cargo, R., Rupp, C. J., Wilson, S., and Klapper, I. (2002a). Biofilm material properties as related to shear-induced deformation and detachment phenomena. *Journal of Industrial Microbiology and Biotechnology*, **29**:361–367.

Stoodley, P., de Beer, D., and Lewandowski, Z. (1994). Liquid flow in biofilm systems. *Applied and Environmental Microbiology*, **60**:2711–2716.

Stoodley, P., Dodds, I., de Beer, D., Scott, H. L., and Boyle, J. D. (2005). Flowing biofilms as a transport mechanism for biomass through porous media under laminar and turbulent conditions in a laboratory reactor system. *Biofouling*, **21**:161–168.

Stoodley, P., Sauer, K., Davies, D. G., and Costerton, J. W. (2002b). Biofilms as complex differentiated communities. *Annual Review of Microbiology*, **56**:187–209.

Stoodley, P., Wilson, S., Hall-Stoodley, L., Boyle, J. D., Lappin-Scott, H. M., and Costerton, J. W. (2001). Growth and detachment of cell clusters from mature mixed-species biofilms. *Applied and Environmental Microbiology*, **67**:5608–5613.

Sturman, P., Stewart, P. S., Cunningham, A. B., Bouwer, E. J., and Wolfram, J. H. (1995). Engineering scale-up of in situ bioremediation processes: a review. *Journal of Contaminant Hydrology*, **19**:171–203.

Tang, H. M., Hwang, S. J., and Wang, W. C. (1997). Degradation of acetone in a biofilter. *Environmental Engineering Science*, **14**:219–226.

Taylor, S. W. and Jaffe, P. R. (1990a). Biofilm growth and the related changes in the physical-properties of a porous-medium. 3. Dispersivity and model verification. *Water Resources Research*, **26**:2171–2180.

Taylor, S. W. and Jaffe, P. R. (1990b). Biofilm growth and the related changes in the physical-properties of a porous-medium. 1. Experimental investigation. *Water Resources Research*, **26**:2153–2159.

Teal, T. K., Lies, D. P., Wold, B. J., and Newman, D. K. (2006). Spatiometabolic stratification of *Shewanella oneidensis* biofilms. *Applied and Environmental Microbiology*, **72**:7324–7330.

Thullner, M. and Baveye, P. (2008). Computational pore network modeling of the influence of biofilm permeability on bioclogging in porous media. *Biotechnology and Bioengineering*, **99**:1337–1351.

Thullner, M., Mauclaire, L., Schroth, M. H., Kinzelbach, W., and Zeyer, J. (2002a). Interaction between water flow and spatial distribution of microbial growth in a two-dimensional flow field in saturated porous media. *Journal of Contaminant Hydrology*, **58**:169–189.

Thullner, M., Schroth, M. H., Zeyer, J., and Kinzelbach, W. (2004). Modeling of a microbial growth experiment with bioclogging in a two-dimensional saturated porous media flow field. *Journal of Contaminant Hydrology*, **70**: 37–62.

Thullner, M., Zeyer, J., and Kinzelbach, W. (2002b). Influence of microbial growth on hydraulic properties of pore networks. *Transport in Porous Media*, **49**:99–122.

Tolker-Nielsen, T. and Molin, S. (2000). Spatial organization of microbial biofilm communities. *Microbial Ecology*, **40**, 75–84.

Tolker-Nielsen, T., Brinch, U. C., Ragas, P. C., Andersen, J. B., Jacobsen, C. S., and Molin, S. (2000). Development and dynamics of *Pseudomonas* sp. biofilms. *Journal of Bacteriology*, **182**:6482–6489.

Tse, S. W. and Yu, J. (2003). Adsorptive immobilization of a Pseudomonas strain on solid carriers for augmented decolourization in a chemostat bioreactor. *Biofouling*, **19**:223–233.

van Loosdrecht, M. C. M., Heijnen, J. J., Eberl, H., Kreft, J., and Picioreanu, C. (2002). Mathematical modelling of biofilm structures. *Antonie Van Leeuwenhoek International Journal of General and Molecular Microbiology*, **81**:245–256.

Vandevivere, P. (1995). Bacterial clogging of porous-media—a new modeling approach. *Biofouling*, **8**:281–291.

Vandevivere, P. and Baveye, P. (1992a). Effect of bacterial extracellular polymers on the saturated hydraulic conductivity of sand columns. *Applied and Environmental Microbiology*, **58**:1690–1698.

Vandevivere, P. and Baveye, P. (1992b). Saturated hydraulic conductivity reduction caused by aerobic-bacteria in sand columns. *Soil Science Society of America Journal*, **56**:1–13.

Vandevivere, P., Baveye, P., Delozada, D. S., and Deleo, P. (1995). Microbial clogging of saturated soils and aquifer materials—evaluation of mathematical-models. *Water Resources Research*, **31**:2173–2180.

VanGulck, J. F. and Rowe, R. K. (2004). Evolution of clog formation with time in columns permeated with synthetic landfill leachate. *Journal of Contaminant Hydrology*, **75**, 115–139.

VanLoosdrecht, M. C. M., Lyklema, J., Norde, W., and Zehnder, A. J. B. (1990). Influence of interfaces on microbial activity. *Microbiological Reviews*, **54**:75–87.

Velicer, G. J. (2003). Social strife in the microbial world. *Trends in Microbiology*, **11**:330–337.

Viamajala, S., Peyton, B. M., Gerlach, R., Sivaswamy, V., Apel, W. A., and Petersen, J. N. (2008). Permeable reactive biobarriers for in situ cr(vi) reduction: bench scale tests using *Cellulomonas* sp. strain ES6. *Biotechnology and Bioengineering*, **101**:1150–1162.

Wan, J., Wilson, J. L., and Kieft, T. L. (1994). Influence of the gas-water interface on transport of microorganisms through unsaturated porous media. *Applied and Environmental Microbiology*, **60**:509–516.

Wani, A. H., Branion, R. M. R., and Lau, A. K. (1997). Biofiltration: A promising and cost-effective control technology for odors, VOCs and air toxics. *Journal of Environmental Science and Health Part A—Environmental Science and Engineering and Toxic and Hazardous Substance Control*, **32**:2027–2055.

Wanko, A., Mose, R., and Beck, C. (2005). Biological processing capacities and biomass growth in waste water treatment by infiltration on two kinds of sand. *Water Air and Soil Pollution*, **165**:279–299.

Wanko, A., Mose, R., Carrayrou, J., and Sadowski, A. G. (2006). Simulation of biodegradation in infiltration seepage—model development and hydrodynamic calibration. *Water Air and Soil Pollution*, **177**:19–43.

Warwood, B., James, G., Horrigan, K., Sturman, P., Cunningham, A. B., Costerton, J. W., and Hiebert, R. (1995). Formation and Persistence of Biobarriers formed from ultra microbacteria in porous media. Bozeman, MT: Center for Biofilm Engineering; Montana State University, DOE Technical Report, Grant 95-C213-CR.

Waybrant, K. R., Blowes, D. W., and Ptacek, C. J. (1998). Selection of reactive mixtures for use in permeable reactive walls for treatment of mine drainage. *Environmental Science & Technology*, **32**:1972–1979.

Werner, E., Roe, F., Bugnicourt, A., Franklin, M. J., Heydorn, A., Molin, S., Pitts, B., and Stewart, P. S. (2004). Stratified growth in *Pseudomonas aeruginosa* biofilms. *Applied and Environmental Microbiology*, **70**: 6188–6196.

Wheeler, L. A. (2009). Establishment of ureolytic biofilms and their influence on the permeability of pulse-flow porous media column systems, M.S. thesis Chemical Engineering, Montana State University.

Wildenschild, D., Hopmans, J. W., Rivers, M. L., and Kent, A. J. R. (2005). Quantitative analysis of flow processes in a sand using synchrotron-based X-ray microtomography. *Vadose Zone Journal*, **4**:112–126.

Wilkin, R. T., Puls, R. W., and Sewell, G. W. (2003). Long-term performance of permeable reactive barriers using zero-valent iron: geochemical and microbiological effects. *Ground Water*, **41**:493.

Williamson, K. and McCarty, P. L. (1976a). Verification studies of biofilm model for bacterial substrate utilization. *Journal Water Pollution Control Federation*, **48**:281–296.

Williamson, K. and McCarty, P. L. (1976b). Model of substrate utilization by bacterial films. *Journal Water Pollution Control Federation*, **48**:9–24.

Willingham, T. W., Werth, C. J., and Valocchi, A. J. (2008). Evaluation of the effects of porous media structure on mixing-controlled reactions using pore-scale modeling and micromodel experiments. *Environmental Science & Technology*, **42**:3185–3193.

Wilson, S., Hamilton, M. A., Hamilton, G. C., Schumann, M. R., and Stoodley, P. (2004). Statistical quantification of detachment rates and size distributions of cell clumps from wild-type (PAO1) and cell signaling mutant (JP1) *Pseudomonas aeruginosa* biofilms. *Applied and Environmental Microbiology*, **70**:5847–5852.

Xu, K. D., Stewart, P. S., Xia, F., Huang, C. T., and McFeters, G. A. (1998). Spatial physiological heterogeneity in *Pseudomonas aeruginosa* biofilm is determined by oxygen availability. *Applied and Environmental Microbiology*, **64**:4035–4039.

Yao, K. M., Habibian, M. T., and O'Melia, C. R. (1971). Water and waste water filtration: concepts and applications. *Environmental Science and Technology*, **5**:1105–1112.

Yarwood, R. R., Rockhold, M. L., Niemet, M. R., Selker, J. S., and Bottomley, P. J. (2002). Noninvasive quantitative measurement of bacterial growth in porous media under unsaturated-flow conditions. *Applied and Environmental Microbiology*, **68**:3597–3605.

Yarwood, R. R., Rockhold, M. L., Niemet, M. R., Selker, J. S., and Bottomley, P. J. (2006). Impact of microbial growth on water flow and solute transport in unsaturated porous media. *Water Resources Research*, **42**. W10405. doi: 10.1029/2005 WR004550.

Yen, T. F. (1990). Microbial Enhanced Oil Recovery: Principle and Practice. CRC Press, Boca Raton, FL.

Zacarias, G. D., Ferreira, C. P., and Velasco-Hernandez, J. (2005). Porosity and tortuosity relations as revealed by a mathematical model of biofilm structure. *Journal of Theoretical Biology*, **233**:245–251.

Zajic, J. E. and Donaldson, E. C. (1985). *Microbes and Oil Recovery.* International Bioresources Journal. Bioresources Publications El Paso, TX.

Zhang, T. C. and Bishop, P. L. (1994). Evaluation of tortuosity factors and effective diffusivities in biofilms. *Water Research*, **28**:2279–2287.

6

Using Porous Media Theory to Determine the Coil Volume Needed to Arrest Flow in Brain Aneurysms

Khalil M. Khanafer
Vascular Mechanics Laboratory, Department of Biomedical Engineering and Section of Vascular Surgery, University of Michigan, Ann Arbor, MI

Ramon Berguer
Vascular Mechanics Laboratory, Department of Biomedical Engineering and Section of Vascular Surgery, University of Michigan, Ann Arbor, MI

CONTENTS

6.1 Nomenclature

d Diameter of the parent vessel
Da Darcy number, K/d^2

Acknowledgment
This work was supported by the Frankel Vascular Research Fund.

F	Forchheimer constant
g	Acceleration due to gravity
J	Unit vector oriented along the pore velocity vector, $\mathbf{J} = \frac{\mathbf{v}_\mathrm{p}}{\lvert\mathbf{v}_\mathrm{p}\rvert}$
K	Permeability of the porous medium
P	Pressure
$\langle P \rangle^f$	Average pressure
Re_m	Mean Reynolds number, $\rho \bar{u}_m d/\mu$
t	Time
$\langle \mathbf{v} \rangle$	Average velocity vector
$\mathbf{v}_\mathrm{p}$	Pore velocity vector
$\mathbf{V}$	Velocity vector in the parent vessel
T_p	Period

Greek symbols

ε	Porosity of the porous medium
μ	Dynamic viscosity
ν	Kinematic viscosity
ρ	Density

Subscripts

f	Fluid

6.2 Introduction

Porous media theory has been utilized to improve our understanding of transport processes. There are numerous practical applications that can be modeled or approximated as a transport through porous media: catalytic reactors, electronic cooling, geothermal systems, thermal insulation, drying technology, and packed-bed heat exchangers. These applications have been discussed by Nield and Bejan [1], Vafai [2], Hadim and Vafai [3], and Vafai and Hadim [4], Vafai and Tien [5,6], and Vafai [7,8]. Vafai and Tien [6] presented a comprehensive analysis of the generalized transport through porous media and developed a set of governing equations utilizing the local volume averaging technique. The use of porous media theory for modeling biomedical phenomena has resulted in significant advances in biofilms, drug delivery, computational biology, brain aneurysms filled with endovascular coil, medical imaging, porous scaffolds for tissue engineering, and diffusion processes in the extracellular space (ECS).

6.3 Physics of Cerebral Aneurysms

Cerebral aneurysms are pathological segmental dilatations of the cerebral arteries. On microscopic examination, cerebral aneurysms show weakening of their vessel walls with thinning of the tunica media and the disruption of the internal elastic lamina. The prevalence of intracranial aneurysms has been estimated at 1%–6% (1–12 million Americans) in large autopsy series. Considerable evidence supports the pathogenesis of aneurysm formation to be related to genetic predisposition and/or environmental factors, including autosomal dominant polycystic kidney disease, Marfan's syndrome, Ehler-Danlos type IV, fibromuscular dysplasia, neurofibromatosis type 1, and age. A common location of intracranial aneurysms is the arteries at the base of the brain, known as the Circle of Willis. The rupture of intracranial aneurysms into the subarachnoid space results in subarachnoid hemorrhage (SAH), a devastating event with high rates of morbidity (15%–30%) and mortality (30%–67%) [9–13]. The worldwide incidence of aneurysmal SAH is approximately 1 per 10,000 people with more than 30,000 Americans suffering from SAH in the United States every year (The American Society of Interventional and Therapeutic Neuroradiology [ASITN]).

Intracranial aneurysms are classified by presumed pathogenesis and geometry. Saccular, berry, or congenital aneurysm represent the majority of all cerebral aneurysms and are spherical expansions of the vessel wall, typically occurring at branch points of major intracranial arteries and most often in the Circle of Willis. The characteristic geometry of a saccular aneurysm is a thin-walled, balloon-like structure or "dome" communicating with the lumen of a parent artery at its base through a "neck." Dolichoectatic, fusiform, or arteriosclerotic aneurysms are extended outpouchings of proximal arteries that account for 7% of all cerebral aneurysms. Infectious or mycotic aneurysms are situated peripherally and comprise 0.5% of all cerebral aneurysms.

Brain aneurysms may be treated by surgery or by less-invasive endovascular techniques [14]. Clipping across the aneurysmal neck excludes the aneurysm from the intracranial circulation yielding excellent long-term efficacy and protection from aneurysm recurrence or rupture. In the last two decades, alternative endovascular coil embolization techniques have been developed to treat intracranial aneurysms. Coil embolization involves the deployment of tiny platinum coils into the aneurysm through an intravascular microcatheter reducing intraaneurysmal blood flow and leading to thrombosis of its sac. In both instances, the aim of the procedure is to occlude the aneurysm with thrombus effectively isolating it from the arterial circulation and preventing its eventual rupture. Recent endovascular advances with bioactive coils, balloon, and stent-assisted techniques have expanded the range of intracranial aneurysms that may be treated via coil embolization. Coil embolization has several advantages over surgical clipping: it produces better survival, freedom from disability, and a lower risk of death than surgery (The International Subarachnoid Aneurysm Trial [ISAT]). Coil embolization, however, cannot be used in wide-necked irregularly shaped aneurysms due to difficulty achieving adequate filling of the aneurysm sac as well as risk of coil protrusion into the parent artery [15]. In wide-neck aneurysms, endovascular stents have been used across the aneurysmal neck in conjunction with coil embolization to contain the coil inside the sac. Placing a stent and coil reduces flow velocity inside the sac and promotes its thrombosis [16–19].

6.4 Background

The dynamics of blood flow dynamics are an important factor in the development and evolution of intracranial aneurysms. A number of specific hemodynamic factors including wall shear stress, pressure, impingement force, flow rate, and particle residence time have been implicated in aneurysm growth and rupture [20]. Therefore, hemodynamic modeling of flow in intracranial aneurysms and their arterial wall properties may help in obtaining quantitative criteria for those parameters that result in expansion or rupture. Furthermore, modeling may provide valuable treatment planning by establishing the minimal coil packing density or the type of intracranial stent that will result in flow arrest within the sac and ultimately in its thrombosis.

6.4.1 Clinical and Experimental Studies Associated with the Treatment of Aneurysms Using Stent Implantation and Coil Placement

Previous studies have analyzed blood flow characteristics of an aneurysm after endovascular treatment using coils and stents. Most of these studies of flow in intracranial aneurysms used idealized *in vitro* models with rigid walls. These idealized models cannot be directly used for relating patient-specific hemodynamics to the treatment plan because they do not reproduce the anatomy or mechanical behavior of the human arteries [21–26]. Yu and Zhao [21] conducted an *in vitro* steady flow study on stented and nonstented side-wall rigid aneurysms using particle image velocimetry (PIV) over a range of Reynolds number from 200 to 1,600. The existing regions of high-wall shear stresses (WSSs) at the distal neck were suppressed by almost 90% in stented aneurysms.

Lieber et al. [22] performed PIV measurements to study experimentally the influence of stent strut size and porosity on the intraneurysmal flow dynamics in a side-wall aneurysm model. Their results showed that stents can significantly reduce both the intraaneurysmal vorticity and the mean flow within the aneurysm. Liou et al. [23] conducted an experimental study to investigate pulsatile flow fields in a saccular brain aneurysm model using helix and mesh stents. Their results showed that the flow features inside the aneurysm changed substantially with the type of stent. Canton et al. [24] conducted an *in vitro* study to quantify the effect of the stents by measuring the changes in the hemodynamic forces acting on a bifurcating aneurysm model (basilar tip configuration) after the placement of flexible neuroform stents. A digital particle image velocimetry (DPIV) system was used to measure the pulsatile velocity and shear stress fields within the aneurysm. Their results showed that peak velocity and strength of vortices inside the aneurysm sac were reduced after placing the stents. Same authors [25] conducted an experimental study to measure the changes in the intraaneurysmal fluid pressure and parent vessel flow characteristics resulting from packing the aneurysmal sac with hydrogel-coated coils. Their results showed that the intraaneurysmal fluid pressure did not increase when packing the aneurysm with hydrogel-coated platinum coils, even with a coil density up to 93%. Gobin et al. [26] observed reduction of inflow and flow stagnation at the dome with coil insertion in their *in vitro* model study.

There are numerous reports of clinical experience with placement of stents and coils in brain aneurysms (see, e.g., Marks et al. [27], Wakhloo et al. [16]). Lanzino et al. [28] reported that stent placement within the parent artery across the aneurysm reduced intraaneurysm flow velocity, which led to intraaneurysm stasis and thrombosis and consequently preventing rupture. Kwon et al. [29] used a new endovascular technique for treatment of cerebral aneurysms. Eight patients with wide-necked aneurysms were successfully treated without complications with detachable coils using the multiple microcatheter technique. Meckel et al. [30] conducted an *in vivo* study to quantify flow velocities and to estimate WSS in patients with cerebral aneurysms using magnetic resonance imaging (MRI). Their results showed a high spatial variation of WSS among different aneurysm geometries reflecting variable flow patterns.

6.4.2 Computational Studies Associated with Combined Use of Stents and Coils for the Treatment of Cerebral Aneurysms

Advanced modeling techniques have enabled multiphysics computations to investigate hemodynamic and other factors contributing to disease progression. In addition, the combination of noninvasive diagnostic tools (e.g., MRI, computed tomography [CT], or ultrasound) and computational fluid dynamics (CFD) techniques provides key hemodynamics feedback for studies of vascular diseases and to plan for therapeutic options [31–41]. Many studies have been conducted in the literature using idealized numerical models to study the flow patterns inside the aneurysms caused by the presence of stents. For example, Aenis et al. [42] used finite element method, pulsatile, Newtonian flows to study the effect of stent placement on a rigid sidewall aneurysm. Their results showed diminished flow and pressure inside the stented aneurysm. Ohta et al. [43] analyzed hemodynamic changes in intracranial aneurysms after stent placement using a finite element modeling approach. Their work illustrated areas with stagnant flow and low-shear rates after stent placement.

To make computational models relevant and useful for clinical applications, it is necessary to ensure their accuracy in capturing important flow features. Some researchers have conducted numerical simulations of the flow in intracranial aneurysms using patient-specific geometries obtained from medical imaging data. Steinman et al. [32] presented image-based computational simulations of the flow dynamics in a giant anatomically realistic, human intracranial aneurysm assuming rigid walls. Their analysis revealed high-speed flow entering the aneurysm at the proximal and distal ends of the neck, promoting the formation of persistent and transient vortices within the aneurysm sac. Stuhne and Steinman [44] conducted a numerical study to analyze the WSS distribution and flow streamlines near the neck of a stented basilar rigid side-wall aneurysm. The numerical simulations were performed assuming constant pressure at the outflow boundary of the model and specifying either steady or pulsatile flow at the inlet. Shojima et al. [45] performed a numerical study in middle cerebral artery (MCA) aneurysms to quantify the magnitude and role of WSS on cerebral aneurysm with the assumption of Newtonian fluid property for blood and the rigid wall property for the parent vessel and the aneurysm. Their results showed that the maximum WSS occurred near the neck of the aneurysm, not in its tip or dome. Appanaboyina et al. [31] conducted a CFD analysis of stented intracranial aneurysms using adaptive embedded unstructured grids. Their results showed that this methodology can be used to model patient-specific anatomies with different stents and makes it possible to explore the effect of different stent designs. Rayz et al. [46] conducted a numerical study to compute the velocity field and WSSs in patient-specific geometries. Flow velocities in the arteries proximal to the aneurysm obtained by MRI scanning were used to specify the inlet flow conditions for each patient. Their results showed a good agreement between the flow fields measured *in vivo* using the in-plane MRV technique and those computed with CFD simulations.

Computational fluid dynamic of coil embolization has been rarely attempted because it is difficult to describe numerically the irregularly shaped coil. A few attempts have been made to simulate aneurysm coiling using different methods. Byun and Rhee [47] used computational methods to analyze the flow fields in partially blocked aneurysm models. These authors modeled the coils as a small solid sphere placed at different locations within the aneurysmal sac. They showed that the intraaneurysmal blood flow motion was smaller when the sphere was placed at the neck compared to when it was placed in the dome of the aneurysm. Groden et al. [48] studied three-dimensional pulsatile flow simulation before and after endovascular coil embolization of a cerebral aneurysm using *in vivo* data obtained by computerized tomographic angiography (CTA). In their CFD model, these authors used cube-shaped cells to represent the coils. Their results showed that a complete cessation of flow through the aneurysm neck was achieved with a 20% filling of blocks. The cube-shaped cells are a rough approximation to model the random tortuosity of the coil. Cha et al. [49] conducted a numerical study to analyze the interaction of coils with the local blood flow after coil embolization of an idealized representation of the basilar bifurcation with a terminal aneurysm using porous media theory. The authors assumed circular tubes of the surrounding arteries with a spherical shape of an aneurysm. The results of that study suggested that there is a complex interaction between the local hemodynamics and intraaneurysmal flow that induces significant forces on the coil mass. Recently, we developed [50] a numerical model to quantify the reduction in blood velocity and pressure resulting from the placement of endovascular coils within a cerebral aneurysm using physiological velocity waveforms. The flow characteristics within the aneurysm sac were modeled using the volume averaged porous media equations. We studied the effects of narrow and wide aneurysmal necks on the velocity fields and pressure within the aneurysmal sac in the absence of the coils. Within the sac at peak systole, wide-neck aneurysms displayed higher velocity and pressure than narrow-neck aneurysms. We showed that velocity fields are significantly affected by the presence of endovascular coil within the aneurysm sac. Moreover, we estimated that a volume density of 20% platinum coil in the aneurysmal sac was sufficient to cause sufficient blood flow arrest in the aneurysm to allow for thrombus formation.

6.5 Mathematical Formulations

We chose to explore the application of porous media theory to the modeling of flow changes in cerebral aneurysms treated by endovascular coils. The conventional CFD technique is not suitable to model flow through a coiled aneurysm because of the difficulties in representing the random geometry of the coils and the large number of nodal points required to model the coil surface. Even if the geometry of the coils could be determined, the density and total number of nodal points required to capture the characteristics of the flow would represent a major limitation. We thought that a CFD-based porous substrate approach should result in a more accurate model of the effect of coiling on the flow and pressure conditions in brain aneurysms. In addition, this technique is computationally efficient because it results in substantial savings in central processing unit (CPU) and memory usage and leads faster to grid-independent solutions. Furthermore, this approach appears to give a more realistic description of the *in vivo* situation since modeling the coils as a porous medium does not act as a solid region as previously modeled but rather slows down the flow activity within the aneurysm that should promote thrombus formation. We modeled the packing of coils in our work as a porous substrate similar to that reported by Srinivasan et al. [51] in his analysis of heat transfer and flow through a spirally fluted tube. The main objective of our work was to model the endovascular coils in the narrow-necked aneurysm sac as a porous medium of decreasing porosity (decreasing as the number of coils increase)

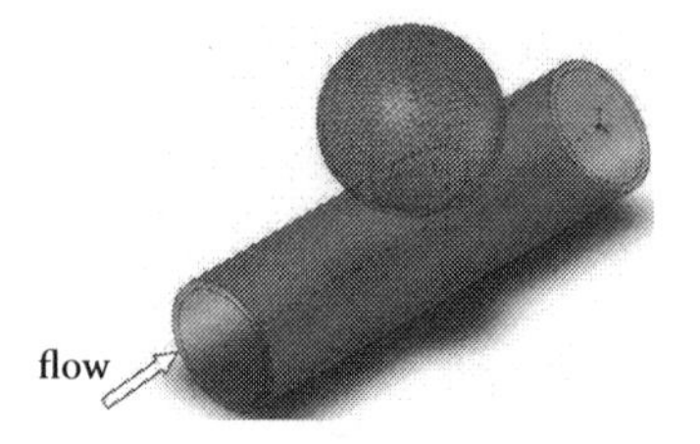

FIGURE 6.1
Schematic of the physical model and coordinate system.

and consequently to show the influence of the coils deployment on the velocity and pressure fields. The endovascular coil filling the sac of the aneurysm was modeled using a porous medium approximation and the transport equations, commonly known as the "generalized model," were solved to determine flow characteristics in an idealized geometry as shown in Figure 6.1. Moreover, the porous medium is viewed as a continuum with the solid and fluid phases in thermal equilibrium, isotropic, homogeneous, and saturated with an incompressible fluid. The generalized model, which was obtained through local volume averaging and matched asymptotic expansions, is also known as the Brinkman–Forchheimer–Darcy model and is described in rigorous detail by Amiri and Vafai [52, 53], Alazmi and Vafai [54], and Khanafer et al. [55].

These equations can be summarized as follows:

Continuity equation:

$$\nabla \cdot \langle \mathbf{v} \rangle = 0 \tag{6.1}$$

Momentum equation:

$$\frac{\rho_f}{\varepsilon}\left[\frac{\partial \langle \mathbf{v} \rangle}{\partial t} + \langle (\mathbf{v} \cdot \nabla)\mathbf{v} \rangle\right] = -\nabla \langle P \rangle^f + \frac{\mu_f}{\varepsilon}\nabla^2 \langle \mathbf{v} \rangle - \frac{\mu_f}{K}\langle \mathbf{v} \rangle - \frac{\rho_f F \varepsilon}{\sqrt{K}}[\langle \mathbf{v} \rangle . \langle \mathbf{v} \rangle] J \tag{6.2}$$

where ε is the porosity, F is the geometric function, K is the permeability, μ_f is the fluid dynamic viscosity, $\mathbf{J} = \mathbf{v}_\mathrm{p}/|\mathbf{v}_\mathrm{p}|$ is the unit vector along the pore velocity vector $\mathbf{v}_\mathrm{p}$, $\langle \mathbf{v} \rangle$ is the average velocity vector, and $\langle P \rangle^f$ is the average pressure. The medium permeability K can be properly modeled as shown by Ergun [56] and Vafai [7,8]. The porosity of the coil may be used as an index to determine the required minimal packing density of a coil and avoid the danger of rupture from unneeded excessive coil packing. More experimental studies are necessary to correlate the porosity of the coil geometry (thickness and length), shape of the coil (i.e., complex-shaped coils or helical coils), and the volume of the aneurysm sac.

The fluid motion within the parent artery is governed by the Navier–Stokes equations with constant density and fluid properties, together with the continuity equation. In a Cartesian coordinate with a fixed reference frame, the conservation of mass and momentum equations for transient, laminar flow without body forces are given by

$$\nabla \bullet \mathbf{V} = 0 \tag{6.3}$$

$$\rho_f \frac{\partial \mathbf{v}}{\partial t} + \rho_f (\mathbf{V} \bullet \nabla \mathbf{V}) = -\nabla P + \mu_f \nabla^2 \mathbf{V} \tag{6.4}$$

where ρ_f is the blood density, μ_f is the blood viscosity, P is the pressure, $\mathbf{V}$ is the velocity vector, and the subscript f refers to the fluid phase. The influence of the non-Newtonian properties of blood on flow is approximated using the Carreau model as

$$\mu = \mu_\infty + (\mu_o - \mu_\infty)(1 + \kappa^2 \dot{\gamma}^2)^{\frac{(n-1)}{2}} \tag{6.5}$$

where μ_o and μ_∞ (used as a reference value for Newtonian case) are the zero and infinite shear rate viscosities, respectively, $\dot{\gamma}$ is the deformation rate, and κ is a time constant. This model was found to fit well the experimental data as shown by Cho and Kensey [57] using the following set of parameters: $\mu_o = 0.056\,\mathrm{N} \cdot \mathrm{sec/m^2}$, $\mu_\infty = 0.00345\,\mathrm{N} \cdot \mathrm{sec/m^2}$, $n = 0.3568$, $\kappa = 3.313\,\mathrm{sec}$. To solve the aforementioned equations (6.1–6.4), the boundary conditions of all computational domains must be formulated. One of the most challenging problems of modeling biological flow is the specification of the boundary flow conditions. Physiological velocity and pressure waveforms must be prescribed.

6.6 Construction of Brain Aneurysm Meshes from CT Scans

Patient-specific geometries or brain aneurysms are essentials for accurate estimation of flow characteristics and wall stresses. Aneurysms can be constructed using DICOM files obtained from magnetic resonance angiography (MRA), CTA, or three-dimensional angiograms of living patients. Mimics software (Materialise, Inc.) is used in this investigation to capture brain aneurysm geometry from CTA images and to create a three-dimensional model for editing and numerical modeling. Mimics software has the capability to improve the quality and speed of finite element analyses through the transformation of irregularly shaped triangles into more or less equilateral triangles. In addition, this software can be used to differentiate thrombus and calcium. Another feature of this software includes the application of assigning variable mechanical properties (i.e., modulus of elasticity and Poisson's ratio) to different volume elements. An example of the applicability of this software in Figure 6.2, shows the geometry of a patient with brain aneurysm.

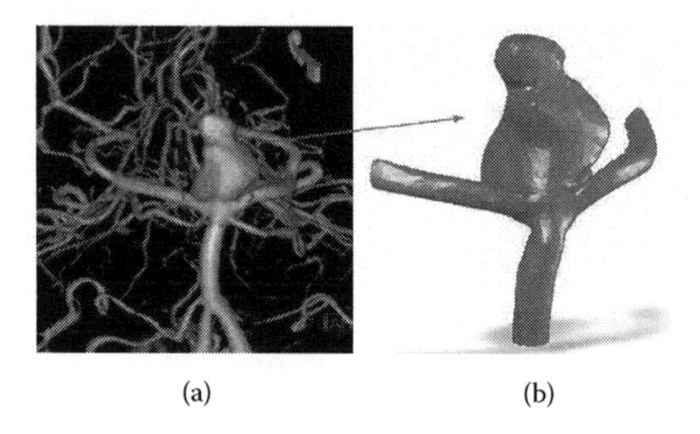

FIGURE 6.2
(a) CTA image of a basilar artery aneurysm, (b) aneurysm model of a basilar artery using Materialise software.

6.7 Results and Discussion

We studied the effect of the neck size on the temporal variations of the flow patterns and average pressure within the aneurysm sac without the presence of the coil as shown in Figures 6.3 and 6.4. Both figures show how blood flows through the lumen of the artery, and impinges on the distal edge of the aneurysmal neck. The impinged flow splits into two streams: one smaller jet enters the aneurysm sac at the distal edge of the neck while the bulk of the blood stream flows though the parent vessel. The blood flow within the aneurysm recirculates and travels in a direction retrograde to the flow in the main arterial lumen until it is entrained into the mainstream flow in the parent vessel. This results in a retrograde vortex forming inside the aneurysm sac with its center near the distal end as shown in Figures 6.3 and 6.4. Although the aneurysm vortex is present throughout the entire cardiac cycle, its magnitude is the greatest at the end of the acceleration phase (peak-flow velocity condition). The strength of the vortex is larger for a wide-necked aneurysm than for a narrow-necked aneurysm. This can be attributed to the fact that an aneurysm with a wide-neck has a larger opening allowing for a larger velocity jet stream to enter the aneurysm sac from the parent vessel than in a narrow-necked aneurysm.

The effect of coil embolization on the flow patterns and pressure within a narrow-necked aneurysm sac is illustrated in Figure 6.5. Owing to the absence of such data in the literature, we assumed a homogenous packing density of the order of 20% (realistic clinical assumption [58]), resulting in uniform porosity, $\varepsilon = \text{const} = 0.8$, everywhere in the aneurysm. Selecting the permeability value and therefore the Darcy number (Da) is more difficult because of the semiempirical nature of the corresponding terms. In most engineering applications, the value of permeability is usually determined experimentally. In our case,

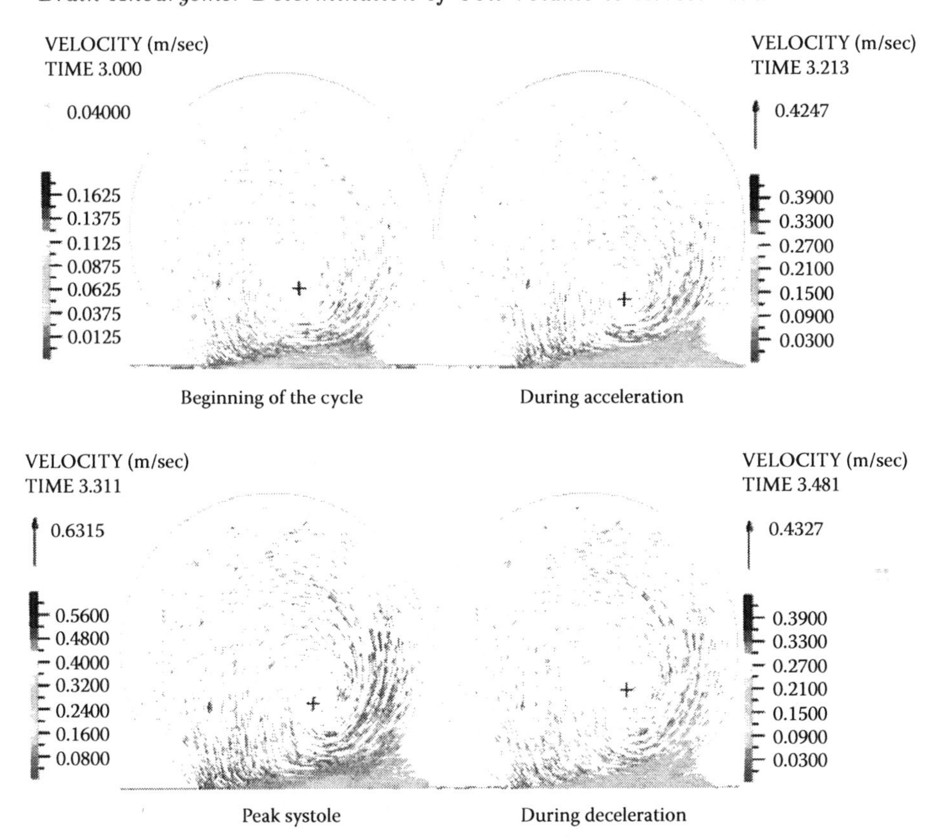

FIGURE 6.3
Temporal variation of the flow patterns within a narrow-necked aneurysm in the absence of coils. Flow from left to right.

due to lack of experimental data, a Darcy number $Da = 10^{-4}$ was assumed to represent high resistance. The Darcy number is a measure of fluid resistance as it flows through the porous matrix. Figure 6.5 shows the flow patterns in the saccular aneurysm and how they are influenced by the flow in the parent vessel and by coil embolization. For small values of the Darcy numbers, the fluid experiences a pronounced large resistance as it flows through the porous matrix, causing the flow to stop within the aneurysm sac. It can be seen from this figure that for a Darcy number of 10^{-4} the narrow-necked aneurysm sac is less permeable to fluid penetration, and, consequently, the convective activities within it are suppressed. Our results showed that the blood flow pulsatility into a side-wall aneurysm was significantly reduced after insertion of a platinum coil, resulting in a packing density of 20% as depicted in Figure 6.5. Our results also were consistent with the *in vitro* findings of Gobin et al. [26].

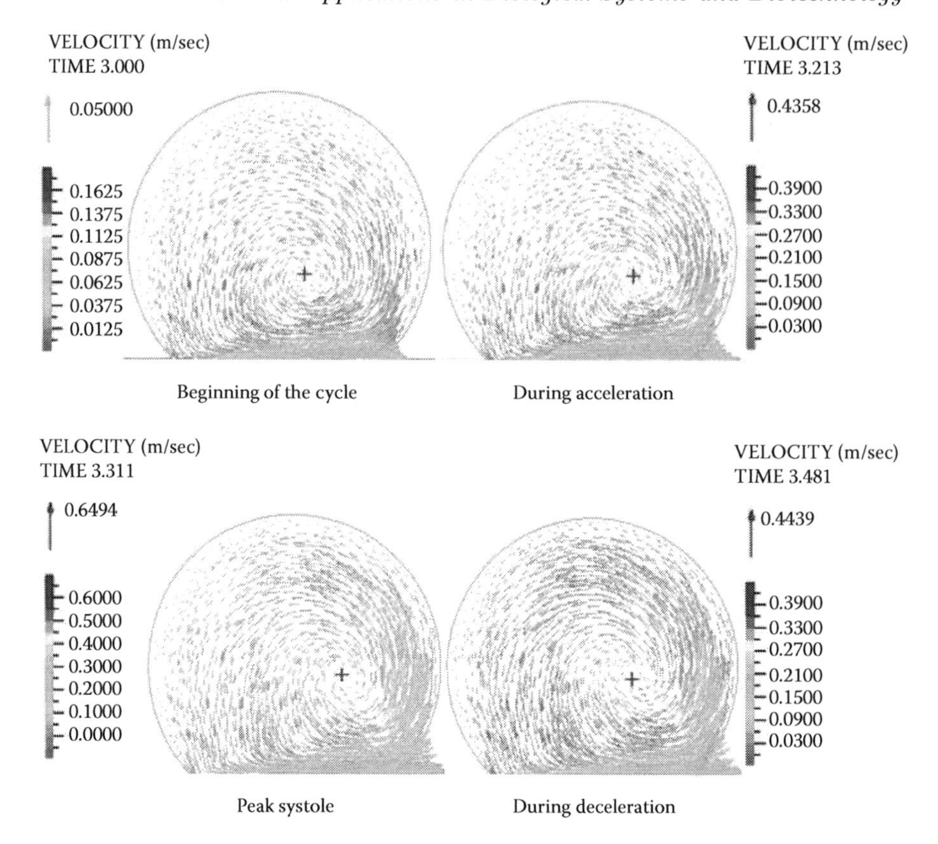

FIGURE 6.4
Temporal variation of the flow patterns within a wide-necked aneurysm in the absence of coils. Flow from left to right.

It can be seen from Figure 6.6 that the inflow to the aneurysm sac was significantly slowed at the aperture plane because of the presence of coils, which occupied a small volume of the aneurysm (20%). This minimum amount of coils is sufficient to arrest flow and promote thrombosis within the sac of the aneurysm, the usual site for rupture.

6.8 Minimum Packing Density of the Endovascular Coil

For aneurysms at high risk of rupture, the minimum length of coil inserted into aneurysm sac needed to arrest blood flow may be estimated from the

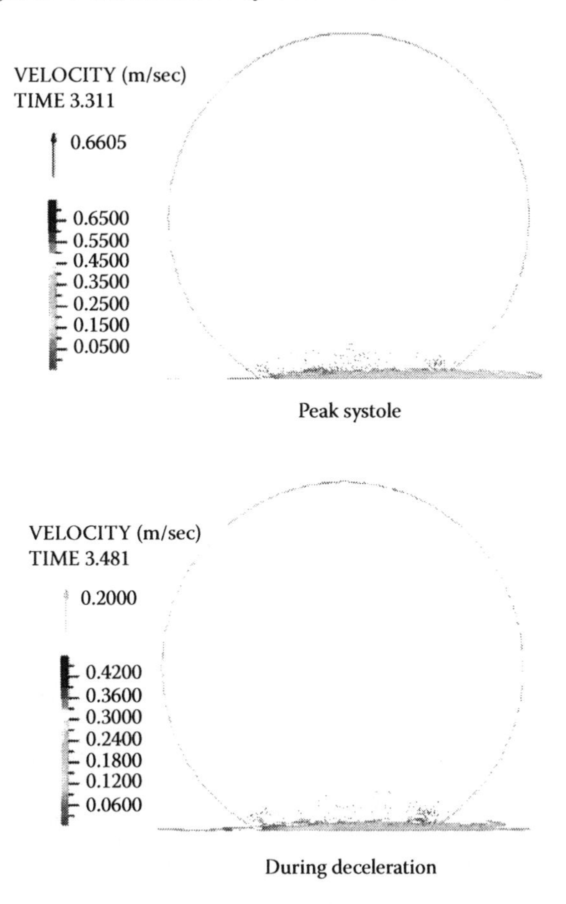

FIGURE 6.5
Temporal variation of the flow patterns a narrow-necked aneurysm with a 20% filling of coils.

aneurysm sac volume (V_{sac}) and the filling density (i.e., volume of the coil) by the following formula:

$$L_{coil} = \frac{V_{coil}}{\pi R_{coil}^2} = \frac{(1-\varepsilon)V_{sac}}{\pi R_{coil}^2} \tag{6.6}$$

This simple formula provides a guideline to the surgeons on the minimum required length of coil that needs to be inserted to arrest flow in the aneurysm sac. Figure 6.7 shows the treatment of a basilar aneurysm using endovascular coil.

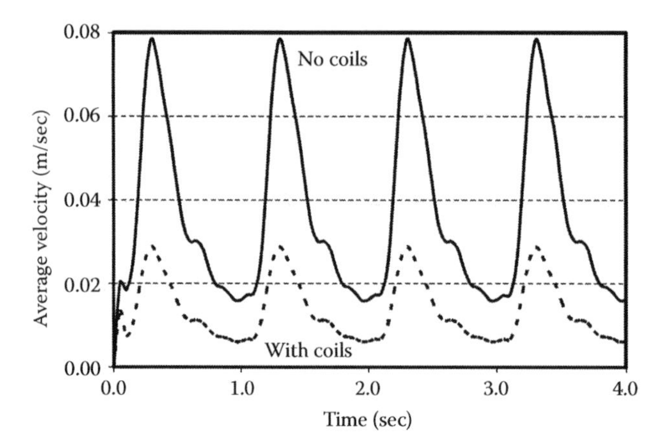

FIGURE 6.6
The effect of the endovascular coil on the temporal variation of the average velocity magnitude at the aperture of a narrow-necked aneurysm.

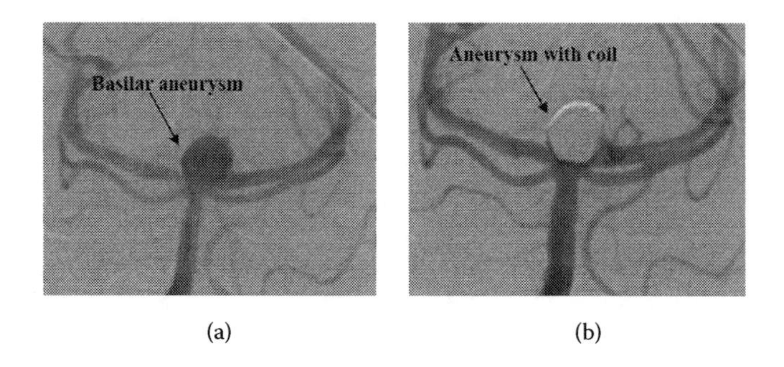

FIGURE 6.7
(a) Basilar aneurysm, (b) angiogram shows complete occlusion of the aneurysm.

6.9 Future Work

One of the major computational challenges in cardiovascular fluid mechanics is accurate modeling of the fluid–structure interactions (FSI) between the blood flow and arterial walls. The flow vectors are determined by the arterial geometry, and the deformation of the arterial wall depends significantly on the accurate values of the mechanical properties of the aneurysm wall (Young's modulus and Poisson's ratio) as well as on the pressure values. Existing publications on computational models of flow and wall mechanics in brain aneurysms have either displayed the fluid field assuming rigid walls or have calculated the mechanical stresses in flexible walls of brain aneurysm assuming same mechanical properties for both aneurysm and intracranial artery.

The ability to perform accurate aneurysm wall stress computations depends significantly on the mechanical behavior of the aneurysm's wall. Our future work will focus on developing a constitutive model for the aneurysm wall capable of reflecting the actual experimental data over a wide range of deformations in a physical model. The mechanical properties of the aneurysm wall will be determined in a tensile testing machine using tissues obtained from autopsy room. The proposed constitutive wall model fitted to mechanical behavior of the brain aneurysms will be entered into the fluid–solid interface model to represent the best approximation to the mechanics of the living artery and aneurysm in a specific patient. We will develop a robust numerical model of the wall stress distribution within brain aneurysms that can assist in the clinical management of patients by predicting the risk of rupture over time and permitting risk/benefit assessment for intervention or observation. Numerical simulations allow for the study of conditions that are difficult or impossible to measure directly in humans or in animal models of brain aneurysm.

6.10 Conclusions

A new model based on the porous media theory is proposed for the study of the effects of coiling in brain aneurysms. In the absence of endovascular coils, a wide-necked aneurysm was found to exhibit higher inflow velocity and average pressure within the aneurysmal sac than a narrow-necked aneurysm. The effect of the inserted platinum coil on the flow fields within an aneurysm sac under pulsatile flow condition was studied numerically. We showed that a 20% filling was sufficient to stop blood flow into aneurysmal sac. Porous media theory permits the study of fluid motion across small spaces of variable and complex geometry.

6.11 References

[1] Nield, D. A. and Bejan, A. (1995). *Convection in Porous Media.* 2nd ed., Springer-Verlag, New York.

Vafai, K. (2000). *Handbook of Porous Media.* 1st ed., Marcel Dekker, Inc., New York.

[2] Vafai, K. (2005). *Handbook of Porous Media.* 2nd ed., Taylor & Francis Group, New York.

[3] Hadim, H. and Vafai, K. (2000). Overview of current computational studies of heat transfer in porous media and their applications-forced convection and multiphase transport. In *Advances in Numerical Heat Transfer.* Taylor & Francis, New York, pp. 291–330.

[4] Vafai, K. and Hadim, H. (2000). Overview of current computational studies of heat transfer in porous media and their applications-natural convection and mixed convection. In *Advances in Numerical Heat Transfer.* Taylor & Francis, New York, pp. 331–371.

[5] Vafai, K. and Tien, C. L. (1981). Boundary and inertia effects on flow and heat transfer in porous media. *International Journal of Heat and Mass Transfer,* **24**:195–203.

[6] Vafai, K. and Tien, C. L. (1982). Boundary and inertia effects on convective mass transfer in porous media. *International Journal of Heat and Mass Transfer,* **25**:1183–1190.

[7] Vafai, K. (1984). Convective flow and heat transfer in variable-porosity media. *Journal of Fluid Mechanics,* **147**:233–259.

[8] Vafai, K. (1986). Analysis of the channeling effect in variable porosity media. *ASME Journal of Energy Resource Technology,* **108**:131–139.

[9] Linn, F., Rinkel, G., Algra, A., and van Gijn, J. (1996). Incidence of subarachnoid hemorrhage: role of region, year, and rate of computed tomography: a meta-analysis. *Stroke,* **27**:625–629.

[10] Weir, B. (2002). Unruptured intracranial aneurysms: a review. *Journal of Neurosurgery,* **96**:3–42.

[11] Tomasello, F., D'Avella, D., Salpietro, F., and Longo, M. (1998). Asumptomatic aneurysms. Literature metanalysis and indications for treatment. *Journal of Neurosurgery,* **42**:47–51.

[12] Stehbens, A. (1972). Intracranial arterial aneurysms. In *Pathology of the Cerebral Blood Vessels.* CV Mosby, St. Louis, MO, pp. 351–470.

[13] Kaminogo, M., Yonekura, M., and Shibata, S. (2003). Incidence and outcome of multiple intracranial aneurysms in a defined population. *Stroke,* **34**:16–21.

[14] Ringer, A., Lopes, D., Boulos, A., Guterman, L., and Hopkins, L. (2001). Current techniques for endovascular treatment of intracranial aneurysms. *Seminar in Cerebrovascular Diseases and Stroke,* **1**:39–51.

[15] Knuckey, N., Haas, R., Jenkins, R., and Epstein, M. (1992). Thrombosis of difficult aneurysms by the endovascular placement of platinum-Dacron microcoils. *Journal of Neurosurgery,* **77**:43–50.

[16] Wakhloo, A. K., Schellhammer, F., de Vries J., Haberstroh, J., Haberstroh, J., and Schumacher, M. (1994). Self-expanding and balloon-expandable stents in the treatment of carotid aneurysms: an experimental study in a canine model. *AJNR American Journal of Neuroradiology,* **15**:493–502.

[17] Wakhloo, A. K., Tio, F. O., Lieber, B. B., Schellhammer, F., Graf, M., and Hopkins, N. (1995). Self-expanding nitinol stents in canine vertebral arteries: Hemodynamics and tissue response. *AJNR American Journal of Neuroradiology,* **16**:1043–1051.

[18] Wakhloo, A. K., Lanzino, G., Lieber, B. B., and Hopkins L. N. (1998). Stents for intracranial aneurysms: the beginning of a new endovascular era? *Neurosurgery,* **43**:377–379.

[19] Turjman, F., Massoud, T. F., Ji, C., Guglielmi, G., Vinuela, F., and Robert, J. (1994). Combined stent implantation and endosaccular coil placement for treatment of experimental wide-necked aneurysms: a feasibility study in swine. *AJNR American Journal of Neuroradiology,* **12**:1087–1090.

[20] Burleson, A. C. and Turitto, V. T. (1996). Identification of quantifiable hemodynamic factors in the assessment of cerebral aneurysm behavior: on behalf of the Subcommittee on Biorheology of the Scientific and Standardization Committee of the ISTH. *Thromb Haemost,* **76**:118–123.

[21] Yu, S. C. M. and Zhao, J. B. (1999). A steady flow analysis on the stented and non-stented sidewall aneurysm models. *Medical Engineering and Physics,* **21**:133–141.

[22] Lieber, R. B., Livescu, V., Hopkins, L. N., and Wakhiloo, A. K. (2002). Particle image velocimetry assessment of stent design influence on intra-aneurysmal flow. *Annals of Biomedical Engineering,* **30**:768–777.

[23] Liou, T. M., Liou, S. M., and Chu, K. L. (2004). Intra-aneurysmal flow with helix and mesh stent placement across side-wall aneurysm pore of a straight parent vessel. *ASME Journal of Biomechanical Engineering,* **126**:36–43.

[24] Canton, G., Levy D. I., and Lasheras, J. C. (2005). Hemodynamic changes due to stent placement in bifurcating intracranial aneurysms. *Journal of Neusurgery,* **103**:146–155, 2005.

[25] Canton, G., Levy, D. I., and Lasheras, J. C. (2005). Changes in the intra-aneurysmal pressure due to hydro*coil* embolization. *AJNR American Journal of Neuroradiology,* **26**:904–907.

[26] Gobin, Y. P., Counord, J. L., Flaud, P., and Duffaux, J. (1994). In vitro study of hemodynamics in a giant saccular aneurysm model: influence

of flow dynamics in the parent vessel and effects of coil embolization. *Neuroradiology,* **36**:530–536.

[27] Marks, M. P., Dake, M. D., Steinberg, G. K, Norbash, A. M., and Lane, B. (1994) Stent placement for arterial and venous cerebrovascular disease: preliminary clinical experience. *Radiology,* **191**:441–446.

[28] Lanzino, G., Wakhloo, A. K., Fessler, R. D., Mericle, R. A., Guterman, L. R., Hopkins, L. N. (1998) Intravascular stents for intracranial internal carotid and vertebral artery aneurysms: preliminary clinical experience. *Neurosurgical Focus,* **5**(4):E3.

[29] Kwon, O. K., Kim, S. H., Oh, C. W., Han, M. H., Kang, H. S., Kwon, B. J., Kim, J. H., and Han, D. H. (2006) Embolization of wide-necked aneurysms with using three or more microcatheters. *Acta Neurochir* (Wien), **148**:1139–1145.

[30] Meckel, S., Stalder, A. F., Santini, F., Ernst-Wilhelm Radü, E. W., Rüfenacht, D. A., Markl, M., and Wetzel, S. G. (2008). In vivo visualization and analysis of 3-D hemodynamics in cerebral aneurysms with flow-sensitized 4-D MR imaging at 3T. *Neuroradiology,* **50**:473–484.

[31] Appanaboyina, S., Mut, F., Löhner, R., Putman, C. M., and Cebral, J. R. (2008). Computational fluid dynamics of stented intracranial aneurysms using adaptive embedded unstructured grids. *International Journal for Numerical Methods in Fluids,* **57**:475–493.

[32] Steinman, D. A., Milner, J. S., Norley, C. J., Lownie, S. P., Holdsworth, D. W. (2003). Image-based computational simulation of flow dynamics in a giant intracranial aneurysm. *American Journal of Neuroradiology,* **24**:559–566.

[33] Hassan, T., Timofeev, E. V., Saito, T., Shimizu, H., Ezura, M., Tominaga, T., Takahashi, A., and Takayama, K. (2004). Computational replicas: Anatomic reconstructions of cerebral vessels as volume numerical grids at three-dimensional angiography. *AJNR American Journal of Neuroradiology,* **25**:1356–1365.

[34] Moore, J. A., Steinman, D. A., and Ethier, C. R. (1998). Computational blood flow modeling: Errors associated with reconstructing finite element models from magnetic resonance images. *Journal of Biomechanics,* **31**:179–184.

[35] Perktold, K., Hofer, M., Rappitsch, G., Loew, M., Kuban, B. D., and Friedman, M. H. (1998). Validated computation of physiologic flow in a realistic coronary artery branch. *Journal of Biomechanics,* **31**: 217–228.

[36] Perktold, K. and Resch, M. (1990). Numerical flow studies in human carotid artery bifurcations: basic discussion of the geometric factor in atherogenesis. *Journal of Biomedical Engineering,* **12**:11–123.

[37] Perktold, K., Resch, M., and Florian, H. (1991b). Pulsatile non-Newtonian flow characteristics in a three-dimensional human carotid bifurcation model. *Journal of Biomechanical Engineering,* **113**:464–475.

[38] Khanafer, K., Bull, J. L., and Berguer, R. (2009). Fluid-structure interaction of laminar and turbulent pulsatile flow within a flexible wall axisymmetric aortic aneurysm models. *European Journal of Mechanics, B/Fluids,* **28**:88–102.

[39] Berguer, R., Bull, J., and Khanafer, K. (2006). Refinements in mathematical models to predict aneurysm growth and rupture. *Annals of the New York Academy of Sciences,* **1085**:110–116.

[40] Khanafer, K., Bull, J. L., and Berguer, R. (2006). Turbulence significantly increases wall pressure and shear stress in an aortic aneurysm model under resting and exercise conditions. *Annals of Vascular Surgery,* **21**:67–74.

[41] Khanafer, K., Gadhoke, P., Berguer, R., and Bull, J. L. (2006). Modeling pulsatile flow in aortic aneurysms: effect of non-Newtonian blood. *Biorheology,* **43**:661–679.

[42] Aenis, M., Stancampiano, A. P., Wakhloo, A. K., and Lieber, B. B. (1997). Modeling of flow in a straight stented and nonstented sidewall aneurysm model. *ASME Journal of Biomechanical Engineering,* **119**:206–212.

[43] Ohta, M., Wetzel, S. G., Dantan, P., Bachelet, C., Lovblad, K. O., Yilmaz, H., Flaud, P., and Rufenacht, D. A. (2005). Rheological changes after stenting of a cerebral aneurysms: a finite element modeling approach. *CardioVascular and Interventional Radiology,* **28**:768–772.

[44] Stuhne, G. R. and Steinman, D. A. (2004). Finite-element modeling of the hemodynamics of stented aneurysm. *ASME Journal of Biomechanical Engineering,* **126**:382–387.

[45] Shojima, M., Oshima, M., Takagi, K., Torii, R., Hayakawa, M., Katada, K., Morita, A., and Kirino, T. (2004). Magnitude and role of wall shear stress on cerebral aneurysm: computational fluid dynamic study of 20 middle cerebral artery aneurysms. *Stroke,* **35**:2500–2505.

[46] Rayz, V. L., Boussel, L., Acevedo-Bolton, G., Martin, A. G., Young, W. L., Lawton, M. T., Higashida, R., and Saloner, D. (2008). Numerical simulations of flow in cerebral aneurysms: comparison of CFD results and in vivo MRI measurements. *Journal of Biomechanical Engineering,* **130**: 1–9.

[47] Byun, H. S. and Rhee, K. (2004). CFD modeling of blood flow following imbolisation of aneurysms. *Medical Engineering and Physics,* **26**:755–761.

[48] Groden, C., Laudan, J., Gatchell, S., and Zeumer, H. (2001). Three-dimensional pulsatile flow simulation before and after endovascular coil embolization of a terminal cerebral aneurysm. *Journal of Cerebral Blood Flow & Metabolism,* **21**:1464–1471.

[49] Cha, K. S., Balaras, E., Liebre, B. B., Sadasivan, C., and Wakhloo, A. K. (2007). Modeling the interaction of coils with the local blood flow after coil embolization of intracranial aneurysms. *Journal of Biomechanical Engineering,* **129**:873–879.

[50] Khanafer, K., Berguer, R., Schlicht, M., and Bull, J. L. (2009). Numerical modeling of coil compaction in the treatment of cerebral aneurysms using porous media theory. *Journal of Porous Media,* **18**:869–886.

[51] Srinivasan, V., Vafai, K., and Christensen, R. N. (1994). Analysis of heat transfer and fluid flow through a spirally fluted tube using a porous substrate approach. *ASME Journal of Heat Transfer,* **116**:543–551.

[52] Amiri, A. and Vafai, K. (1994). Analysis of dispersion effects and non-thermal equilibrium, non-Darcian, variable porosity incompressible flow through porous media. *International Journal of Heat and Mass Transfer,* **37**:939–954.

[53] Amiri, A. and Vafai, K. (1998). Transient analysis of incompressible flow through a packed bed. *International Journal of Heat Mass Transfer,* **41**:4259–4279.

[54] Alazmi, B. and Vafai, K. (2002). Constant wall heat flux boundary conditions in porous media under local thermal non-equilibrium conditions. *International Journal of Heat and Mass Transfer,* **45**:3071–3087.

[55] Khanafer, K. M., Bull, J. L., Pop, I., and Berguer, R. (2007). Influence of pulsatile blood flow and heating scheme on the temperature distribution during hyperthermia treatment. *International Journal of Heat and Mass Transfer,* **50**:4883–4890.

[56] Ergun, S. (1952). Fluid flow through packed columns. *Chemical Engineering Progress,* **48**:89–94.

[57] Cho, Y. and Kensey, K. (1991). Effects of the non-Newtonian viscosity of blood on flows in a diseased arterial vessel: steady flows. *Biorhelogy,* **28**:241–262.

[58] Slob, M. J., van Rooij, W. J., and Sluzewski, M. (2005). Coil thickness and packing of cerebral aneurysms: A comparative study of two types of coils. *AJNR American Journal Neuroradiology,* **26**:901–903.

7

Lagrangian Particle Methods for Biological Systems

Alexandre M. Tartakovsky

Pacific Northwest National Laboratory, Computational Mathematics Group, Richland, WA

Zhijie Xu

Idaho National Laboratory, Energy Resource Recovery & Management, Idaho Falls, ID

Paul Meakin

Idaho National Laboratory, Center for Advanced Modeling and Simulation, Idaho Falls, ID

Physics of Geological Processes, University of Oslo, Oslo, Norway

Multiphase Flow Assurance Innovation Center, Institute for Energy Technology, Kjeller, Norway

CONTENTS

Acknowledgments
This research was supported by the Office of Science of the U.S. Department of Energy under Scientific Discovery through Advanced Computing. Pacific Northwest National Laboratory is operated for the U.S. Department of Energy by Battelle under contract DE-AC06-76RL01830, and the Idaho National Laboratory is operated for the U.S. Department of Energy by the Battelle Energy Alliance under contract DE-AC07-05ID14517.

7.1 Introduction

Solute transport coupled with biomass growth and/or mineral precipitation/dissolution is a complex and challenging nonlinear problem. Important applications in biomedical systems include tumor growth (Zheng et al. 2005); infection of prosthetic devices (Bandyk et al. 1991) and stents (Speer et al. 1988); dental plaque (Thomas and Nakaishi 2006); physiologic mineralization (Hartgerink et al. 2001) and demineralization (Holliday et al. 1997) in vertebrate bones (including cartilage), teeth, and otoconia; and ectopic calcification that occurs when mineral precipitates pathologically in soft tissues (Azari et al. 2008). In geological systems, microorganisms play an important role in the formation of iron mineral deposits in acid mine drainage (Karamanev 1991), in the precipitation of carbonates in hot springs (Riding 2000), in the growth of stromatolites (Reid 2000), and in weathering leading to the release of nutrients to the environment (Leyval and Berthelin 1991). Microorganisms also play an important role in corrosion (Beech and Gaylarde 1999), waste water treatment (Wagner et al. 1996), and blockage of water pipes (Brigmon et al. 1997) and heat exchangers. In the subsurface, microorganisms may significantly reduce permeability (Rittmann 1993), catalyze redox reactions relevant to contaminant remediation (Lensing et al. 1994), decompose organic contaminants (Zhang et al. 1995), and improve oil recovery (Van Hamme et al. 2003). Owing to the importance of these applications, there is a strong incentive to develop predictive numerical models.

The modeling and simulation of coupled biogeochemical processes in porous materials on the continuum (Darcy) scale relies heavily on phenomenological descriptions that complicate error analysis and reduce the predictive ability of the models. Phenomenology is required to describe the relationship between the macroscopic properties of porous media (e.g., permeability, dispersion coefficient, and effective surface area), the growth of precipitated minerals and/or biomass and the interactions between different solid and fluid phases. Pore-scale reactive transport models are based on fundamental conservation laws, such as the conservation of mass, momentum, and energy, and they have a much higher predictive ability (Knutson et al. 2005; Tartakovsky et al. 2007a,b). However, the high computational cost of pore-scale models does not allow them to completely replace Darcy-scale models in the entire computational domain. When mineral precipitation and/or biomass growth is highly localized, hybrid/multiphysics algorithms can provide an efficient computational tool for combining micro- and macroscale descriptions of physical phenomena. In the following sections we present two Lagrangian particle models, dissipative particle dynamics (DPD) and smoothed particle hydrodynamics (SPH), and their applications to pore-scale reactive transport, biomass growth, and mineral precipitation.

Dissipative particle dynamics, a stochastic Lagrangian approach introduced by Hoogerbrugge and Koelman in 1992 is based on the idea that particles can be used to represent clusters of atoms or molecules instead of single atoms or molecule to provide a simple and robust way of coarse graining the molecular dynamics (MDs) of dense fluids and soft condensed matter systems. Because of the internal degrees of freedom associated with individual DPD particles, the DPD particle–particle interactions include dissipative and fluctuating interactions that are related by the fluctuation–dissipation theorem (Kubo 1966; Espanol and Warren 1995) in addition to the conservative particle–particle interactions, and the dissipative and fluctuating interactions function as a thermostat for the model. The grouping of atoms or molecules into a single DPD particle (coarse graining) leads to averaged effective conservative interaction potentials (soft repulsive-only potentials in the standard DPD model) between the DPD particles. Consequently, the computational cost is substantially lowered due to the soft potentials as well as the coarse graining. The computational advantage of DPD over MD is about $1{,}000N_m^{5/3}$, where N_m is the number of atoms represented by a single DPD particle (Pivkin and Karniadakis 2006). This makes DPD an effective mesoscale particle simulation technique on length and timescales, that are larger than those accessible to fully atomistic MD simulations.

Smoothed particle hydrodynamics is a fully Lagrangian particle method (Monaghan 2005). In SPH models, fluid phases are represented by a set of particles, and the particle positions serve as interpolation points for solving partial differential equations, such as the Navier–Stokes (NS) and advection–diffusion equations. The particles move with velocities given by the NS equations and the composition (concentrations) of the particles change according to the advection–diffusion equation (diffusion equation in the Lagrangian coordinate system). The size (or number density) of the particles sets the numerical resolution of SPH simulations, and the time step is limited by the standard Courant–Friedrichs–Lewy (CFL) conditions. Consequently, SPH models can be regarded as continuum models with an unstructured grid, and they operate on much larger time and length scales than mesoscale (between the atomistic and continuum scale) DPD models. SPH can be used to model biogeochemical systems that can be adequately described by conservation laws in the form of deterministic partial differential equations.

In the final section, we describe a hybrid algorithm for modeling a general class of diffusion-reaction systems and its application to mixing-induced precipitation.

7.2 DPD Models for Biological Applications

When DPD models are used to simulate liquids, the particles overlap extensively (there are tens of particles in the interaction volume of $4\pi r_0^3/3$, where r_0 is the cut-off range of the particle–particle interactions). As the number of atoms, N_m, represented by a DPD particle increases, the thermal energy ($k_BT/2$ per degree of freedom, where k_B is the Boltzmann constant and T is the temperature), remains constant while the DPD particle–particle interactions increase in magnitude. The decrease in the thermal energy relative to the interaction energy drives the system through a Kirkwood–Alder transition (Kirkwood 1939; Alder and Wainwright 1962) from a fluid to a solid, and this limits the size of the cluster of atoms or molecules that the DPD particles can represent. For DPD simulations with interaction parameters that have been selected so that the DPD fluid properties match the properties of real liquids the DPD particle mass can be no more than 10–100 times the atomic or molecular mass (Dzwinel and Yuen 2000). Consequently, speedups (relative to MDs) greater than about 10^5 cannot be achieved using the standard DPD model. DPD can be regarded as thermostatted nonequilibrium MDs with soft particle–particle interactions, and the DPD thermostat has been used in nonequilibrium MDs simulations (Guo et al. 2002). The soft particle–particle interactions allow much larger time steps to be taken, and this is a more important advantage than the coarse graining, relative to standard MDs.

Owing to its computational efficiency compared to MDs, while a significant molecular level detail is retained, DPD has been extensively applied to investigate the effects of the size, shape, and rigidity of large molecules, and their intermolecular interactions on the behavior of soft condensed matter. This makes DPD an attractive approach for the simulation of biological systems on supraatomic length scales and on timescales beyond the MD timescale, but where the effects of the thermal fluctuations and thermal (entropic) forces are still important. DPD has been used quite extensively to simulate the dynamic behavior of cell membranes, and lipid bilayers, which play very important roles in living cells (Venturoli and Smit 1999; Groot and Rabone 2001; Shillcock and Lipowsky 2002; Yamamoto et al. 2002; Kranenburg et al. 2004; Ortiz, Nielsen et al. 2005; Shillcock and Lipowsky 2005; Gao et al. 2007; Revalee et al. 2008; Wu and Guo 2008). As the computational cost of full MDs simulations of the self-assembly of membranes is very high, Venturoli and Smit (1999) developed a DPD model to study the lateral pressure and density variation across biological membranes. In this coarse-grained DPD model, the hydrophilic head of the surfactant was represented by one DPD particle, the hydrophobic tail was represented by five particles, and the particles were connected by harmonic springs. When the surfactant was dissolved in a solvent consisting of water-like hydrophilic DPD particles, the surfactant spontaneously assembled into a bilayer. The collective behavior of molecules forming the membrane is properly retained while other unnecessary atomistic level details (not of interest in the investigation [Venturoli and Smit 1999]) were discarded to minimize the computational cost.

Later, DPD was applied to the dynamic behavior of biological membranes (membrane damage, rupture, and morphology changes) that are exposed to nonionic surfactants (Groot and Rabone 2001). In principle, biological processes take place over a wide range of timescales, and simulations must be run for long enough to allow phenomena such as phase transition, phase separation, and the uptake of surfactant by the biological system to occur. Therefore, DPD provides a very promising computationally efficient mesoscale approach to the simulation of biological systems.

The successful application of DPD to specific molecular systems depends on the parameters associated with the interactions among the various types of DPD particles that represent the distinct molecular species that constitute the system. These interaction parameters may be obtained by matching experimental and observed properties (Groot and Rabone 2001). For small molecules that can be represented by a single DPD particle, or a group of identical particles, the interactions between particles representing the same chemical component may be obtained by matching the experimental and simulated compressibility, and solubilities are often used to calibrate the interactions between DPD particles that represent different chemical components. Many larger molecules are represented by a number of different particles, and calibration of the particle–particle interactions is more challenging. Simple surfactants may often be represented by a "head" consisting of one or more particles of type A and a "tail" consisting of one or more particles of type B. In this case, neutron scattering and/or x-ray scattering may be used to obtain information about the density profile of the head or tail groups at a liquid/air interface, and this information may be used to (partially) calibrate the particle–particle interactions. MD simulations may be used to obtain more detailed information, and this information may be used to calibrate or validate DPD models (Groot and Rabone 2001). In more complex systems, a larger number of distinct DPD particles is needed, and the number of particle–particle interactions is $n(n+1)/2$, where n is the number of distinct types of DPD particles, and this is a lower limit on the number of interaction parameters. Consequently, the challenge of calibrating DPD models grows rapidly with the complexity of the system, and there is rarely, if ever, sufficient relevant experimental information. An alternative approach, largely under development, is to base the DPD model on a MD model by means of a systematic coarse-graining procedure. If DPD simulations are used to investigate generic behaviors, there is no need for system-specific interaction parameters, and a realistic set of interaction parameters can often be obtained by using simple theoretical concepts.

In general, mesoscale DPD simulations make it possible to study the assembly, equilibrium properties, and dynamics of synthetic and biological lipid bilayers and other large molecular assemblies. The equilibrium structure and mechanical properties of membrane patches obtained from DPD simulations are in agreement with experimental results (Gao et al. 2007).

In addition to computational efficiency, DPD provides a consistent way of simulating the effects of thermal fluctuations on biological systems, which can be very important at small scales (Sugii et al. 2007). For instance, lipid bilayer vesicles usually exhibit a much larger surface area than the spheres of same volume. A large portion of the lipid bilayer vesicles surface area is folded because of the thermal fluctuation in the system. This excess surface area could be very important in certain situations, for example, the deformation of lipid bilayer vesicles in shear flow (de Haas et al. 1997), and this can be investigated by DPD simulation.

Dissipative particle dynamics has also been applied to other important biological phenomena. Examples include platelet-mediated thrombus formation in blood vessels; the motion, collision, aggregation, and adhesion of activated platelets (Filipovic et al. 2006, 2007, 2008); modeling of red blood cells in shear flow (Richardson et al. 2008); and modeling of the conformations of DNA molecules suspended in a fluid flowing in microchannels (Fan et al. 2006).

Here we describe a DPD model for the growth and deformation of biofilm in a flowing fluid. The hydrodynamic interactions between the biofilm and liquid flow are expected to be important in this application, and an important advantage of DPD, relative to Brownian dynamics simulations, is that the hydrodynamic interactions are naturally included in DPD models. Biofilm can be defined as a microbial community composed of either single or multiple species embedded in extracellular biopolymer, which adheres to a solid substrate. The modeling of biofilm provides important insight that contributes to a better understanding of the mechanisms of biofilm formation, growth, and cell death.

Modeling the effects of nutrient and metabolic waste transport and fluid flow on biofilm development is very challenging because of the complexity of the underlying fundamental physical, chemical, and biological processes occurring at various scales (Picioreanu et al. 2000). The interplay and coupling between those physical, chemical, and biological processes governs the evolution of biofilm structure. Biofilm structure development can be conceptualized as a competition between "positive" and "negative" processes (Picioreanu et al. 2000). Here "positive" processes refer to processes that lead to biofilm volume expansion, while "negative" processes refer to processes that lead to biofilm volume reduction. Typical "positive" processes include cell attachment, cell division, and extracellular polymeric substance (EPS) production as a result of the transport of dissolved substrate via both advection and diffusion from or to the biofilm, while typical "negative" processes include cell death, erosion (the removal of single cells or small cell clusters from the biofilm), and fracturing or sloughing (the removal of a large number of living and/or dead microorganisms in a single event) due to liquid–biofilm hydrodynamic interactions. The effects of possible physiological and biochemical factors were not considered in the current model. However, in principle, those effects can be incorporated in more comprehensive models.

In the DPD model, we use a mathematical description that is widely employed in the biofilm literature (Picioreanu et al. 2000; Eberl et al. 2001). The entire computation domain is divided into three phases, namely the liquid phase, the biofilm phase, and the biofilm support (substratum) phase. In general, the presence of two types of nutritional substrate, electron donors and electron acceptors, are required for biofilm growth. If there is an unlimited supply of one substrate (either electron donors or electron acceptors) then the biofilm growth is limited by the concentration of only one nutritional substrate, S (Picioreanu et al. 2000). Then the biofilm growth, and the advection and diffusion of nutritional substrate, S, are governed by the advection–diffusion-reaction equation:

$$\partial C_s/\partial t + \mathbf{V} \cdot \nabla C_s = D_s \nabla^2 C_s - r_s \tag{7.1}$$

where $\mathbf{V}$ is the liquid velocity, $C_s(\mathbf{x},t)$ is the concentration of substrate, S, at position x and time t, and D_s is the diffusion coefficient of the substrate. The substrate consumption rate (kg m^{-3} s^{-1}) in the biofilm phase, r_s, represents substrate consumption due to biofilm growth. Here we assume that for a single substrate, r_s, the substrate consumption rate is given by a Monod function (Picioreanu et al. 2000):

$$r_s = \frac{\mu_m}{Y_{bs}} \frac{C_b C_s}{(K_s + C_s)} \tag{7.2}$$

where μ_m is the maximum biomass growth rate (s^{-1}), K_s is the substrate saturation constant (kg m^{-3}), Y_{bs} is the dimensionless biofilm yield (kg biomass/kg substrate) for substrate S, and $C_b(\mathbf{x},t)$ is the biomass density.

The liquid velocity field, $\mathbf{V}$, is given by the incompressible NS equations

$$\nabla \cdot \mathbf{V} = 0 \tag{7.3}$$

$$\partial \mathbf{V}/\partial t + \mathbf{V} \cdot \nabla \mathbf{V} = -\nabla P/\rho + \nu \nabla^2 \mathbf{V} \tag{7.4}$$

Equations (7.3) and (7.4) describe the mass and momentum conservation in the liquid phase, where P is the pressure field, ρ is the liquid density, and ν is the kinematic viscosity. The kinetic equation describing the biofilm growth and/or decay in the biofilm phase is written as

$$dC_b/dt = Y_{bs}(r_s - m_s C_b) \tag{7.5}$$

where m_s is a maintenance coefficient (kg substrate/[kg biomass s]) representing the biomass simultaneous decay effect. Spreading is an important characteristic in the biofilm kinetics model. The biomass density has a maximum value, C_{bm}, and whenever the biomass density, C_b, grows larger than the threshold value, C_{bm}, the extra biomass is redistributed giving rise to biofilm volume expansion. Specific examples of growth algorithm of this type are described in the following DPD simulation.

The original set of equations (7.1–7.5) can be rewritten in a dimensionless form for the convenience of simulation and modeling by introducing the characteristic length, l_c, and velocity, v_c, as the units of length and velocity and substituting equation (7.2) into equation (7.5). The resulting dimensionless equations are

$$\partial c_s/\partial t' + \mathbf{v}' \cdot \nabla c_s = \nabla^2 c_s/P_e - k_1 k_4 c_b c_s/(k_2 + c_s) \tag{7.6}$$

$$\nabla \cdot \mathbf{v}' = 0 \tag{7.7}$$

$$\partial \mathbf{v}'/\partial t' + \mathbf{v}' \cdot \nabla \mathbf{v}' = -\nabla p' + \nabla^2 \mathbf{v}'/R_e \tag{7.8}$$

$$dc_b/dt' = k_1 c_b \{c_s/(k_2 + c_s) - k_3\} \tag{7.9}$$

where the dimensionless numbers $k_1 - k_4$ are defined as, $k_1 = \mu_m l_c/v_c$, $k_2 = K_s/C_{sm}$,

$k_3 = Y_{bs} m_s/\mu_m$, and $k_4 = C_{bm}/(Y_{bs} C_{sm})$. C_{sm} is the maximum substrate concentration in the system, and it is used for normalization. The Reynolds number is defined as, $R_e = v_c l_c/\nu$ and the Peclet number is defined as $P_e = v_c l_c/D_s$. The dimensionless velocity and pressure fields are defined as $\mathbf{v}' = \mathbf{V}/v_c$ and $p' = P/\rho v_c^2$. The dimensionless substrate concentration and biofilm density are normalized by $c_s = C_s/C_{sm}$ and $c_b = C_b/C_{bm}$, and they both lie in the range 0–1. This set of dimensionless equations (7.6–7.9) can be further reduced by introducing the new timescale $t'' = t' k_1$ and velocity scale $\mathbf{v} = \mathbf{v}'/k_1$. This gives

$$\partial c_s/\partial t'' + \mathbf{v} \cdot \nabla c_s = \nabla^2 c_s/P_{eb} - k_4 c_b c_s/(k_2 + c_s) \tag{7.10}$$

$$\nabla \cdot \mathbf{v} = 0 \tag{7.11}$$

$$\partial \mathbf{v}/\partial t'' + \mathbf{v} \cdot \nabla \mathbf{v} = -\nabla p + \nabla^2 \mathbf{v}/R_{eb} \tag{7.12}$$

$$dc_b/dt'' = c_b \{c_s/(k_2 + c_s) - k_3\} \tag{7.13}$$

where the new set of equations contains only five dimensionless numbers k_2, k_3, k_4, $R_{eb} = \mu_m l_c^2/\nu$, and $P_{eb} = \mu_m l_c^2/D_s$. The corresponding dimensionless time, velocity, and pressure fields are defined as, $t'' = \mu_m t$, $\mathbf{v} = \mathbf{V}/(\mu_m l_c)$ and $p = P/\rho(\mu_m l_c)^2$.

Appropriate velocity, concentration, and pressure boundary conditions associated with equations (7.10–7.13) must be provided to complete the definition of the model. The no-slip boundary conditions for the velocity are applied on the interface between the liquid phase and the biofilm or solid substratum phase. A variety of numerical approaches have been applied to solve the coupled differential equations in biofilm modeling studies. A continuum model similar to the phase-field approach has been developed by treating both the liquid and the biofilm phase as continuous media, but with different diffusion

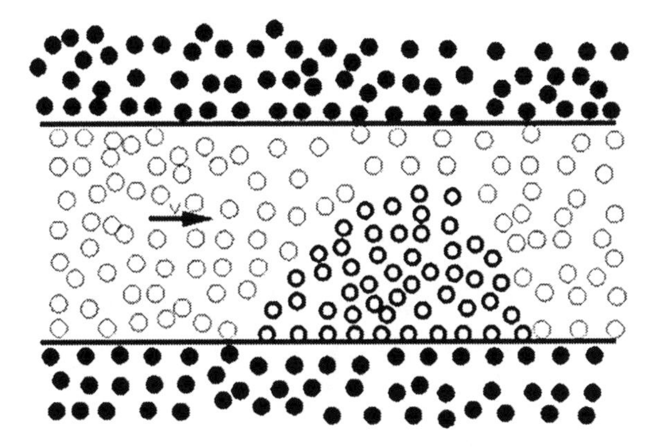

FIGURE 7.1
A schematic representation of the DPD biofilm model: Open circles represent liquid flowing through a channel, filled circles represent the solid substratum, and partially filled circles represent the biomass.

coefficients, D_s (Eberl et al. 2001). Special attention to D_s is needed in the biofilm phase to account for the fact that biofilm spreading is only significant when the biomass density, C_b, is close to the maximum biofilm density, C_{bm}. Grid-based cellular automata models have also been developed to solve the coupled differential equations presented earlier (Picioreanu et al. 1998; Picioreanu et al. 2000; Picioreanu et al. 2001). The hydrodynamic interactions between flow and biofilm structure can be further considered by using the finite-element method to solve the stress and strain in the biofilm structure at each time step (Picioreanu et al. 2001).

In the DPD model used in this work, three types of DPD particles are used to represent the liquid, biomass, and solid substratum phases (Figure 7.1).

In standard DPD simulations (Hoogerbrugge and Koelman 1992), a fluid is represented by an ensemble of particles that move because of the combined effects of conservative (nondissipative), $\mathbf{f}^C$, dissipative, $\mathbf{f}^D$, fluctuating (random) $\mathbf{f}^R$, and external, $\mathbf{f}^{ext}$, forces, and the equation of motion is

$$dm_i\mathbf{v}_i/dt = \mathbf{f}_i^{int} + \mathbf{f}_i^{ext} = \mathbf{f}_i^C + \mathbf{f}_i^D + \mathbf{f}_i^R + \mathbf{f}_i^{ext} = \mathbf{f}_i^{ext} + \sum_{j\neq i}(\mathbf{f}_{ij}^C + \mathbf{f}_{ij}^D + \mathbf{f}_{ij}^R) \tag{7.14}$$

where $\mathbf{v}_i$ is the velocity of particle i and m_i is its mass. In models for single-phase fluid flow, the conservative forces between particles are usually given by a simple and purely repulsive form such as $\mathbf{f}_{ij}^C = S(1 - r_{ij}/r_0)\hat{\mathbf{r}}_{ij}$ for $r_{ij} = |\mathbf{r}_{ij}| = |\mathbf{r}_i - \mathbf{r_j}| < r_0$ and $\mathbf{f}_{ij}^C = 0$ for $r_{ij} \geq r_0$, where S is the strength of the particle–particle interaction, r_0 is the cut-off range of the particle–particle interactions, and $\hat{\mathbf{r}}_{ij}$ is the unit vector pointing from particle j to particle i

($\hat{\mathbf{r}}_{ij} = (\mathbf{x}_i - \mathbf{x}_j)/|\mathbf{x}_i - \mathbf{x}_j|$). The total force acting on particle i due to conservative forces is

$$\mathbf{f}_i^C = \sum_{j \neq i} \mathbf{f}_{ij}^C = \sum_{j \neq i, r_{ij} < r_0} S_{ij}(1 - r_{ij}/r_0)\hat{\mathbf{r}}_{ij} \tag{7.15}$$

where S_{ij} is the strength of the interaction between particle i and particle j. The dissipative particle–particle interactions are given by $\mathbf{f}_{ij}^D = \gamma W^D(r_{ij})(\mathbf{r}_{ij} \cdot \mathbf{v}_{ij})\hat{\mathbf{r}}_{ij}$, where γ is a viscosity coefficient and $\mathbf{v}_{ij} = \mathbf{v}_j - \mathbf{v}_i$, for $r_{ij} < r_0$ and $\mathbf{f}_{ij}^D = 0$ for $r_{ij} > r_0$ so that

$$\mathbf{f}_i^D = \sum_{j \neq i} \mathbf{f}_{ij}^D = - \sum_{j \neq i, r_{ij} < r_0} \gamma W^D(r_{ij})(\mathbf{r}_{ij} \cdot \mathbf{v}_{ij})\hat{\mathbf{r}}_{ij} \tag{7.16}$$

and the random forces are given by $\mathbf{f}_{ij}^R = \sigma W^R(r_{ij})\zeta \hat{\mathbf{r}}_{ij}$ for $r_{ij} < r_0$ and $\mathbf{f}_{ij}^R = 0$ for $r_{ij} > r_0$, where σ is the fluctuation strength coefficient and ζ is a random variable selected from a Gaussian distribution with a zero mean and a unit variance so that

$$\mathbf{f}_i^R = \sum_{j \neq i} \mathbf{f}_{ij}^R = \sum_{j \neq i, r_{ij} < r_0} \sigma W^R \zeta \hat{\mathbf{r}}_{ij} \tag{7.17}$$

In practice, ζ can be selected randomly from a uniform distribution with a zero mean and a unit variance (the sum of a quite small number of uniformly distributed random numbers over a number of time steps quite accurately approximates a Gaussian random variable). The random and dissipative particle–particle interactions are related through the fluctuation–dissipation theorem (Espanol and Warren 1995), which requires that $\gamma = \sigma^2/2k_BT$, where k_B is the Boltzmann constant, T is the prescribed temperature, and $W^D(r) = (W^R(r))^2$, where $W^D(r)$ and $W^R(r)$ are r-dependent weight functions, both vanishing for $r \geq r_0$. These are the detailed balance condition required for DPD. In standard DPD models, the simple weighting function, $W^D(r) = \left[W^R(r)\right]^2 = (1 - r/r_0)^2$ (for $r < r_0$), is used. The combination of dissipative and fluctuating forces, related by the fluctuation–dissipation theorem (Kubo 1966), acts as a thermostat, which maintains the temperature of the system, measured through the average kinetic energy of the particles at a temperature of T, provided that the time step used in the simulation is small enough. This idea can be taken one step further (Lowe 1999) by integrating the equations of motion with only the conservative forces $\{\mathbf{f}_i^C\}$ over the time step, Δt, and then "thermalizing" the relative velocities of a fraction, f_T, of the particle pairs separated by a distance of r_0 or less by randomly selecting the relative velocities from a Maxwell–Boltzmann distribution and multiplying by $\sqrt{2}$ to convert the Maxwell–Boltzmann single particle velocity distribution at temperature T to a relative velocity distribution function and preserving the average velocity of the particle pairs to conserve momentum. The fraction, f_T, of particles that are "thermalized" at each step in the simulation can then be varied to change the effective viscosity of the fluid. An advantage of this approach is that it eliminates the problems associated with integration of stochastic differential equations, and the temperature, measured via the particle kinetic energy, deviates less from the temperature used in the DPD algorithm (Nikunen et al. 2003).

In the current DPD biofilm model, the solution of the NS equations (7.10–7.11) is approximated by the low Mach number flow of a slightly compressible fluid represented by the DPD particles. The velocities and positions of the DPD particles are found from equation (7.14) using a modified velocity Verlet algorithm to integrate the equation of motion (Groot and Warren 1997). The no-slip boundary condition on the interface between the liquid and substratum/biofilm phases is implicitly implemented through the particle–particle interactions, which are strong enough to prevent the penetration of the liquid particles into the region occupied by the biofilm particles. This interaction can also be tuned to allow the liquid to penetrate into the biofilm phase, which is important for diffusion of the nutrient substrate into the biofilm phase. In practice, the fluid velocity field will penetrate into the biofilm and there will be slip at the fluid–solid interface. However, both of these effects are very small (the permeability of biofilm is very small and the slip length is typically on the order of 1 nm).

The mass of substrate and biomass carried by DPD particle i is specified by the substrate concentration, $c_{s,i}$, and the biomass density, $c_{b,i}$, associated with particle i. The changes of substrate concentration in liquid and biofilm DPD particles is given by the advection–diffusion equation (7.12). The DPD representation of the advection–diffusion equation is similar to the representation of the heat-conduction equation (Avalos and Mackie 1997; Espanol 1997; Ripoll and Espanol 1998), and it has the form

$$\frac{dc_{s,i}}{dt} = \sum_{j \neq i} \lambda_{ij} W^R(r_{ij})(c_{s,j} - c_{s,i}) + \sum_{j \neq i} \zeta^c \sqrt{2\alpha^{-1}\lambda_{ij} W^R(r_{ij}) c_{s,i} c_{s,j} / \Delta t} - \frac{k_4 c_{b,i} c_{s,i}}{k_2 + c_{s,i}} \tag{7.18}$$

where λ_{ij} is the interparticle diffusion constant between particles i and j, which can be related to the continuum molecular diffusion coefficient, and Δt is the time step used in the DPD simulations. In equation (7.18), α is a material constant representing the magnitude of the concentration fluctuation, ζ^c is a random variable of the same type as ζ in equation (7.17), but it is uncorrelated with ζ. The term $\Delta t^{-1/2}$ appears in equation (7.18) because the average of a random force acting over the time step of length Δt is proportional to $\Delta t^{-1/2}$ (Groot and Warren 1997). The last term in equation (7.18) is a particle formulation of the rate of substrate consumption.

Because of the soft particle–particle interactions used in DPD models, the DPD particles have a relatively large intrinsic self-diffusion coefficient (the cage effect that makes the momentum diffusion coefficient much larger than the molecular diffusion coefficient in liquids is much weaker in DPD fluids) and the contribution of the DPD particle self-diffusion to the diffusive transport of the nutrient substrate is significant. Consequently, the real nutrient substrate diffusion coefficient in a DPD model must be determined for given λ_{ij} and T by calibrating the DPD model using the known analytical solutions for one-dimensional diffusion equation.

The biofilm density, $c_{b,i}$, is found from equation (7.13):

$$dc_{b,i}/dt = c_{b,i}\{c_{s,i}/(k_2 + c_{s,i}) - k_3\} \quad (7.19)$$

Equation (7.19) should be solved for biofilm DPD particles only where $c_{b,i} > 0$. Once the biomass density, c_b, exceeds the maximum biomass density (1.0 for the normalized biomass density), the excess biomass is transferred to the nearest fluid DPD particle in the cut-off range with $c_b = 0$, and this fluid DPD particle is spontaneously changed to a biofilm DPD particle. In the rare event of an absence of fluid DPD particles in the cut-off region of particle i (this may happen if particle i lies deep inside biofilm domain), the excess biomass is assumed to be lost. In practice, this "unphysical" procedure is highly unlikely to occur because biomass growth is concentrated mainly on the interface, and decay dominates inside the biomass domain. This biomass spreading mechanism is similar to the discrete rules used to redistribute the biomass in cellular automata models (Picioreanu et al. 1998), where a search for a "free-space" element among the nearest-neighbor elements is performed.

Lennard-Jones or bi-harmonic potentials have been used in MD studies of crack propagation in solids (Abraham et al. 1997; Buehler et al. 2003). Similarly, a simple harmonic potential with an equilibrium distance, r_e, that is slightly smaller than the cut-off distance, r_0, is used to model the interactions between biofilm DPD particles, and this results in a soft solid-like biofilm structure. The harmonic potential energy and force between biofilm particles i and j are given by

$$e_{ij} = -\frac{S_{bb}}{2r_e}(r_{ij} - r_e)^2 \quad (7.20)$$

$$\mathbf{f}_{ij} = -\mathbf{f}_{ji} = S_{bb}(1 - r_{ij}/r_e)\hat{\mathbf{r}}_{ij} \quad (7.21)$$

When the distance, r_{ij}, exceeds the cut-off distance, r_0, the force between particles i and j falls to zero because the "bond" between these particles is ruptured.

The biofilm–biofilm particle interaction strength, S_{bb}, and the equilibrium distance, r_e, control the mechanical properties of the biofilm, such as the elastic modulus and the maximum strain at failure. In the model, the interaction between a liquid DPD particle and a biofilm DPD particle is assumed to be equal to the purely repulsive liquid–liquid interaction or it can be assumed to have a harmonic functional form with an equilibrium distance of $r_e = r_0$, and the strength, S_{lb}, can be tuned for various liquid–biofilm interaction strengths to satisfy the no-slip boundary condition at the liquid–biofilm interface. Here S_{lb} and S_{ll} are the strength coefficients used in equation (7.15) to compute the conservative forces between liquid–biofilm particles and liquid–liquid particles.

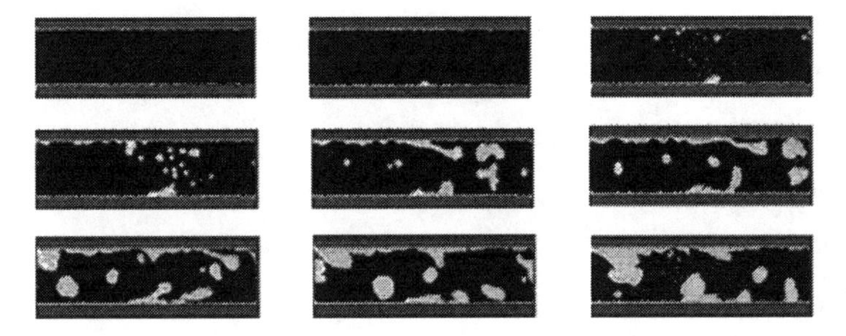

FIGURE 7.2
DPD simulation of the growth deformation and fragmentation of a relatively soft biofilm. Medium gray represents the substratum, dark gray represents the flowing fluid, and light gray represents the biofilm. Snapshots are taken at simulation times of $t = 500$, $550 \ldots 900$.

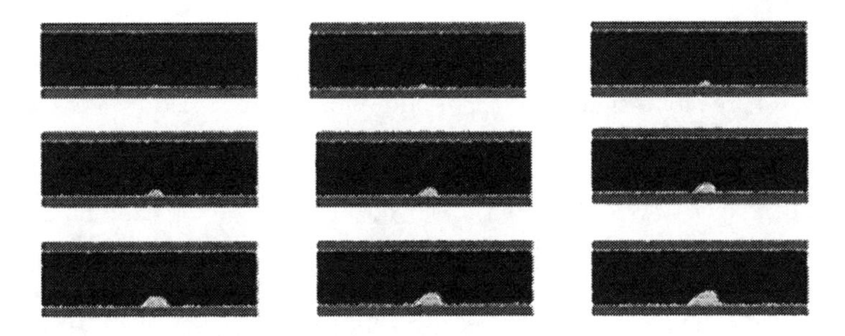

FIGURE 7.3
DPD simulation model of the growth and deformation of a medium stiffness and strength biofilm. Medium gray represents the substratum, dark gray represents the flowing fluid, and light gray represents the biofilm. Snapshots are taken at simulation times $t = 500$, $550 \ldots 900$.

Preliminary results from a DPD model incorporating nutrient substrate transport, biofilm growth, and hydrodynamic interactions are shown in Figures 7.2, 7.3, and 7.4 at the same time. The simulation was started with a seed biofilm DPD particle attached to the bottom surface of a channel, and gravity drove the fluid flow is from left to right. In the figures, medium gray represents the substratum, dark gray represents the flowing fluid, and light gray represents the biofilm. Periodic boundary conditions were applied at the vertical boundaries: fluid particles exiting the right boundary were reinserted into the computational domain through the left boundary with a fixed substrate concentration prescribed at the boundary. Three types of biofilm, with various rigidities, were investigated, namely soft, medium, and rigid biofilms. For a given geometry and gravitational acceleration, the rigidity of the biofilms

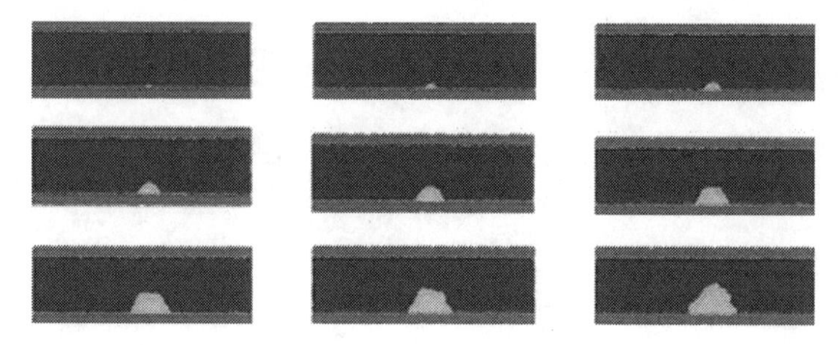

FIGURE 7.4
DPD simulations of the growth and deformation of a relatively rigid and strong biofilm. Medium gray represents the substratum, dark gray represents the flowing fluid, and light gray represents the biofilm. Snapshots are taken at simulation times $t = 500, 550 \ldots 900$.

is controlled by the magnitude of biofilm–biofilm DPD particle–particle interaction. For all three investigations, the liquid–biofilm interaction was assumed to be the same as the liquid–liquid interaction, and the biofilm–biofilm interaction was tuned to simulate soft, medium, and rigid biofilms. In all simulations, the only length scale of relevance is the cut-off distance $r_0 = 1$. The DPD particle density was set to $\rho = 4$, and the liquid–liquid interaction strength was set to $S_u = 18.75$ to match the compressibility of water (Groot and Warren 1997). A dimensionless force, $g = 0.02$, was applied on each DPD particle to sustain the fluid flow from left to right. The liquid–biofilm interaction was the same as the liquid–liquid interaction, and the biofilm–biofilm interaction strengths were set to $S_{bb} = 0.25S_u$, S_u, and $4.0S_u$, respectively, for soft, medium, and rigid biofilms.

For soft biofilm, the biofilm DPD particle–particle interaction strength is small. The biofilm structure is strongly deformed and stretched in the fluid flow direction by the shear stress exerted by the flowing liquid, and the biofilm masses can very easily be detached. The biofilm spreads over the substratum rather than growing toward the middle of the channel. For medium modulus biofilm, the structure is still strongly influenced by the liquid flow, and it forms an asymmetric shape. Medium strength biofilm can grow into the liquid, but not as far as the stiff biofilm. This results in a lower biofilm production rate compared to stiffer biofilms, because the substrate concentration is lower close to the substratum and higher deep in the liquid. As expected for rigid biofilms, the hydrodynamic interactions do not have a significant effect on the biofilm morphology. The biofilm grows preferentially toward the left (up stream direction) because more substrate is available on the left-hand side. For both medium and strong biofilms, detachment did not occur in the simulations, but longer simulations would lead to larger biofilm masses and larger stresses at the biofilm–substratum interface, which could lead to detachment.

7.3 SPHs Models for Biofilm Growth

Smoothed particle hydrodynamics was introduced by Lucy (1977) and by Gingold and Monaghan (1977) to simulate fluid dynamics in the context of astrophysical applications. SPH models have been successfully used to simulate a variety of multiphase flow and transport processes in porous media including microscale unsaturated fluid flow (Tartakovsky and Meakin 2005a,b), saturated flows (Morris et al. 1997; Zhu et al. 1999), multiphase flows (Tartakovsky and Meakin 2005c, 2006), and nonreactive and reactive solute transport (Zhu and Fox 2001, 2002; Tartakovsky and Meakin 2005c; Tartakovsky et al. 2009) in fractured and porous media. The SPH approach to the simulation of biomass (Tartakovsky et al. 2009), described here, is conceptually similar to the particle-based approach of Picioreanu et al. (2004) who represented biomass as discrete particles moving continuously (not constrained to a grid), but calculated solute transport and reactions on a separate grid using a multigrid PDE solution approach. Because SPH uses a well-established meshless numerical framework, it is able to directly incorporate the full range of relevant processes (e.g., hydrodynamic flow, solute transport, and chemical and biological reactions) into simulation of biomass growth with an arbitrary geometry, using a single consistent formulation. That is, no secondary grids or other methods are required for simulation of flow, transport, or reactions (except for an underlying grid that is used to rapidly locate pairs of particles that are located near enough to each other to interact), and all pertinent processes are simulated within the unified SPH framework. The Lagrangian particle nature of SPH allows complex biomass-fluid and biomass-solid interactions to be modeled through simple pair-wise interaction forces.

In the SPH biomass model (Tartakovsky et al. 2009), pair-wise particle–particle interactions are used to simulate interactions within the biomass, and interactions between biomass and fluid and between biomass and soil grains. A model-fractured porous medium was generated by randomly inserting nonoverlapping particles with radii selected randomly from a truncated Gaussian distribution on either side of the gap between two self-affine fractal curves representing a microfracture, and biomass was randomly distributed on the soil grains to initialize the simulations. The injection of two solutions, the first containing electron donors and the second containing electron acceptors, into different halves of the domain was simulated.

This work assumes that the changes in the biomass are governed by dual Monod kinetics (Knutson et al. 2005):

$$\frac{dM}{dt} = Yk_s M \frac{A}{K_A + A} \frac{B}{K_B + B} - k_d M \tag{7.22}$$

where M is the biofilm concentration; A and B are the concentrations of electron donors and electron acceptors; Y is the yield coefficient; k_S is the maximum growth rate; k_d is the decay rate; and K_A and K_B are the half-saturation constants for A and B. The advection and diffusion of electron donors and acceptors is governed by the continuity and momentum conservation equations (NS equations) coupled with diffusion equations for A and B:

$$\frac{dA}{dt} = D_A \nabla^2 A - k_s M \frac{A}{K_A + A} \frac{B}{K_B + B} \tag{7.23}$$

and

$$\frac{dB}{dt} = D_B \nabla^2 B - k_s M \frac{A}{K_A + A} \frac{B}{K_B + B} \tag{7.24}$$

The NS, diffusion and biomass evolution equations were solved using a Lagrangian particle method based on SPH. SPH uses a meshless discretization of the computational domain and an interpolation scheme, $A(\mathbf{r}) = \sum_i (A_i/n_i) W(\mathbf{r} - \mathbf{r}_i, h)$, allowing approximation of a continuous field $A(\mathbf{r})$ using the values of A at a set of discretization points. Here, W is the bell-shaped SPH weighting function with compact support of scale h, $\mathbf{r}_i$ is the positions of discretization point i, $A_i = A(\mathbf{r}_i)$, $n_i = \rho_i/m_i$ is the particle number density, ρ_i and m_i are the density and the mass of the phase associated with point i. Because each point possesses a mass and volume, it is natural to think of discretization points as physical particles. The SPH approximation of continuous fields allows the mass and momentum conservation equations to be written in the form of a system of ordinary differential equations (ODEs) (Tartakovsky et al. 2005c),

$$n_i = \sum_j W(\mathbf{r}_j - \mathbf{r}_i, h) \tag{7.25}$$

and

$$m_i \frac{d\mathbf{v}_i}{dt} = \mathbf{F}_i^{N-S} \tag{7.26}$$

where

$$\mathbf{F}_i^{N-S} = -\sum_j \left(\frac{P_j}{n_j^2} + \frac{P_i}{n_i^2} \right) \nabla_i W(\mathbf{r}_i - \mathbf{r}_j, h) + \sum_j \frac{4\mu_i \mu_j}{(\mu_i + \mu_j)} \frac{(\mathbf{v}_i - \mathbf{v}_j)}{n_i n_j (\mathbf{r}_i - \mathbf{r}_j)^2} (\mathbf{r}_i - \mathbf{r}_j) \cdot \nabla_i W(\mathbf{r}_i - \mathbf{r}_j, h) + \mathbf{g} \tag{7.27}$$

In the SPH model fluid and solid (e.g., soil grains) are represented by particles. In equations (7.25) and (7.27), $\sum_j$ indicates summation over all particles. Particles representing soil grains are frozen in space, their velocity is set to zero, and they enter into the calculation of the densities of fluid particles (Equation 7.25) and forces acting on the fluid particles (equation 7.7). The force $\mathbf{F}_i^{N-S}$, defined by equation (7.27), is the hydrodynamic force acting on the fluid particles (calculated from the NS equation), $\mathbf{v}$ is the fluid velocity vector, P is the pressure, $\mathbf{g}$ is the gravitational acceleration vector, and μ is the dynamic viscosity of the solution. In general, the mass m_i associated with fluid particle i and the local viscosity μ_i depend on the fluid composition, which may change with time. In this work it was assumed that the biomass density is equal to the fluid density and that the viscosity of fluid containing biomass is ten times greater than the viscosity of fluid containing no biomass.

The equation of state, $P_i = (P_{eq}/n_{eq})n_i$ was used to close the system of equations (7.25–7.27). In the equation of state n_{eq} is the average particle density and P_{eq} is the fluid pressure in the system at dynamic equilibrium.

Following Tartakovsky et al. (2007a) the system of diffusion/reaction equations (7.22–7.24) can be cast in the form of a system of ODEs:

$$\frac{dA_i}{dt} = \frac{1}{m_i} \sum_{j \in \text{fluid}} \frac{D_A (m_i n_i + m_j n_j)(A_i - A_j)}{n_i n_j (\mathbf{r}_i - \mathbf{r}_j)^2} (\mathbf{r}_i - \mathbf{r}_j) \cdot \nabla_i W(\mathbf{r}_i - \mathbf{r}_j, h) - k_s M \frac{A_i}{K_A + A_i} \frac{B_i}{K_B + B_i} \tag{7.28}$$

$$\frac{dB_i}{dt} = \frac{1}{m_i} \sum_{j \in \text{fluid}} \frac{D_B (m_i n_i + m_j n_j)(B_i - B_j)}{n_i n_j (\mathbf{r}_i - \mathbf{r}_j)^2} (\mathbf{r}_i - \mathbf{r}_j) \cdot \nabla_i W(\mathbf{r}_i - \mathbf{r}_j, h) - k_s M \frac{A_i}{K_A + A_i} \frac{B_i}{K_B + B_i} \tag{7.29}$$

and

$$\frac{dM_i}{dt} = Y k_s M_i \frac{A_i}{K_A + A_i} \frac{B_i}{K_B + B_i} - k_d M_i \tag{7.30}$$

where ∇_i denotes the gradient with respect to the position vector $\mathbf{r}_i$ and the symbol $\sum_{j \in \text{fluid}}$ indicates summation over all the fluid particles. Excluding solid particles from the summations enforces no-diffusion no-reaction conditions at solid–fluid interfaces. Initially biomass was randomly distributed on the surfaces of the porous medium by randomly selecting fluid particles near the solid particles and assigning a biomass concentration M_0 to these particles. Zero biomass concentration was assigned to the rest of the fluid particles. Evolution of the biomass was calculated according to equation (7.30). Once the concentration of the biomass at any particle i exceeded M_0, the excess biomass was moved to the nearest fluid particle with a biomass concentration less than M_0. Biomass growth is also influenced by the forces exerted on the biomass by the flowing fluid. In this chapter we investigated two models to account for the effect of fluid shear stress on the spreading of the biofilm.

7.3.1 Model 1

Model 1 is similar to the cellular automaton model used by Knutson et al. (2005). Particles with nonzero biomass concentration are assumed to be immobile, and excess biomass is allowed to move only to the nearest particles with a shear stress, $\tau = \mu(\partial v_x/\partial y + \partial v_y/\partial x)$, less than the critical stress τ_{cr}. If no particles satisfy the above requirements, excess biomass is assumed to be lost.

This model also disregards detachment and deformation of the biomass as a result of the shear stresses.

7.3.2 Model 2

Model 2 is based on the ideas that biomass is held together by attractive forces mediated by biopolymer, and that attractive forces between biomass and soil grains keep biomass attached to the solid surface. In Model 2, the excess biomass is transferred to a neighboring fluid particle with $M < M_0$, and the behavior of the biomass (deformation, attachment, detachment, and splitting) is modeled through a combination of hydrodynamic forces and a combination of short-range repulsive and medium-range attractive pair-wise forces that act between all pairs of particles including particles representing soil grains and fluid particles with and without biomass. The short-range component of the pair-wise force prevents particles from approaching each other too closely, and the medium-range attractive component of the force represents the attractive forces within the biomass and between biomass and soil grains. The exact form of the particle–particle interactions is not critical to the success of the simulations as long as the interactions satisfy the requirements described above and have symmetric forms to conserve linear momentum.

In this work, we used the pair-wise interaction forces:

$$\mathbf{F}_{ij} = \begin{cases} s_{ij} \cos\left(\dfrac{1.5\pi}{h}|\mathbf{r}_i - \mathbf{r}_j|\right) \dfrac{\mathbf{r}_i - \mathbf{r}_j}{|\mathbf{r}_i - \mathbf{r}_j|}, & |\mathbf{r}_i - \mathbf{r}_j| \leq h \\ 0, & |\mathbf{r}_i - \mathbf{r}_j| > h \end{cases} \tag{7.31}$$

where s_{ij} is the strength of the force that depends on the properties of materials and is different for interaction between different types of particles. The interaction strength, s_{ij}, between pairs of particles containing biomass was set to be larger than s_{ij} for the interaction between pairs of particles with and without biomass. If the interaction strength, s_{ij}, for fluid particles containing biomass and solid particles is greater than s_{ij} for interaction between two particles containing biomass, the model produces biomass growth in the form of continuous biofilm. Otherwise, patchy biofilm is preferentially formed. In this work the strength, s_{ij}, for interactions between fluid particles containing biomass and solid particles was set to be greater than s_{ij} for interactions between two particles containing biomass. The interaction strength, s_{ij}, between fluid particles with zero biomass and solid particles was set to be the same as s_{ij} for two fluid particles containing zero biomass. Since solid particles are immobile, interactions between pairs of solid particles were not considered, and s_{ij} for the forces between pairs of solid particles were not specified.

The total force acting on every fluid particle with or without biomass is a combination of hydrodynamic forces and particle interaction forces, and the motion of fluid particles is controlled by the momentum conservation equation

$$m_i \frac{d\mathbf{v}_i}{dt} = \mathbf{F}_i^{N-S} + \sum_j \mathbf{F}_{ij} \tag{7.32}$$

7.3.3 Implementation of the SPH Model

At each time step in a simulation, the particle number densities, n_i, at each of the particles are calculated using equation (7.25) and the pressure at each particle is obtained using the equation of state, $P_i = P_{eq} n_i / n_{eq}$. In Model 1, particles with nonzero biomass, $M_i > 0$ are assumed to be immobile. The positions of fluid particles with $M_i = 0$ and new concentrations A, B, and M are found from (7.26) and (7.28–7.30) using the explicit "velocity Verlet" algorithm (Allen and Tildesley 2001). In Model 2, all fluid particles with or without biomass are considered to be mobile, and the new positions of the fluid particles and new concentrations are found by explicit time integration of equations (7.32) and (7.28–7.30). An M_6 spline function (Schoenberg 1946) was used for the SPH weighting function.

7.3.4 Numerical Results

To initialize the simulations, particles were placed randomly into a 16 × 32 box (in units of h), and the SPH equation (7.27), with $\mathbf{g} = 0$, and periodic boundary conditions in all directions were used to bring the system into an equilibrium state. The equilibrium particle density was $n_{eq} = 19h^{-2}$ (19 particles in an area of h^2). The coefficient P_{eq} in the equation of state was $P_{eq} = 20$ and the viscosity was $\mu_i = 1$. Model units of time, length, and mass are used in the descriptions of the simulations presented later. After equilibrium was reached, the particles at positions $\mathbf{r}_i$ covered by the discs representing soil grains were "frozen" to form impermeable boundaries to the flow. A fractured porous medium was generated by randomly inserting nonoverlapping discs with random radii on either side of the gap between two self-affine fractal curves representing a microfracture. The initial concentrations of A and B were set to zero. A body force was then applied in the y direction. No-flow boundary conditions were imposed at the boundaries of the computational domain in the x direction, and periodic flow boundary conditions were used in the y direction. Particles exiting the flow domain at $y = 0$ were returned into the flow domain at $y = 32$.

The injection of electron donors and acceptors in different halves of the computational domain was simulated by assigning concentrations $A = 1$, $B = 0$, $M = 0$ to the particles entering at $y = 32$ in the left part of domain and $A = 0$, $B = 1$, $M = 0$ in the right part of the domain. The parameters $D_A = D_B = 0.5$, $\mathbf{g} = [0, -0.01]$, $K_A = 0.2$, $K_B = 0.1$, $Y = 0.07$, $k_S = 0.1$, and $k_d = 0.001$ were used in the simulations. Figure 7.5 shows the steady-state distribution of biomass resulting from continuous injection of solutions containing electron donors and acceptors for two different values of the critical stress, τ_{cr}, when Model 1 was used. For the smaller value of τ_{cr} the biomass grew preferentially near the entrance of the fracture where the solutes were injected. From the profiles of the product of concentrations A and B, AB, depicted in Figure 7.5, it can be seen that the nutrient substrate becomes rapidly depleted

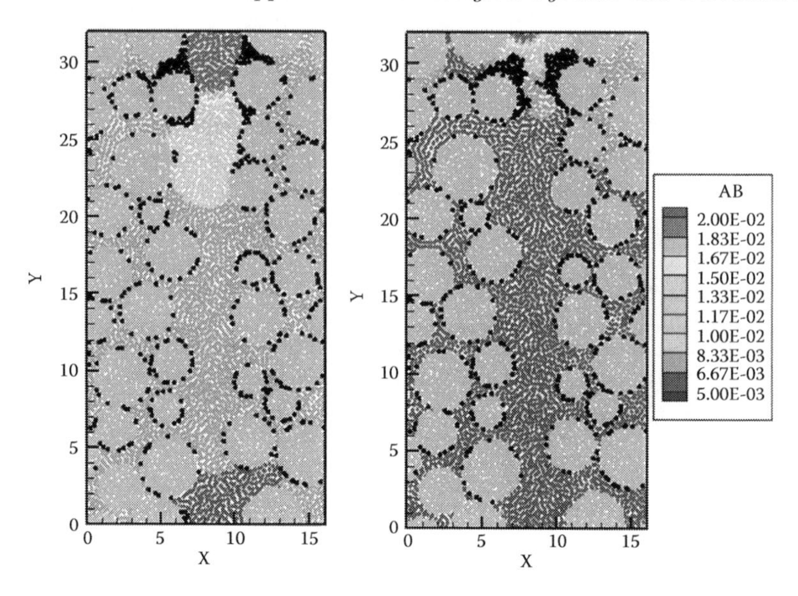

FIGURE 7.5
Distribution of the biomass produced by SPH Model 1 for two different values of τ_{cr}. Gray particles denote soil grains, black particles represent fluid particle with nonzero biomass concentration, and the quasicontinuous gray scale indicates the concentration product, *AB*, in the fluid particles with zero biomass concentration (Tartakovsky et al. 2009). (Tartakovsky, A.M., Scheibe, T.D., and Meakin, P., *J. Por. Med.*, 12, 5, 2009. Copyright 2009 Bagell House. Modified from Bagell House.)

near the fracture entrance preventing biomass growth further along the fracture walls. The biomass grew in the form of bridges between the soil grains oriented in the direction of the flow. The mixing zone between solutes A and B did not extend far enough into the porous matrix to facilitate substantial growth of the biomass in the porous matrix. For the larger critical stress, τ_{cr}, the biomass extended further toward the middle of the fracture but still grew only near the fracture entrance.

Figure 7.6 shows the distribution of biomass resulting from Model 2. It can be seen that, as in Model 1 with smaller τ_{cr}, the biomass grew between soil grains where the fluid stresses were the smallest. However, owing to the attachment/detachment mechanism included in the Model 2 the biomass spreads along the whole length of the fracture, completely sealing the fracture walls. Comparison of Figures 7.5 and 7.6 shows similar distributions of the concentrations of A and B but very different distributions of the biomass, indicating the *dominant* role of the attachment/detachment mechanism in the biomass growth and spreading. The difference between the results of Model 1 and 2 demonstrates the importance of realistic treatment of biomass that is mostly

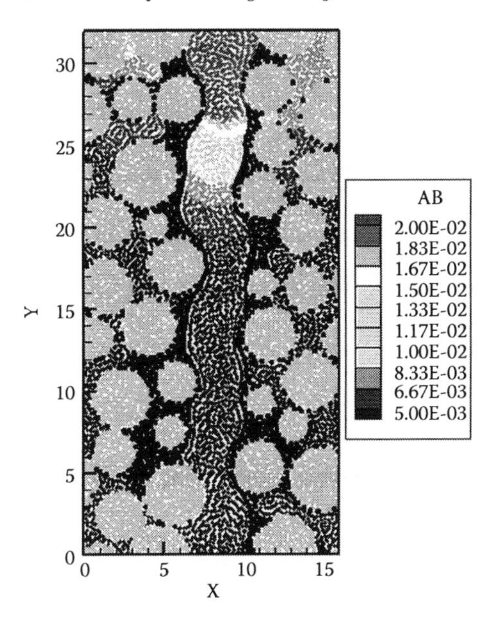

FIGURE 7.6
Distribution of the biomass produced by Model 2. Gray particles denote soil grains, black particles represent fluid particle with nonzero biomass concentration, and the quasicontinuous gray scale indicates the concentration product AB in the fluid particles with zero biomass concentration (Tartakovsky et al. 2009). (Tartakovsky, A.M., Scheibe, T.D., and Meakin, P., *J. Por. Med.*, 12, 5, 2009. Copyright 2009 Bagell House. Modified from Bagell House.)

missing in cellular automata models, the most common approach to simulating biofilm. In Model 2, biomass was modeled as a viscous fluid phase with a viscosity ten times greater than the viscosity of the fluid containing no biomass. Depending on the nature of the biomass, it is possible to treat the biomass as a non-Newtonian or visco-elastic material that will increase predictive ability of simulations.

7.4 An SPH Model for Mineral Precipitation

SPH models were developed to simulate mineral precipitation in porous and fractured media (Tartakovsky et al. 2007a,b, 2008a). The SPH discretization of the advection–diffusion-reaction equation with heterogeneous (dissolution or precipitation) reaction kinetics of the form $k(C - C_{eq})^{\beta}$, where C is the solute concentration at the liquid–solid interface, C_{eq} is the solute concentration at equilibrium with the solid, k is the rate constant, and β is the order of the heterogeneous reactions, is

$$\frac{dC_i}{dt} = \frac{1}{m_i} \sum_{j \in \text{fluid}} \frac{(D_i n_i m_i + D_j n_j m_j)(C_i - C_j)}{n_i n_j (\mathbf{r}_i - \mathbf{r}_j)^2} (\mathbf{r}_i - \mathbf{r}_j) \cdot \nabla_i W(\mathbf{r}_i - \mathbf{r}_j, h)$$
$$- k(C_i - C_{eq})^\beta \sum_{k \in \text{solid}} \frac{W(\mathbf{r}_i - \mathbf{r}_k, h)}{n_k^\gamma \sum_{l \in \text{solid}} n_l W(\mathbf{r}_k - \mathbf{r}_l, h)} \tag{7.33}$$

Here, $\gamma = 2/3$ for three-dimensional simulations, $\gamma = 1/2$ for two-dimensional simulations, and D is the diffusion coefficient. In equation (7.33) the symbol $\sum_{j \in \text{fluid}}$ indicates summation over all fluid particles and $\sum_{j \in \text{solid}}$ indicates summation over all solid particles. The last term in equation (7.33) is proportional to the rate of the mass loss/gain (per unit mass of solution) due to the precipitation/dissolution. Mass conservation requires the rate of solid mass gain/loss due to precipitation or dissolution to be

$$\frac{dm_k}{dt} = \frac{k}{n_k^\gamma \sum_{i \in \text{fluid}} n_l W(\mathbf{r}_k - \mathbf{r}_i, h)} \sum_{i \in \text{fluid}} (C_i - C_{eq})^\beta W(\mathbf{r}_i - \mathbf{r}_k, h) \tag{7.34}$$

Homogeneous reactions can be included by adding terms to the right-hand side of equation (7.33). For example, if the solute is unstable or reacts with itself, the term $-rC_i^\alpha$, where α is the order of the reaction, can be added to the right-hand side of equation (7.33).

The masses, m_i, of the solid particles are tracked, and once the mass of a solid particle, m_k, exceeds $2m_k^0$, where m_k^0 is the mass of the mineral phase within a volume of $1/n_k$ (where n_k is the SPH particle number density of the solid particle i), the nearest fluid particle "precipitates," becoming a new solid particle, and the mass of the new solid particle is set to $m_i - m_k^0$ while the mass of the old solid particle becomes m_k^0. Similarly, if the mass of a solid particle reaches zero, the solid particle becomes a new fluid particle. Since the fluid velocity adjacent to a solid surface is very small, the velocity of the new fluid particle is set to zero and the concentration of the fluid particle is set to the equilibrium concentration.

Figure 7.7 illustrates an SPH simulation of the injection of a supersaturated solution into a fractured porous medium that was initially saturated with a solution at equilibrium with the solid. In the simulation, the precipitate eventually seals the fracture walls completely.

Figure 7.8 illustrates simulations of multicomponent reactive transport in a two-dimensional porous medium. Solutions of A and B were injected into two different halves of a porous medium, and the product, C, of the homogeneous

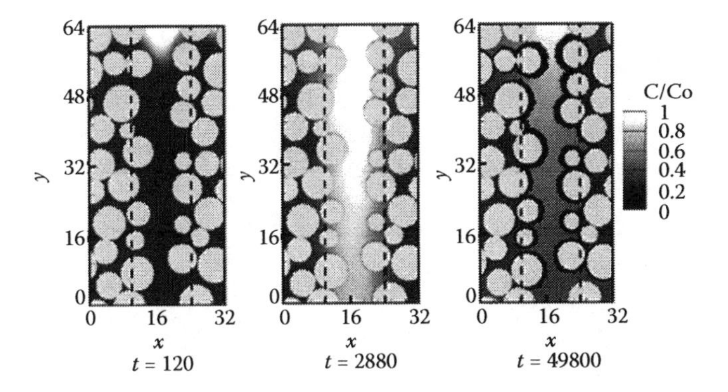

FIGURE 7.7
SPH simulation of reactive transport and precipitation in a porous medium with a microfracture. The Damkohler number was $Da = 2.548$ and the Peclet numbers before precipitation were $Pe_m = 0.082$ for the matrix and $Pe_f = 22.82$ for the fracture. The light gray particles represent mineral grains, and the black particles represent precipitated mineral. The gray scale denotes the dimensionless concentration, C/C_0. A solution with a solute concentration of ($C_0 = 4C_{eq}$) was injected via the upper boundary, located at $y = 64$, into the fractured porous medium, which was initially filled with a solution having a solute concentration of C_{eq} [(Tartakovsky et al. 2007a,b) (Tartakovsky, A.M., Meakin, P., Scheibe, T., and Wood, B., *Water Res. Res.*, 3, 2007, doi:10.1029/2005WR004770. Copyright 2007 American Geophysical Union. Reproduced/modified by permission of American Geophysical Union.)

reaction

$$A + B \leftrightarrow C \tag{7.35}$$

precipitated on the soil grains once the product concentration, C, exceeded the equilibrium concentration, C_{eq}. The simulations showed that the precipitate formed a thin layer in the mixing zone, and that the precipitate layer significantly reduced the mixing between the two reactants, A and B. To prevent dissolution due to reduced mixing, the injected solutions contained the reaction product, C, with a concentration of C_{eq}. The rate of precipitation increases with increasing Peclet number, and the width of the precipitation zone decreases with increasing Damkohler number.

Because the precipitate separating the two solutions has a width on the order of the grain diameter, continuum scale simulations would provide misleading results (the amount of precipitation would be seriously overestimated). On the other hand, pore scale simulations cannot be used to simulate large-scale processes. In this case, a multiscale multiresolution model that uses small

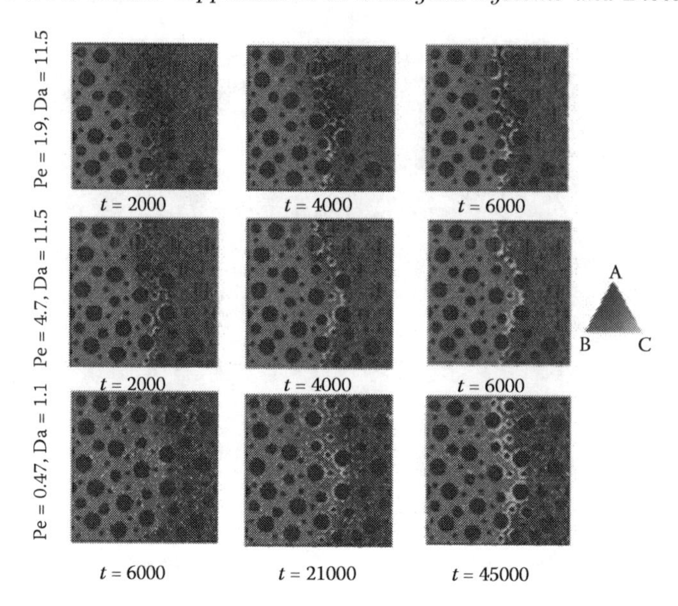

FIGURE 7.8
Two-dimensional SPH simulations of precipitation resulting from the mixing of two reactive solutions in a porous medium at three different Peclet numbers and two different Damkohler numbers. The solutes, A (medium gray) and B (dark gray), were injected at the same rate into the right and left halves of the computational domain. As the solutions mix, A and B react and produce the product C, which precipitates on mineral surfaces. The gray scale indicates the concentrations of A, B, and C. The gray particles represent mineral grains and the light gray particles are precipitated solid C (Tartakovsky et al. 2008a). (Tartakovsky, A.M., Redden, G., Lichtner, P., Scheibe, T., and Meakin, P., *Water Res. Res.*, 44, 2008, doi:10.1029/2006WR005725. Copyright 2008 American Geophysical Union. Reproduced/modified by permission of American Geophysical Union.)

SPH particles to simulate the precipitation zone and large SPH particles to simulate solute dispersion can be used.

7.5 Hybrid Models for Diffusion-Reaction Systems

Mineral precipitation and biofilm growth are amenable to mathematical descriptions on a multiplicity of scales that range from the atomistic scale to the continuum (Darcy) scale. While pore-scale and finer-scale models can be used to simulate such phenomena with a high degree of fidelity, they often require computational capabilities that are not currently available and they are typically replaced with coarse-scale Darcy descriptions that can often provide an acceptable degree of accuracy at a fraction of the computational cost. Hybrid (multiscale/multiphysics) numerical algorithms are rapidly gaining acceptance as a method of choice when coarse-scale continuum models fail to accurately describe a physical phenomenon in a small part of the computational domain. In the region or regions in which the continuum methods fail, they are replaced with their finer-scale counterparts.

The coupling of two (or more) mathematical models operating on vastly different spatial and/or temporal scales remains a major theoretical and computational challenge. Apart from fundamental issues, such as the propagation of errors between the discrete and continuum scales (Alexander et al. 2002, 2005a,b), a key to the success of a hybrid methods is a computationally efficient implementation of the continuity and/or jump conditions on the interface(s) between the different models (e.g., Albuquerque et al. 2006).

While the general challenges posed by hybrid methods are common to most multiscale/multiphysics models, their practical implementation is highly problem specific and depends, in large part, on the scale and nature of each mathematical component of a given hybrid. The goal of the work presented here is to simulate mixing-induced mineral precipitation in porous media. Macroscopic or Darcy-scale descriptions of this process, which rely on a coupled system of averaged reaction-diffusion equations (RDEs), fail to account for the highly localized precipitation that occurs in the mixing zone (Tartakovsky et al. 2008a). At the same time, pore-scale solutions of systems of RDEs are computationally expensive and require detailed knowledge of the pore geometry at a level that is rarely available on the macroscopic scales of interest. This scenario calls for a hybrid methods in which microscopic pore-scale simulations are used in a narrow reaction zone, while macroscopic Darcy-scale simulations are used in the rest of the computational domain.

Here we describe a hybrid algorithm (Tartakovsky et al. 2008b), in which SPH is used to solve RDEs in the part of the domain that must be represented on the pore scale. SPH is also used to solve the averaged RDEs to model diffusion and reactions at the macroscale (Darcy-scale). Since the resulting hybrid does not require an iterative procedure to establish the proper coupling between the solutions of RDEs on the micro- and macroscales, it is computationally efficient.

7.5.1 Hybrid Formulation for Reaction-Diffusion Systems in Porous Media

Consider a reaction-diffusion process in a volume, Ω^T, of a fully saturated porous medium. Diffusion and reaction take place inside the fluid-filled pores, $\Omega_p^T \subset \Omega^T$. Pore-scale descriptions of this process solve the appropriate RDEs defined over Ω_p^T and these equations are subject to boundary conditions on its boundary, $\partial\Omega_p^T$. Darcy-scale models treat the porous medium as a continuum, that is, the corresponding RDEs are defined at every point $\mathbf{x} \in \Omega^T$. In a hybrid formulation, pore-scale simulations are carried out in Ω_d, a (small) portion of the porous medium, Ω^T, while the Darcy-scale model is solved in the remainder of the porous medium, Ω_c ($\Omega^T = \Omega_c \cup \Omega_d$). While the hybrid algorithm developed in this study is applicable to a large class of multi-component reaction-diffusion systems in porous media, we formulate it in terms of mixing-induced heterogeneous precipitation. Here, we assume that mixing-induced heterogeneous precipitation involves a homogeneous reaction between two mixing solutes A and B, which forms a reaction product C that grows heterogeneously from a supersaturated solution. For simplicity, we disregard both the reverse homogeneous reaction and homogeneous nucleation. (The latter approximation implies that the supersaturation index is not large enough to support precipitation in the liquid phase.) Furthermore, we assume that precipitation occurs only as a dense overgrowth on solid surfaces. In practice, the chemistry of precipitation and dissolution can be quite complex. A high supersaturation may be required to initiate heterogeneous precipitation, and a number of surface and solution species may play significant roles in the precipitation process. A simple alternative to the formation of intermediate C, which precipitates heterogeneously on the surface, would be the separate incorporation of A and B, in equal amounts, into the solid at the interface. The nonequilibrium thermodynamics approach to interfacial chemistry (Onsager 1931a,b; Prigogine 1947) indicates that the rate of precipitation will depend on the supersaturation index, and both simple models for the precipitation chemistry can be expected to give similar results.

7.5.2 Pore-Scale Description and Its SPH Formulation

The subdomain, $\Omega_d = \Omega_p(t) \cup \Omega_s(t)$, in which the pore-scale simulations are conducted, consists of the fluid-filled pore space, $\Omega_p(t)$, and the solid matrix, $\Omega_s(t)$, with $F(t) = \Omega_s(t) \cap \Omega_p(t)$ denoting the corresponding (multiconnected) fluid–solid interface. Precipitation/dissolution causes the pore geometry, that is, $\Omega_p(t)$, $\Omega_s(t)$, and $F(t)$, to change with time t.

Let $C^A(\mathbf{x},t)$, $C^B(\mathbf{x},t)$, and $C^C(\mathbf{x},t)$ denote the concentrations in the solvent of solutes A and B and the reaction product, C. The concentrations are defined as the mass dissolved in a unit volume of fluid; the concentrations, C^A and C^B, are normalized with the corresponding initial concentrations $C^{A,0}$ and $C^{B,0}$; and the concentration, C^C, is normalized with $C^{A,0} + C^{B,0}$. The precipitation process can be described by a system of coupled RDEs:

$$\frac{\partial C^I}{\partial t} = \nabla \cdot (D^I \nabla C^I) - k^{AB} C^A C^B, \quad I = A, B, \ \mathbf{x} \in \Omega_p \tag{7.36}$$

$$\frac{\partial C^C}{\partial t} = \nabla \cdot (D^C \nabla C^C) + k^{AB} C^A C^B - k \int_F (C^C - C_{eq}) \delta(\mathbf{x} - \mathbf{x}_f) da, \quad \mathbf{x} \in \Omega_p \tag{7.37}$$

where $D^I > 0$ $(I = A, B, C)$ are the molecular diffusion coefficients of species I in the solvent, $k^{AB} > 0$ and $k > 0$ are the rate coefficients of the homogeneous and heterogeneous reactions, $\mathbf{x}_f$ is a point on the interface, F, and da is an infinitesimally small surface element of F. It is important to recognize that the formulation (7.36–7.37) reflects several physical assumptions. First, it neglects diffusion in the solid phase, which is usually many orders of magnitude slower than its counterpart in the liquid phase. Second, it disregards the reverse reaction $C(\text{aqueous}) \to A + B$. Third, (7.37) implies that precipitation/dissolution of the soluble reaction product, C, is described by a first-order kinetic-reaction model at the fluid–solid interface,

$$D^C \nabla C^C \cdot \mathbf{n} = k(C^C - C_{eq}) \tag{7.38}$$

where C_{eq} is the concentration of C in equilibrium with the solid matrix, and $\mathbf{n}$ is the unit vector in the direction normal to the interface pointing toward the fluid. The normal velocity, v_n, with which the fluid–solid interface at a point $\mathbf{x}_s$ advances into the liquid, is given by

$$v_n(\mathbf{x}_s) = \theta D^C \nabla C^C \cdot \mathbf{n}, \quad \theta = \frac{C^{A,0} + C^{B,0}}{\rho_s} \tag{7.39}$$

where $\rho_s > 0$ is the density of the precipitated solid phase. The parameters D^I $(I = A, B, C)$, k^{AB}, k, and ρ_s are measurable quantities that are assumed to be known.

7.5.3 SPH Representation of the Pore-Scale RDEs

Using the SPH interpolation scheme, equations (7.36–7.37) can be discretized as (Tartakovsky et al. 2008b):

$$\frac{\partial C_a^I}{\partial t} = 4 \sum_{b \in \Omega_p} V_b \frac{D_a^I D_b^I}{D_a^I + D_b^I} \frac{C_a^I - C_b^I}{x_{ab}^2} \mathbf{x}_{ab} \cdot \nabla_a x_{ab} \frac{\partial W(x_{ab}, h)}{\partial x_{ab}} - k^{AB} C_a^A C_a^B, I = A, B \tag{7.40}$$

$$\begin{aligned} \frac{\partial C_a^C}{\partial t} &= 4 \sum_{b \in \Omega_p} V_b \frac{D_a^C D_b^C}{D_a^C + D_b^C} \frac{C_a^C - C_b^C}{x_{ab}^2} \mathbf{x}_{ab} \cdot \nabla_a x_{ab} \frac{\partial W(x_{ab}, h)}{\partial x_{ab}} + k^{AB} C_a^A C_a^B \\ &- k \sum_{b \in \Omega_s} \Delta_b (C_a^C - C_{eq}) W_b^{-1} W(x_{ab}, h_r), \qquad W_b = \sum_{j \in \Omega_p} V_J \, W(x_{jb}, h_r) \end{aligned} \tag{7.41}$$

Here, subscripts a and b denote properties and positions associated with SPH particles a and b, $\mathbf{x}_{ab} = \mathbf{x}_a - \mathbf{x}_b$; $x_{ab} = |\mathbf{x}_{ab}|$; $\sum_{b \in \Omega_p}$; and $\sum_{b \in \Omega_s}$ indicate summation over fluid and solid particles; V_a is the volume (area in two-dimensional simulations) of particle a and Δ_b denotes the reactive surface area (length in two-dimensional simulations) associated with solid particle b. The third term on the right-hand side of (7.41) represents the heterogeneous reaction term obtained by discretization of the integral in equation (7.37), where $W(x,h_r)$ was used to approximate the Dirac delta function. In general, the support lengths, h_r and h, of the smoothing function, W, in the reactions and diffusion terms are not the same, and there is often good reason why they should be different. Criteria that could be used to select the problem-specific values for h_r and h are provided in the following section. The normalization factor, W_b, guarantees that the total change of concentration due to the heterogeneous reaction is equal to $k\int_F(C^C - C_{eq})da$.

The SPH particles representing solids are frozen in space. We neglect the movement of fluid particles due to mineral precipitation, so that the hydrostatic conditions assumed in this study render fluid particles immobile (otherwise, their dynamics can be described by an SPH discretization of the NS equations).

To describe the evolution of the fluid–solid interfaces due to precipitation and dissolution, we introduce "ghost" particles whose initial mass is zero and whose initial locations coincide with those of fluid particles. The masses of the ghost particles, m_a, change because of the heterogeneous precipitation reaction according to (7.41) so that

$$\frac{dm_a}{dt} = k(C^{A,0} + C^{B,0}) \sum_{b\in\Omega_s} V_b W_b^{-1} \Delta_b (C_a^C - C_{eq}) W(x_{ab}, h_r) \tag{7.42}$$

Once m_a reaches the prescribed solid particle mass, m_0, the ghost particle is converted into a solid particle and the corresponding fluid particle is removed. Dissolution is modeled in a similar fashion by tracking the mass of solid particles according to (7.42). When the mass of a solid particle reaches zero, the solid particle is reclassified as a new fluid particle.

7.5.4 Darcy-Scale (Continuum) Description

On the Darcy scale, the porous medium is treated as a continuum, and equations (7.36–7.39) are replaced by a system of averaged coupled RDEs

$$\frac{\partial \phi C^I}{\partial t} = \nabla\cdot(\phi D^I \nabla C^I) - \phi k_{\text{eff}}^{AB} C^A C^B, \quad I = A,B, \quad \mathbf{x}\in\Omega_c \tag{7.43}$$

$$\frac{\partial \phi C^C}{\partial t} = \nabla\cdot(\phi D^C \nabla C^C) + \phi k_{\text{eff}}^{AB} C^A C^B - \phi k_{\text{eff}}(C^C - C_{eq}), \quad \mathbf{x}\in\Omega_c \tag{7.44}$$

where D^I $(I = A,B,C)$ are the effective diffusion coefficients; ϕ is the porosity of the porous medium; and k_{eff}^{AB} and k_{eff} are the effective rate coefficients for the homogeneous and heterogeneous reactions, respectively. In principle, the parameters ϕ, D^I, k_{eff}^{AB}, and k_{eff} are measurable quantities, which are assumed to be known as long as the internal structure and topology of a porous medium are not significantly affected by precipitation and/or dissolution.

When the effects of precipitation and/or dissolution cannot be ignored, it is common to resort to ad hoc constitutive relationships that describe the dependence of macroscopic parameters, such as the permeability on the changing porosity, $\phi = \phi(\mathbf{x}, t)$. The hybrid algorithm described later obviates the necessity of using constitutive relationships of this kind, since it relies on the Darcy-scale description only for the portion of the porous medium where changes in the pore geometry are insignificant and the parameters ϕ, k_{eff}^{AB}, and k_{eff} remain constant.

7.5.5 SPH Representation of Averaged Darcy-Scale RDEs

Since the Darcy-scale description does not explicitly account for the solid and liquid phases, only one kind of particle is used to discretize the computational domain, Ω_c. The SPH discretization of the Darcy-scale RDEs is given by

$$\frac{\partial \phi_a C_a^I}{\partial t} = 4 \sum_{b \in \Omega_c} V_b \frac{\phi_a \phi_b D_a^I D_b^I}{\phi_a D_a^I + \phi_b D_b^I} \frac{C_a^I - C_b^I}{x_{ab}^2} \mathbf{x}_{ab} \cdot \nabla_a x_{ab} \frac{\partial W(x_{ab}, h)}{\partial x_{ab}} - \phi_a k_{\text{eff}}^{AB} C_a^A C_a^B, \quad I = A, B \tag{7.45}$$

and

$$\frac{\partial \phi_a C_a^C}{\partial t} = 4 \sum_{b \in \Omega_c} V_b \frac{\phi_a \phi_b D_a^C D_b^C}{\phi_a D_a^C + \phi_b D_b^C} \frac{C_a^C - C_b^C}{x_{ab}^2} \mathbf{x}_{ab} \cdot \nabla_a x_{ab} \frac{\partial W(x_{ab}, h)}{\partial x_{ab}} + \phi_a k_{\text{eff}}^{AB} C_a^A C_a^B - \phi_a k_{\text{eff}} (C_a^C - C_{eq}) \tag{7.46}$$

Darcy-scale (continuum) descriptions of reaction-diffusion processes in porous media are based on the averaging of microscopic RDEs over a representative volume of the porous media. Darcy-scale descriptions break down when the concentration gradients are large enough for the concentrations to change significantly on the scale of the representative volume (Whitaker 1999). If this occurs in only a small region, Ω_p, of the computational domain, Ω^T, hybrid simulations, which combine pore-scale simulations in Ω_p with Darcy-scale simulations in Ω_c, become attractive. The efficiency of a typical hybrid algorithm increases as the ratio $||\Omega^T||/||\Omega_d||$ increases, where $||\Omega||$ indicates the volume of the domain, Ω. Whether a small region in which pore scale simulation is required, Ω_d develops and whether the ratio $||\Omega^T||/||\Omega_d||$ is large is determined by the physical process(es) under consideration, and by the initial and boundary conditions for the advection–diffusion-reaction equations. In this study we are concerned with reaction-diffusion processes with localized reaction fronts, which are formed, for example, when a solution with concentrations $C^A = 1$ and $C^B = 0$ is brought instantaneously into contact with a solution with concentrations $C^A = 0$ and $C^B = 1$ in the same solvent.

7.5.6 Hybrid Formulation

A hybrid pore-scale/Darcy-scale algorithm was constructed by combining the SPH representation of the pore-scale RDEs defined on the domain Ω_d with the SPH representation of the Darcy-scale RDEs defined on the domain Ω_c. The two components of the hybrid are coupled by imposing the continuity of normal fluxes of each species along the interface, $\Gamma = \Omega_c \cap \Omega_d$, separating the two models,

$$D^I \mathbf{n} \cdot \nabla C^I{}_{|\Gamma_d} = \phi D^I \mathbf{n} \cdot \nabla C^I{}_{|\Gamma_c}, \qquad I = A, B, C \tag{7.47}$$

where the subscripts d and c indicate the side (discrete [pore] or continuum [Darcy]) of the hybrid interface, Γ, on which the relevant quantities are evaluated, and $\mathbf{n}$ is the unit vector normal to Γ pointing outside of the pore-scale domain.

7.5.7 Numerical Implementation of the Hybrid Algorithm

The coupling of the two components of the pore-scale/Darcy-scale hybrid algorithm, that is, the enforcement of the continuity conditions (7.47), is facilitated by employing SPH to numerically discretize both components of the hybrid.

7.5.8 Coupling of the Pore-Scale and Darcy-Scale Simulations

An SPH discretization of both continuum and discrete components of the hybrid provides a seamless way to couple the pore-scale and Darcy-scale descriptions of reaction-diffusion processes in porous media. Effective coupling of the continuum and discrete parts of the SPH hybrid model was achieved by combining the two sets of equations, (7.40–7.41) and (7.45–7.46), into one set that is valid over the entire computational domain, Ω^T.

$$\begin{aligned}\frac{\partial \omega_a C_a^I}{\partial t} &= 4 \sum_{b \in \Omega_c \cup \Omega_p} V_b \frac{\omega_a \omega_b d_a^I d_b^I}{\omega_a d_a^I + \omega_b d_b^I} \frac{C_a^I - C_b^I}{x_{ab}^2} \mathbf{x}_{ab} \\ &\quad \cdot \nabla_a x_{ab} \frac{\partial W(x_{ab}, h)}{\partial x_{ab}} - \omega_a r_a^{AB} C_a^A C_a^B, \qquad I = A, B \end{aligned} \tag{7.48}$$

$$\begin{aligned}\frac{\partial \omega_a C_a^C}{\partial t} &= 4 \sum_{b \in \Omega_c \cup \Omega_p} V_b \frac{\omega_a \omega_b d_a^C d_b^C}{\omega_a d_a^C + \omega_b d_b^C} \frac{C_a^C - C_b^C}{x_{ab}^2} \mathbf{x}_{ab} \\ &\quad \cdot \nabla_a x_{ab} \frac{\partial W(x_{ab}, h)}{\partial x_{ab}} + \omega_a r_a^{AB} C_a^A C_a^B \\ &\quad - \omega_a k_a^c (C_a^C - C_{eq}) - \omega_a k_a^d \sum_{b \in \Omega_s} \Delta_b (C_a^C - C_{eq}) W_b^{-1} W(x_{ab}, h_r) \end{aligned} \tag{7.49}$$

Here

$$d_a^I = \begin{cases} D^I, & a \in \Omega_p \\ \mathcal{D}^I, & a \in \Omega_c \end{cases}, \quad r_a^{AB} = \begin{cases} k^{AB}, & a \in \Omega_p \\ k_{eff}^{AB}, & a \in \Omega_c \end{cases}$$
$$k_a^c = \begin{cases} 0, & a \in \Omega_p \\ k_{eff}, & a \in \Omega_c \end{cases}, \quad k_a^d = \begin{cases} k, & a \in \Omega_p \\ 0, & a \in \Omega_c \end{cases}, \quad \omega_a = \begin{cases} 1, & a \in \Omega_p \\ \phi, & a \in \Omega_c \end{cases} \tag{7.50}$$

The last term in (7.49) vanishes in the Darcy-scale domain, since the soil grains and mineral surfaces are not explicitly represented there.

The symmetric form of the interactions between "pore scale" and "Darcy scale" particles in equations (7.48–7.49) ensure the continuity of mass fluxes across the interface between the Darcy-scale and pore-scale domains and implicitly imposes the boundary condition (7.47). The parameters in the hybrid model, defined by (7.50), characterize the continuum properties of the porous media and/or properties of the solutions. These parameters can be measured by standard laboratory methods or field experiments, and they are tabulated for the wide class of soils and chemical compounds. The hybrid model does not require any additional parameters beyond those used in the Darcy-scale or pore-scale models and, hence, parameterization of the hybrid model can be easily achieved.

7.5.9 Multiresolution Implementation of the Hybrid Algorithm

To increase the computational efficiency of the pore-scale/Darcy-scale hybrid algorithm, a lower spatial resolution (increased spacing between adjacent particles in SPH models) was used in the continuum domain relative to its counterpart in the pore-scale domain. Particles in both domains were placed on the nodes of square lattices. In the discrete domain, Ω_d, particles were placed on a square lattice with a size $\Delta x = 0.25$ (in model length units). In the continuum domain, Ω_c, the spacing between particles was increased from $\Delta x = 0.25$, near the boundary, Γ, to $\Delta x = 1$, away from Γ. The support length for each particle was set at

$$h_i = 4\Delta x_i \tag{7.51}$$

where Δx_i is the size of the lattice on which particle i is located. This defines a smoothing length over which each particle interacts with approximately 50 neighbors. Finally, h_r was set to

$$h_r = \sqrt{2}\Delta x + 0.01\Delta x \tag{7.52}$$

so that only one layer of fluid particles can react with the solid particles and vice versa. Upon setting $h_{ab} = (h_a + h_b)/2$, this leads to the following multiresolution SPH formulation:

$$\frac{\partial \omega_a C_a^I}{\partial t} = 4 \sum_{b \in \Omega_c \cup \Omega_p} V_b \frac{\omega_a \omega_b d_a^I d_b^I}{\omega_a d_a^I + \omega_b d_b^I} \frac{C_a^I - C_b^I}{x_{ab}^2} \mathbf{x}_{ab} \cdot \nabla_a x_{ab} \frac{\partial W(x_{ab}, h_{ab})}{\partial x_{ab}}$$
$$- \omega_a r_a^{AB} C_a^A C_a^B, \qquad I = A, B \tag{7.53}$$

$$\frac{\partial \omega_a C_a^C}{\partial t} = 4 \sum_{b \in \Omega_c \cup \Omega_p} V_b \frac{\omega_a \omega_b d_a^C d_b^C}{\omega_a d_a^C + \omega_b d_b^C} \frac{C_a^C - C_b^C}{x_{ab}^2} \mathbf{x}_{ab} \cdot \nabla_a x_{ab} \frac{\partial W(x_{ab}, h_{ab})}{\partial x_{ab}}$$
$$+ \omega_a r_a^{AB} C_a^A C_a^B - \omega_a k_a^c (C_a^C - C_{eq}) - \omega_a k_a^d \sum_{b \in \Omega_s}$$
$$\Delta_b (C_a^C - C_{eq}) W_b^{-1} W(x_{ab}, h_r) \tag{7.54}$$

An extensive discussion of the multiresolution SPH method can be found in Kitsionas and Whitworth (2002) and references therein.

7.5.10 Time Integration

Integration of the SPH equations (7.53), (7.54) can be carried out using various explicit (Monaghan 2005) or fully implicit schemes (Chaniotis et al. 2003). To improve the algorithm's efficiency, adaptive particle time stepping can be used (Kitsionas and Whitworth 2002).

In this study, we employed an explicit Euler time stepping integration method,

$$C_a^I(t + \Delta t) = C_a^I(t) + \Delta t \frac{dC_a^I(t)}{dt} \tag{7.55}$$

where the time step, Δt, satisfies the following conditions:

$$\Delta t < \min\left(\varepsilon_1 \frac{\Delta x_a^2}{d_a^I}, \frac{\varepsilon_2}{k_a^{AB}}, \varepsilon_3 \frac{\Delta x_a}{k}\right) \tag{7.56}$$

Our numerical experiments have shown that setting the constants ε_1, ε_2, and ε_3 to $\varepsilon_1 = \varepsilon_2 = \varepsilon_3 = 0.25$ provides a stable and accurate solution (see also Monaghan [2005] where the first of these inequalities, $\varepsilon_1 = 0.25$, was postulated).

7.5.11 Numerical Example

We consider mixing-induced precipitation in the porous medium depicted in Figure 7.9. A hybrid simulation of this process was conducted and validated against numerical solution obtained with the single scale pore-scale model.

7.5.12 Pore-Scale SPH Simulations

First pore-scale simulations were conducted over the whole computational domain, $\Omega_T = [0, L] \times [0, B]$, in which diffusion of solutes A and B and the

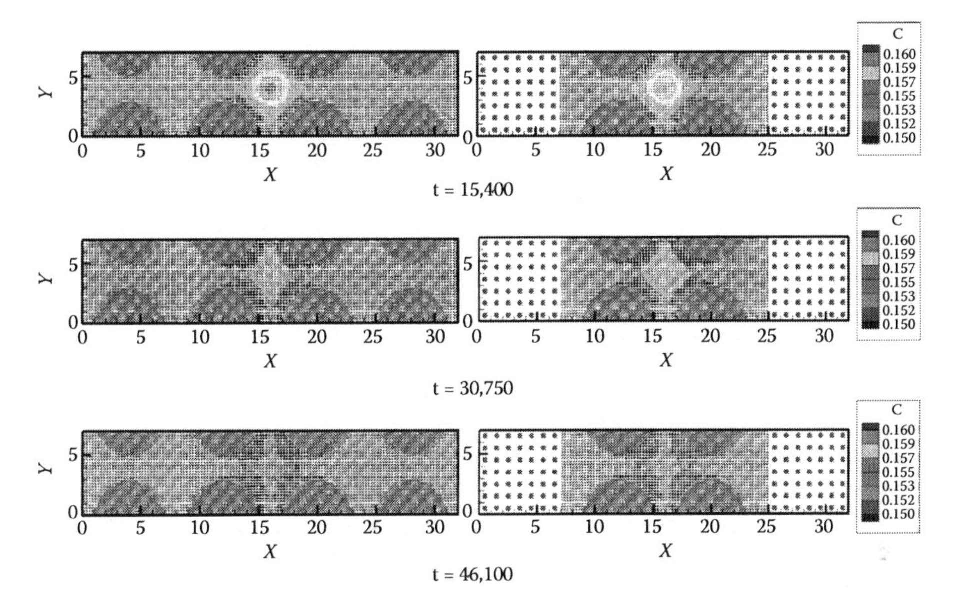

FIGURE 7.9
Snapshots of the distribution of precipitates (the dark gray circles) and the concentration of solute C (the quasicontinuous gray scale) at three different times. The medium gray circles denote particles representing soil grains. The left and right columns depict results of the pore-scale and hybrid simulations, respectively (Tartakovsky et al. 2008b). (Tartakovsky, A. M., Tartakovsky, D. M., Scheibe, T. D., and Meakin, P., *SIAM J. Sci. Comput.*, 30, 6, 2008b. Copyright 2008 Society for Industrial and Applied Mathematics. Reproduced/modified by permission of Society for Industrial and Applied Mathematics.)

production and precipitation of the reaction product, C, are governed by equations (7.36–7.39) subject to the boundary conditions:

$$\begin{aligned} C^A(0,y,t) = & C^B(L,y,t) = 1, \quad \frac{\partial C^A}{\partial x}(L,y,t) = \frac{\partial C^B}{\partial x}(0,y,t) = 0 \\ & \frac{\partial C^C}{\partial x}(0,y,t) = \frac{\partial C^C}{\partial x}(L,y,t) = 0 \end{aligned} \tag{7.57}$$

and the initial conditions

$$\begin{aligned} & C^A(x,y,0) = \begin{cases} 1, & x \leq 16 \\ 0, & x > 16, \end{cases} \quad C^B(x,y,0) = \begin{cases} 0, & x \leq 16 \\ 16, & x > 16 \end{cases} \\ & C^C(x,y,0) = \begin{cases} 0, & x \leq 16 \\ 0, & x > 16 \end{cases} \end{aligned} \tag{7.58}$$

The left column in Figure 7.9 shows three temporal snapshots of the pore-scale SPH simulations of this reaction-diffusion system for $L = 31$, $B = 8$, $D^A = D^B = D^C = 0.5$, $k^{AB} = 1.5$, and $k = 20$. The precipitation of the reaction product, C, (the black particles in Figure 7.9) modifies the structure of the porous medium, thus leading to changes in its effective properties, including porosity, effective diffusion coefficient, and effective reaction rates. As should be expected, these processes occur only in the narrow reaction zone separating the solutions of the two reactants, A and B.

7.5.13 Hybrid Simulations

The presence of a small region within a computational domain in which a coarse-grain description breaks down is, of course, the *raison d'être* for a hybrid algorithm. Since evolution of the pore geometry cannot be accurately described on the Darcy (continuum) scale, we employed the hybrid pore-scale/Darcy-scale algorithm. It combines a pore-scale simulation in the central part of the computational domain with a continuum reaction-diffusion description elsewhere (the right column in Figure 7.9).

To parameterize the Darcy-scale model, we obtained a steady-state finite-element solution of the one-dimensional version of the Darcy-scale diffusion equation subject to the boundary conditions $C^A(0) = 1$, $C^A(L) = 0$, $C^B(0) = 0$, $C^B(L) = 1$, and $C^C(0) = C^C(L) = 0$. This solution was fitted to the y-averaged steady-state SPH solution of the corresponding pore-scale model with $L = 31$, $B = 8$, $D = 0.5$, $k^{AB} = 1.5$, and $k = 20$ to yield the effective transport parameters $D^I = 0.31$ ($I = A, B, C$), $k^{AB}_{\text{eff}} = 0.98$, and $k_{\text{eff}} = 3.3$.

The right column in Figure 7.9 presents the results of the hybrid pore-scale/Darcy-scale simulations. Figure 7.10 shows the temporal evolution of the porosity of the portion of the porous medium affected by precipitation, $x \in [8, 24]$, obtained with the pore-scale and hybrid simulations. It can be seen that the two are in a close agreement.

In the examples presented here, the hybrid algorithm allowed the number of SPH particles to be reduced from 4,096 in the pore-scale simulations over the whole computational domain to 2,416 in the hybrid pore-scale/Darcy-scale simulations. In these simulations, the domain of the pore-scale simulations occupied half of the total computational domain. In practical applications of the hybrid model, this ratio is expected to be orders of magnitude smaller, so that the savings in the number of particles is expected to be substantially larger. The savings can be increased further by reducing the number of particles used to discretize the continuum (Darcy-scale) subdomains as the distance from the nearest pore-scale domain increases. Since the number of operations in SPH simulations increases linearly with the number of particles, the hybrid algorithm allows for a significant reduction in computational time.

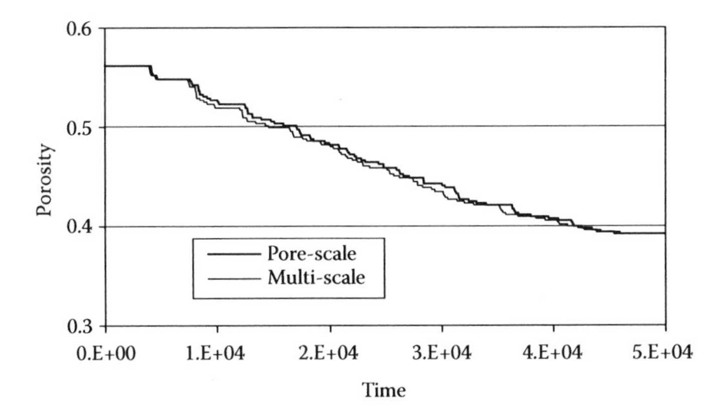

FIGURE 7.10
The change in porosity of the portion of a porous medium located between $x = 8$ and $x = 24$ computed with the pore-scale (the solid line) and hybrid pore-scale/Darcy-scale (the broken line) simulations. (Tartakovsky, A.M., Tartakovsky, D.M., Scheibe, T.D., and Meakin, P., *SIAM J. Sci. Comput.*, 30, 6, 2008b. Copyright 2008 American Geophysical Union. Reproduced/modified by permission of American Geophysical Union.)

7.6 Summary

In this chapter we reviewed two Lagrangian particle methods, DPD and SPH, and their application to modeling biofilm growth and mineral precipitation in porous media. Both biofilm growth and mineral precipitation play a critical role in a wide range of biological systems. The simulation results presented here show that mineral precipitation and biofilm growth lead to changes in pore-geometry and pore flow, and affect the continuum (effective) properties of porous media such as porosity, permeability, dispersion coefficient, and effective reaction rates. Existing continuum (Darcy scale) models use phenomenological relationships to describe changes in the effective properties due to mineral precipitation and dissolution. This significantly reduces the predictive power of Darcy-scale models. To increase the predictive ability of numerical models, a number of pore-scale models were developed in recent years. The two models, described here, operate on two different scales. On smaller scales, where the effects of random thermal fluctuations on biogeochemical processes are important, DPD models provide a good alternative to MD models. The DPD approach to biofilm simulation consistently incorporates thermally driven fluctuations into models of biomass growth, while significantly extending the range of typical MD simulations. This allows the effects of fluid flow on biomass growth and spreading to be studied—a phenomena that is out of reach of all but impractically large MD simulations. On larger scales, where thermodynamic fluctuations are not important and biogeochemical processes can be accurately described by deterministic partial differential equations (NS and advection–diffusion-reaction equations), SPH is an efficient tool. Due to the presence of moving interfaces, the PDEs describing biofilm growth and mineral precipitation are highly nonlinear and present a significant challenge to traditional grid-based methods. Owing to their Lagrangian nature, SPH models handle moving boundaries without any complications. Also, SPH, MD, and DPD allow complex interactions between biomass, fluid, and substratum to be simulated using simple pair-wise interactions that significantly simplify the biofilm and mineral precipitation models (in some cases, there are advantages to using more complex interactions that depend on the local particle density).

Despite the superior accuracy of pore-scale models, their high computational cost does not allow them to completely replace pore network and Darcy-scale models in a large computational domain. We presented a hybrid (multiphysics) numerical model that employs a SPH approach as the numerical engine for a scenario in which mineral precipitation is highly localized and coarse-scale continuum models fail to accurately describe the reactive transport in only a small part of a computational domain. The main advantages of the hybrid model are as follows: (1) Effective parameters in the hybrid model characterize the continuum properties of the porous media and/or properties of the solutions. They can be measured by standard laboratory or field experiments, and they are tabulated for a wide class of soils and chemical compounds. Parameterization of the hybrid model does not require any additional parameters beyond those used in the Darcy-scale or pore-scale models and hence can be easily achieved; (2) The continuum representation and effective parameters are used only in the part of the computational domain where no significant precipitation occurs. As a result, the effective parameters used in the continuum model do not change during a simulation, and this significantly increases the fidelity of the hybrid model predictions.

7.7 References

Abraham, F. F., Brodbeck, D., Rudge, W. E., and Xu, X. P. (1997). A molecular dynamics investigation of rapid fracture mechanics. *Journal of the Mechanics and Physics of Solids*, **45**(9):1595–1619.

Albuquerque, P., Alemani, D., Chopard, B., and Leone, P. (2006). A hybrid lattice Boltzmann finite difference scheme for the diffusion equation. *International Journal of Multiscale Computational Engineering*, **4**:209–219.

Alder, B. J. and Wainwright, T. E. (1962). Phase transition in elastic disks. *Physical Review,* **127**(2):359–361.

Alexander, F. J., Garcia, A. L., and Tartakovsky, D. M. (2002). Algorithm refinement for stochastic partial differential equations: 1. Linear diffusion. *Journal of Computational Physics,* **182**:47–66.

Alexander, F. J., Garcia, A. L., and Tartakovsky, D. M. (2005a). Algorithm refinement for stochastic partial differential equations: II. Correlated systems. *Journal of Computational Physics,* **207**:769–787.

Alexander, F. J., Garcia, A. L., and Tartakovsky, D. M. (2005b). Noise in algorithm refinement Methods. *Computing in Science & Engineering,* **7**: 32–38.

Allen M. P. and Tildesley, D. J. (2001). *Computer Simulation of Liquids.* Oxford University Press, Oxford, 1987.

Avalos, J. B. and Mackie, A. D. (1997). Dissipative particle dynamics with energy conservation. *Europhysics Letters,* **40**(2):141–146.

Azari, F., Vali, H., Guerquin-Kern, J.-L., Wu, T.-D., Croisy, A., Sears, S. K., Tabrizian, M., McKee, M. D. (2008). Intracellular precipitation of hydroxyapatite mineral and implications for pathologic calcification. *Journal of Structural Biology,* **162**:468–479.

Bandyk, D. F., Bergamini, T. M., Kinney, T. M., Kinney, E. V., Seabrook, G. R., and Towne, J. B. (1991). *In situ* replacement of vasculat prosthesis infected by bacterial biofilms. *Journal of Vascular Surgery,* **13**(5):575–583.

Beech, I. B. and Gaylarde, C. C. (1999). Recent advances in biocorrosion. *Revista de Microbiologica,* **30**(3):177–190.

Brigmon, R. L., Martin, H. W., and Aldrich, H. C. (1997). Biofouling of groundwater systems *Thiorix* spp. *Current Microbiology,* **35**(3):169–174.

Buehler, M. J., Abraham, F. F., and Gao, H. J. (2003). Hyperelasticity governs dynamic fracture at a critical length scale. *Nature,* **426**(6963):141–146.

Chaniotis, A. K., Frouzakis, C. E., Lee, J. C., Tomboulidies, A. G., and Poulikakos, D. A. B. (2003). Remeshed smoothed particle hydrodynamics for the simulation of laminar chemically reactive flows. *Journal of Computational Physics,* **191**:1–17.

Cottet, G.-H. and Koumoutsakos, P. (2000). *Vortex Methods: Theory and Applications.* Cambridge University Press, Cambridge.

de Haas, K. H., Blom, C., van den Ende, D., Duits, M. H. G., and Mellema, J. (1997). Deformation of giant lipid bilayer vesicles in shear flow. *Physical Review E,* **56**(6):7132–7137.

Dzwinel, W. and Yuen, D. A. (2000). Matching macroscopic properties of binary fluids to the interactions of dissipative particle dynamics. *International Journal of Modern Physics C,* **11**(1):1–25.

Eberl, H. J., Parker, D. F., and van Loosdrecht, M. C. M. (2001). A new deterministic spatio-temporal continuum model for biofilm development. *Journal of Theoretical Medicine,* **3**:161–175.

Espanol, P. (1997). Dissipative particle dynamics with energy conservation. *Europhysics Letters,* **40**(6):631–636.

Espanol, P. and Warren, P. (1995). Statistical-mechanics of dissipative particle dynamics. *Europhysics Letters,* **30**(4):191–196.

Fan, X. J., Phan-Thien, N., Chen, S., Wu, X. H., and Ng, T. Y. (2006). Simulating flow of DNA suspension using dissipative particle dynamics. *Physics of Fluids,* **18**(6):063102.

Filipovic, N., Kojic, M., and Tsuda, A. (2006). Modeling of thrombosis by dissipative particle dynamics (DPD). *International Journal of Artificial Organs,* **29**(5):514–514.

Filipovic, N., Kojic, M., and Tsuda, A. (2007). A multiscale modeling of thrombosis using coupling of dissipative particle dynamics with continuum finite element method. *International Journal of Artificial Organs,* **30**(8):709–709.

Filipovic, N., Kojic, M., and Tsuda, A. (2008). Modelling thrombosis using dissipative particle dynamics method. *Philosophical Transactions of the Royal Society A—Mathematical Physical and Engineering Sciences,* **366**(1879):3265–3279.

Gao, L. H., Shillcock, J., and Lipowsky, R. (2007). Improved dissipative particle dynamics simulations of lipid bilayers. *Journal of Chemical Physics,* **126**(1):015101.

Gingold, R. A. and Monaghan, J. J. (1977). Smoothed particle hydrodynamics—theory and application to non-spherical stars. *Monthly Notices of the Royal Astronomical Society,* **181**(2):375–389.

Groot, R. D. and Rabone, K. L. (2001). Mesoscopic simulation of cell membrane damage, morphology change and rupture by nonionic surfactants. *Biophysical Journal,* **81**(2):725–736.

Groot, R. D. and Warren, P. B. (1997). Dissipative particle dynamics: bridging the gap between atomistic and mesoscopic simulation. *Journal of Chemical Physics,* **107**(11):4423–4435.

Guo, H. X., Kremer, K., and Soddemann, T. (2002). Nonequilibrium molecular dynamics simulation of shear-induced alignment of amphiphilic model systems. *Physical Review E,* **66**(6):061503.

Hartgerink, J. D., Beniash, E., and Stupp, S. I. (2001). Self-asssembly and mineralization of peptide-amphiphile nanofibers. *Science,* **294**: 1684–1688.

Holliday, L. S., Welgus, H. G., Fliszar, C. J., Veith, G. M., Jeffrey, J. J., and Cluck, S. L. (1997). Initiation of osteoclast bone resorption by interstitial collagenase. *Journal of Biological Chemistry,* **272**:22053–22058

Hoogerbrugge, P. J. and Koelman, J. M. V. A. (1992). Simulating microscopic hydrodynamic phenomena with dissipative particle dynamics. *Europhysics Letters,* **19**(3):155–160.

Karamanev, D. G. (1991). Model of biofilm structure of thiobacillus-ferrooxidans. *Journal of Biotechnology,* **20**(1):51–64.

Kirkwood, J. G. (1939). Molecular distribution in liquids. *Journal of Chemical Physics,* **7**:919–925.

Kitsionas, S. and Whitworth, A. P. (2002). Smoothed particle hydrodynamics with particle splitting, applied to self-gravitating collapse. *Monthly Notices of the Royal Astronomical Society,* **330**:129–136.

Knutson, C. E., Werth, C. J., and Valocchi, A. J. (2005). Pore-scale simulation of biomass growth along the transverse mixing zone of a model two-dimensional porous medium. *Water Resources Research,* **41**:W07007, doi:10.1029/2004WR003459.

Kranenburg, M., Nicolas, J. P., and Smit, B. (2004). Comparison of mesoscopic phospholipid-water models. *Physical Chemistry Chemical Physics,* **6**(16):4142–4151.

Kubo, R. (1966). Fluctuation-dissipation theorem. *Reports on Progress in Physics,* **29**:255–282.

Lensing, H. J., Vogt, M., and Herrling, B. (1994). Modeling of biologically mediated redox processes in the subsurface. *Journal of Hydrology,* **159** (1–4):124–143.

Leyval, C. and Berthelin, J. (1991). The weathering of a mica by roots and rhizospheric microorganisms of pine. *Soil Science Society of America Journal,* **55**(4):1009–1016.

Lowe, C. P. (1999). An alternative approach to dissipative particle dynamics. *Europhysics Letters,* **47**(2):145–151.

Lucy, L. B. (1977). Numerical approach to testing of fission hypothesis. *Astronomical Journal,* **82**(12):1013–1024.

Monaghan, J. J. (2005). Smoothed particle hydrodynamics. *Reports on Progress in Physics,* **68**:1703–1759.

Morris J. P., Fox, P. J., and Zhu, Y. (1997). Modeling low Reynolds number incompressible flows using SPH. *Journal of Computational Physics,* **136**:214–226.

Nikunen, P., Karttunen, M., and Vattulainen, I. (2003). How would you integrate the equations of motion in dissipative particle dynamics simulations? *Computer Physics Communications,* **153**(3):407–423.

Onsager, L. (1931a). Reciprocal processes in irreversible processes I. *Physical Review,* **37**:405–426.

Onsager, L. (1931b). Reciprocal processes in irreversible processes II. *Physical Review,* **38**:2265–2279.

Ortiz, V., Nielsen, S. O., Discher, D. E., Klein, M. L., Lipowsky, R., and Shillcock, J. (2005). Dissipative particle dynamics simulations of polymersomes. *Journal of Physical Chemistry B,* **109**(37):17708–17714.

Picioreanu, C., van Loosdrecht, M. C. M., and Heijnen, J. J. (1998). A new combined differential-discrete cellular automaton approach for biofilm modeling: application for growth in gel beads. *Biotechnology and Bioengineering,* **57**(6):718–731.

Picioreanu, C., van Loosdrecht, M. C. M., and Heijnen, J. J. (2000). Effect of diffusive and convective substrate transport on biofilm structure formation: a two-dimensional modeling study. *Biotechnology and Bioengineering,* **69**(5):504–515.

Picioreanu, C., van Loosdrecht, M. C. M., and Heijnen, J. J. (2000). A theoretical study on the effect of surface roughness on mass transport and transformation in biofilms. *Biotechnology and Bioengineering,* **68**(4): 355–369.

Picioreanu, C., van Loosdrecht, M. C. M., and Heijnen, J. J. (2001). Two-dimensional model of biofilm detachment caused by internal stress from liquid flow. *Biotechnology and Bioengineering,* **72**(2):205–218.

Picioreanu, C., Kreft, J.-U., and van Loosdrecht, M. C. M. (2004). Particle-based multidimensional multispecies biofilm model. *Applied and Environmental Microbiology,* **70**(5):3024–3040.

Pivkin, I. V. and Karniadakis, G. E. (2006). Coarse-graining limits in open and wall-bounded dissipative particle dynamics systems. *Journal of Chemical Physics,* **124**(18):184101.

Prigogine, I. (1947). *Etude Thermodynamique des Phenomenes Irreversibles.* Liege, Desoer.

Reid, R. P., Visscher, P. T., Decho, A. W., Stolz, J. F., Bebout, B. M., Dupraz, C., Macintyre, I. G., Paerl, H. W., Pinckney, J. L., Prufert-Bebout, L., Steppe, T. F., and DesMarais, D. J. (2000). The role of microbes in accretion, lamination and early lithification of modern marine stromatolites. *Nature,* **406**:989–992.

Revalee, J. D., Laradji, M., and Kumar, P. B. S. (2008). Implicit-solvent mesoscale model based on soft-core potentials for self-assembled lipid membranes. *Journal of Chemical Physics*, **128**(3):035102.

Richardson, P. D., Pivkin, I. V., and Karniadakis, G. E. (2008). Red cells in shear flow: dissipative particle dynamics modeling. *Biorheology*, **45** (1–2):107–108.

Riding, R. (2000). Microbial carbonates: the geological record of calcified bacterial-algal mats and biofolms. *Sedimentology*, **47:**179–214.

Ripoll, M. and Espanol, P. (1998). Dissipative particle dynamics with energy conservation: heat conduction. *International Journal of Modern Physics C*, **9**(8):1329–1338.

Rittmann, B. E. (1993). The significance of biofoilm in porous-media. *Water Resources Research,* **29**(7):2195–2202.

Schoenberg, I. J. (1946). Contributions to the problem of approximation of equidistant data by analytical functions: part A. *Quart. Appl. Math.*, **IV**: 45–99.

Shillcock, J. C. and Lipowsky, R. (2002). Equilibrium structure and lateral stress distribution of amphiphilic bilayers from dissipative particle dynamics simulations. *Journal of Chemical Physics*, **117**(10):5048–5061.

Shillcock, J. C. and Lipowsky, R. (2005). Tension-induced fusion of bilayer membranes and vesicles. *Nature Materials*, **4**(3):225–228.

Speer, A. G., Cotton, P. B., and Rode, Y. (1988). Biliary blockage with bacterial biofilm—a light and electron microscopy study. *Annals of Internal Medicine,* **108**(4):546–553.

Sugii, T., Takagi, S., and Matsumoto, Y. (2007). A meso-scale analysis of lipid bilayers with the dissipative particle dynamics method: thermally fluctuating interfaces. *International Journal for Numerical Methods in Fluids,* **54**(6–8):831–840.

Tartakovsky, A. M. and Meakin, P. (2005a). Modeling of surface tension and contact angles with smoothed particle hydrodynamics. *Physical Review E,* **72**:026301.

Tartakovsky, A. M. and Meakin, P. (2005b). Simulation of free-surface flow and injection of fluids into fracture apertures using smoothed particle hydrodynamics. *Vadose Zone Journal,* **4**:848–855.

Tartakovsky, A. M. and Meakin, P. (2005c). A smoothed particle hydrodynamics model for miscible flow in three-dimensional fractures and the two-dimensional Rayleigh–Taylor instability. *Journal of Computational Physics,* **207**:610–624.

Tartakovsky, A. M. and Meakin, P. (2006). Pore-scale modeling of immiscible and miscible fluid flows using smoothed particle hydrodynamics. *Advanced Water Resources,* **29**:1464–1478.

Tartakovsky, A. M., Meakin, P., Scheibe, T., Wood, B. (2007a). A smoothed particle hydrodynamics model for reactive transport and mineral precipitation in porous and fractured porous media. *Water Resources Research,* **3**: W05437, doi:10.1029/2005WR004770.

Tartakovsky, A. M., Meakin, P., Scheibe, T., and Eichler West, R. M. (2007b). Simulations of reactive transport and precipitation with smoothed particle hydrodynamics. *Journal of Computational Physics,* **222**:654–672.

Tartakovsky, A. M., Redden, G., Lichtner, P., Scheibe, T., and Meakin, P. (2008a). Mixing-induced precipitation: experimental study and multi-scale numerical analysis. *Water Resources Research,* **44**, W06S04, doi:10.1029/2006WR005725.

Tartakovsky, A. M., Tartakovsky, D. M., Scheibe, T. D., and Meakin, P. (2008b). Hybrid simulations of reaction-diffusion systems in porous media. *SIAM Journal of Scientific Computing,* **30**(6):2799–2816.

Tartakovsky, A. M., Scheibe, T. D., and Meakin, P. (2009). Pore-scale model for reactive transport and biomass growth. *Journal of Porous Media,* **12**(5):417–434.

Thomas, J. G. and Nakaishi, L. A. (2006). Managing the complexity of a dynamic biofilm. *Journal of the American Dental Association (1939),* **137**:10S–15S.

Van Hamme, J. D., Singh, A., and Ward, O. P. (2003). Recent advances in petroleun microbiology. *Microbiology and Molecular Biology Reviews,* **67**(4):503–549.

Venturoli, M. and Smit, B. (1999). Simulating the self-assembly of model membranes. *Phys. Chem. Comm,* **10**:1–5.

Wagner, M., Rath, M., Kopps, H.-P., Flood, J., and Amann, R. (1996). In situ analysis of nitrifying bacteria in sewage treatment plants. *Water Science and Technology,* **34**(1–2):237–244.

Whitaker, S. (1999). *The Method of Volume Averaging,* Springer-Verlag, New York.

Wu, S. G. and Guo, H. X. (2008). Dissipative particle dynamics simulation study of the bilayer–vesicle transition. *Science in China Series B-Chemistry,* **51**(8):743–750.

Yamamoto, S., Maruyama, Y., and Hyodo, S. (2002). Dissipative particle dynamics study of spontaneous vesicle formation of amphiphilic molecules. *Journal of Chemical Physics,* **116**(13):5842–5849.

Zhang, W., Bouwer, E., Wilson, L., and Durnat, N. (1995). Biotransformation of aromatic-hydrocarbons in subsurface biofilmls. *Water Science and Technology,* **31**(1):1–14.

Zheng, X., Wisea, S. M., and Cristinia, V. (2005). Nonlinear simulation of tumor necrosis, neo-vascularization and tissue invasion via an adaptive finite-element/level-set method. *Bulletin of Mathematical Biology,* **67**:211–259.

Zhu, Y. and Fox, P. J. (2001). Smoothed particle hydrodynamics model for diffusion through porous media. *Transport in Porous Media,* **43**:441–471.

Zhu, Y. and Fox, P. J. (2002). Simulation of pore-scale dispersion in periodic porous media using smoothed particle hydrodynamics. *Journal of Computational Physics,* **182**:622–645.

Zhu, Y., Fox, P. J., and Morris, J. P. (1999). A pore-scale numerical model for flow through porous media. *Int. J. Num. Analyt. Meth. Geomech,* **23**: 881–904.

8

Passive Mass Transport Processes in Cellular Membranes and their Biophysical Implications

Armin Kargol

Department of Physics, Loyola University, New Orleans, LA

Marian Kargol

Department of Physics, The Jan Kochanowski University of Humanities and Sciences, Kielce, Poland

CONTENTS

8.1 Introduction

Water is the most abundant and also one of the smallest molecules in living organisms. It plays a fundamental role in many biological and physiological processes. A living cell, whether it exists as a separate organism or as a part of a multicellular organ or organism, in the course of all its physiological functions continuously exchanges with the extracellular medium water, nutrients, and metabolic waste products. This exchange takes place across cellular membranes and is controlled by the transport properties of the membranes.

Physical foundations of mechanisms of water and solute transport across cell membranes have been subjects of numerous research and review papers (Disalvo et al. 1989; Zeuthen and MacAulay 2002; Walsh et al. 2004; Kargol 2007, Elmoazzen et al. 2008). Initially, the processes of transport of non-electrolytic substances across cell membranes have been described using the Fick's law of diffusion. In 1932, Jacobs and Stewart (Elmoazzen et al. 2008) gave quantitative estimates of membrane permeability and developed a differential equation describing the rate of water and solute permeation as a function of concentration, cell size, and the membrane permeation coefficient. They have made certain simplifying assumptions, such as the constancy of membrane thickness and the extracellular concentration. Also, assuming that the osmotic pressure for a given substance is proportional to its concentration, they de facto introduced an assumption that solutions are dilute. This is also an assumption made in a vast majority of papers on membrane transport.

Another assumption was that the transport is passive, that is, it is driven by thermodynamic forces, such as concentration or pressure difference. This is in contrast to active transport, which requires energy input from some source, typically from the adenosine triphosphate (ATP) hydrolysis. Kedem and Katchalsky (1958) (Katchalsky and Curran 1965) developed a formalism describing transport properties of membranes using three parameters: the coefficients of filtration L_p, permeation ω, and reflection σ. Starting from the laws of linear thermodynamics of irreversible processes, they derived equations for the volume flux and the solute flux induced by the osmotic pressure and hydraulic pressure gradients. These equations, known as the Kedem-Katchalsky (KK) equations, have been widely used in studies of passive membrane transport processes. Coefficients of filtration, permeation, and reflection have been found experimentally for numerous membranes and solutes.

Despite its successes, the KK formalism has also certain limitations. In particular, often questioned is the reflection coefficient introduced to account for interaction of water and solute fluxes permeating via the same channels. Many authors believe that this coefficient is frequently incorrectly interpreted and computed. To solve this problem Kleinhans (1998) developed a two-parameter (2P) formalism that uses only two parameters: filtration and permeation. The comparison of 2P and KK formalisms can be found in several papers (Katkov 2000; Chuenkhum and Cui 2006).

Both the KK and 2P formalisms describe global transport across membranes and do not take into account possible inhomogeneities in the membrane. Both methods are based on thermodynamical approach and they do not consider the microscopic structure of the membrane. They yield only very limited and indirect information on the details of transport mechanisms of water and various solutes. On one hand that means that they are very general and apply to a variety of membranes and membrane permeation mechanisms. On the other hand, as our understanding of membrane structure and biophysical mechanisms of permeation of various substances improves, it makes it difficult to relate the macroscopic description of membrane transport with the microscopic details of transport mechanisms. The mechanistic formalism (ME) proposed recently (Kargol 2002; Kargol and Kargol 2003, 2006) uses several concepts of the KK formalism; however, it is based on a specific model of a membrane, in which permeation of water and solutes takes place through membrane pores. The aim of this formalism was to link the macroscopic transport equations with the microscopic properties of a membrane described by a specific, although very simplified, model.

In this chapter we describe the KK, 2P, and ME formalisms. We show that the KK and ME formalisms are equivalent in the global form, but the latter predict effects not detectable by the KK formalism, which may have physiological significance. We discuss two applications of the KK and ME equations in biology.

8.2 Thermodynamic KK Equations

Cell membrane forms a barrier separating the cytoplasm from the external environment. It permits a controlled exchange of water and various chemical compounds, as required for cell homeostasis and for all physiologic processes the cell participates in. At the same time the membrane maintains proper gradients of concentration, pressure, temperature, and electric potential between the cytoplasm and the extracellular medium. These two fundamental functions of the cell membrane can be described in terms of transport processes, such as diffusion or osmosis.

Membrane transport processes, like most natural processes, are irreversible. On the basis of the linear thermodynamics of irreversible processes, in 1958 Kedem and Katchalsky developed a formalism describing passive membrane transport processes. It has been widely accepted and is commonly used to analyze membrane phenomena. This formalism defines phenomenological parameters describing transport properties of a membrane, and expresses the fluxes of solvent and solute in terms of existing thermodynamic forces. The derivation of the KK equations is well known but for completeness we present its outline.

8.2.1 Derivation of Phenomenological KK Equations

For a thermodynamic system, the dissipation function ψ is defined as

$$\psi = \delta T = \sum_{k} X_k J_k \tag{8.1}$$

where δ expresses entropy production d_iS/dt in irreversible processes in the system and T is the temperature. The function ψ is the sum of products of thermodynamic stimuli X_k and flux densities J_k conjugated with them, and is a measure of energy consumption in irreversible processes.

We consider a membrane system shown in Figure 8.1. A membrane M of thickness Δx separates two nonelectrolytic solutions of different concentrations. If the solutions are well mixed (e.g., by mechanical stirrers m), and the volumes are sufficiently large, a stationary concentration profile develops on the membrane (Table 8.1).

Assuming isothermal conditions, the dissipation function for an infinitesimal membrane thickness element dx can be expressed as

$$\varphi = \sum_{i=1}^{n} J_i \cdot \text{grad}\,(-\mu_i) \tag{8.2}$$

where J_i is the flux density for the ith solution component and μ_i is its chemical potential. Since the flux density is constant across the thickness of

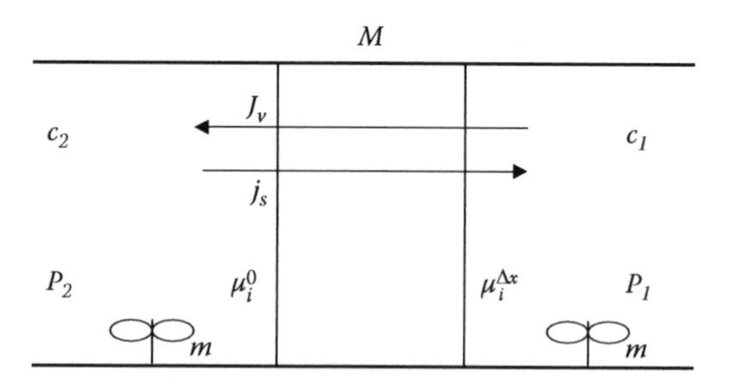

FIGURE 8.1
Membrane system (M is the membrane of thickness Δx; m is the mixing devices; P_1, P_2 is the pressures; c_1, c_2 is the concentrations; μ_i^0, $\mu_i^{\Delta x}$ is the chemical potentials; J_v is the volume flux; j_s is the solute flux).

TABLE 8.1
List of Symbols

J	**Flux density**
j_s	Solute flux density
J_v	Volume flux density
J_{vw}, J_{vs}	Water/solute volume flux density
μ	Chemical potential
ψ	Dissipation function
$\overline{V}_i$	Molar volume
P	Hydrostatic pressure
Π	Osmotic pressure
c	Concentration
L_p, L_D, L_{pD}, L_{Dp}	Onsager coefficients
L_p	Filtration coefficient
σ	Reflection coefficient
ω	Permeation coefficient (KK formalism)
ω_d	Permeation coefficient (ME formalism)
v_s, v_w	Velocities of molecules of solute and water
Subscript "a"	Quantities referring to part (a) of the membrane (cf. Figure 8.3)
Subscript "b"	Quantities referring to part (b) of the membrane (cf. Figure 8.3)
Subscript "M"	Fluxes obtained in ME formalism

the membrane, the dissipation function for the entire membrane is

$$\begin{aligned}\psi_{\Delta x} &= \int_0^{\Delta x} \sum_i J_i \text{grad}(-\mu_i) dx = \sum_i J_i \int_0^{\Delta x} \left(-\frac{d\mu_i}{dx}\right) dx \\ &= \sum J_i \left(\mu_i^0 - \mu_i^{\Delta x}\right) = \sum_i J_i \Delta\mu_i,\end{aligned} \tag{8.3}$$

where $\Delta\mu_i = \mu_i^0{}_i - \mu_i^{\Delta x}$ is the chemical potential difference across the membrane. If we assume binary solutions (i.e., solutions consisting of the solvent and one solute only), then we get

$$\psi_{\Delta x} = j_s \Delta\mu_s + j_w \Delta\mu_w \tag{8.4}$$

where j_s, j_w are the flux densities for the solute (s) and the solvent (w), and $\Delta\mu_s$, $\Delta\mu_w$ are chemical potential differences for the two components of the solution. In what follows we will refer to js as simply fluxes.

The chemical potential μ_i of the ith component can be expressed as

$$\mu_i = \mu_i^{st} + \overline{V}_i P + \mu_i^c \tag{8.5}$$

where μ_i^{st} is the standard chemical potential, $\overline{V}_i P$ is the pressure component of the chemical potential ($\overline{V}_i$ is the molar volume, and P is the pressure), and μ_i^c is the concentration component of the potential. In particular, for water the chemical potential difference across the membrane equals

$$\begin{aligned}\Delta\mu_w &= \mu_w^0 - \mu_w^{\Delta x} = \overline{V}_w \left(P^0 - P^{\Delta x}\right) + \left(\mu_w^c\right)^0 - \left(\mu_w^c\right)^{\Delta x} \\ &= \overline{V}_w \Delta P + \Delta\mu_w^c = \overline{V}_w \left(\Delta P - \Delta\Pi\right)\end{aligned} \tag{8.6}$$

where $\Delta P = P^0 - P^{\Delta x}$, the osmotic pressure difference $\Delta\Pi$ is given by the van't Hoff formula $\Delta\Pi = RT(c_s^0 - c_s^{\Delta x})$, R is the universal gas constant, T is the temperature, and $c_s^0, c_s^{\Delta x}$ are the concentrations on both sides of the membrane. For well-mixed solutions $c_s^0 = c_1$ and $c_s^{\Delta x} = c_2$. For the solute (s)

$$\Delta\mu_s = \overline{V}_s \Delta P + \Delta\mu_s^c = \overline{V}_s \Delta P + \frac{\Delta\Pi}{\overline{c}_s} \tag{8.7}$$

where the mean concentration is $\overline{c}_s = \frac{\Delta\Pi}{\Delta\mu_s^c}$. For diluted solutions we can use an approximation:

$$\overline{c}_s = \frac{c_s^0 + c_s^{\Delta x}}{2} \tag{8.8}$$

and the dissipation function can be written as

$$\psi = \left(j_w \overline{V}_w + j_s \overline{V}_s\right) \Delta P + \left(\frac{j_s}{\overline{c}_s} - \overline{V}_w j_w\right) \Delta\Pi. \tag{8.9}$$

The term $j_w \overline{V}_w + j_s \overline{V}_s$ is the volume flux:

$$J_v = j_w \overline{V}_w + j_s \overline{V}_s = J_{vs} + J_{vw} \tag{8.10}$$

The mean volume fraction of the solute and the solvent (water) in a solution can be written as

$$\begin{aligned} y_s &= \overline{c}_s \overline{V}_s \\ y_w &= \overline{c}_w \overline{V}_w \approx 1 \end{aligned} \tag{8.11}$$

Then the second term in equation (8.9) can be written as

$$\frac{j_s}{\overline{c}_s} - \overline{V}_w j_w = \frac{j_s}{\overline{c}_s} - \frac{j_w}{\overline{c}_w} \tag{8.12}$$

The flux j_i of the ith component of the solution is

$$j_i = \frac{\Delta\overline{n}_i \Delta x}{A \Delta x \Delta t} = \frac{\Delta\overline{n}_i}{\Delta V} \frac{\Delta x}{\Delta t} = \overline{c}_i v_i \tag{8.13}$$

where A is the active membrane surface, $\Delta\overline{n}_i$ is the number of moles of the ith component, ΔV is the volume of solution permeating the membrane in time Δt, Δx is the membrane thickness, $\overline{c}_i = \frac{\Delta\overline{n}_i}{\Delta V}$ is the mean concentration, and $v_i = \frac{\Delta x}{\Delta t}$ is the permeation speed of ith component molecules across the membrane. Equation (8.12) can be rewritten as

$$\frac{j_s}{\overline{c}_s} - \overline{V}_w j_w = v_s - v_w = J_D \tag{8.14}$$

and it expresses the diffusion flux J_D of solute (s), that is, the diffusion speed of this solute relative to the solvent. Equation (8.9) can now be interpreted as

$$\psi = J_v \Delta P + J_D \Delta\Pi \tag{8.15}$$

In a system in near-equilibrium, every flux is a linear function of all driving forces:

$$J_i = \sum_{k=1}^{n} L_{ik} X_k \tag{8.16}$$

where L_{ik} is the proportionality coefficients. Having identified the fluxes and driving forces present in the system in equation (8.15), we can write this condition (8.16) as

$$\begin{aligned} J_v &= L_p \Delta P + L_{pD} \Delta\Pi \\ J_D &= L_{Dp} \Delta P + L_D \Delta\Pi \end{aligned} \tag{8.17}$$

These equations are known as the phenomenological KK equations. The four terms in the equations represent processes of filtration ($L_p\Delta P$), osmosis ($L_{pD}\Delta\Pi$), ultrafiltration ($L_{Dp}\Delta P$), and diffusion ($L_D\Delta\Pi$). Accordingly, the coefficients are called the coefficients of filtration, osmotic permeation, ultrafiltration, and diffusion. Additionally, the Onsager relation requires

$$L_{Dp} = L_{pD} \tag{8.18}$$

8.2.2 Practical KK Equations

Equation (8.17) for a volume flux in a membrane system shown in Figure 8.1 can be rewritten as

$$J_v = L_p \Delta P - L_p \sigma \Delta\Pi \tag{8.19}$$

where σ denotes the so-called reflection coefficient, defined as $\sigma = -L_{pD}/L_p$ (Staverman 1951). It may assume values $0 \leq \sigma \leq 1$. If $\sigma = 1$, the membrane is semipermeable, that is, it fluxes only the solvent and generates maximal osmotic pressure. For $0 \leq \sigma \leq 1$ the membrane is selective, that is, it fluxes solvent better than the solute, and for $\sigma = 0$ the membrane is called permeable. The solute and the solvent permeate equally, and such a membrane is osmotically inactive.

From equations (8.17) we can write

$$J_D + J_v = (L_p + L_{Dp})\Delta P + (L_{pD} + L_D)\Delta\Pi \tag{8.20}$$

On the other hand $J_D + J_v$ can be expressed using equations (8.10), (8.14), as

$$J_D + J_v = \frac{j_s}{\bar{c}_s} - \overline{V}_w j_w + \overline{V}_w j_w + \overline{V}_s j_s = \frac{j_s}{\bar{c}_s} + \overline{V}_s j_s = \frac{j_s}{\bar{c}_s}(1 + y_s) \approx \frac{j_s}{\bar{c}_s} \tag{8.21}$$

Comparing these two expressions we obtain

$$j_s = \omega\Delta\Pi + \bar{c}_s(1 - \sigma)J_v \tag{8.22}$$

where

$$\omega = \frac{\bar{c}_s(L_p L_D - L_{pD^2})}{L_p} \tag{8.23}$$

is the permeation coefficient. Equations (8.19) and (8.23) are known as the practical KK equations.

8.2.3 Transport Parameters $\mathbf{L_p}$, σ, and ω

Membrane properties with regard to passive transport are described by Kedem and Katchalsky using three phenomenological transport parameters. The reflection coefficient σ was first defined by Staverman (1951) and is dependent on both the membrane and the permeating solute. Similarly, the filtration coefficient L_p, and the permeation coefficient ω are defined in terms of the Onsager coefficients in equations (8.17). The practical KK equations give physical interpretation of these coefficients. If we assume $\Delta\Pi = 0$ in equation (8.19), we obtain

$$L_p = \left(\frac{J_v}{\Delta P}\right)_{\Delta\Pi=0} \tag{8.24}$$

that is, the filtration coefficient is the volume flux per unit pressure in the absence of osmotic pressure. Similarly, putting $J_v = 0$ we get

$$\sigma = \left(\frac{\Delta P}{\Delta\Pi}\right)_{J_v=0} \tag{8.25}$$

If experimentally we obtain $\Delta P = \Delta\Pi$, then $\sigma = 1$, and the membrane is selective. If $\Delta P = 0$, than the membrane is osmotically inactive. Assuming $J_v = 0$ in equation (8.22) we get

$$\omega = \left(\frac{j_s}{\Delta\Pi}\right)_{J_v=0} \tag{8.26}$$

that is, the permeation coefficient expresses the solute flux generated by the unit osmotic pressure in the absence of the net volume flux.

The KK equations were derived from the laws of thermodynamics and apply to generic membrane systems, without regard to the details of solute and solvent permeation across the membrane. However, it has been shown (Kleinhans 1998) that when water and solute permeate through membrane pores, the role of the reflection coefficient is to describe the mutual interaction between water and solute fluxes in the same pores. When the fluxes flow through separate pores, the KK formalism can be simplified; Kleinhans developed a 2P formalism requiring only the filtration and permeation coefficients. It can be shown that both approaches agree in experiments where water and solutes have distinct permeation mechanisms.

8.3 Porous Membranes

A membrane acts as a barrier to various chemical compounds and maintains a difference of concentration, pressure, temperature, and the electric potential. If the solutions on both sides are mixed well, then the gradients exist only across the membrane. These gradients are the thermodynamic forces that generate fluxes in the system. In a thermodynamic KK formalism described in the previous section, fluxes across the membrane are described from a macroscopic point of view and we do not consider internal structure of the membrane and the microscopic mechanisms of permeation. A membrane is considered homogeneous with respect to its transport properties, which can be summarized in terms of three phenomenological parameters, that is, coefficients of filtration L_p, reflection σ, and permeation ω.

Recently, our understanding of membrane structure and permeation mechanisms of both artificial and biological membranes has improved significantly. Artificial membranes, porous and nonporous, can be manufactured by different methods. By a porous membrane we mean a thin impermeable barrier containing pores through which the solvent and, to a different degree, various solutes can permeate. Typical examples are cellophane, ceramic, metal, or polymer membranes that differ in the dimensions and geometry of their pores. An example of a nonporous membrane can be, for instance, a gel membrane on an appropriate support providing its stability. Such membranes do not have stable pores, and permeation is based on dissolving and diffusion of different solutes in the gel matrix. Biological membranes are phospholipid cell membranes and organelle membranes. They allow the cells or organelles to fulfill their physiological functions by controlling the exchange of various substances with the external environment. This exchange may be active, that is, requiring energy from some source, such as the ATP hydrolysis, or passive, where the transport is induced by thermodynamic forces, such as the concentration or pressure gradient. The microscopic details of permeation vary from the diffusion across the phospholipids bilayer to transport by specialized membrane proteins acting as transporters or forming pores. Among the latter are aquaporins forming pores permeable to water, ion channels for permeation of selected ions, or pores formed by certain antibiotics, such as nystatin. These pores have hydrophilic internal walls and are filled with water. They all differ in their dimensions and specialize in fluxing water and/or various nonelectrolytic and electrolytic solutes (Hejnowicz 1996; Elmoazzen et al. 2002; Zeuthen and MacAulay 2002). With respect to their transport properties we divide porous membranes into homogeneous and inhomogeneous. A homogeneous porous membrane has all pores of identical sizes, while pores of an inhomogeneous porous membrane have different geometry and sizes.

8.3.1 Homogeneous and Inhomogeneous Porous Membranes

Examples of homogeneous porous membranes are so-called nucleopore membranes. They are manufactured by bombarding very thin solid foils with high-velocity particles. If one type of particle is used, we obtain a membrane with cylindrical pores of the same diameter. Depending on the size of the particles used in manufacturing, the membrane may be permeable to water and some solutes. If such a membrane has pores of radius r larger than the radius of the water molecule r_w, then it is permeable to water. When the size of a given solute molecule r_{s1} is larger than the pore radius r, the membrane is impermeable, and the KK formalism has a reflection coefficient $\sigma = 1$ (a semipermeable membrane). For a solute with molecular radius $r_w < r_{s2} < r$, the membrane has a reflection coefficient $0 < \sigma < 1$ (a selective membrane), and for a solute with $r_{s3} < r_w$, the coefficient is $\sigma = 0$ (a permeable membrane).

A membrane with pores differing in their dimensions is called inhomogeneous. Such membranes usually have irregular pore geometry, and they differ locally in their transport properties. For simplicity, however, we assume here that all pores have cylindrical geometry and are perpendicular to the membrane surface. An example is a nucleopore membrane obtained with a beam of different particles. In real membranes the pores are randomly distributed, but for modeling purposes we can order them according to their radii. We assume that the smallest pores are larger than the dimensions of a water molecule ($r_w < r_1$), and for a solute with molecular radius r_s we have

$$r_1 < r_2 < \cdots < r_s < \cdots < r_N \tag{8.27}$$

that is, out of N pores in the membrane a certain number, n_a, of sufficiently small pores is impermeable to the solute, while the remaining $n_b = N - n_a$ pores are permeable. To describe the degree of selectivity of individual pores we introduce a coefficient σ_p. We postulate that a single pore with pore radius r_p can have the coefficient $\sigma_p = 1$ or 0 expressing the fact that it is either permeable to the solute molecules (if $r_s < r_p$) or not. By similarity with the KK formalism, we will call it the reflection coefficient but it needs to be understood that its definition is different from Staverman's. A definition of the reflection coefficient of the entire membrane will be given later but if a given membrane has only impermeable pores we treat it as a semipermeable membrane and assign it the reflection coefficient $\sigma = 0$. If there are only permeable pores then we say the entire membrane is permeable and has $\sigma = 1$. In case of a membrane with both permeable and impermeable pores we divide the membrane into two parts: (a) and (b). For a given solute, all impermeable pores are in part (a) of the membrane and all permeable pores are in part (b). We can also allow a number of unstable pores that will change their dimensions and change from being impermeable to permeable and vice versa. We assume, however, that statistically the numbers of permeable and impermeable are constant.

8.3.2 Poiseuille's Equation for Individual Pores and for the Membrane

Detailed models of water permeation in pores, especially biological, have been difficult to develop. When the pores are very narrow, with sizes comparable to the dimensions of water molecules, a molecular description is required, but is not available. For larger pores, standard techniques of hydrodynamics can be used. For the lack of better description, here we are using hydrodynamical concepts to describe permeation through individual pores; however, it needs to be understood that the validity of the approximation becomes questionable as pores get more and more narrow (Figure 8.2).

Since membrane pores have hydrophilic walls and are filled with water, motion of water molecules in the pores is determined by cohesion and adhesion forces. The narrowest pores are the least permeable, and, as the pore diameter increases, the molecules move with increasing ease. We assume the Poiseuille's

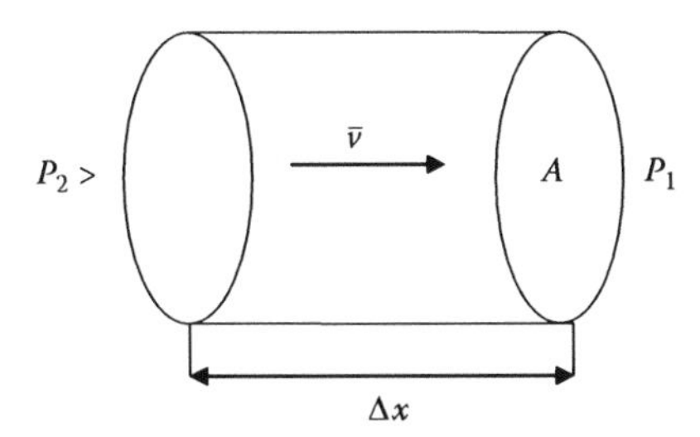

FIGURE 8.2
A cylindrical pore (A is the cross-section; P_1, P_2 is the pressures on both sides of the membrane; Δx is the pore length; v is the average flow speed in the pore).

Law can be applied to each pore. For a pore of length Δx, radius r, and cross-section $A = \pi r^2$ we get, because $J_v = \overline{v}$:

$$J_v = \frac{A}{8\eta\pi\Delta x}\Delta P \tag{8.28}$$

where η denotes viscosity, $\Delta P = P_2 - P_1$ is the pressure difference, J_v is the volume flux. Since

$$J_v = L_p \Delta P \tag{8.29}$$

we get

$$L_p = \frac{A}{8\eta\pi\Delta x} \tag{8.30}$$

For an inhomogeneous membrane of thickness Δx we can write

$$\begin{aligned} J_{v_1} &= \frac{A_1}{8\eta\pi\Delta x} \Delta P \\ &\vdots \\ J_{v_i} &= \frac{A_i}{8\eta\pi\Delta x} \Delta P \\ &\vdots \\ J_{v_N} &= \frac{A_N}{8\eta\pi\Delta x} \Delta P \end{aligned} \tag{8.31}$$

or after summation as

$$J_{v_t} = \frac{A_t}{8\eta\pi\Delta x}\Delta P \tag{8.32}$$

where $J_{v_t} = \sum_1^N J_{v_i}$ is the total volume flux across the membrane, $A_t = \sum_1^N A_i = \sum_1^N \pi r_i^2$ is the total cross-section area of all pores. From equation (8.32) we obtain

$$L_{pt} = \frac{A_t}{8\eta\pi\Delta x} \tag{8.37}$$

where L_{pt} denotes the total filtration coefficient of all pores. Equation (8.36) also gives

$$\overline{J}_v = \frac{\overline{A}}{8\eta\pi\Delta x}\Delta P \tag{8.38}$$

where $\overline{A} = \frac{1}{N} A_t$ is the mean cross-section area of one pore, and $\overline{J}_v = \frac{1}{N} J_{vt}$ is the mean volume flux through one pore.

8.4 Mechanistic Equations of Membrane Transport

We consider a membrane system shown in Figure 8.3. An inhomogeneous porous membrane separates two compartments, containing solutions of the

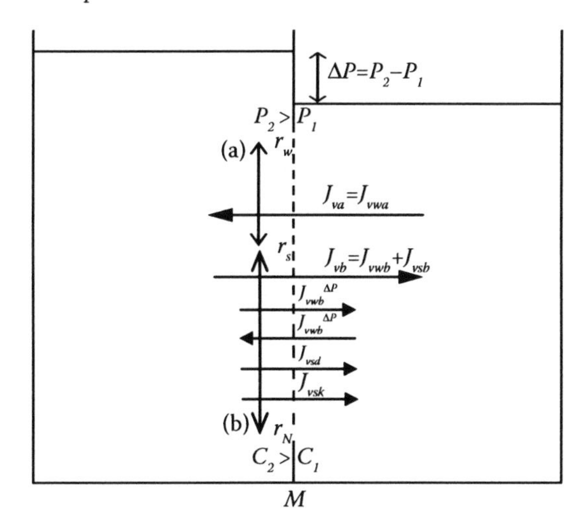

FIGURE 8.3
Membrane system (M is the membrane; m is the stirrers; c_1, c_2 are the concentrations; P_1, P_2 are the pressures; J_v is the volume fluxes of solute; J_{vw} is the volume fluxes of water; J_{vs} is the volume fluxes of solute; $J^{\Delta P}$, $J^{\Delta \Pi}$, are the volume fluxes of water generated by ΔP and $\Delta \Pi$, respectively; J_{vsd}, J_{vsk} is the volume fluxes of solute induced by ΔP and $\Delta \Pi$). The membrane pores are arranged according to sizes from the smallest (at the top—part [a]) to the largest (at the bottom—part [b] of the membrane).

same solute, with concentrations c_1, c_2 ($c_1 < c_2$) and under pressures P_1, P_2 ($P_1 < P_2$). The membrane has N pores permeable to water. We assume the pores are cylindrical, are perpendicular to the membrane surface, and are ordered according to sizes, with the smallest pores in the top part of the membrane. As described in the previous paragraph, for a given solute with molecular radius r_s the membrane can be divided into part (a) (top part in Figure 8.3), containing only semipermeable pores, and part (b) (bottom part) with only permeable pores.

8.4.1 Equation for the Volume Flux

The hydraulic pressure, ΔP, and the osmotic pressure, $\Delta \Pi$, across the membrane generate volume fluxes, J_{va} and J_{vb}, in both parts of the membrane:

$$J_{va} = L_{pa}\Delta P - L_{pa}\Delta \Pi \tag{8.39}$$

$$J_{vb} = L_{pb}\Delta P \tag{8.40}$$

where $\Delta P = P_2 - P_1$, $\Delta\Pi = RT(c_2 - c_1)$, and L_{pa}, L_{pb} are the filtration coefficients of both parts of the membrane. They are given as

$$L_{pa} = \left(\frac{J_{va}}{\Delta P}\right)_{\Delta\Pi=0} \tag{8.41}$$

$$L_{pb} = \left(\frac{J_{vb}}{\Delta P}\right)_{\Delta\Pi=0} \tag{8.42}$$

The total volume flux across the membrane is obviously

$$J_{vM} = (L_{pa} + L_{pb})\Delta P - L_{pa}\Delta\Pi = \mathrm{L}_p\Delta P - L_p\sigma\Delta\Pi \tag{8.43}$$

where

$$L_p = L_{pa} + L_{pb} \tag{8.44}$$

is the total filtration coefficient of the membrane. The quantity

$$\sigma = \frac{L_{pa}}{L_p} \tag{8.45}$$

which also satisfies

$$\sigma = \left(\frac{\Delta P}{\Delta\Pi}\right)_{J_{vM}=0} \tag{8.46}$$

is called the reflection coefficient, in analogy with the KK equations. If $L_{pa} = L_p$, then $\sigma = 1$ and the membrane is semipermeable. If $L_{pa} = 0$, then $\sigma = 0$ (a permeable membrane). For $L_p > L_{pa} > 0$ the membrane is selective and $1 > \sigma > 0$.

8.4.2 Equation for the Solute Flux

Contrary to water, the solute molecules permeate only through part (b) of the membrane. The flux J_{vb} is the sum of water and solute volume fluxes

$$J_{vb} = J_{vwb} + J_{vsb} \tag{8.47}$$

Each of these in turn can be written as the flux induced by the hydrostatic pressure, ΔP, and osmotic pressure, $\Delta\Pi$

$$J_{vwb} = J_{vwb}^{\Delta P} + J_{vwb}^{\Delta\Pi} \tag{8.48}$$

$$J_{vsb} = J_{vsb}^{\Delta P} + J_{vsb}^{\Delta\Pi} \tag{8.49}$$

where the subscripts ΔP and $\Delta\Pi$ denote the generating forces for the fluxes. We consider two cases.

8.4.2.1 Case 1

Assume that $\Delta\Pi = 0$, $\Delta P > 0$, and the solutions on both sides of the membrane are the same and equal c_s. Then also $\overline{c}_s = c_s$. The hydraulic pressure generates a flow of solvent and by convection also of the solute. The convective solute flux can be written as

$$j_{sk} = \overline{c}_s J_{vb} \tag{8.50}$$

where J_{vb} is given by equation (8.40). Hence

$$j_{sk} = (1-\sigma)\overline{c}_s L_p \Delta P \tag{8.51}$$

8.4.2.2 Case 2

Assume $\Delta P = 0$ and $\Delta\Pi > 0$. Then from (8.47–8.49) we have:

$$J_{vb} = J_{vwb}^{\Delta\Pi} + J_{vsb}^{\Delta\Pi} = 0 \tag{8.52}$$

since $\sigma_b = 0$ (σ_b is the filtration coefficient of part [b] of the membrane, see Figure 8.3). The fluxes $J_{vwb}^{\Delta\Pi}$ and $J_{vsb}^{\Delta\Pi}$ have opposite directions, as shown in Figure 8.3. We denote them as $J_{vsb}^{\Delta\Pi} = J_{vsd}$, and $J_{vwb}^{\Delta\Pi} = J_{vwd}$, where the latter is the diffusive volume flux of water relative to the membrane. To show that J_{vsd} is the diffusive volume flux of solvent, also measured relative to the membrane, we notice that $J_{vsd} = j_s\overline{V}_s = \overline{c}_s\nu_s\overline{V}_s$ and $J_{vwd} = j_w\overline{V}_w = \overline{c}_w\nu_w\overline{V}_w$.

Since

$$J_{vsd} = -J_{vwd} \tag{8.53}$$

we get

$$\overline{c}_s\nu_s\overline{V}_s = -\overline{c}_w\nu_w\overline{V}_w \tag{8.54}$$

where v_s, v_w are the velocities of molecules of the solute and water, respectively. For diluted solutions $\overline{c}_s \ll \overline{c}_w \approx 1$; hence we can conclude $\nu_s \gg \nu_w$. In other words, the velocity of the water relative to the membrane is negligibly small and the convection of solute by moving water can be disregarded.

The fluxes J_{vwd} and J_{vsd} can be written as

$$J_{vwd} = L_{pbw}\overline{c}_w\overline{V}_w\Delta\Pi \tag{8.55}$$

$$J_{vsd} = L_{pbs}\overline{c}_s\overline{V}_s\Delta\Pi \tag{8.56}$$

where L_{pbw} and L_{pbs} are the hydraulic conductivities for the diffusive permeation of water and the solute through part (b) of the membrane. The latter formula is also written as

$$j_{sd} = L_{pbs}\overline{c}_s\Delta\Pi = \omega_d\Delta\Pi \tag{8.57}$$

where

$$\omega_d = L_{pbs}\overline{c}_s \tag{8.58}$$

is called the coefficient of diffusive permeation of solute.

When both ΔP and $\Delta\Pi$ exist on the membrane, then the total solute flux $j_{sM} = \frac{\Delta m}{A\Delta t}$ across the membrane is the sum of the diffusive and convective flux

$$j_{sM} = \omega_d \Delta\Pi + (1-\sigma)\bar{c}_s L_p \Delta P \tag{8.59}$$

Equations (8.43) and (8.59) are the mechanistic (ME) transport equations. The coefficient ω_d given by equation (8.58) also satisfies:

$$\omega_d = \left(\frac{j_{sd}}{\Delta\Pi}\right)_{\Delta P=0} \tag{8.60}$$

Equation (8.59) for the solute flux can also be expressed as the solute volume flux

$$J_{vsM} = \omega_d \overline{V}_s \Delta\Pi + (1-\sigma) c_s \overline{V}_s L_p \Delta P \tag{8.61}$$

As we can notice in equations (8.48 and 8.49) the total volume flux

$$J_{vM} = J_{va} + J_{vb} = \underbrace{J_{vwa}}_{J_{va}} + \underbrace{\overbrace{J_{vwb}^{\Delta P} + J_{vwb}^{\Delta\Pi}}^{J_{vw}} + \overbrace{J_{vsd} + J_{vsk}}^{J_{vs}}}_{J_{vb}} = J_{vw} + J_{vs} \tag{8.62}$$

is the sum of J_{vwa}, $J_{vsb}^{\Delta P}$, $J_{vsb}^{\Delta\Pi}$, J_{vsd}, and J_{vsk}.

8.4.3 Correlation Relation for Parameters $\mathbf{L_p}$, σ, and ω_d

It is known that the three phenomenological transport parameters are not independent. For instance, this allowed elimination of the reflection coefficient in the 2P formalism. Using the ME described in the previous section we can derive a correlation relation for the transport parameters in a form:

$$\omega_d = (1-\sigma)\bar{c}_s L_p \tag{8.63}$$

To that end we consider equation (8.52), where the fluxes J_{vM} and J_{vs} are given by equations (8.43) and (8.61). In a case when $\Delta P = -\Delta\Pi$ these equations assume forms

$$J_{vM} = -L_p \Delta\Pi - L_p \sigma \Delta\Pi \tag{8.64}$$

$$J_{vs} = \omega_d \overline{V}_s \Delta\Pi - (1-\sigma)\bar{c}_s \overline{V}_s L_p \Delta\Pi \tag{8.65}$$

The water volume flux J_{vwa} is

$$J_{vw} = J_{vwa} + J_{vwb}^{\Delta P} + J_{vwb}^{\Delta\Pi} \tag{8.66}$$

where, according to equations (8.39) and (8.45), J_{vwa} equals

$$J_{vwa} = J_{va} = L_p \sigma \Delta P L_p \sigma \Delta\Pi \tag{8.67}$$

Also, for dilute solutions

$$J_{vwb}^{\Delta P} \approx J_{vb} = (1-\sigma)L_p\Delta P \tag{8.68}$$

Again, assuming $\Delta P = -\Delta\Pi$ we can write equation (8.66) in a form

$$J_{vw} = -L_p\Delta\Pi - L_p\sigma\Delta\Pi + J_{vwb}^{\Delta\Pi}$$

When $\Delta P = -\Delta\Pi$ flux $J_{vwb}^{\Delta\Pi}$ satisfies $|-L_p\Delta\Pi - L_p\sigma\Delta\Pi| \gg |J_{vwb}^{\Delta\Pi}|$ and can be disregarded, yielding

$$J_{vw} = -L_p\Delta\Pi - L_p\sigma\Delta\Pi \tag{8.69}$$

Now from equations (8.62), (8.64), (8.65), and (8.69) we obtain the correlation relation for membrane transport parameters, that is, equation (8.59).

8.4.4 2P Form of the Mechanistic Equations

Using the correlation relation (8.59) in transport equations (8.43) and (8.59) we can reduce them to the following 2P form:

$$J_{vM} = L_p\Delta P - L_p\sigma\Delta\Pi \tag{8.70}$$

$$j_{sM} = (1-\sigma)\bar{c}_s L_p(\Delta\Pi + \Delta P) \tag{8.71}$$

The latter can also be written as the solute volume flux equation

$$J_{svM} = (1-\sigma)\bar{c}_s\overline{V}_s L_p(\Delta\Pi + \Delta P) \tag{8.72}$$

8.4.5 Corrected Form of the Mechanistic Transport Equations

In both the KK and the ME formalisms the volume and solute fluxes are defined as $J_v = \frac{\Delta V}{A\Delta t}$, $j_s = \frac{\Delta m}{A\Delta t}$, $J_{vM} = \frac{\Delta V}{A\Delta t}$, and $j_{sM} = \frac{\Delta m}{A\Delta t}$, where A is the active surface area of the membrane, that is, the area of the two-sided contact of the membrane with solutions. However, porous membranes have pores in otherwise totally impermeable matrix and transport takes place only through the pores. The important thing is not the entire surface of the membrane but the total cross-section area of the pores A_t. Therefore the definition of fluxes can be corrected as

$$J_{vMt} = \frac{\Delta V}{A_t\Delta t} \tag{8.73}$$

and

$$j_{sMt} = \frac{\Delta m}{A_{bt}\Delta t} \tag{8.74}$$

where A_{tb} is the total cross-section area of permeable pores, that is, pores in part (b) of the membrane. The mechanistic transport equations can then be written as

$$J_{vMt} = L_{pt}\Delta P - L_{pt}\sigma\Delta\Pi \tag{8.75}$$

$$j_{sMt} = \omega_{dt}\Delta\Pi + (1-\sigma)\bar{c}_s L_{pt}\Delta P \tag{8.76}$$

where the corrected coefficients are defined as

$$L_{pt} = \left(\frac{J_{vtM}}{\Delta P}\right)_{\Delta\Pi=0} \tag{8.77}$$
$$\sigma = \left(\frac{\Delta P}{\Delta\Pi}\right)_{J_{vMt}=0, J_{vM}=0}$$
$$\omega_{dt} = \left(\frac{j_{sMt}}{\Delta\Pi}\right)_{\Delta P=0}$$

Similarly, the 2P mechanistic equations can be corrected to a form:

$$J_{vMt} = L_{pt}\Delta P - L_{pt}\sigma\Delta\Pi \tag{8.78}$$
$$j_{sMt} = (1-\sigma)\bar{c}_s L_{pt}(\Delta P + \Delta\Pi) \tag{8.79}$$

since

$$\omega_{dt} = (1-\sigma)\bar{c}_s L_{pt} \tag{8.80}$$

Let us recall that the coefficient L_{pt} is defined as (Equation [8.37]):

$$L_{pt} = \frac{A_t}{8\eta\pi\Delta x}$$

but

$$L_{pt} = L_{pta} + L_{ptb} = \frac{A_{at}}{8\eta\pi\Delta x} + \frac{A_{bt}}{8\eta\pi\Delta x}$$

where L_{pta} and L_{ptb} represent the total filtration coefficients of pores n_a and n_b, respectively. Then the reflection coefficient is defined as

$$\sigma = \frac{A_{at}}{A_t} \tag{8.81}$$

8.4.6 Equivalence of KK and ME Equations

The KK equations are based on thermodynamics and the ME equations relay on a specific model, a porous membrane, and allow at least partial microscopic interpretation of the transport parameters. It is obvious, however, that they should yield the same numerical results in situations where both are applicable. The volume flux equations in both formalism, equations (8.19) and (8.43) are formally identical but the transport parameters are defined differently. In the KK formalism the reflection coefficient, σ, is given by the Staverman's definition, $\sigma = -L_{pD}/L_p$, while in ME equations it is defined as $\sigma = \sigma_M = L_{pa}/L_p$. In the Staverman's definition the coefficient L_{pD} is negative, but if we put $|L_{pD}| = L_{pa}$ then both definitions are equivalent.

The ME solute flux equation

$$j_{sM} = \omega_d \Delta\Pi + (1-\sigma)\overline{c}_s L_p \Delta P$$

and the KK equation:

$$j_s = \omega\Delta\Pi + \overline{c}_s(1-\sigma)J_v$$

have different forms. The former was written in the following 2P form:

$$j_{sM} = (1-\sigma)\overline{c}_s L_p(\Delta P + \Delta\Pi) = (1-\sigma)\overline{c}_s L_p \Delta\Pi + (1-\sigma)\overline{c}_s L_p \Delta P \quad (8.82)$$

where $(1-\sigma)\overline{c}_s L_p \Delta\Pi = j_{sd}$ is the diffusion flux and $(1-\sigma)\overline{c}_s L_p \Delta P = j_{sk}$ is the convection flux of the solute. To bring the KK to the analogous 2P form, we derive first a correlation relation for the KK transport parameters L_p, σ, and ω. Let us recall that the total volume flux is given by

$$J_v = j_w \overline{V}_w + j_s \overline{V}_s = J_{vw} + j_s \overline{V}_s$$

Considering equations (8.59) and (8.60) we obtain

$$J_v = j_{vw} + \omega\overline{V}_s\Delta\Pi + (1-\sigma)\overline{c}_s\overline{V}_s(L_p\Delta P - L_p\sigma\Delta\Pi) \quad (8.83)$$

For dilute solutions ($c_w \gg c_s$) we can assume $J_v \approx J_{vw}$, hence

$$0 = \omega\Delta\Pi + (1-\sigma)\overline{c}_s L_p \Delta P - (1-\sigma)\overline{c}_s L_p \sigma\Delta\Pi \quad (8.84)$$

Assuming $\Delta P = -\Delta\Pi$ we get the correlation equation

$$\omega = (1-\sigma^2)\overline{c}_s L_p \quad (8.85)$$

Substituting in the KK equation we get its 2Ps form

$$j_s = (1-\sigma)\overline{c}_s L_p(\Delta P + \Delta\Pi) = (1-\sigma)\overline{c}_s L_p \Delta\Pi + (1-\sigma)\overline{c}_s L_p \Delta P \quad (8.86)$$

which is identical with the reduced ME equation for the solute flux. Also, considering correlation formulas (8.59), (8.85) we can find the connection between the transport parameters in the form:

$$\omega = (1+\sigma)(1-\sigma)\overline{c}_s L_p = \omega_d(1+\sigma) \quad (8.87)$$

8.5 Water Exchange between Aquatic Plants and the Environment

Every living organism, whether multicellular or a single cell, needs to continuously exchange with the environment water and various chemicals. It uptakes water and nutrients and expels water and metabolic waste products. In case of multicellular land plants, water uptake takes place mostly osmotically through the root system. Water is then expelled in leaves. It transpires to the atmosphere, mainly from the walls of mesophyll cells and from leaf cuticles. Water can diffuse across cell membranes and cuticles; however, in many cases we may observe a significantly faster permeation through water pores formed in membranes by specialized proteins or water pores that spontaneously form in cuticles. The majority of studies of such transport have only qualitative character, and the details of transport mechanisms are not well understood (Schönherr 2006).

In case of aquatic plants that live entirely submerged, the mechanisms of water exchange with the surroundings are simpler but still not entirely understood. Aquatic plants do not uptake water through roots, since they have smaller roots that serve mostly to attach the plant. Water also cannot transpire from leaf surfaces. Therefore we assume water is exchanged not by specialized organs within a plant, but by each plant cell individually. This exchange takes place across the cell membrane and is induced by thermodynamic forces, such as osmotic pressure, $\Delta\Pi$, or hydraulic pressure, ΔP. One of the main microscopic mechanisms of water transport is permeation through a variety of pores, such as aquaporins or ion channels. Therefore this transport can be described using practical KK equations (Chapter 2) or the ME equations (Chapter 4). In this section, we describe biophysical foundations of water exchange with the surroundings on the example of two aquatic plants: *Nitella translucens* and *Chara Corallina* (Kedem and Katchalsky 1965; Hertel and Steudle 1997; Dąbska 1964).

8.5.1 KK Equations Applied to Water Exchange by Aquatic Plants

We consider a model of an aquatic plant cell in an aqueous environment as shown in Figure 8.4. The cell surrounded by a selective membrane contains an osmotically active nonelectrolytic solute (s). The concentration of the same solute in the environment, c_{so}, is smaller than intracellular concentration, c_{si}.

According to the KK formalism, the transport properties of the membrane are characterized by transport parameters L_p, σ, and ω. The reflection coefficient has a value $0 < \sigma < 1$ for a selective membrane. Therefore, the permeation coefficient is $\omega > 0$. The concentration difference across the membrane $\Delta c = c_{si} - c_{so}$ induces a solute efflux, given by the volume solute flux J_{vs}.

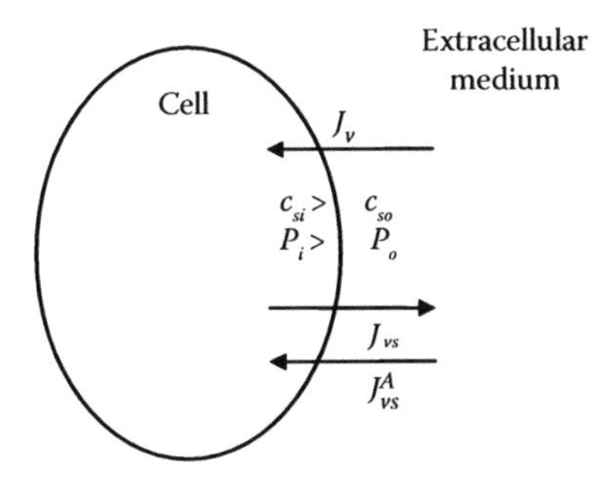

FIGURE 8.4
A model of an aquatic plant cell (c_{si}, c_{so} is the concentrations; P_i, P_o is the pressures, J_v, J_{vs}, J_{vs}^A is the volume fluxes).

To maintain a steady concentration gradient there must be an active influx of the same solute, described as J_{vs}^A in the figure. In such a stationary state there is a constant osmotic pressure, $\Delta\Pi = RT(c_{si} - c_{so})$, and a constant turgor pressure, $\Delta P = P_i - P_o$. The KK equations for the water and solute transport have the form

$$J_v = L_p \Delta P - L_p \sigma \Delta\Pi \tag{8.88}$$

$$j_s = \omega \Delta\Pi + (1 - \sigma)\bar{c}_s J_v \tag{8.89}$$

The latter can be rewritten as the solute volume flux

$$J_{vs} = \omega \overline{V_s} \Delta\Pi + (1 - \sigma)\bar{c}_s \overline{V_s} J_v \tag{8.90}$$

where $\overline{V}_s$ is the molar volume. The total volume flux is then given as

$$J_v^t = J_v + J_{vs}^A \tag{8.91}$$

In the KK formalism, the flux J_v is defined as $J_v = J_{vw} + J_{vs}$; hence

$$J_v^t = \mathrm{J}_{vw} + \mathrm{J}_{vs} + J_{vs}^A \tag{8.92}$$

In a stationary state $J_v^t = 0$, $J_{vs} + J_{vs}^A = 0$, and the water flux is also zero ($J_{vw} = 0$). The cell does not change its volume and it does not absorb or reject water.

8.5.2 Water Exchange Described by Mechanistic Equations

We consider a model of a cell including fluxes relevant to the ME shown in Figure 8.5 (Kargol and Kargol 2005; Kargol et al. 2005). For illustration purposes we assume that the osmotically active surface of the cell containing all

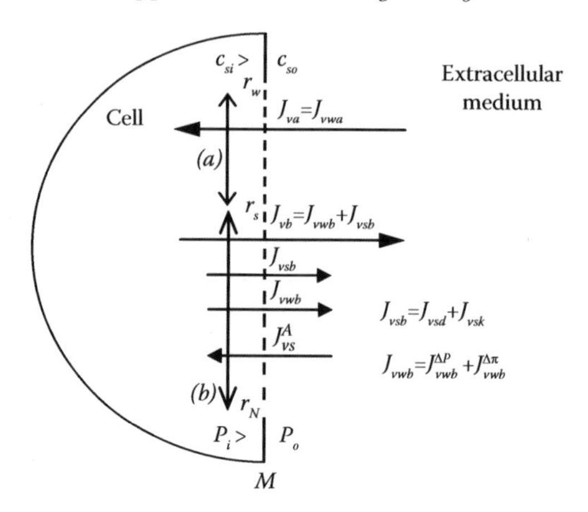

FIGURE 8.5
Model of an aquatic plant cell (M is the cell membrane; c_{si}, c_{so} is the concentrations; P_i, P_o is the pressures; $J_{va}, J_{vb}, J_{vwb}, J_{vsb}$, $J_v^A s$ is the volume fluxes). (Reprinted from publication, Kargol, A., Przestalski, M., and Kargol, M., *Cryobiol.*, 50, 2005. ©2005, with permission from Elsevier.)

pores is represented as membrane M and that the pores are ordered according to sizes from top to bottom. All pores with radii $r < r_s$, that is, pores impermeable to the solute are in part (a) of the membrane and all pores larger than the size of solute molecule, that is, permeable pores are in part (b) of the membrane. Membrane M has a filtration coefficient L_p, reflection coefficient σ, and the diffusive permeation coefficient ω_d. Parts (a) and (b) of the membrane have filtration coefficients L_{pa} and L_{pb} and reflection coefficients $\sigma_a = 1$ and $\sigma_b = 0$, respectively. We also assume there is an active solute volume flux, J_{vs}^A. Again, in a stationary state there is a constant osmotic pressure, hydraulic pressure, and concentration gradient across the membrane.

The net volume flux across membrane M is

$$J_v^t = J_{vM} + J_{vs}^A = 0 \tag{8.93}$$

where $J_{vM} = J_{va} + J_{vb}$ is the passive volume flux, and J_{va}, J_{vb} are the volume fluxes across parts (a) and (b) of the membrane, respectively. From the mechanistic volume flux equation (8.43) we can find the following expression for the turgor pressure:

$$\Delta P = \sigma \Delta \Pi - \frac{J_{vs}^A}{L_p} \tag{8.94}$$

where $\Delta \Pi = RT(c_{si} - c_{s0})$. In the stationary state, the solute volume flux is zero as well, that is, $J_{vs}^t = J_{vsM} + J_{vs}^A = 0$. From the mechanistic solute equation (8.59) we get:

$$-J_{vs}^{A} = J_{vsM} = (1-\sigma)\bar{c}_s\overline{V}_s L_p \Delta\Pi + (1-\sigma)\bar{c}_s\overline{V}_s L_p \Delta P \tag{8.95}$$

From (8.94) and (8.95) we see that the turgor pressure satisfies

$$\Delta P = \bar{\sigma}\Delta\Pi \tag{8.96}$$

where

$$\bar{\sigma} = \frac{\sigma + (1-\sigma)\bar{c}_s\overline{V}_s}{1-(1-\sigma)\bar{c}_s\overline{V}_s} \tag{8.97}$$

8.5.3 Numerical Results for *Nitella translucens* and *Chara Corallina*

In the model cell shown in Figure 8.5, the water volume influx, J_{vwa}, across part (a) of the membrane is given as

$$J_{vwa} = L_p\sigma(\bar{\sigma}-1)\Delta\Pi = L_p\sigma(\bar{\sigma}-1)RT(c_{si}-c_{so}) \tag{8.98}$$

where we used equation (8.96) and the volume flux equation (8.43). Similarly, the volume flux through part (b) can be expressed as

$$J_{vwb} + J_{vsb} = J_{vb} = L_{pb}\Delta P \tag{8.99}$$

Using equation (8.96) and the solute flux equation (8.59) we can find the following expression for the water volume efflux:

$$J_{vwb} = (1-\sigma)[(1-\bar{c}_s\overline{V}_s)\bar{\sigma} - \bar{c}_s\overline{V}_s]L_pRT(c_{si}-c_{so}) \tag{8.100}$$

Equations (8.98) and (8.100) show that in a stationary state the cell can simultaneously absorb and reject water through different parts of the membrane. In Table 8.2 we show numerical results for two particular plants, *Nitella translucens* and *Chara Corallina.* Values of transport parameters were obtained from available literature. The table shows values of water influx and efflux calculated from equations (8.98) and (8.100), respectively. Water influx is osmotically driven and takes place in part (a) of the membrane and water efflux is driven by the turgor pressure and occurs through pores in part (b) of the membrane.

8.6 Passive Transport through Cell Membranes of Human Erythrocytes

Fundamental functions of human erythrocytes are related to transport of oxygen throughout the organism and removal of carbon dioxide. Contrary to most

TABLE 8.2
Transport Properties of *Nitella translucens* and *Chara Corallina* Plants and Calculated Water Volume Fluxes

No.	Membrane	Solute (s)	L_p*10^{12} [m^3/N·s]	σ	$\overline{V}_s$*10^3 [m^3/mol]	Source	J_{vwa}*10^8 [m/s]	J_{vwb}*10^8 [m/s]
1	*Nitella translucens*	Ethanol	1.1	0.44	0.058	Katchalsky and Curran (1965)	−6.69	6.69
2	*Nitella translucens*	Isopropanol	1.1	0.40	0.076		−5.60	5.60
3	*Chara corallina*	Ethanol	1.6	0.36	0.058	Hertel and Steudle (1997)	−9.11	9.11
4	*Chara corallina*	Isopropanol	1.6	0.35	0.076		−8.97	8.97

Other data: $C_{si} = 150$ [mol/m^3]; $C_{so} = 50$ [mol/m^3]; $\bar{c}_s = 100$ [mol/m^3]; $R = 8.3$ [N·m/mol·K]; $T = 300$ [K].
Source: Kargol, M. and Kargol, A. *Acta Physiol. Planta.*, 27, 2005. With permission.

human cells the erythrocytes undergo relatively large changes in volume resulting from blood pressure changes, in particular in the heart and the arteries. However, in veins the changes in external pressure are significantly smaller and we can treat water and solute exchange with blood plasma as stationary. We can describe these processes applying methods similar to those used in Section 8.5 for water transport in aqueous plants.

8.6.1 Regulation of Water Exchange between Erythrocytes and Blood Plasma

To describe water permeation across erythrocyte membranes (Sha'afi and Gary-Bobo 1973; Kargol et al. 2005) we use the mechanistic equations. We replace the entire membrane by an equivalent membrane that contains all water permeable pores ordered according to their sizes. Following the procedure outlined in Section 8.5 we write the water influx through the equivalent membrane as

$$J_{vwa} = L_p\sigma(\overline{\sigma} - 1)\Delta\Pi = L_p\sigma(\overline{\sigma} - 1)RT(c_{si} - c_{s0}) \qquad (8.101)$$

and water efflux as

$$J_{vwb} = (1 - \sigma)[(1 - \overline{c}_s\overline{V}_s)\overline{\sigma} - \overline{c}_s\overline{V}_s]L_pRT(c_{si} - c_{s0}) \qquad (8.102)$$

We illustrate these equations using values of transport parameters for two solutes (Table 8.3).

For concentrations c_{si} = 150 [mol/m^3], c_{so} = 50 [mol/m^3], $\overline{c}_s$ = 100 [mol/m^3], and taking $R = 8.3$ [N·m/mol·K], $T = 300$ [K] we calculated the fluxes. For ethyl glycol we obtained $J_{vwa} = -5.29 \times 10^{-8}$ [m/sec], $J_{vwb} = 5.29 \times 10^{-8}$ [m/sec], and for urea: $J_{vwa} = -7.77 \times 10^{-8}$ [m/sec], $J_{vwb} = 7.77 \times 10^{-8}$ [m/sec] (Figure 8.6).

As these graphs show that the fluxes depend linearly on the internal solute concentration. This concentration on the other hand depends on the active transport, J_{vs}^A, required to maintain the stationary state. Similarly, the dependence on the filtration coefficient is linear while the largest changes of the fluxes as functions of the reflection coefficient occur for very small (near 0) and very large (near 1) values of σ.

TABLE 8.3
Transport Parameters for Membranes of Human Erythrocytes

Solute	L_p **[m^3/Ns]**	σ	$\overline{V}_s$ **[m^3/mol]**	**References**
Ethyl glycol	0.92×10^{-12}	0.63	0.0566×10^{-3}	Katchalsky and Curran (1965)
Urea	1.27×10^{-12}	0.55	0.042×10^{-3}	Sha'afi and Gary-Bobo (1973)

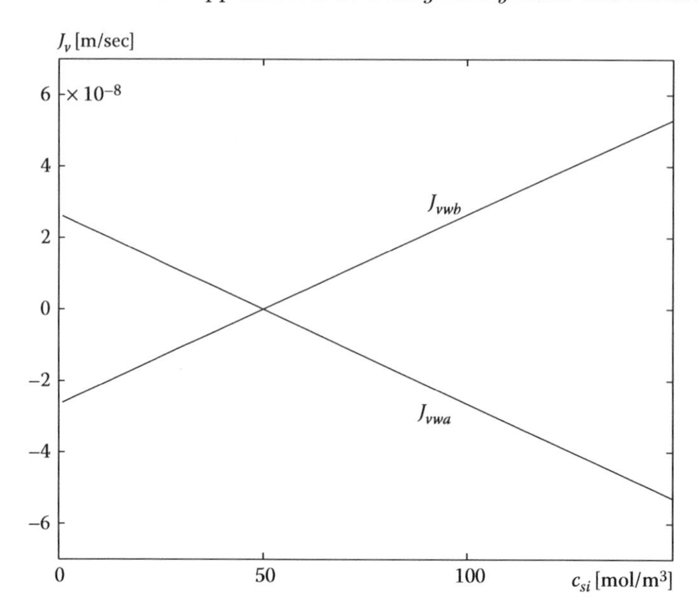

FIGURE 8.6
Relations: $J_{vwa} = f(c_{si})$ and $J_{vwb} = f(c_{si})$, for human erythrocytes. (Reprinted from publication, Kargol, A., Przestalski, M., and Kargol, M., *Cryobiol.*, 50, 2005. ©2005, with permission from Elsevier.)

For these relationships to have any physiological significance one must identify possible mechanisms for changes of these parameters for a give erythrocyte. To that end we recall the mechanistic definition of the reflection coefficient, equation (8.45):

$$\sigma = \frac{L_{pa}}{L_p}$$

where L_{pa} is the filtration coefficient of permeable pores of the equivalent membrane (Figure 8.7).

According to this definition any change in the number of permeable or impermeable pores due, for instance, to gating of ion channels or blockage of aquaporins and other pores contributing to this transport, not only changes the filtration coefficient of the erythrocyte membrane but also its reflection coefficient. It then affects water influx and efflux across the membrane (Figure 8.8).

8.6.2 Distribution of Pore Sizes

Another aspect of passive transport across porous membranes is the determination of the pore sizes. In previous sections we implicitly assumed knowledge

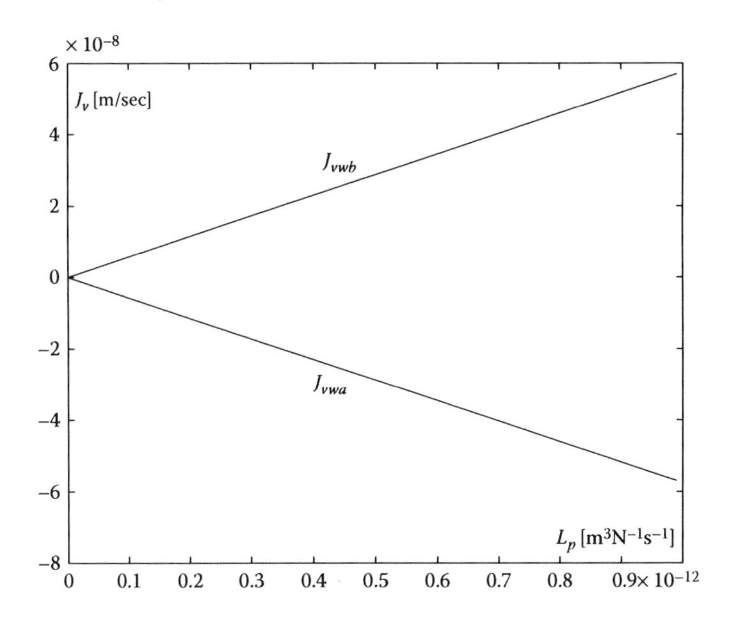

FIGURE 8.7
Relations: $J_{vwa} = f(L_p)$ and $J_{vwb} = f(L_p)$, for human erythrocytes. (Reprinted from publication, Kargol, A., Przestalski, M., and Kargol, M., *Cryobiol.*, 50, 2005. ©2005, with permission from Elsevier.)

of pore size distribution for a particular membrane and derived membrane transport parameters from this knowledge. In practice it is the reverse, that is, the transport parameters can be determined experimentally and we would like to deduce some information about the pore numbers and sizes. In this section we illustrate a proposed method for determining pore distribution from known membrane transport properties and illustrate it on the example of human erythrocytes (Kargol et al. 2005).

In previous paragraphs we assumed pore sizes are random. Let us also assume that the distribution of pore cross-section areas is Gaussian (Kargol et al. 2005):

$$f(s) = \frac{N}{h\sqrt{2\pi}} \exp\left(-\frac{(s-\overline{s})^2}{2h^2}\right) \tag{8.103}$$

where $\overline{s}$ is the distribution mean, h is the standard deviation, and N is a normalization constant. In the ME the reflection coefficient for individual pores can be either 0 or 1. However, the reflection coefficient for the entire membrane can be any number between 0 and 1, depending on the ratio of permeable and impermeable pores for a given solute. Let us choose a particular solute (s) with molecular radius r_s. Let A_{at}, A_{bt} denote the total cross-section

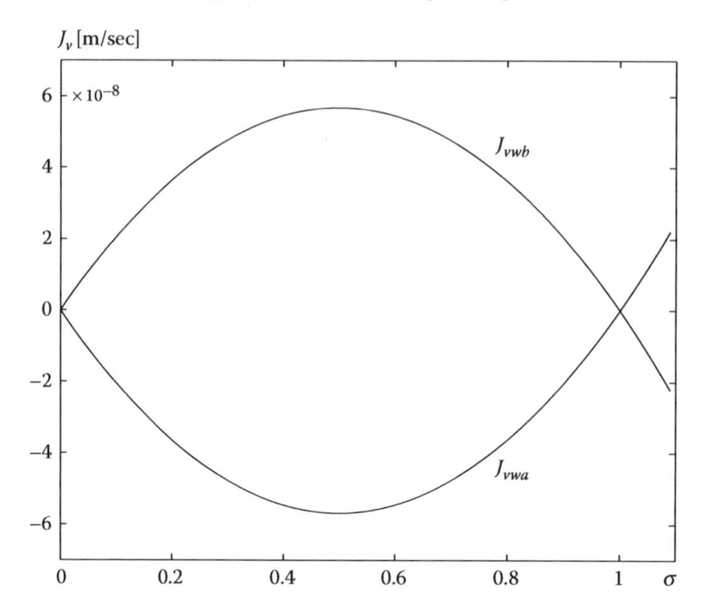

FIGURE 8.8
Relations: $J_{vwa} = f(\sigma)$ and $J_{vwb} = f(\sigma)$, for human erythrocytes. (Reprinted from publication, Kargol, A., Przestalski, M., and Kargol, M., *Cryobiol.*, 50, 2005. ©2005, with permission from Elsevier.)

areas of all pores impermeable and permeable to this solute, respectively. For instance, if $A_{at} = A_{bt}$ then the corresponding reflection coefficient, σ, is 0.5. In general the ratio of A_{at} to the total cross-section area of all pores A_t equals the reflection coefficient (equation 8.81). However, the dependence of σ on A_{at} is nonlinear and to find parameters of this Gaussian pore distribution we need to find explicit dependence of the reflection coefficient on the solute dimensions, based on equation (8.81). The area A_{at} can be expressed as

$$A_a = \int_0^{s_x} f(s)ds \tag{8.104}$$

where $s_x = \pi r_s^2$ and similarly the area A_{bt} is

$$A_b = \int_{s_x}^{\infty} f(s)ds \tag{8.105}$$

If we assume that the area under the Gaussian tail from $-\infty$ to 0 is negligible, we have approximately:

$$A_t = A_{at} + A_{bt} \approx \int_{-\infty}^{\infty} f(s)ds \tag{8.106}$$

Therefore, we obtain the following formula for the reflection coefficient, σ, of a given solute:

$$\sigma = \frac{1}{2}\left[erf\left(\frac{s_x - \overline{s}}{h\sqrt{2}}\right) + erf\left(\frac{\overline{s}}{h\sqrt{2}}\right)\right] \tag{8.107}$$

Although, equation (8.107) expresses the dependence of σ on the solute dimensions, s_x, it also involves other unknown distribution parameters (h and $\overline{s}$). We can find the latter using experimental data for the membrane (the reflection coefficients and molecular dimensions of various solutes). We plot the reflection coefficient, σ, as a function of s_x. Next we find a function of the form (8.107) that fits these experimental data (the plot $\sigma = f(s_x)$). Data for human erythrocytes collected from available literature is given in Table 8.4, and the fitting was done using a simulated annealing algorithm to minimize the χ^2 error. In our computations we had an additional constraint that the majority of the pores ought to be permeable to water, that is, $\overline{s} - s_w > 3h$. Using data shown in Table 8.4 we obtained $\overline{s} = 0.137$ [nm^2] and $h = 0.024$.

When transport is described using the KK equation where the volume flux is defined as the rate of volume flow, $\Delta V/\Delta t$, per unit surface area of the membrane, then assuming $\Delta\Pi = 0$, we get

$$J_v = L_p \Delta P = \frac{\Delta V}{A_{tm}\Delta t} \tag{8.108}$$

where A_{tm} is the membrane surface area. For porous membranes, an alternative definition of the volume flux can be considered: the volume flow per unit

TABLE 8.4
Transport Data for Human Erythrocytes and Selected Solutes

Solute	Water	Formamide	Urea	Acetamide	Propionamide
Molecular radius r [nm]*	0.15	0.207	0.211	0.23	0.231
Molecular cross-section area s_x [nm^2]*	0.07	0.134	0.134	0.166	0.168
L_p [m^3/(Ns)]†		1.27×10^{-12}	1.27×10^{-12}	1.27×10^{-12}	1.27×10^{-12}
$\sigma^\dagger$		0.58	0.53	0.80	0.55

Source: *Levitt 1974, Goldstein and Solomon 1961; †Sha'afi and Gary-Bobo 1973.

effective area of pores

$$\tilde{J}_v = \frac{\Delta V}{A_{tp}\Delta t} \qquad (8.109)$$

where $A_{tp} = \sum_{i=1}^{N} A_i$ is the total effective area of pores in the membrane. In analogy to (8.109) the dependence of the flux $\tilde{J}_v$ on the pressure stimulus can be expressed as

$$\tilde{J}_v = \tilde{L}_v \Delta P \qquad (8.110)$$

where $\tilde{L}_p$ is the total filtration coefficient of the pores (a measure of the total hydraulic conductivity of pores in the membrane). Experimentally measurable are the rate of the volume flow $\Delta V/\Delta t$ and the membrane surface A_{tm} from which the values of J_v and L_p can be determined. The values of filtration coefficients given in literature refer thus to quantities defined by equations (8.108). On the other hand a comparison of equations (8.108–8.109) shows that

$$\tilde{L}_p = \frac{A_{tm}}{A_{tp}} L_p \qquad (8.111)$$

From equations (8.111) and (8.37) we have

$$A_{tp} = \sqrt{8\eta\pi l A_{tm} L_p} \qquad (8.112)$$

where L_p is the filtration coefficient from the KK formalism. This formula allows computations of the total pore area, A_{tp}, provided η, l, A_{tm}, and L_p are known. On the other hand, knowing values of $\overline{s}$ and A_{tp} the number of pores can be found as $N = A_{tp}/\overline{s}$. We computed the values based on the following data: viscosity of human blood plasma, $\eta = 1.81 \times 10^{-21}$ [Ns/nm^2] (Kane and Sternheim 1988); filtration coefficient of human erythrocytes membrane, $L_p = 1.27 \times 10^{-12}$ [m^3/Ns] (Sha'afi and Gary-Bobo 1973); total surface area of a human erythrocyte, $A_{tm} = 167 \times 10^6$ [nm^2] (Sha'afi et al. 1970; Milgram and Solomon 1977); thickness of a human erythrocyte membrane, $l = 7.5$ [nm] (Stryer 2000); mean pore cross-section area, $\overline{s} = 0.137$ [nm^2]. We found that for such an erythrocyte the total pore cross-section area is $A_{tp} = 268.9$ [nm^2], which gives the number of pores at $N = 1.9 \times 10^3$. That is, on the average there is one pore per 87×10^3 [nm^2] of cell surface area.

8.7 Comparison of Transport Formalisms: KK, ME, and 2P

The main aim of this chapter is to compare various formalisms for membrane transport of binary solutions. As mentioned in the introduction they were derived from different assumptions and thus they apply in somewhat different situations. The KK and 2P formalisms are of thermodynamic nature and do not consider microscopic details of membrane transport of water and solute molecules. On one hand this makes them more general and more widely applicable, on the other hand they provide only limited knowledge of the relation between macroscopically observable flows and the underlying microscopic transport mechanisms. In contrast, the ME formalism is based on a specific model of membrane permeation—diffusion through pores. It uses concepts borrowed from the KK formalism but it aims at linking the microscopic transport properties of membranes with the macroscopic description of mass transport. We conclude this chapter with a brief summary of the three formalisms, including their main equations.

1. Formalism KK (Kedem and Katchalsky 1958; Katchalsky and Curran 1965)

Formalism KK is based on the three phenomenological coefficients describing membrane transport properties: coefficients of filtration, permeation, and reflection. They are defined as $L_p = (J_V/\Delta P)_{\Delta\Pi=0}$, $\omega = (j_S/\Delta\Pi)_{J_v=0}$ and $\sigma = (\Delta P/\Delta\Pi)_{J_v=0}$. The volume flux density (i.e., the flux of water and permeable solute) is given as

$$J_v = L_p\Delta P - L_p\sigma\Delta\Pi \tag{8.113}$$

and the solute flux density as

$$j_s = \omega\,\Delta\Pi + (1-\sigma)\bar{c}_s J_v \tag{8.114}$$

The formalism applies to a variety of problems, independently of the nature of membrane permeation mechanism, and has been widely used. Nevertheless, there have been questions raised as to certain aspects of its interpretation. One of the questions regarding the reflection coefficient introduced to describe coupling between water and solute fluxes, for example, when they occur through the same pathways (Kleinhans 1998). Another question is about the interpretation of the permeation coefficient (Kargol 2002; Kargol and Kargol 2003, 2006). The parameter is measured with $J_v = 0$, that is, in the presence of two stimuli satisfying $|\Delta P| = |-\sigma\Delta\Pi|$. For a membrane with $\sigma = 1$ the solute transport is driven by both of these stimuli; hence, ω does not describe diffusive solute permeation only.

2. Formalism 2P (Kleinhans 1998)

The 2P formalism has been developed for situations where water and solute fluxes are independent. In such cases the reflection coefficient is redundant and the transport equations can be simplified to the following form. Water flux density is (assuming $\Delta P = 0$):

$$J_{vw} = -L_p RT\Delta c \tag{8.115}$$

and solute flux density is

$$j_s = P_s \Delta c \tag{8.116}$$

where $P_s = \omega \mathrm{RT}$ is the solute permeability. The solute flux equations are identical and the difference in volume flux equations appears to formally reduce to a substitution $\sigma = 1$. However, as emphasized in Kleinhans (1998); Elmoazzen et al. (2008), this would be incorrect since for $\sigma = 1$ there is no solute transport. Equation (8.115) describes water flux only. To directly compare with (8.113) one must compute the volume flux of water and solute (Elmoazzen et al. 2008):

$$J_v = J_{vw} + J_{vs} = -L_p RT \left(1 - \frac{P_s v_s}{L_p RT}\right) \Delta c \tag{8.117}$$

Now equations (8.113) and (8.115) are almost identical. The only difference is the expression $1 - (P_s v_s / L_p RT)$ instead of σ in the 2P equation. According to the 2P formalism this expression is the upper limit of σ. It is reached when water and solute permeate through separate pathways. In such case the 2P and KK formalisms are equivalent.

3. ME formalism (Kargol 2002; Kargol and Kargol 2003, 2006)

This formalism has been developed based on a specific microscopic model of permeation, described in detail in the chapter. The transport equations are given as (equations 8.43 and 8.59):

$$J_{vm} = L_p \Delta P - L_p \sigma \Delta \Pi \tag{8.118}$$

$$j_{sM} = \omega_{\mathrm{d}} \Delta \Pi + (1 - \sigma) \bar{c}_s L_p \Delta P \tag{8.119}$$

where J_{vM} is the volume flux density and j_{sM} is the solute flux density. Both fluxes are driven by pressure differences, ΔP and $\Delta \Pi$. Parameters L_p, σ, and ω_d are defined as

$$L_p = \left(\frac{J_{vM}}{\Delta P}\right)_{\Delta \Pi = 0}, \quad \sigma = \left(\frac{\Delta P}{\Delta \Pi}\right)_{J_{vM} = 0}, \quad \text{and} \quad \omega_d = \left(\frac{j_{sM}}{\Delta \Pi}\right)_{\Delta P = 0}$$

where $J_{vM} = J_{va} + J_{vb}$, $L_p = L_{pa} + L_{pb}$, $\sigma = L_{pa}/L_p$, $L_{pa} = (J_{va}/\Delta P)_{\Delta \Pi = 0}$, and $L_{pb} = (J_{vb}/\Delta P)_{\Delta \Pi = 0}$. Different terms in (8.118–8.119) describe the phenomena of filtration, osmosis, diffusion, and convection, respectively.

In the ME equations, the permeation coefficient is defined differently from the KK formalism, and it refers to the diffusive solute permeation only. In Section 8.4.6 we derived correlation relations for the KK and ME formalisms in the forms $\omega = (1 - \sigma^2) \bar{c}_s L_p$ and $\omega_d = (1 - \sigma) \bar{c}_s L_p$, respectively. Using these relations one can show the equivalence of both formalisms (see Section 8.4.6).

8.8 References

Chuenkhum, S. and Cui, Z. (2006). The parameter conversion from the Kedem–Katchalsky model into the two-parameter model. *CryoLetters*, **27**:185–99.

Dbska, J. (1964). *Charophyta-Ramienice, Flora Sodkowodna Polski [Poland's Freshwater Flora]*. PWN, Warszawa.

Disalvo, A., Siddiqi, F. A., and Ti Tien, H. (1989). Membrane transport with emphasis on water and nonelectrolytes in experimental lipid bilayers and biomembranes. In *Water Transport in Biological Membranes*, ed. G. Benga, CRC, Boca Raton, FL, pp. 41–75.

Elmoazzen, H. Y., Elliott, J. A. W., and McGann, L. E. (2008). Osmotic transport across cells in nondilute solutions. *Biophysical Journal* (to appear).

Elmoazzen, H. Y., Elliott, J. A. W., and McGann, L. E. (2002). The effect of temperature on membrane hydraulic conductivity. *Cryobiology*, **45**:68–79.

Goldstein, D. A. and Solomon, A. K. (1961). Determination of equivalent pore radius for red cells by osmotic pressure measurement. *The Journal of General Physiology*, **44**:1–17.

Hejnowicz, Z. (1996). Aquaporin water channels in plant and animal cells. *Postępy Biologii Komórki*, **23**:529–546 (in Polish).

Hertel, A. and Steudle, E. (1997). The function of water channels in *Chara*: the temperature dependence of water and solute flows provides evidence composite transport for a slippage of small organic solutes across water channels. *Planta*, **202**:324–35.

Jacobs, M. H. and Stewart, D. R. (1932). A simple method for the quantitative measurement of cell permeability. *Journal of Cellular and Comparative Physiology*, **1**:71–82.

Kane, J.W. and Sternheim, M. M. (1988). *Physics for Natural Scientists, 2.* PWN, Warsaw.

Kargol, A. (2002). A mechanistic model of transport processes in porous membranes generated by osmotic and hydrostatic pressures. *Journal of Medical Screening*, **191**:61–69.

Kargol, M. (2007). *Mass Transport Processes in Membranes and Their Biophysical Implications.* Kielce: WSTKT Kielce, Poland.

Kargol, M. and Kargol, A. (2003). Mechanistic equations for membrane substance transport and their identity with Kedem–Katchalsky equations. *Biophysical Chemistry,* **103**:117–27.

Kargol, M. and Kargol, A. (2003). Mechanistic formalism for membrane transport generated by osmotic and mechanical pressure. *General Physiology and Biophysics,* **22**:51–68.

Kargol, M. and Kargol, A. (2005). Biophysical mechanisms of physiological water exchange with the surroundings by the cells of the *Nitella translucens* and *Chara corralina* plants. *Acta Physiologiae Plantarum,* **27**:71–79.

Kargol, M. and Kargol, A. (2006). Investigation of reverse osmosis on the basis of the Kedem–Katchalsky equations and mechanistic transport equations. *Desalination,* **190**:267–276.

Kargol, M., Kargol, A., and Przestalski, S. (2001). Studies on the structural properties of porous membranes: measurement of linear dimensions of solutes. *Biophysical Chemistry,* **91**:263–271.

Kargol, A., Przestalski, M., and Kargol, M. (2005). A study of porous structure of cellular membranes in human erythrocytes. *Cryobiology,* **50**:332–337.

Kargol, M., Suchanek, G., Przestalski, M., Siedlecki, J., and Kargol, A. (2005). The problem of water exchange by the living cells in the light of mechanistic transport equations. *Polish Journal of Environmental Studies,* **14**:605–611.

Katchalsky, A. and Curran, P. F. (1965). *Nonequilibruim Thermodynamics in Biophysics.* Harvard University Press, Cambridge, MA.

Katkov, I. (2000). A two-parameter model of cell membrane permeability for multisolute systems. *Cryobiology,* **40**:64–83.

Kedem, O. and Katchalsky, A. (1958). Thermodynamic analysis of the permeability of biological membranes to non-electrolytes, *Biochimica et Biophysica Acta,* **27**:229–246.

Kleinhans, F.W. (1998). Membrane permeability modeling: Kedem–Katchalsky vs a two-parameter formalism. *Cryobiology,* **37**:271–289.

Levitt, D.G. (1974). A new theory of transport for cell membrane pores. *Biochimica et Biophysica Acta,* **373**:115–131.

Milgram, J. H. and Solomon, A. K. (1977). Membrane permeability equations and their solutions for red cells. *The Journal of Membrane Biology,* **34**: 103–144.

Schönherr, J. (2006). Characterization of aqueous pores in plant cuticles and permeation of ionic solutes. *Journal of Experimental Botany,* **57**:2471–2491.

Sha'afi, R. J. and Gary-Bobo, C. M. (1973). Water and nonelectrolytes permeability in mammalian red cell membranes, In *Progress in Biophysics and Molecular Biology,* ed. J. A. V. Butler and D. Noble, 26. Pergamon Press, Oxford and New York, pp. 106–145.

Sha'afi, R. J., Rich, G. T., Mickulecky, D. C., and Solomon, A. K. (1970). Determination of urea permeability in red cells by minimum method. *The Journal of General Physiology,* **55**:427–450.

Staverman, A. J. (1951). The theory of measurement of osmotic pressure. *Recueil des Travaux Chimiques des Pays-Bas,* **70**:344–352.

Stryer, L. (2000). *Biochemistry.* PWN, Warsaw.

Walsh, J. R., Diller, K. R., and Brand, J. J. (2004). Measurement and simulation of water and methanol transport in algal cells. *Journal of Biomechanical Engineering,* **126**:167–179.

Zeuthen, T. and MacAulay, N. (2002). Passive water transport in biological pores. In *Molecular Mechanisms of Water Transport Across Biological Membranes,* ed., T. Zeuthen and W. D. Stein. Academic Press, San Diego, London, pp. 203–230.

9

Skin Electroporation: Modeling Perspectives

S. M. Becker
Institute for Thermo-Fluid Dynamics, Hamburg University of Technology Hamburg, Germany

A. V. Kuznetsov
Department of Mechanical and Aerospace Engineering, North Carolina State University, Raleigh, NC

CONTENTS

Acknowledgments
Special thanks to the Alexander von Humboldt Foundation for its generous Fellowship awarded to Dr. Sid Becker. The authors also greatly appreciate the generosity of Prof. Heinz Herwig who provided facilities at the Institute for Thermo-Fluid Dynamics, Hamburg University of Technology, and whose insights and encouragement greatly benefited the authors.

9.1 Introduction

The subject of this chapter concerns modeling and treatment of mass transport through biological tissues, specifically the skin. To the reader unfamiliar with the physiological aspects of the skin, a basic description of the skin and each of its composite layers is provided. The treatment of the skin as a porous medium is then addressed. The chapter continues by introducing current methods used to describe nondestructive transdermal transport of low-molecular weight hydrophilic and hydrophobic solutes. This provides a glimpse of the current approaches of the porous media perspective of transdermal transport, and presents access to a list of these models.

The second half of the chapter covers the subject of electroporation (a structure altering electrokinetic means of transport in which the skin is exposed to a series of intense, short-duration electric pulses). A review of experimental findings and observations regarding the electrically induced creation of local transport regions (LTRs) is presented. This is done to familiarize the reader with clinical and experimental results that link the thermal aspects of the applied pulse to the thermal behavior of the skin. Furthermore the review points out a thermodynamic basis to the creation of these local sites of increased permeability. A description of various methods used to describe electroporation of the skin (both empirical and mechanistic) is also provided.

9.2 Transdermal Drug Delivery

Traditional systemic drug delivery methods often result in low efficacy and unintentional treatment of healthy tissue. Oral intake may result in drug denaturation in the body's gastro-intestinal tract: drugs consisting of water-soluble molecules may degrade in the digestive tract before reaching intended locations of treatment. Injection of drugs into the circulatory system results in a lower selectivity level of the region of drug influence. For drugs with a high toxicity, for example, those used in chemotherapy, systemic delivery is associated with a high risk of unintentional damage to healthy tissues. Targeted drug delivery refers to any method of drug administration designed to ensure and enhance a drug's transfer rate to specific location (rather than the entire body as in systemic delivery). Transdermal drug delivery is one method that allows for localized delivery by transporting the drug directly through the skin to the drug's desired target.

This chapter begins with a general physical description of the skin and its barrier properties. This is followed by a review of current porous–media-based descriptions of passive transdermal transport. The second half of this chapter focuses on one such method called electroporation in which the skin is exposed to short (µs to ms) pulses that act to disrupt the skin's barrier structure to allow increased permeability. Then the process of electroporation is detailed to describe the primary and secondary effects of exposing the skin to these intense electric pulses. Experimental skin electroporation findings are given to add insight into the mechanics underlying current hypotheses.

The following section is devoted to providing a basic physical description of the skin and the obstacles that must be overcome to transport drugs through its impermeable nature.

9.3 The Skin as a Composite

Although the skin is the body's first defense against infection, in general, the skin should not be treated as a homogenous medium. Even in a very basic description, human skin is made of multiple constituent composite layers, each providing specific function with varying thicknesses (see Figure 9.1).The reader unfamiliar with the physiology of the skin may find the introductory text (Millington and Wilkinson 1983) helpful.

The outer skin layer, the epidermis, (0.05–1.5 mm), is without vasculature and acts as a protective barrier preventing molecular transport. The epidermis is basically an assembly line of viable living cells at its base that transition to flat dead cells at its outer surface. The outer layer of the epidermis provides the skin's primary barrier to transport.

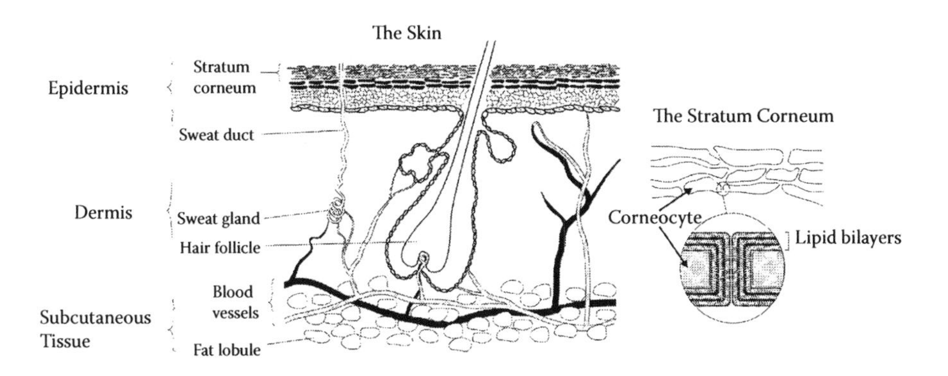

FIGURE 9.1
The skin.

Below the epidermis is the highly vascular inner skin layer, the dermis (0.3–3 mm). An important characteristic of the dermis is the large network of capillaries with high blood flow rates exceeding several times that of metabolic requirements. The primary reason of such high-perfusion rates in the epidermis is to regulate body temperature: perfusion rates decrease to conserve body heat and increase to cool the body. The dermis is also characterized by its high collagen content, which provides the skin with its structural support.

Below the dermis lies a thermally insulative fatty layer that acts to conserve body heat. The skin, at most sites on the body, is perforated by appendageal pathways in the form of sweat glands and hair follicles.

In any model of *in vivo* skin (living, as in skin attached to a living body) that focuses on thermoelectric response of the skin, it is important to consider the adequacy of the model's composite layer description.

9.4 Stratum Corneum and the Lipid Barrier

The thin (10–50 μm) outermost layer of the epidermis is called the *stratum corneum* (*SC*). This is the layer most resistive to transport through the skin. Although the *SC*'s barrier function is vitally important to healthy skin (by keeping harmful molecules from passing into the skin and providing an initial defense against infection), it is this high resistance to permeability that presents a major obstacle for successful transdermal delivery. This layer is dynamic and is constantly renewing itself: as cells are degraded by age and friction on the outer *SC*, they are replaced by new ones originating at the base of the *SC*. The renewal process takes approximately 30–50 days.

The *SC* is composed of 15–20 layers of corneocytes (flat dead cell shells), which are interconnected by a lipid lamellar bilayer structure in a crystalline-gel phase. Transport circumventing the barrier function of the *SC* is primarily associated to occur within the lamellar lipid structure of the *SC* (Bouwstra 2003; Madison 2003); thus it is the lipid structures within the interstitial space (space between the corneocytes) that are important for transdermal delivery success. The thin *SC* corneocytes-lipid matrix architecture may be conceptualized as a brick and mortar structure in which the highly impermeable corneocytes are represented by the brick and the slightly less impermeable lipid sheets are represented by the mortar-filled spaces between the lipid sheets (see Figure 9.1).

9.5 Nondestructive Transport Modeling: The SC as a Porous Medium

Nondestructive transdermal drug delivery methods that rely on pure diffusion or very low voltage electrically enhanced diffusion (iontophoresis) are typically associated with solutes of low molecular weight. The basic equations governing the transport of solute through the skin are for pure diffusion:

$$\frac{\partial C}{\partial t} = \nabla \cdot (D_{\text{eff}} \nabla C) \tag{9.1}$$

and, for nondestructive electrically enhanced diffusion:

$$\frac{\partial C}{\partial t} = \nabla \cdot (D_{\text{eff}} \nabla C) + \nabla \cdot (m_{\text{eff}} C \nabla \phi) - U_{\text{eff}} \cdot \nabla C \tag{9.2}$$

where C is the solute concentration, D_{eff} is the effective diffusion coefficient, ϕ is the electric potential resulting from the applied electric field, m_{eff} is the effective electrophoretic mobility coefficient, which describes the ability of the field to move the solute, and U_{eff} is the effective electroosmotic flow.

The challenge to researchers has been the way to use the porous medium approach to define the effective coefficients (diffusion, electrophoretic mobility, and electroosmotic flow) within the skin based on the architecture and the physics of the skin. This section describes some of the methods that are currently being used to describe these coefficients—primarily with respect to pure diffusion within the *SC*.

It is important to note that the following examples are almost exclusively modeled as *in vitro* situations in which the skin acts as a barrier membrane between a donor and receiver reservoir. Although these do not necessarily closely resemble transport through living skin into the subcutaneous tissue, important information can be extracted from these types of models and experiments. The permeability, P, is defined as the steady dimensionless flux of solute transported though the *SC*:

$$P = D_{\text{eff}} \frac{1}{C_I} \frac{C_I - C_O}{L} \tag{9.3}$$

where C_I is the concentration of solute on the donor side of the *SC*, C_O is the concentration of solute on the receiver side of the *SC*, and L is the thickness of the *SC*.

A particular solute's permeability within the *SC* is often described by a dependency on chemical and geometric parameters, and it is these parameter values that are used to validate a model's accuracy by comparing the results with those of the experiment (often a Franz diffusion cell).

There are two mechanistic approaches used to describe the permeability within the *SC* architecture. One takes into account the known structure of the lipid–corneocyte matrix, while the other incorporates more traditional membrane modeling strategies by a representation of the *SC*, which is permeated by microscopic tortuous pores.

9.5.1 Brick and Mortar Models

The underlying concept behind the brick and mortar model is that the lipid-filled space is permeable, while the corneocytes are either completely or highly impermeable to the solute. This is the basic idea upon which most porous medium representation of the SC has been developed. The representation of the SC architecture consisting of a symmetric brick and mortar model was introduced by Michaels et al. (1975) in which the interstitial horizontal lengths between vertical spaces are all of the same length (see Figure 9.2(a)). Variations of this model have been proposed to better represent the SC geometry. A nonsymmetric concept of the SC has been introduced by Johnson et al. (1997), which more accurately describes the lateral diffusive path by using an offset that is defined as the ratio of long to short lateral diffusion path lengths (see Figure 9.2(b)). A trapezoidal representation of the corneocyte is presented by Wang et al. (2006), which attempts to more realistically represent the SC microstructure and includes diffusion through the corneocytes. The study finds that this route does play a minor role. Further developments of this model include those of Frasch and Barbero (2005) and Barbero and Frasch (2005), which use irregular description of the SC architecture based on actual micrograph of mouse SC with the commercial software package ANSYS to more accurately capture transient transdermal diffusion.

An excellent presentation and comparison of the models based on the geometry of Figures 9.2(a) and 9.2(b) is presented by Kushner et al. (2007a). This study derives a two-tortuosity model that accounts for the total lipid-filled space as well as lateral diffusion pathways along. The study compares the model results with the more traditional model descriptions of porosity and tortuosity (see Table 9.1). In the descriptions of Table 9.1, N refers to the number of layers within the SC, and the other geometric parameters of the SC representation are shown in Figures 9.2(a) and 9.2(b). The corneocyte offset, ω, is defined as the ratio of long to short lateral paths $\omega = d_L/d_S$. The

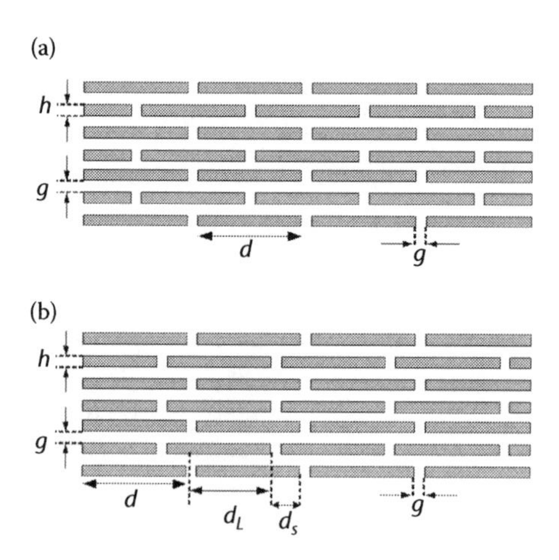

FIGURE 9.2
Brick and mortar representations of SC geometry: (a) symmetric, (b) asymmetric.

TABLE 9.1
Description of *SC* Tortuosity and Porosity Based on Brick and Mortar Structures of Figure 9.2.

Reference	Porosity (ε)	Tortuosity (τ)
Michaels et al. (1975)	$\frac{2g}{d}$	$1+\frac{\frac{d}{h}+\frac{g}{h}}{2\left(1+\frac{g}{h}\right)}$
Cussler et al. (1988)	$\frac{g}{g+d}$	$\frac{Nh+(N-1)\left(\frac{d+g}{4}\right)}{L}$
Lange-Lieckfeldt and Lee (1992)	$\frac{\left(\frac{g+d}{2}\right)^2-\left(\frac{d}{2}\right)^2}{\left(\frac{d}{2}\right)^2}$	$\frac{L+(N-1)\left(\frac{d+g}{2}\right)}{L}$
Johnson et al. (1997)	$\frac{g}{(g+d)}$	$\frac{Nh+(N-1)\left(\frac{\omega}{(1+\omega)^2}\right)d}{L}$
Kusner et al. (2007)	$\frac{g}{(g+d)}$	$\tau_{\text{flux}}=\frac{Nh+(N-1)g+(N-1)\frac{\omega}{(1+\omega)^2}d}{Nh+(N-1)g}$ $\tau_{\text{volume}}=\frac{Nh+(N-1)g+(N-1)d}{Nh+(N-1)g}$

Source: Adapted from Compendium provided in Kushner et al. 2007.

study compares the model predictions of D_b, the solute–*SC* diffusion coefficient, and K_b, the solute–lipid bilayer partition coefficient (a measure of the solute's solubility within the lipid bilayers). To do this, the permeability of equation (9.3) is defined as

$$P=\left(\frac{\varepsilon}{\tau}\right)\frac{D_bK_b}{L} \tag{9.4}$$

where ε and τ are, respectively, the porosity and the tortuosity of the intercellular paths (lipid-filled spaces between the corneocytes).

Should the porosity and tortuosity be known, it is then possible to find accurate values for the D_b, the solute–*SC* diffusion coefficient, and K_b, the solute–lipid bilayer partition coefficient. These values are found by combining analytic solutions with experimental results (Kushner et al. 2007a). This is done using experimental transient SC diffusion flux data in combination with the analytic solution to the one-dimensional transient flux through the SC barrier. The following equations depicting solute flux are developed by Kushner et al. (2007a), which incorporates the porosity and tortuosity for single tortuosity models:

$$Q(t) = (\varepsilon K_b C_2 \tau L)\left[\frac{D_b}{(\tau L)^2} - \frac{1}{6} - \frac{2}{\pi^2}\sum_{n=1}^{\infty}\frac{(-1)^n}{n^2}\exp\left(\frac{-D_b n^2 \pi^2 t}{(\tau L)^2}\right)\right] \tag{9.5}$$

and for the two-tortuosity model:

$$Q(t) = (\varepsilon K_b C_2 \tau_{\text{volume}} L) \times \left[\frac{D_b}{\tau_{\text{flux}}\tau_{\text{volume}}L^2} - \frac{1}{6} - \frac{2}{\pi^2}\sum_{n=1}^{\infty}\frac{(-1)^n}{n^2}\exp\left(\frac{-D_b n^2 \pi^2 t}{\tau_{\text{flux}}\tau_{\text{volume}}L^2}\right)\right] \tag{9.6}$$

The study finds that the two-tortuosity model most closely predicts D_b and K_b in a comparison between the various models listed in Table 9.1, and compares well with experimental results when predicting D_b and K_b of hydrophobic permeants.

It should be noted that the models listed in Table 9.1 can produce varying results. For example, using the following input values suggested for normal SC by Kushner et al. (2007a): $N = 15$, $h = 1$ µm, $g = 0.1$ µm, $w = 3$, $d = 40$ µm,the values of ε/τ of the studies are as follows: 2.60×10^{-4} (Michaels et al. 1975), 2.93×10^{-4} (Cussler et al. 1988), 2.78×10^{-4} (Lange-Lieckfeldt and Lee 1992), 3.92×10^{-4} (Johnson et al. 1997), and $\varepsilon/\tau_{\text{volume}}$ of the (Kushner et al. 2007a) two-tortuosity model has a value of 3.37×10^{-4}.

9.5.2 Models Based on Lipid Microstructure: Free Volume Diffusion

For very small hydrophobic solutes ($MW < 400$ Da), it is believed that diffusion transport takes place within the lipid bilayers where the fluctuations of the bilayer lipids provide transport pathways through "free volume pockets." These fluctuations produce free pockets with radii estimated at 4 Å that exist for 1.6 µs, while the time for the molecule to "jump" through the free volume is on the order of ns (Mitragotri 2003). The steady state free volume permeability is defined as

$$P = \frac{1}{\tau^*}\frac{D_b K_b}{L} \tag{9.7}$$

The representation here of the tortuosity, τ^*, includes contributions from the diffusion path length within the intercellular spaces as well as contributions from the area fraction occupied by the lipid bilayers (a value of $\tau^* = 3.6$ cm) is used in this study as referenced from Johnson et al. (1997).

Using scaled particle theory based on statistical mechanics of lipid chains within the bilayers, expressions of the partition coefficient and solute diffusion coefficient within the lipid bilayers are presented by Mitragotri (2002). The solute partition coefficient within the *SC* is related to the more easily experimentally measured water octanol partition coefficient by

$$K_b = K_{O/W}^{0.7} \tag{9.8}$$

The study provides a detailed formulation of the average diffusion coefficient within the *SC* by assuming a spherical solute of radius r, which diffuses through a continuous spherocylindrical path that is free of lipid molecules. By integrating over the lipid bilayers using structural parameter values based on dipalmitoyl phosphatidycholine and the Wilke–Chang model of solute diffusion in an isotropic hydrocarbon, the study arrives at the following representation of diffusion coefficient:

$$D_b = 2 \times 10^{-5} \exp\left(-0.46r^2\right) \tag{9.9}$$

When solute molecular weight, MW, is known and solute molecular radius, r, is unknown, the radius may be approximated by:

$$4/3\pi r^3 = 0.9087MW \tag{9.10}$$

The permeability of equation (9.7) can now be represented as

$$P = 5.6 \times 10^{-6} K_{O/W}^{-0.7} \exp\left(-0.46r^2\right) \tag{9.11}$$

This equation has been found to closely reflect experimental results exceeding empirical model accuracy for small hydrophobic solutes (Lian et al. 2008).

9.5.3 Aqueous Pore-Membrane Models

For hydrophilic solutes an aqueous pore model has been proposed in which it is postulated that the solute travels through water-filled spaces within the bilayers. The basis behind the aqueous pore model is that there exist nm-sized voids within the *SC* that exist as a result of defects within the lipid bilayers. These defects or imperfections "manifest themselves as separation of grain boundaries, lattice vacancies, multi-molecular voids due to missing lipids, or defects caused by steric constraints placed by keratinocytes on intercellular lipid bilayers" (Mitragotri 2003). The aqueous pore-membrane model represents these defects by a system of cylindrical tortuous pores (see Figure 9.3) that traverse the barrier layer of the skin and are described in terms of radial size and tortuosity, thereby allowing the *SC* to be described as a porous membrane.

Although indirect physical experimental evidence supports the presence of these water-filled defect pathways, it should be recognized when implementing

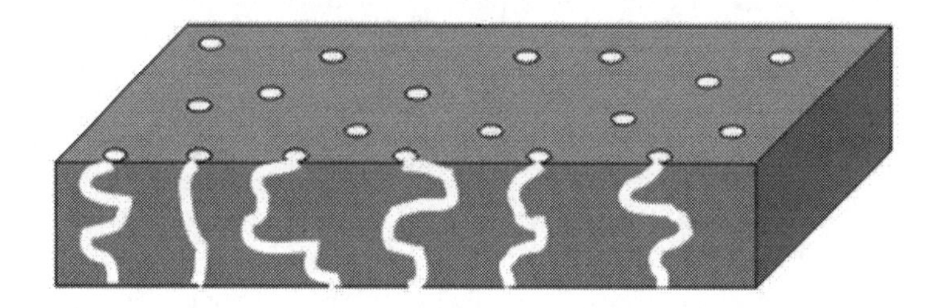

FIGURE 9.3
Aqueous pore-membrane representation of the *SC*.

this method that there is not conclusive evidence that this route actually exists:

> "While interpreting the model predictions it should be remembered that no direct physical evidence has been proposed on the existence of pores within the skin." (Tezel and Mitragotri 2003)
>
> "While has been no direct evidence for the physical existence of aqueous pore channels in the skin to date, some researchers have claimed that the presence of an aqueous pore pathway can be inferred from evidence of connected lacunar domains within the lipid lamellar bilayers." (Kushner et al. 2007b)

The concept has been used to describe pure diffusion through the *SC* (Mitragotri 2003; Kushner et al. 2007b; Tezel et al. 2003), or electrically enhanced diffusion (nonstructure altering) in which electrophoretic and electroosmotic contributions are considered (Li et al. 2001, 2004).

This class of solute transport description relies on a hindrance factor, which is based on a spherical solute transported through a fluid-filled cylindrical pore. In the presence of an applied electric field, the flux can be represented by the modified Nernst–Plank equation:

$$J = \frac{\varepsilon}{\tau}\left\{-H'D\left(\frac{dC}{dx} - \frac{CzF}{RT}\frac{d\phi}{dx}\right) + W'vC\right\} \tag{9.12}$$

Here the passive diffusion component across the membrane is represented by the term $D\frac{dC}{dx}$. The electrophoretic component is $\frac{CzF}{RT}\frac{d\phi}{dx}$, where z is the charge of the ionic permeant (solute), F is Faraday's number, R is the gas constant, T is the absolute temperature, and ϕ is the electric potential. The electroosmotic-induced convective flow is vC, where v is the average flow velocity. Here the term H' is the diffusion hindrance factor, which is associated with Brownian motion and electrophoretic migration within the resistive membrane, while the term W' is a convective flow hindrance factor associated with the bulk electroosmotic convective flow within the membrane. The hindrance factors are defined with a dependency on the ratio of ionic solute radius to the preexisting pore radius, $\lambda = r/r_P$.

Often, for simplification, studies rely on a pore model in which the pores are represented by a single average pore radius value (Higuchi et al. 1999; Li et al. 2004). For instance, Higuchi et al. (1999) assumed that the lipid bilayers membranes of the SC have small pores of preexisting radii in the range $10 \leq r_P \leq 20\,\text{Å}$. Under this approximation, direct descriptions of the diffusive hindrance and hydrodynamic hindrance are applicable (Deen 1987). Two regimes based on the magnitude of this ratio are commonly put into practice.

For medium and small molecule transport ($\lambda < 0.4$), hindrance factors are represented by

$$H(\lambda) = (1-\lambda)^2 \left(1 - 2.104\lambda + 2.09\lambda^3 - 0.948\lambda^5\right) \tag{9.13}$$

$$W(\lambda) = (1-\lambda)^2 \left(2 - (1-\lambda)^2\right)\left(1 - 0.667\lambda^2 - 0.163\lambda^3\right) \tag{9.14}$$

while large solute particles ($0.4 \leq \lambda$) have the associated hindrance factor relations:

$$H(\lambda) = \frac{6\pi (1-\lambda)^2}{K_t} \tag{9.15}$$

$$W(\lambda) = \frac{(1-\lambda)^2 (2 - (1-\lambda)^2) K_s}{2K_t} \tag{9.16}$$

where

$$\begin{pmatrix} K_s \\ K_t \end{pmatrix} = \frac{9}{4}\pi^2 \sqrt{2}(1-\lambda)^{-5/2} \left[1 + \sum_{n=1}^{2} \begin{pmatrix} a_n \\ b_n \end{pmatrix} (1-\lambda)^n \right] + \sum_{n=0}^{4} \begin{pmatrix} a_{n+3} \\ b_{n+3} \end{pmatrix} \lambda^n \tag{9.17}$$

and

$$\begin{array}{llll} a_1 = -1.217; & a_2 = 1.534; & a_3 = -22.51; & a_4 = -5.612; \\ a_5 = -0.3363; & a_6 = -1.216; & a_7 = 1.647; & b_1 = 0.1167; \\ b_2 = -0.04419; & b_3 = 4.018; & b_4 = -3.979; & b_5 = -1.922; \\ b_6 = 4.392; & b_7 = 5.006 & & \end{array}$$

However, in the case of a distribution of pore radii within the skin, a distributive function, $\gamma(r_P)$, is applied and the hindrance factor is defined as

$$\bar{H} = \int_0^\infty \gamma(r_P) H(\lambda)\, dr_P \tag{9.18}$$

The distributive function is based on the exponential form (Tezel et al. 2003)

$$\gamma(r_P) = \chi \exp(-\varsigma r_P^2) \tag{9.19}$$

The constants χ and ς may be derived from singular experiment, for example, the values of these constants as measured for pig *SC* are given as follows: $\chi = 0.024$; $\varsigma = 0.00045$ (Tezel et al. 2003). The distributive function is explicitly defined as the normalized distributive function in which $\chi = 2/\sigma\sqrt{2\pi}$; $\varsigma = 2\sigma^2$, where σ is the standard deviation, whose value is also determined experimentally (Kushner et al. 2007b).

The methods that rely on nondestructive transport (low voltage electrically assisted and pure diffusion) are successful only for small molecular transport. To transport larger drugs into the skin, researchers have focused on altering the barrier function of the *SC*. To facilitate transdermal transport of "large molecule," the barrier structure of the *SC* and the lamellar lipid bilayers must be altered. In the following section a method called electroporation, which relies on electrical and thermal destruction of the *SC* barrier function, is introduced.

9.6 Skin Electroporation

Skin electroporation is an approach used to enhance localized transdermal transport in which the skin is exposed to a series of electric pulses (Regnier et al. 1999; Denet et al. 2004). Typically, in skin electroporation a section of skin coated with an applicator gel is pinched between two electrodes, which deliver the electric pulses to the skin (see Figure 9.4[a]). Electroporation of the *SC* barrier dramatically increases the permeability and electrical conductivity of the *SC* by creating microscopic pores through which agents are able to pass through the outer barrier (Pliquett et al. 1995; Prausnitz 1996; Vanbever and Preat 1999; Pliquett and Gusbeth 2000). Electroporation pulses can be classified in one of the two regimes: short pulse and long pulse, and because the physical effects of the two regimes differ greatly, the choice of which regime to use is dependant largely on the drug size (Denet et al. 2004).

9.6.1 Short Pulse (Nonthermal)

Small molecules require lower permeability increases, and once this minimum *SC* permeability is met, solutes of low molecular weight can access deeper skin layers purely by diffusion. Electroporation pulse times that are classified as short duration are typically $\leq 100\,\mu s$. In this regime nonthermal skin

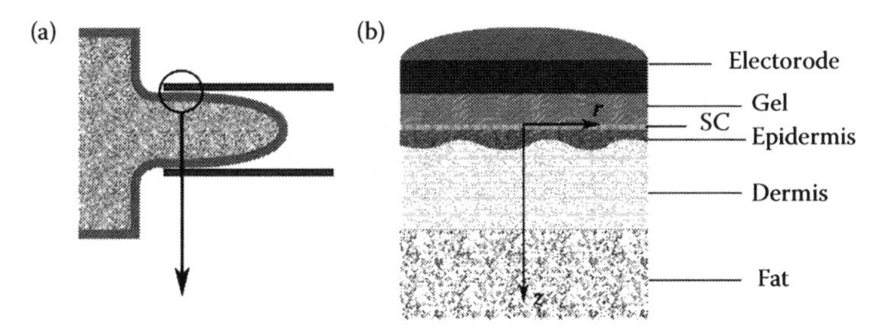

FIGURE 9.4
Electroporation overview: (a) skin fold, (b) close-up of composite representation of skin electroporation.

electroporation takes place in which nm-sized pores develop within the lipid bilayer membranes (and under the right conditions, through the corneocytes). Although this phenomenon has been documented in countless experimental studies (both *in vitro* and *in vivo*), the exact mechanism behind the sudden development of these pores is not fully understood. The general theoretical description of this process has been developed for single lipid bilayers and is believed to be related to pulse magnitude and lipid–lipid interactions. It is understood that within an individual lipid bilayer membrane, electropores are created initially hydrophobic: at this early stage of pore formation the edges of the pore consist of exposed hydrophobic lipid tails. As the pulse progresses, the electroporation pore makes a transition to hydrophilic: the hydrophilic lipid heads position themselves to the edge of the pore (thus covering the hydrophobic tails). This formation of individual nm-sized hydrophilic pores within the individual lipid bilayer membrane is electroporation in its strictest sense. When approaching electroporation of the skin it is important to consider that the individual electropores do not cross the entire *SC*, but only single bilayers. The concept of single lipid bilayer electroporation should be contrasted with skin electroporation. The *SC* cannot be represented by a single bilayer because it consists of a lamellar network of lipid bilayers (about 100) that surround the corneocytes. Thus, for drug molecules to pass through the *SC*, they must pass through an electropore in each of the 100 bilayers. Only small drugs are capable of being transdermally transmitted through the small, tortuous pathways created during short-pulse electroporation.

Because the pulse duration is short, there is no significant rise in temperature associated with Joule heating. This means that the *SC* lipid structure does not experience a thermal moderate scale fluidization at the mm scale, although it has been postulated that nm-scale heating contributes to what is considered to be nonthermal electroporation (Pliquett et al. 2008). In the following section attention is directed to transdermal transport of larger molecules by longer pulses that have been associated with large temperature rises.

9.6.2 Long Pulse (Thermal)

For larger molecule transport, longer duration pulses (up to several hundred ms) are required. Longer pulses make possible two important secondary effects: electrophoresis and Joule heating.

Electrophoresis is the electric–field-induced motion of charged particles through a solution. Even with increased permeation of the *SC*, larger molecules experience greater viscous resistance to transport through the tortuous *SC*. Because simple diffusion is not adequate to overcome this barrier, the electric field provides the force to overcome resistive viscous forces and drive the large charged molecule through the layers of the *SC*. Thus, longer pulsing times are used to increase the distance that the electrophoretic forces move the solute.

Also important to skin permeation studies are the Joule–heating-induced local temperature rises. It is understood that the effects of long duration skin electroporation pulses (>25 ms) are Joule heating and the resulting localized temperature rises, which result in thermotropic phase transitions within the lipid bilayer matrix of the *SC* (Denet et al. 2004). At temperatures around 70°C the barrier function of the *SC* is dramatically reduced (Golden et al. 1987; Potts and Francoeur 1990) because the *SC* lipid lamellar structure experiences a fluidizing phase transition (Golden et al. 1987; Cornwell et al. 1996; Al-Saidan et al. 1998). It is recognized that localized Joule heating associated with electroporation contributes to increased permeability of the *SC* by lipid chain melting (Pliquett et al. 1998, 2005; Vanbever et al. 1999; Denet et al. 2004). The temperature rises are proportional to Joule heating, Q_J, and pulsing times, τ_P, by the relation

$$\Delta T \propto \frac{\tau_P Q_J}{\rho C_p} \tag{9.20}$$

The thermal alterations associated with these longer pulses result in μm- to mm-sized regions of increased permeability. Early researchers of skin electroporation have used the phrase LTR to describe these concentrated electrically induced regions of increased permeability (Pliquett et al. 1996, 1998). The development of LTRs is always associated with thermal effects (Pliquett et al. 2008), and it is stressed that this increase in permeability due to Joule heating should not be mistaken for pure electroporation (which does not require large-scale temperature rises.) To the reader unfamiliar with skin electroporation, the distinction between short- and long-pulse electroporation is stressed; the reader must also recognize the distinction between the nonthermal primary nm-sized pores that occur within a single bilayer and the μm- to mm-sized LTRs that pass through the entire *SC* and are associated with Joule heating. Also the reader should not confuse the term *SC* thermal LTR formation with actual macroscopic voids within the skin. This type of electroporation of the skin provides a measure of disruption of the *SC* architecture (the lamellar lipid sheets), and it is this disruption that allows for increased permeability. These disruptions do not generally occur homogenously throughout the skin, but at localized spots. Skin electroporation causes local regions of high permeability on a membrane that in its unperturbed state acts as a barrier. To reiterate: skin electroporation does not create large voids through the *SC*, but local regions of high permeability.

9.6.3 LTR: Experimental Observation

Experimentally it has been shown that in some cases the high-temperature contours associated with Joule heating may originate near skin appendages (sweat glands or hair follicles) and as the pulse is applied the LTRs propagate (spread) radially outward along the *SC* (Pliquett and Gusbeth 2000). Figure 9.5 depicts a close-up of the *SC* near a preexisting pore. As the lipid-phase transition temperatures are attained and move radially outward, the lipid sheets become fluidized as in the lower panel of Figure 9.5. Depending on pulse intensity, the scale of the distances covered by these high-temperature fronts is on the order of 100 μm occurring on a timescale of 10–100 msec (Vanbever et al. 1999). As this heat front moves through the *SC*, it supplies the energy to activate lipid-phase transitions. Thus the lipid-phase transition (barrier function breakdown) can be thought of as a propagating heat front through the *SC*. Adding support to this idea are the results of *in vitro* studies in which it is shown that within the front the electrical and mass permeabilities may be many orders of magnitude higher than outside of this front (Vanbever et al. 1999; Pliquett and Gusbeth 2004).

Direct evidence of the localized moving heat front and localized regions of transport is not available *in vivo* (within living tissue) studies, although remarkable findings are given from *in vitro* studies that are conducted in which human *SC* is removed and electroporated under observation (Pliquett et al. 1995, 2005; Prausnitz 1996; Vanbever and Preat 1999; Vanbever et al. 1999; Pliquett and Gusbeth 2000, 2004). Using fluorescence microscopy and time-resolved freeze fracture electron microscopy during applied pulses of 80 V at various durations, structural changes occurring within the *SC* (human and porcine) are found as a result of localized Joule heating, and it is shown that these structural changes are highly localized, taking up less than 1% of the skin's surface area (Pliquett et al. 2005). A fluorescence microscope matched to a very sensitive camera is used to capture the progression of the growing heat front associated with Joule heating (Pliquett and Gusbeth 2000). For pulses shorter than 1 msec, Joule heating is negligible; however, when pulses of 200 V intensity and 200 msec length are used, striking thermal phenomena are captured. In fact, in that study, the moving thermal front is captured in a series of microscopic photographs at 40, 80, 160, and 240 msec by tracking isotherms of 40°C and 60°C that grow radially outward from a point of origin. By 240 msec the front diameter has grown to 0.4 mm. A comparison between local

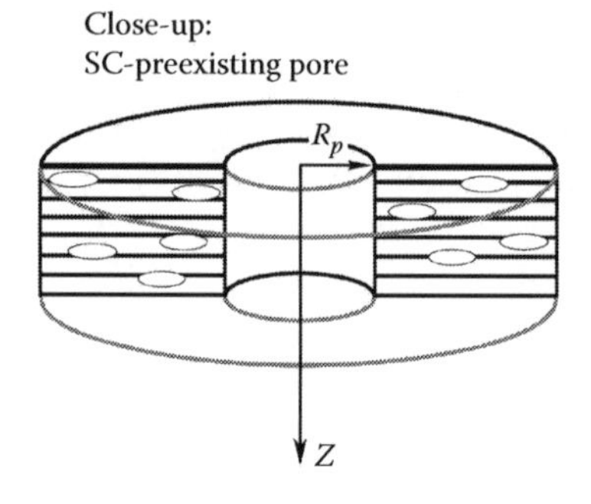

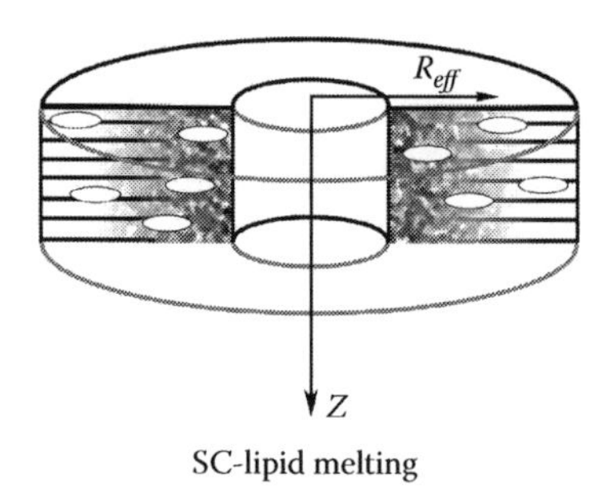

FIGURE 9.5
Thermal destruction of lipid architecture in the vicinity of preexisting pathway.

site transport and pulse protocol was made by Vanbever et al. (1999). Here the excised skin is placed between two reservoirs (donor and receptor) during electroporation, and using fluorescence microscopy relations between several pulse parameters and size and number of the LTRs are recorded. No regions were recorded for 20 pulses at 100 V and of duration 100 msec, although by tripling the duration to 300 msec large, rare regions up to 0.8 mm in diameter are found. A train of 20 pulses at 200 V and 100 ms (doubling the voltage) results in a few pores of diameter between 0.4 and 0.8 mm. Increasing both the voltage and duration to 300 V and 300 msec results in 0.2–0.6 mm diameter pores occurring more often at about 6 per 0.1 cm^2.

9.6.4 Lipid Thermal Phase Transitions

One of the most important characteristic phenomena of longer-pulse-type skin electroporation are the microscale thermal effects that also help change the structure of the lipid sheets. While in pure diffusion models of transdermal transport these thermal effects are not really considered, these influences are

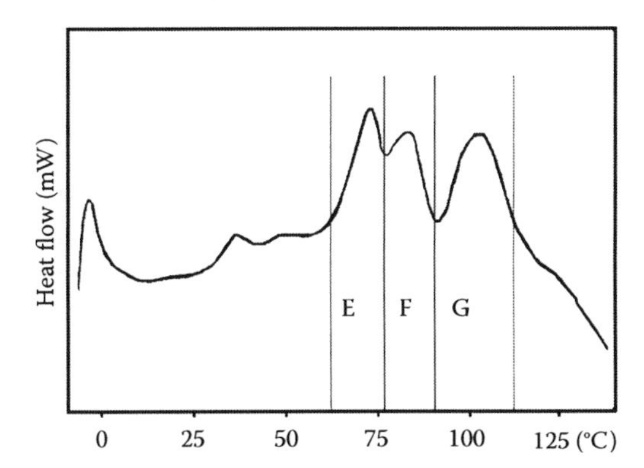

FIGURE 9.6
Representative heat versus temperature curve of *SC* lipids in thermal phase transition.

important in a thermodynamic-based understanding of the *SC*. Here we provide a short description of lipid thermal behavior to better understand the physics underlying the thermal influences during the electroporation LTR development.

The lipid structure is known to become destabilized at elevated temperatures. It is generally accepted that there exist four main endothermic transitions in the *SC* in the temperature range 40°C–130°C, which have been independently confirmed by differential scanning calorimetry (Golden et al. 1986, 1987; Cornwell et al. 1996; Al-Saidan et al. 1998; Tanojo et al. 1999). Figure 9.6 shows a representative heat versus temperature in which the *SC* endothermic transitions are evident as the peaks and have been identified using the convention followed by Silva et al. (2006a,b). Of great interest to thermally assisted destabilization of the *SC*'s lipid barrier function is phase change E, which has primarily been attributed to the disordering of the lamellar lipid phase (Cornwell et al. 1996). Phase change E has been documented in numerous differential scanning calorimetry studies at endothermic peak temperatures from 65°C to 72°C (Silva et al. 2006b). Using X-ray diffraction microscopy it has been shown that the lipid lamellar structure in the *SC* is evident up to temperatures about 60°C and then disappears within 10°C (White et al. 1988; Bouwstra et al. 2003). Polarized light thermal microscopy of lipids extracted from the *SC* directly shows an overall fluidization of lipid structures at 60°C (Silva et al. 2006a). Further evidence of lipid chain melting and increased permeation at this phase change is provided in the reports by Golden et al. (1987) and Potts and Francoeur (1990), where it is shown that the *SC* becomes more permeable to water flux abruptly at 70°C.

The remaining two high-temperature phase transitions, F and G, are of less direct importance to skin permeability. The enthalpic peaks of phase change G occur at temperatures above 90°C and have been associated with protein denaturation (Cornwell et al. 1996; Tanojo et al. 1999; Silva et al. 2006a). Secondary lipid melting has been associated with phase change, F, at temperatures around 80°C and has been linked to lipids covalently bonded to corneocytes (Cornwell et al. 1996; Tanojo et al. 1999).

So it is clear then that there exists a specific temperature range within which the lipid barrier architecture is destroyed, and that the lipid thermal behavior should be considered in any electroporation study in which the pulse parameters are such that significant temperature rises are possible.

9.7 Skin Electroporation Models (Nonthermal)

The following section displays the models currently used to depict electroporation. The reader should keep in mind that the physics underlying electroporation (even of a single bilayer) are not completely understood and that the models used to describe electroporation are experimentally based (either empirically or hypothetically). Whatever the models basis, its purpose it to relate permeability (measure of porosity) to electroporation pulse parameters, typically applied voltage, V_{app}, and pulse time, τ_P. With the structural changes of electroporated *SC* come large increases in electrical conductivity of the *SC*, σ_{SC}. Because electrical characteristics (conductivity, current, and voltage drop) can be easily and precisely measured, they have traditionally been used as parameters to show the transient increases in permeability. Furthermore, because the current electroporation models have origins tied to experimental findings (which monitor electrical behavior), it should come as no surprise that the numeric and theoretic models used to describe skin electroporation often use electric parameters to interpret the degree of permeability.

9.7.1 Single Bilayer Electroporation Modeling

We begin the discussion of using *SC* electrical behavior to describe electroporation with a description of single bilayer pore formation (recall that the *SC* architecture consists of about 100 lamellar sheets made of these single bilayer membranes.) The single bilayer models track the creation and growth of small pores (<10 nm) in which the increase in permeability is based on the local potential drop across the membrane, V_m. To model single bilayer electroporation, investigators keep track of each individual pore's transient growth or decay. It is not difficult to imagine what a task this would become when dealing with 100 bilayers interconnected by the corneocytes.

The models used to describe single lipid bilayer membrane electropore creation are usually some extension of the Smoluchowski equation that links pore creation and growth to V_m and lipid–lipid interaction energies within the bilayer.

It is believed that most pores begin as tiny ($r < 1\,\mathrm{nm}$) hydrophobic pores. If these pores are to last, they must grow to become hydrophilic by allowing the hydrophilic heads to fill in the pore sides. However, many newly created pores are so small that they are quickly destroyed by lipid molecular fluctuations. If the pore radii are smaller than some minimum pore radius $r < r^*$ they may safely be neglected as they are quickly destroyed by these lipid fluctuations. If lipids are created at a radius $r \geq r^*$ then they are able to transition to sustainable hydrophilic and grow slightly to a minimum sustainable hydrophilic radius, r_m.

In studies by Smith et al. (2004) the transient evolution of electropores is described in such a way as to neglect the creation of the quickly destroyed pores of $r < r^*$:

$$\frac{dN}{dt} = \alpha\left(\exp[V_m/V_{ep}]^2 - \frac{N}{N_O \exp[r_m/r^*]}\right) \tag{9.21}$$

where N is pore density, N_O is equilibrium pore density, V_{ep} is characteristic electroporation voltage, and α is a rate coefficient.

The important aspect of equation (9.21) to keep in mind is that the rate at which pores are created is exponentially related to the square of the transmembrane voltage.

This method then keeps track of the time rate of growth of each of the n pore radii by the equation

$$\frac{dr_j}{dt} = \frac{D}{kT}\left(\frac{V_m^2 F_{\max}}{1 + r_h/(r + r_t)}\right) + 4\beta\left(\frac{r^*}{r}\right)^4 \frac{1}{r} - 2\pi\gamma + 2\pi\sigma_{\mathrm{eff}} r, \quad j = 1, 2, \ldots, n \tag{9.22}$$

where D is the diffusion coefficient; k is the Boltzmann constant; T is the absolute temperature; $F_{\max}$, r_h, r_t, β; and γ are constants; and σ_{eff} is a geometric function relating the membrane tension to the pore size.

In the rate equation, equation (9.22), the first term accounts for the transmembrane contribution to bilayer energy, and the second and third and fourth terms denote geometric and spatial membrane effects (repulsion of lipid heads, membrane tension, and the pore perimeter energy, respectively). The dominating influence of pore growth during the application of the electroporation pulse is the first term, which is a function of the square of the transmembrane potential.

Typically, in membrane electroporation studies, lateral variations of permeability are neglected, and only transient effects are modeled. In the study by Smith et al. (2004) the membrane is assumed to be a homogenous perfusion and the transient behavior of the transmembrane potential is approximated by the transient circuit representation:

$$C\frac{dV_m}{dt} + \left(R_m^{-1} + R^{-1}\right) \cdot V_m + \sum_{j=1}^{K} \frac{V_m}{R_i + R_P} = \frac{V_O}{R_P} \tag{9.23}$$

Clearly this is a simplified representation of the potential distribution and does not account for any lateral variations within the membrane. It should be noted that in single-cell electroporation studies, the spatial variation of transmembrane potential and pore evolution have been captured by calculating the electrical potential of the solution domain on both sides of the membrane (Stewart et al. 2004; Krassowska and Filev 2007).

Perhaps the most important conclusion that should be drawn from the current models of single-membrane electroporation is that there exists a strong tie between the transmembrane electric potential and the degree of permeation (as seen in concentration of pores, N; and pore size, r). It is important to note that even if a study were to account for lateral variations (not represented in equation [9.23]), this theoretical method may become quite involved as the total number of different pore sizes, K, must be accounted for within each of the up to 100 SC bilayers at each lateral location. Thus modeling individual pore creation, while informative, is not practical for skin electroporation modeling.

9.7.2 Empirical Models

Empirical models allow the researcher to model skin electroporation without keeping track of the permeability of each of the 100 bilayers by incorporating experimental observations that monitor the increase in electrical permeability as the SC undergoes electroporation. The local SC permeability is captured by the local SC electrical conductivity value, σ_{SC}. The capacitive charging time associated with non-Ohmic behavior of the SC is very short and may be neglected at pulsing times greater than 1 msec (Chizmandzhev et al. 1998). This behavior is typically neglected in studies that are less concerned with individual electropores but focus on describing the behavior of the SC as a whole. The electric potential distribution is usually solved from the Laplace equation:

$$\nabla \cdot (\sigma_{SC} \nabla \phi) = 0 \tag{9.24}$$

where ϕ is the electric potential.

The simplest representation of the increase in electrical conductivity with electroporation is a time-dependant step function. In (Becker and Kuznetsov 2007a,b) numerical studies focusing on thermal damage assessment are conducted to represent a section of living skin clamped between two electrodes and exposed to a series of electroporation pulses. These studies provide a very crude, but simple relation between SC electrical conductivity and degree of electroporation. The concept is based on the experimental results reported by Pliquett et al. (1995), Prausnitz (1996), and Pliquett and Gusbeth (2000) in which it is shown that when exposed to a high-voltage electric field the electrical resistance of the skin drops dramatically in a time period of μsec. To model the breakdown of the *SC* electrical resistance, the conductivity of the *SC* is prescribed as follows:

$$\sigma_{SC}(t) = \begin{cases} t \leq 5\,\mathrm{ms}: & \sigma_{SC} = 10^{-5}\ \mathrm{S/m} \\ t > 5\,\mathrm{ms}: & \sigma_{SC} = 10^{-3},\ \sigma_{SC} = 10^{-2}\ \mathrm{S/m} \end{cases} \tag{9.25}$$

The increase in electrical conductivity represents the permeation of the *SC*: basically during the first 5 msec of pulsing the *SC* is treated as completely unperturbed, and after the first 5 msec the *SC* is treated as fully permeabilized. This method takes advantage of the short time period required to electroporate the *SC* at high-intensity applied voltages. This description is fully empirical, neglecting *SC* thermal and voltage dependencies.

Recalling that the current understanding of single-membrane electroporation relies on V_m to determine the degree of electroporation, it follows that a macroscale model perhaps should also make use of this dependency. The idea here is to use experimental data in which degree of electroporation is represented by directly measured electrical current and local potential drop. The increase in transdermal current that is associated with electroporation is represented by an increase in electrical conductivity. It follows that the degree of electroporation may be similarly represented by an increase in electrical conductivity.

A novel empirical approach that takes into account the local electric field influence on the increased permeability is presented by Pavselj et al. (2005, 2007). In these studies the degree of permeability, as represented by the electrical conductivity, is related to the local electric. The electric field is defined as

$$E = \nabla \phi \tag{9.26}$$

The study (Pavselj et al. 2007) conducts a finite element analysis of a skin fold undergoing electropermeabilization. This is a parametric study that uses experimentally determined electrical conductivity relation to magnitude of electric field for three tissue types: subcutaneous, *SC*, and the combined dermis and epidermis (excluding *SC*). This article uses experimental data of skin electroporation of electrical conductivity change of the skin composite layers. The drop in conductivity is modeled as a step function of the magnitude of the electric field consisting of four different steps. For instance, the *SC* conductivity is represented as

$$\sigma_{SC}\,(\mathrm{S/m}) = \begin{cases} E\ (\mathrm{V/m}) < 600: & \sigma_{SC} = 0.0005 \\ 600 < E < 800: & \sigma_{SC} = 0.0165 \\ 800 < E < 1000: & \sigma_{SC} = 0.06 \\ 1000 < E < 1200: & \sigma_{SC} = 0.178 \\ 1000 < E: & \sigma_{SC} = 0.5 \end{cases} \tag{9.27}$$

This offers a much more accurate parametric description of the permeabilization of the tissues during electroporation. The study allows for a macroscopic representation of the current-applied voltage relation. The study compares well with experimental findings relating the electrical behavior of the skin fold at various applied voltages.

The concept of using the electric field magnitude to define the degree of permeability of the skin and underlying tissue is used the studies by Pavselj et al. (2005) in which a single value of postelectroporated tissue electrical conductivity increase is used. The basis of this study is that experimentally it has been shown that electroporation begins at a specific transmembrane voltage, E_0, but that when a certain voltage is exceeded, E_1, irreversible cellular damage begins. The study relies on experimentally derived values of E_0 and E_1 to represent the threshold values of various tissues undergoing electroporation. The study models the degree of electroporation by an increase in local tissue electrical conductivity. Electroporation is modeled as a function of local electric field—that is the degree of electroporation is represented by the local increase in magnitude of the electric field.

The electrical conductivity, $\sigma(E)$, is related to degree of electroporation by some relationship between electrical conductivity before permeabilitization, σ_0, and a maximum value of electrical conductivity due to electroporation, σ_1.

The novel concept of this study is the variety of functional dependencies that are incorporated to depict the relation between conductivity and potential drop:

Step function:

$$\sigma(E) = \begin{cases} \sigma_0 & E < E_0 \\ \sigma_1 & E \geq E_0 \end{cases} \tag{9.28}$$

Linear dependence:

$$\sigma(E) = \frac{\sigma_1 - \sigma_0}{E_1 - E_0} E + \sigma_0 \tag{9.29}$$

Exponential dependencies:

$$\sigma(E) = (\sigma_1 - \sigma_0) \frac{1 - \exp\left[\dfrac{E - E_0}{B}\right]}{1 - \exp\left[\dfrac{E - E_1}{B}\right]} + \sigma_0 \tag{9.30}$$

$$\sigma(E) = (\sigma_0 - \sigma_1) \frac{\exp\left[\dfrac{E - E_1}{B}\right] - 1}{\exp\left[\dfrac{E - E_0}{B}\right] - 1} + \sigma_1 \tag{9.31}$$

Sigmoid dependence:

$$\sigma(E) = \sigma_0 + \frac{\sigma_1 - \sigma_0}{1 + \exp\left[\dfrac{(E_1 + E_0 - E)}{2B}\right]} \tag{9.32}$$

where B is a shape parameter.

Comparing the simplicity of the empirical model to the complexity of the single bilayer models, it is clear that the benefit of using experimental results to model electroporation lies in the ability to match the physical phenomenon measured in the electrical skin behavior. Furthermore, these studies cleverly use the experimental data to capture lateral variations in the *SC*. The limiting factor of empirical models is that they require detailed data regarding tissue electrical behavior during electroporation, which can be a problem when this data does not exist. That is the motivation of the recently introduced thermodynamic approach to modeling skin electroporation.

9.8 Thermodynamic Approach

The increase in permeability associated with the formations of the LTR has been linked to sudden local rises in temperature. We recall the findings described in Section 9.6.3, which point toward the following trends associated with long-pulse electroporation:

- The electroporated *SC* is not homogenous laterally but experiences localized regions of higher permeability.
- The skin experiences local heat fronts that propagate radially outward forming regions of high permeability (for medium pulse amplitudes these originate in skin appendageal ducts).
- These local heating regions have been observed to reach temperatures above 70°C.
- That at temperatures above 60°C *SC* lipids have been shown to experience thermal phase transitions that destroy the lamellar barrier architecture.

Figure 9.5 shows a representation of the *SC* prior to electroporation in which a preexisting pore of radius, R_P, at the axial center passes through the *SC*. Before electroporation, the lipid structure connecting the corneocytes of the *SC* is lamellar and uniform. During electroporation a large current density through the pore causes local temperature rises within the *SC*. Figure 9.5 shows the lamellar structure in phase transition and that the effective pore radius, R_{eff}, extends to the region at which *SC* lipids are unaffected by the temperature rises.

These are the primary observed phenomena that the thermodynamic approach focuses on in its theoretical development. The preceding descriptions of *SC* electroporation have either neglected these points entirely or they rely entirely on empirical evidence. Thus, these do not capture the local thermal kinetics of the creation of the LTR. A recent publication from the group that helped introduce the description and the underlying physics of LTR formation in long-pulse skin electroporation builds a hypothesis based on experimental observations in which the development of the LTR is explicitly linked to local temperature rises associated with Joule heating (Pliquett et al. 2008). This study makes the thermal connection to increased *SC* electroporation based on the experimental observation that the minimum transdermal potential required to initiate electroporation is inversely related to local temperature rise: at 4°C the required voltage to induce electroporation is 80–100 V, while at 60°C the required transdermal potential drop is in the range 10–20 V. A numerical model is presented in which a step function method is used to define the *SC* behavior: once the condition for electroporation for one element is reached, the element is classified as electroporated, and its properties are switched from that of lipid to that of saline.

9.8.1 Fully Thermodynamic Approach

As the electroporation pulses cause local Joule heating within the *SC*, nearby lipid architecture experiences a transition from being a highly organized lamellar structure to a highly disorganized one. As the lipid structure is destroyed, the *SC* experiences dramatic increases in mass permeability and electrical conductivity. On the basis of the key points of the experimentally observed phenomena, a thermodynamic model of skin electroporation, which seeks to capture the evolution of the LTR, must provide a logical connection between temperature rise and permeability (both ionic and molecular.) With this in mind Becker and Kuznetsov (2008a,b) propose a thermally based function that seeks to capture the degree of *SC* lipid disorder. Recalling the heat versus temperature curve of Figure 9.6, the thermally influenced disorder is attributed to phase change E, which lies in the temperature range (65°C–75°C). To describe the degree of disorder, these studies borrow from methods that have been traditionally designed to model melting and solidification processes occurring over a temperature range (Voller and Brent 1989; Ozisik 1993), the analogy being that the thermally influenced transition from structured to disordered is represented by the solid structure transitioning from solid to liquid.

9.8.2 LTR Lipid Thermal Phase Change

In light of the previously discussed *SC* lipid thermal behavior and permeability studies, this model uses phase transition E to model *SC* lipid melting, during which the lipid melt fraction is used to describe the degree of lipid disorder

TABLE 9.2
SC Enthalpic Phase Transition Data

Phase Change	Description	T_1 (°C)	T_2 (°C)	ΔH (J/kg)
E	Fluidization of lamellar lipids	65	75	5,300
F	Fluidization of protein associated lipids	75	90	4,100
G	Denaturation of proteins	90	110	5,100

Source: Taken from data provided in Cornwell et al. 1996.

and is defined as

$$\varphi = \frac{(H - c_{SC}T)}{\Delta H_E} \tag{9.33}$$

where H is total enthalpy and is defined:

$$H = \int_{T_{E1}}^{T_{E2}} c_{SC,APP} dT \tag{9.34}$$

where ΔH_E is the latent heat associated with phase transition E, and T_{E1} and T_{E2} are the temperatures over which transition E takes place. The values have been taken from experimental studies (see Table 9.2).

To simplify the description considerably, a rectangular shaped specific heat versus temperature curve is used to model phase transitions (Bart and VanderLaag 1990; Roy and Avanic 2001).

The apparent SC specific heat, $c_{SC,APP}$, is defined as

$$c_{SC,APP} = c_{SC} + c_{SC,L} \tag{9.35}$$

where $c_{SC,L}$ is the latent specific heat, which is represented as

$$c_{SC,L} = \frac{\Delta H_{PC}}{T_{PC2} - T_{PC1}} \tag{9.36}$$

where ΔH is the latent heat, T_2 and T_1 are the representative phase change end and beginning temperatures, and the subscript PC refers to one of the three phase transitions: E, F, or G. Table 9.2 lists the values used in computations for the three SC phase change transitions. It is noted that long-pulse duration is linked to increased postpulse electrical resistance recovery times (>5 minutes) (Dujardin et al. 2002; Pliquett et al. 2005). The electric and mass transport barrier properties of SC regions experiencing lipid thermal phase transition by electroporation show very long recovery times (>30 minutes)(Pliquett and Gusbeth 2004). *SC* lipid structural changes associated with localized Joule heating during electroporation remain for some time after cooling of the *SC* (Pliquett et al. 2005). Polarized light thermal microscopy of *SC* and extracted lipid samples heated to 130°C show that, as the *SC* samples return to room temperature, the restructuring of the lipids shows evidence of aggregate variation (Silva et al. 2006a). Because of these findings and in light of the relatively short time periods that transience is modeled (ms to s) upon cooling it may not be necessary to consider lamellar restructuring of the lipid bilayers and subsequent electrical recovery. Thus, the lipid melt fraction may be described as

$$\varphi^{t+1} = \begin{cases} \max(\varphi^t, 0): & T < T_{E,1} \\ \max(\varphi^t, EQ(p17)): & T_{E,1} < T < T_{E,2} \\ 1: & T_{E,2} < T \end{cases} \tag{9.37}$$

where the superscript t and $t+1$ refer to the previous and current time steps, respectively.

At this point we have a thermodynamically derived function that represents the degree of lipid disorder, which is in the range $0 \leq \varphi \leq 1$, where $\varphi = 0$ corresponds to the insulative unaltered lipid organization and $\varphi = 1$ corresponds to permeable, fully disrupted *SC* lipids architecture.

The next task of the model is to quantify the *SC* permeability based on the lipid melt fraction. The lamellar structure of the bilayer matrix greatly reduces ionic permeability, and with the fluidization the lamellar lipid extracellular sheets of phase transition E comes increased *SC* ionic permeability. Thermal dependence of electrical resistance has been shown experimentally to drop by two orders of magnitude in the temperature range of this phase change (Craane van Hinsberg et al. 1995). To capture this effect, the local *SC* effective electrical conductivity increase is linearly related to lipid melt fraction by the relation:

$$\sigma^*_{SC} = \sigma_{SC} + \varphi(\sigma_{SC,MELT} - \sigma_{SC}) \tag{9.38}$$

where σ_{SC} is the normal electrical conductivity and $\sigma_{SC,MELT}$ is the electrical conductivity associated with the *SC* after full lipid melting. In the studies by Becker and Kuznetsov (2008a) the values $\sigma_{SC} = 10^{-5}$ S/m and $\sigma_{SC,MELT} = 10^{-3}$ S/m are used.

9.8.3 Transport

Electrically driven transdermal delivery is negotiated by the three modes of transport: electrophoresis, electroosmosis, and diffusion. Studies focusing on electrically driven transport of large charged molecules show that especially for short-duration pulses, electroosmotic effects are negligible compared to electrophoresis forces (Zaharoff et al. 2002). Homogenous tissue *in vivo* electroporation studies show that electrophoretic forces dominate in the transport of large molecules (Bureau et al. 2000; Satkauskas et al. 2005). In skin electroporation the pulse parameters are short (compared to iontophoresis); thus electroosmosis plays a negligible role, while electrophoresis effects are the primary effects in transdermal delivery of charged molecules (Regnier et al. 1999; Denet et al. 2004).

The transport of solute through the *SC* and into the underlying domain is formulated from the Nernst–Planck equation (Planck, 1890) as follows:

$$\frac{\partial C}{\partial t} = \nabla \cdot (m_i C \nabla \phi) + \nabla \cdot (D_i \nabla C) \tag{9.39}$$

where D is the diffusion coefficient; m is the electromobility; and the subscript i refers to one of the composite layers g, *SC*, *ED*, *DERM*, or *FAT;* and C is the solute concentration.

Similar to the electrical conductivity the enhanced electrokinetic mobility of the *SC* is related to lipid melt fraction by the relation

$$m_{SC}^{*} = m_{SC} + \varphi(m_{SC,MELT} - m_{SC}) \tag{9.40}$$

where m_{SC} is the unperturbed mobility of the *SC* and $m_{SC,MELT}$ is the mobility associated with the *SC* after full lipid melting. In the studies by Becker and Kuznetsov, (2008a) the values $m_{SC} = 10^{-17} m^2/Vs$ and $m_{SC,\ MELT} = 5 \times 10^{-11}\ m^2/Vs$ are used.

The SC diffusion coefficient is related to the lipid melt fraction by

$$D_{SC}^{*} = D_{SC} + \varphi(D_{SC,MELT} - D_{SC}) \tag{9.41}$$

where D_{SC} corresponds to the initial diffusion coefficient of the solute in the *SC* and $D_{SC,MELT}$ is the diffusion coefficient associated with the *SC* after full lipid melting. In the studies by Becker and Kuznetsov (2008a) The values $D_{SC} = 10^{-17}\text{m}^2/\text{sec}$ and $D_{SC,MELT} = 10^{-13}\text{m}^2/\text{sec}$ are used.

The values used to represent the mass transport coefficients associated with the fully thermally altered *SC* have been chosen to approximate the very resistive conditions a large DNA molecule would experience within the tortuous route between the unaltered corneocytes of the *SC*: the approximated diffusion and electrophoretic mobility coefficients used are much lower than those associated with those of even the high collagen content tissue of the epidermis. These are the coefficients that should be modeled using porous media perspectives. As yet, however, this has not been accomplished.

9.8.4 Thermal Energy

The lipid melt fraction of equation (9.37) and associated electrical and mass permeability rises of equations (9.38), (9.40), and (9.41) require an accurate description of the temperature distribution within the *SC*. Should the study attempt to reflect a skin fold it is important to consider that the process of skin electroporation involves additional tissues and materials to the *SC* that are exposed to electric fields. In the studies by Becker and Kuznetsov (2008a)

A physical model is developed of a skin fold which includes the electrode, applicator gel, and skin composite layers (see Figure 9.4[a]). Figure 9.4(b) shows the composite representation of this configuration consisting of electrodes (e), gel (g), and skin. The skin model consists of four sections: *SC*, epidermis (ED), dermis ($DERM$), and subcutaneous fat (FAT).

The Pennes' bioheat equation (Pennes 1948) is used along with an additional source term to describe thermal energy:

$$\rho_i c_i \frac{\partial T}{\partial t} = \nabla \cdot (k_i \nabla T) - \omega_i c_b (T - T_a) + q_i''' + Q_J \tag{9.42}$$

where ρ is the density, c is the specific heat, k is the thermal conductivity, T is the temperature, T_a is the arterial temperature, q''' is the metabolic volumetric heat generation, ω_m is the nondirectional blood flow associated with perfusion, and t is the time. The parameter c_b in the perfusion term is the specific heat of blood, which is assigned a value of $c_b = 3{,}800\,\mathrm{J/kgK}$ (Duck 1990). The subscript i refers to the gel, electrode, or one of the four composite skin layers *SC*, *ED*, *DERM*, or *FAT*.

The second and third terms on the right-hand side (RHS) of equation (9.42), which denote perfusion and metabolic heat generation are not present in the nonbiological electrode and gel layers. The term, Q_J, on the RHS of equation (9.42) is the Joule heat generated from the induced electric field. Joule heating occurs during the applied electric pulse in the gel and tissue composite layers, and it is defined as

$$Q_J = \sigma_i |\nabla \phi|^2 \tag{9.43}$$

where ϕ is the electric potential from the applied electric field, which is solved from the Laplace equation

$$\nabla \cdot (\sigma_i \nabla \phi) = 0 \tag{9.44}$$

where ϕ is the electric potential and σ_i is the composite electrical conductivity, and the subscript i refers to one of the composite layers g, *SC*, *ED*, *DERM*, or *FAT*.

Equation (9.44) neglects non-Ohmic behavior as well as the minor influence of the charged molecules of the solute on the electrical distribution of the large magnitude pulse.

9.9 Conclusions

The link between the theory of porous media and transport by the electroporation of the *SC* must be made. A model that represents large molecule transdermal transport should include electrophoretic effects. These effects occur during the pulse and that is also the time during which the skin *SC* structure is undergoing radical structural rearrangements. In this chapter we have covered some of the porous media models that are being used to describe passive transport through the *SC*. The chapter has covered some of the most current methods used to describe skin electroporation and has shown the methods that researchers are using at this stage to estimate the extent of magnitude of electroporation. We have concluded with a description of moderate long-pulse electroporation protocols in which localized microscopic regions of high permeability are created within the *SC* as a result of localized Joule heating. Supporting experimental findings have been provided to illuminate what phenomena the electroporation model should be expected to capture. What the field still lacks is a link between the porous–media-focused models used in the passive transport descriptions and the models that describe the degree of electroporation-induced *SC* permeation. Further work must be done that allows the description of the transient structural behavior of the *SC* to be considered in terms of the porosity and tortuosity of the *SC* architecture.

9.10 References

Al-Saidan, S. M., Barry, B. W., and Williams, A. C. (1998). Differential scanning calorimetry of human and animal stratum corneum membranes. *International Journal of Pharmaceutics,* **168**(1):17–22.

Barbero, A. M. and Frasch, H. F. (2005). Modeling of diffusion with partitioning in stratum corneum using a finite element model. *Annals of Biomedical Engineering,* **33**(9):1281–292.

Bart, G. C. J. and VanderLaag, P. C. (1990). Modeling of arbitrary-shaped specific and latent heat curves in phase-change storage simulation routines. *Journal of Solar Energy Engineering-Transactions of the ASME,* **112**(1):29–33.

Becker, S. M. and Kuznetsov, A. V. (2007a). Numerical assessment of thermal response associated with in vivo skin electroporation: the importance of the composite skin model. *Journal of Biomechanical Engineering-Transactions of the ASME,* **129**(3):330–340.

Becker, S. M. and Kuznetsov, A. V. (2007b). Thermal damage reduction associated with in vivo skin electroporation: a numerical investigation justifying aggressive pre-cooling. *International Journal of Heat and Mass Transfer,* **50**(1–2):105–116.

Becker, S. M. and Kuznetsov, A. V. (2008a). Thermal in-vivo skin electroporation pore development and charged macromolecule trans-dermal delivery: a numerical study of the influence of chemically enhanced lower lipid phase transition temperatures. *International Journal of Heat and Mass Transfer,* **51**(7–8):2060–2074.

Becker, S. M. and Kuznetsov, A. V. (2008b). Thermally induced pore growth associated with in vivo skin electroporation: a numerical model of lipid phase transition influence. *Journal of Bio-Mechanical Engineering-Transactions of the ASME,* **129**(5):712–721.

Bouwstra, J. A., Honeywell-Nguyen, P. L., Gooris, G. S., and Ponec, M. (2003). Structure of the skin barrier and its modulation by vesicular formulations. *Progress in Lipid Research,* **42**(1):1–36.

Bureau, M. F., Gehl, J., Deleuze, V., Mir, L. M., and Scherman, D. (2000). Importance of association between permeabilization and electrophoretic forces for intramuscular DNA electrotransfer. *Biochimica et Biophysica Acta-General Subjects,* **1474**(3):353–359.

Chizmadzhev, Y. A., Indenbom, A. V., Kuzmin, P. I., Galichenko, S. V., Weaver, J. C., and Potts, R. O. (1998). Electrical properties of skin at moderate voltages: contribution of appendageal macropores. *Biophysical Journal,* **74**(2):843–856.

Cornwell, P. A., Barry, B. W., Bouwstra, J. A., and Gooris, G. S. (1996). Modes of action of terpene penetration enhancers in human skin differential scanning calorimetry, small-angle X-ray diffraction and enhancer uptake studies. *International Journal of Pharmaceutics,* **127**(1):9–26.

Craane van Hinsberg, W. H. M., Verhoef, J. C., Junginger, H. E., and Bodde, H. E. (1995). Thermoelectrical analysis of the human skin barrier. *Thermochimica Acta,* **248**:303–318.

Cussler, E., Hughes, S., Ward, W., and Aris, R. (1988). Barrier membranes, *Journal of Membrane Science*, **38**(2):161–174.

Deen, W. M. (1987). Hindered transport of large molecules in liquid-filled pores. *AIChe Journal,* **33**(9):1409–1425.

Denet, A. R., Vanbever. R., and Préat, V. (2004). Skin electroporation for transdermal and topical delivery. *Advanced Drug Delivery Reviews,* **56**(5): 659–674.

Duck, F. A. (1990). *Physical Properties of Tissue: A Comprehensive Reference Book.* Academic Press, London.

Dujardin, N., Staes, E., Kalia, Y., Clarys, P., Guy, R., and Preat, V. (2002). In vivo assessment of skin electroporation using square wave pulses. *Journal of Controlled Release,* **79**(1–3):219–227.

Frasch, H. F. and Barbero, A. M. (2005). Steady-state flux and lag time in the stratum corneum lipid pathway: results from finite element models. *Journal of Pharmaceutical Sciences,* **92**(11):2196–2207.

Golden, G. M., Guzek, D. B., Harris, R. R., Mckie, J. E., and Potts, R. O. (1986). Lipid thermotropic transitions in human stratum corneum. *Journal of Investigative Dermatology,* **86**(3):255–259.

Golden, G. M., Guzek, D. B., Kennedy, A. H., Mckie, J. E., and Potts, R. O. (1987). Stratum corneum lipid phase-transitions and water barrier properties. *Biochemistry,* **26**(8):2382–2388.

Higuchi, W. I., Li, K. S., Abdel-Halim, G., Honggang, Z., and Yang, S. (1999). Mechanistic aspects of iontophoresis in human epidermal membrane. *Journal of Controlled Release,* **62**(1–2):13–23.

Johnson, M. E., Blankschtein, D., and Langer, R. (1997). Evaluation of solute permeation through the stratum corneum: lateral bilayer diffusion as the primary mechanism. *Journal of Pharmaceutical Sciences,* **86**(10): 1162–1172.

Kitson, N. and Thewalt, J. L. (2000). Hypothesis: the epidermal permeability barrier is a porous medium. *Acta Dermato-Venereologica,* **208**:12–15.

Krassowska, K. and Filev, P. (2007). Modeling electroporation in a single cell. *Biophysical Journal,* **92**(2):404–417.

Kushner, J., Blankschtein, D., and Langer, R. (2007b). Evaluation of the porosity, the tortuosity, and the Hindrance factor for the transdermal delivery of hydrophyllic permeants in the context of the aqueous pore pathway hypothesis using dual-radiolabeled permeability experiments. *Journal of Pharmaceutical Sciences,* **96**(12):3263–3282.

Kushner, J., Deen, W., Blankschtein, D., and Langer, R. (2007a). First-principles, structure-based transdermal transport model to evaluate lipid partition and diffusion coefficients of hydrophobic permeants solely from stratum corneum permeation experiments. *Journal of Pharmaceutical Sciences,* **96**(12):3236–3251.

Lange-Lieckfeldt, R. and Lee, G. (1992). Use of a model lipid matrix to demonstrate the dependence of the stratum–corneum barrier properties on its internal geometry. *Journal of Controlled Release,* **20**(3):183–194.

Li, S. K., Ghanem, A. H., Teng, C. L., Hardee, G. E., and Higuchi, W. I. (2001). Iontophoretic transport of oligonucleotides across human epidermal membrane: a study of the Nernst–Planck Model. *Journal of Pharmaceutical Sciences,* **90**(7): 915–931.

Li, S. K., Higuchi, I., Kochambilli, R. P., and Zhu, H. (2004). Mechanistic studies of flux variability of neutral and ionic permeants during constant current dc iontophoresis with human epidermal membranre. *International Journal of Pharmaceutics,* **273**(1–2):9–22.

Lian, G., Chen, L., and Han, L. (2008). An evaluation of mathematical models for predicting skin permeability. *Journal of Pharmaceutical Sciences,* **97**(1):584–598.

Madison, K. C. (2003). Barrier function of the skin: "la raison d'Être" of the epidermis. *Journal of Investigative Dermatology,* **121**(2):231–241.

Michaels, A. S., Chandraskeran, S. K., and Shaw, J. E. (1975). Drug permeation through human skin: theory and in vitro experimental measurement. *American Institute of Chemical Engineering,* **21**(5):985–996.

Mitragotri, S. (2002). A theoretical analysis of permeation of small hydrophobic solutes across the stratum corneum based on scaled particle theory. *Journal of Pharmaceutical Sciences,* **91**(3):744–752.

Mitragotri, S. (2003). Modeling skin permeability to hydrophilic and hydrophobic solutes based on four permeation pathways. *Journal of Contolled Release,* **86**(1):69–92.

Millington, P. F. and Wilkinson, R. (1983). *Skin.* Cambridge University Press, Cambridge.

Ozisik, M. N. (1993). *Heat Conduction.* John Wiley & Sons, Inc., New York, NY, USA.

Pavselj, N., Bregar, Z., Cukjati, D., Batiuskaite, D., Mir, L. M., and Miklavcic, D. (2005). The course of tissue permeabilization studied on a mathematical model of a subcutaneous tumor in small animals. *IEEE Transactions on Bio-Medical Engineering,* **52**(8):1373–1381.

Pavselj, N., Preat, V., and Miklavcic, M. (2007). A numerical model of skin electropermeabilization based on *in vivo* experiments. *Annals of Biomedical Engineering,* **35**(12):2138–21144.

Pennes, H. H. (1948). Analysis of tissue and arterial blood temperatures in the resting forearm. *Journal of Applied Physiology,* **1**:93–122.

Planck, M. (1890). Über die erregung von elektrizität und wärme in elektrolyten. *Annalen der Physik und Chemie,* **39**:161–186.

Pliquett, U. and Gusbeth, C. (2004). Surface area involved in transdermal transport of charged species due to skin electroporation. *Bioelectrochemistry,* **65**(1):27–32.

Pliquett, U. F. and Gusbeth, C. A. (2000). Perturbation of human skin due to application of high voltage. *Bioelectrochemistry,* **51**(1):41–51.

Pliquett, U., Langer, R., and Weaver, J. C. (1995). Changes in the passive electrical properties of human stratum corneum due to electroporation. *Biochimica et Biophysica Acta-Biomembranes,* **1239**(2):111–121.

Pliquett, U. F., Vanbever, R., Preat, V., and Weaver, J. C. (1998). Local transport regions (LTRs) in human stratum corneum due to long and short 'high voltage' pulses. *Bioelectrochemistry and Bioenergetics,* **47**(1): 151–161.

Pliquett, U., Zewert, T. E., Chen, T., Langer, R., and Weaver, J. C. (1996). Imaging of fluorescent molecule and small ion transport through human stratum corneum during high voltage pulsing: localized transport regions are involved. *Biophysical Chemistry,* **58**(1–2):185–204.

Pliquett, U., Gallo, S., Hui, S. W., Gusbeth, C., and Neumann, E. (2005). Local and transient changes in stratum corneum at high electric fields: contribution of joule heating. *Bioelectrochemistry,* **67**(1):37–46.

Pliquett, U., Gusbeth, C., and Nuccitelli, R. (2008). A propagating heat wave model of skin electroporation. *Journal of Theoretical Biology,* **251**(2): 195–201.

Potts, R. O. and Francoeur, M. L. (1990). Lipid biophysics of water loss through the skin. *Proceedings of the National Academy of Science, USA,* **87**(10):3871–3873.

Prausnitz, M. R. (1996). Do high voltage pulses cause changes in skin structure? *Journal of Controlled Release,* **40**(3):321–326.

Regnier, V., De Morre, N., Jadoul, A., and Préat, V. (1999). Mechanisms of a phosphorothioate oligonucleotide delivery by skin electroporation. *International Journal of Pharmaceutics,* **84**(2):147–156.

Roy, S. K. and Avanic, B. L. (2001). Turbulent heat transfer with phase change material suspensions. *International Journal of Heat and Mass Transfer,* **44**(12):2277–2285.

Satkauskas, S., Andre, F., Bureau, M. F., Scherman, D., Miklavcic, D., and Mir, L. M. (2005). Electrophoretic component of electric pulses determines the efficacy of *in vivo* DNA electrotransfer. *Human Gene Therapy,* **16**(10):1194–1201.

Silva, C. L., Nunes, S. C. C., Eusebio, M. E. S., Pais, A. A. C. C., and Sousa, J. J. S. (2006a). Study of human stratum corneum and extracted lipids by thermomicroscopy and DSC. *Chemistry and the Physics of Lipids,* **140**(1–2):36–47.

Silva, C. L., Nunes, S. C. C., Eusebio, M. E. S., Sousa, J. J. S., and Pais, A. A. C. C. (2006b). Thermal behaviour of human stratum Corneum- a differential scanning calorimetry study at high scanning rates. *Skin Pharmacology Physiology,* **19**(3):132–139.

Stewart, D. A., Gowrishankar, T. R., and Weaver, J. C. (2004). Transport lattice approach to describing cell electroporation: use of a local asymptotic model. *IEEE Transactions on Plasma Science,* **32**(4):1696–1708.

Tanojo, H., Bouwstra, J. A., Junginger, H. E., and Bodde, H. E. (1999). Thermal analysis studies on human skin and skin barrier modulation by fatty acids and propylene glycol. *Journal of Thermal Analisys and Calorimitry,* **57**(1):313–322.

Tezel, A., Sens, A., and Mitragotri, S. (2003). Description of transdermal transport of hydrophilic solutes during low-frequency sonophoresis based on a modified porous pathway model. *Journal of Pharmaceutical Sciences,* **92**(2):381–393.

Vanbever, R. and Preat, V. (1999). In vivo efficacy and safety of skin electroporation. *Advanced Drug Delivery Reviews,* **35**(1):77–88.

Vanbever, R., Pliquett, U. F., Preat, V., and Weaver, J. C. (1999). Comparison of the effects of short, high-voltage and long, medium-voltage pulses on skin electrical and transport properties. *Journal of Controlled Release,* **60**(1): 35–47.

Voller, V. R. and Brent, A. D. (1989). The modeling of heat, mass and solute transport in solidification systems. *International Journal of Heat and Mass Transfer,* **32**(9):1719–1732.

Wang, T., Kasting, G. B., and Nitsche, J. M. (2006). A multiphase microscopic diffusion model for stratum corneum permeability. I. Formulation, solution, and illustrative results for representative compounds. *Journal of Pharmaceutical Sciences,* **95**(3):620–648.

White, S. H., Mirejovsky, D., and King, G. I. (1988). Structure of lamellar lipid domains and corneocyte envelopes of murine stratum corneum. An x-ray diffraction study. *Biochemistry,* **27**(10):3725–3732.

Zaharoff, D. A., Barr, R. C., Li, C. Y., and Yuan, F. (2002). Electromobility of plasmid DNA in tumor tissues during electric field-mediated gene delivery. *Gene Therapy,* **9**(19):1286–1290.

10

Application of Porous Media Theories in Marine Biological Modeling

Arzhang Khalili[†]
Max Planck Institute for Marine Microbiology, Bremen, Germany
Earth and Space Sciences, Jacobs University, Bremen, Germany

Bo Liu, Khodayar Javadi, Mohammad R. Morad, Kolja Kindler
Max Planck Institute for Marine Microbiology, Bremen, Germany

Maciej Matyka
Max Planck Institute for Marine Microbiology, Bremen, Germany
Institute for Theoretical Physics, University of Wrocław, Wrocław, Poland

Roman Stocker
Department of Civil and Environmental Engineering, Massachusetts Institute of Technology, Cambridge, MA

Zbigniew Koza
Institute of Theoretical Physics, University of Wrocław, Wrocław, Poland

CONTENTS

[†]Corresponding author: Max Planck Institute for Marine Microbiology, Celsiusstr. 1, 28359 Bremen, Germany, Email: akhalili@mpi-bremen.de, Tel.: +49 421 20 28 636, Fax.: +49 421 20 28 690

10.1 Introduction

Theories initially developed to describe transport phenomena through the classical porous medium "soil" and "ground" (Darcy 1856) are encountered literally everywhere in everyday life, in nature, and in technical applications. The reason is that except metals, some plastics and dense rocks, almost all solids and semisolid materials can be considered as "porous" to varying degrees. Hence, there exist many types of different technology that depend on or make use of theories in porous media. The most prominent examples are given in the field of (1) hydrology, which deals with the water movement in earth and soil structures (e.g., dams, wells, filter beds, sewage), (2) petroleum engineering, which deals with exploration and production of oil and gas, and (3) chemical engineering (e.g., heterogeneous catalysis, chromatography, in particular, gel chromatography, separation processes using porous polymers, biological, and inorganic membranes). Also it has been long discovered that granular material sintering (Chen et al. 2005) is a very large tonnage technology, where pore structures are significant, and finds application in manufacturing ceramic products, paper, textile, and so forth.

However, the use of porous media theories in the field of marine microbiology is rather new, and was initiated by the discovery of the role of the seabed in regulation of the chemical composition of water masses in world oceans, and with it the role of oceans in the global cycles (Pamatmat and Banse 1969; Smith Jr. and Teal 1973; Sayles 1979; Emerson et al. 1980; Berelson et al. 1982; Glud et al. 1996, 2007; Ivey et al. 2000; Nikora et al. 2002; Oldham et al. 2004).

It has been found that at the bottom of rivers, lakes, sees, and oceans an enhanced transport of solutes and particulate matter can be encountered in a thin layer, which comprises of a tiny portion of the seawater layer from top and a tiny portion of the porewater layer from below, referred to as the benthic boundary layer (BBL). The BBL has been found to have a direct impact on all physical, chemical, biological, and biogeochemical processes occurring in aquatic systems (Boudreau and Jørgensen 2001).

Most direct denitrification rate measurements for continental shelves have been made on fine-grained, muddy sediments, which cover only 30% of global shelf area. The remaining 70% of continental shelf area is covered by sandy sediments. These sandy sediment environments are generally characterized by low organic matter and high pore water dissolved oxygen concentrations, properties typically considered unfavorable for heterotrophic denitrification

(Emery 1968; Vance-Harris and Ingall 2005). However, it is believed that N_2 production in high-permeability coastal sediments may play an important role in the global nitrogen cycle (Rao et al. 2007). This is another important evidence for the significant role of porous media theories in understanding global cycles.

When seabed sediments are permeable, the advective flux predominates the diffusive one (Huettel and Gust 1992a) drastically. In the context of permeable sediments, a variety of interesting phenomena exists in the field of marine microbiology, which can benefit from the knowledge available in porous media. Examples include, but are not limited to, topography effects in nutrient transport into deeper sediment layers (Huettel and Gust 1992b), enhanced bottom transport by gravity waves (Shum 1992a), reactive solute transport below rippled beds (Shum 1992b), and tide-driven deep pore-water flow in intertidal sands flats (Røy and Lee 2008).

Also the classification of different sediment types providing a habitat for marine species depend on how well the physical properties of the sediment have been described. It has been found that, beside permeability and porosity, the knowledge over tortuosity plays a significant role, for example, in exchange processes in the porewater (Iversen and Jørgensen 1993).

Furthermore, a variety of microorganisms inhabit the seabed, which have the ability of altering or influencing the ongoing interfacial exchange. Prominent examples of this group are burrowing animals, which construct U-, V-, or L-shaped tubes into the seabed, and ventilate the overlying seawater and generate an enhanced mixing. Using peristaltic or oscillatory motions, larvae are able to transfer oxygen into deeper sediment layers, and perform an ecologically significant interfacial nutrient exchange (Riisgård and Larsen 2005). Theoretically it seems obvious that the seawater ventilated by the larvae might also penetrate laterally into the ambient sediment, and generate, in addition to the currently accepted diffusive transport, yet another new mode of transport, namely the advective one. Modeling studies considering flow through a composite region made of saturated sediment and pure-fluid layers can provide useful hints bringing more light into this complex and important phenomena of bioirrigation.

Also in the water column of world oceans, a great deal of situations arise where porous media theories can be applied. An interesting example is that of marine aggregates. It has been found that particle settling has a significant effect on the biogeochemistry and ecology of the oceans due to the fact that particles are the key factor for carbon sequestration, and indirectly responsible for the amount of CO_2 that is released into the atmosphere from the seabed (Chisholm 2000; Azam and Long 2001).

When marine particles coagulate, bigger aggregates are formed that sink from the ocean surface down to the seabed within several hours or days depending on their sinking velocity and the ocean depth. The release of nutrients from sinking aggregates into the ambient seawater or vice versa plays an important role for the marine life. Although some simple models exist in which aggregates were considered as solid bodies (Kiørboe et al. 2001), transmission electron microscopy images (Leppard et al. 2004) clearly reveal that aggregates are rather porous organisms. Hence, implementing porous media theories can enhance the current quantitative estimations of the nutrient exchange mediated by the aggregates from one side, and provide an improved picture of biological consequences. Certainly there exist more biogeochemical problems, which are treated by the means of porous media theories, however, we settle for the examples mentioned to not explode the given framework.

This manuscript is organized as follows. First, a brief description of the mathematical model is brought. In the next sections, some recent examples are given with application in the field of marine microbiology. Finally, some concluding remarks and examples of future applications of porous media in marine microbiology and biogeochemistry have been mentioned.

10.2 Description of the Mathematical Model

For the numerical solution of the porous media equations different techniques such as finite difference method, finite element method, and finite volume method have been suggested. However, lattice Boltzmann model (LBM) has proved to be a promising technique to be applied in porous domains (Guo et al. 2002; Jue 2003; Wu et al. 2005). Compared to other numerical methods, LBM has the advantage of being most suitable for parallel algorithms. Besides, LBM is known to have a simple structure, which makes it most attractive for program coding. Being based on lattices, LBM has the ability of tackling complex meshes, dealing with multiphase, multicomponent fluids or domains (Succi 2001), which frequently occur in the field of marine biogeochemistry. For the sake of completeness, only a brief description of the LBM is brought here. The interested reader may refer to Succi (2001) for further details.

10.2.1 BGK Model

The Boltzmann equation describes a fluid from a microscopic viewpoint as an ensemble of discrete particles following the distribution $f = f(\mathbf{u}, \mathbf{x}, t)$, where f is the probability of finding a particle with velocity (or momentum) in the range $(\mathbf{u},\ \mathbf{u} + d\mathbf{u})$ and position in the range $(\mathbf{x},\ \mathbf{x} + d\mathbf{x})$ at time t. Then the discretized Boltzmann equation in D2Q9 lattice is expressed as follows:

$$f_i(\mathbf{x} + \mathbf{e}_i\delta_t, t + \delta_t) - f_i(\mathbf{x}, t) = -\Omega_i(f(\mathbf{x}, t)) \tag{10.1}$$

where the subscript i is the direction of the velocity. Furthermore, δt is the time increment and Ω_i denotes the collision operator. The discrete velocities

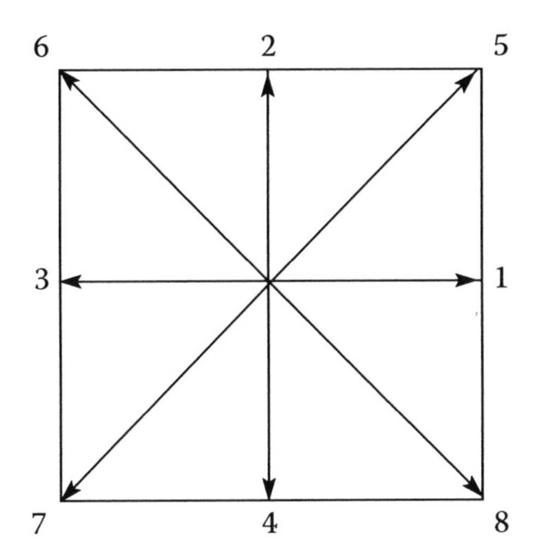

FIGURE 10.1
The lattice direction system for D2Q9 model.

are given by $\mathbf{e}_0 = 0$ and $\mathbf{e}_i = \lambda_i(\cos\theta_i, \sin\theta_i)c$, with $\lambda_i = 1$, $\theta_i = (i-1)\pi/2$ for $i = 1-4$, and $\lambda_i = \sqrt{2}$, $\theta_i = (i-5)\pi/2 + \pi/4$ for $i = 5-8$ (Figure 10.1).

The hydrodynamic variables include mass density (ρ), momentum ($\mathbf{j}$), and flux tensor ($\mathbf{\Pi}$), and are computed by the following:

$$\delta\rho = \sum_i f_i \tag{10.2}$$

$$\mathbf{j} = \rho\mathbf{u} = \sum_i \mathbf{e}_i f_i \tag{10.3}$$

$$\mathbf{\Pi} = \sum_i \mathbf{e}_i\mathbf{e}_i f_i \tag{10.4}$$

Its simplest and by now most popular form is the Bhatnagar–Gross–Krook (BGK) model, which expresses the collision as a relaxation toward a local equilibrium, $\Omega_i = -\frac{1}{\tau}(f_i - f_i^{eq})$, where τ is the nondimensional relaxation time directly related to the kinematic fluid viscosity $\nu = c_s^2\left(\tau - \frac{1}{2}\right)\delta_t$ and $f_i^{(eq)}$ is the equilibrium distribution function (Zou et al. 1995; He and Luo 1997; Dellar 2003):

$$f_i^{(eq)} = \omega_i\rho\left[1 + \frac{\mathbf{e}_i\cdot\mathbf{u}}{c_s^2} + \frac{\mathbf{uu}:(\mathbf{e}_i\mathbf{e}_i - c_s^2\mathbf{I})}{2c_s^4}\right] \tag{10.5}$$

in which ω_i is a weight factor, c_s is the speed of sound (set as $c_s^2 = 1/3$), and $\mathbf{I}$ is the unit tensor. The weights are given by $\omega_0 = 4/9$, $\omega_i = 1/9$ for $i = 1–4$, and $\omega_i = 1/36$ for $i = 5–8$.

The Navier–Stokes equations can be derived from the Chapman–Enskog procedure (Chopard and Droz 1998), which leads to

$$\frac{\partial \rho}{\partial t} + \nabla \cdot (\rho \mathbf{u}) = 0 \tag{10.6}$$

$$\frac{\partial (\rho \mathbf{u})}{\partial t} + \nabla \cdot (\rho \mathbf{u}\mathbf{u}) = -\nabla p + \nu \nabla \cdot [\rho(\nabla \mathbf{u} + \mathbf{u}\nabla)] \tag{10.7}$$

where $p = c_s^2 \rho$ is the pressure, and the effective viscosity is defined as

$$\nu = \tfrac{1}{3}\left(\tau - \tfrac{1}{2}\right)\delta_t \tag{10.8}$$

10.2.2 LBM for Incompressible Flows in Porous Media

Flow in porous media is usually modeled by some semiempirical models because of the complex structure of a porous medium based on the volume averaging at the scale of representative elementary volume (REV). Several widely used models have been introduced in the literature, such as the Darcy, the Brinkman-extended Darcy, and the Forchheimer-extended Darcy models. A recent achievement in modeling flow in porous media is the so-called generalized model, in which all fluid forces and the solid drag force are considered in the momentum equation given by:

$$\nabla \cdot (\mathbf{u}) = 0 \tag{10.9}$$

$$\frac{\partial (\mathbf{u})}{\partial t} + \nabla \cdot \left(\frac{\mathbf{u}\mathbf{u}}{\phi}\right) = -\frac{1}{\rho}\nabla(\phi p) + \nu_e \nabla^2 \mathbf{u} + \mathbf{F} \tag{10.10}$$

In the above equation, ν_e is the effective viscosity and $\mathbf{F}$ represents the total body force given by the following:

$$\mathbf{F} = -\frac{\phi \nu}{K}\mathbf{u} - \frac{\phi F_\phi}{\sqrt{K}}|\mathbf{u}|\mathbf{u} + \phi \mathbf{G} \tag{10.11}$$

in which the three terms on the right side represent Darcy, Forchheimer, and gravity force, respectively. The geometric function F_ϕ and permeability K can be expressed as follows:

$$F_\phi = \frac{1.75}{\sqrt{150\phi^3}} \tag{10.12}$$

$$K = \frac{\phi^3 d_p^2}{\sqrt{150(1-\phi)^2}} \tag{10.13}$$

where d_p is the solid particle diameter.

In the LBM notation, the momentum equation for the fluid flow in a porous medium can be expressed as:

$$f_i(\mathbf{x} + \mathbf{e}_i\delta_t, t + \delta_t) - f_i(\mathbf{x}, t) = -\frac{1}{\tau}\left[f_i(x,t) - f_i^{(eq)}(x,t)\right] + \delta_t F_i \tag{10.14}$$

in which the equilibrium distribution function has been modified as

$$f_i^{(eq)} = \omega_i \rho \left[1 + \frac{\mathbf{e} \cdot \mathbf{u}}{c_s^2} + \frac{\mathbf{uu} : (\mathbf{e}_i \mathbf{e}_i - c_s^2 \mathbf{I})}{2\phi c_s^4} \right] \tag{10.15}$$

and force term as

$$F_i = \omega_i \rho \left(1 - \frac{1}{2\tau} \right) \left[\frac{\mathbf{e}_i \cdot \mathbf{F}}{c_s^2} + \frac{\mathbf{uF} : (\mathbf{e}_i \mathbf{e}_i - c_s^2 \mathbf{I})}{2\phi c_s^2} \right] \tag{10.16}$$

Accordingly, the fluid density and velocity are given by

$$\rho = \sum_i f_i \tag{10.17}$$

$$\mathbf{u} = \frac{\mathbf{v}}{c_0 + \sqrt{c_0^2 + c_1 |\mathbf{v}|}} \tag{10.18}$$

where $\mathbf{v}$ is an auxiliary velocity and is defined as

$$\rho \mathbf{v} = \sum_i \mathbf{e}_i f_i + \frac{\delta_i}{2} \phi \rho \mathbf{G} \tag{10.19}$$

The two parameters c_0 and c_1 are given by

$$c_0 = \frac{1}{2} \left(1 + \phi \frac{\delta_t}{2} \frac{\nu}{K} \right), \quad c_1 = \phi \frac{\delta_t}{2} \frac{F_\phi}{\sqrt{K}} \tag{10.20}$$

By a similar procedure described above, from equation (10.14) one can obtain the extended Darcy equation for a porous medium containing the Brinkman and Forchheimer suggestions as

$$\frac{\partial \rho}{\partial t} + \nabla \cdot (\rho \mathbf{u}) = 0 \tag{10.21}$$

$$\frac{\partial (\rho \mathbf{u})}{\partial t} + \nabla \cdot \left(\frac{\rho \mathbf{uu}}{\phi} \right) = -\nabla (\phi p) + \nabla \cdot [\rho \nu_e (\nabla \mathbf{u} + \mathbf{u} \nabla)] + \rho \mathbf{F} \tag{10.22}$$

where $p = c_s^2 \rho / \phi$ is the pressure, while the effective viscosity is defined as

$$\nu_e = c_s^2 \left(\tau - \frac{1}{2} \right) \delta_t \tag{10.23}$$

10.2.3 LBM for Concentration Release in Porous Media

The LBM for concentration release can be expressed as

$$g_i(\mathbf{x} + \mathbf{e}_i \delta_t, t + \delta_t) - g_i(\mathbf{x}, t) = -\frac{1}{\tau_g} \left[g_i(x, t) - g_i^{(eq)}(x, t) \right] \tag{10.24}$$

where τ_g is the relaxation time and g_i the distribution function for concentration. The equilibrium distribution function was modified as

$$g_i^{(eq)} = \omega_i C \left[\phi + \frac{\mathbf{e} \cdot \mathbf{u}}{c_s^2} \right] \tag{10.25}$$

Accordingly, the concentration and velocity are given by

$$\phi C = \sum_i g_i \tag{10.26}$$

$$\mathbf{u} C = \sum_i g_i c_i \tag{10.27}$$

By a similar procedure described above, from equation (10.14) one can obtain the concentration equation for a porous medium

$$\phi \frac{\partial C}{\partial t} + (\mathbf{u} \cdot \nabla) C = \nabla \cdot [D_m \nabla C] \tag{10.28}$$

with the effective diffusion coefficient

$$D_m = \phi c_s^2 \left(\tau_g - \tfrac{1}{2} \right) \delta_t. \tag{10.29}$$

10.3 Application of Porous Media in Marine Microbiology

As mentioned earlier, in marine microbiology there exists a great deal of situations in which porous media theories apply. From different examples given above, in this section following problems will be discussed: (1) shear-stress control at seabed bottom, (2) tortuosity of marine sediments, (3) oscillatory flows over permeable seabed ripples, (4) nutrient release from sinking marine aggregates, and (5) enhanced nutrient exchange by burrowing macrozoobenthos species.

10.3.1 Shear-Stress Control at Bottom Sediment

In a variety of marine microbiological or environmental issues, generating uniform shear stress planes are desired. An example is given by sampling devices applied in marine sciences—known as microcosms—in which a controlled flow is generated to minimize the erosion threshold by producing a uniform shear stress on the sediment surface.

Recently, shear stress control devices have been considered in technologies for integration of cell separation and protein isolation from mammalian cell cultures. In filtration systems a few circular disks are designed below the rotating cone. This way, due to the fact that shear force, pressure generation, and the specific hydrodynamics of the system are decoupled, shear rates can be easily optimized and precisely controlled to maximize filtration performance while viability of the shear sensitive animal cells is maintained (Vogel et al. 2002).

So far two different categories of devices have been suggested for uniforming the bottom shear stresses. The first one is suggested by Gust (1990), and is composed of a rotating disk in a cylindrical housing. Through the central section of the disk fluid is pumped out and is reentered into the container via the peripheral zone. The disk has optional skirts attached to it (see top image in Figure 10.2).

In the second category, the rotating disk has a conical shape (either flat or curved), and is suggested by Kroner and Vogel (2001) and in modified versions by Sun and Lee (2005) and Ting and Chen (2008). The geometrical configuration of the second category device—termed as shear inducer—has been shown in the middle image in Figure 10.2.

As shown by Khalili et al. (2008), the shear stress uniformity generated by microcosms covers only 72% of the bottom area. A further disadvantage of the microcosm is that it cannot be miniaturized for biological applications such as cell culturing. Also in the case of shear inducers a shear stress uniformity of 84% can be achieved under restricted conditions (very small cone tip-substrate distances, very small Reynolds number). For more details on this issue, the reader is referred to Javadi and Khalili (2009).

As an alternative to the devices in both categories, Khalili and Javadi (2009) have suggested a new device (see the bottom image in Figure 10.2) with which a shear stress uniformity over 94% or larger sections of the bottom area may be generated. As shown in the figure, the system composes of a central rotating disk, surrounded by a number of rings that rotate with lower angular velocities. For the sake of comparison, the shear stress uniformity achieved by all three devices are shown in Figure 10.3. Note that the calculations made for the multiring device contain a central disk and three rotating rings.

As clearly demonstrated, the multiring device performs best. The real advantage of the multiring device is that it can be applied to any flow and size constraints, and can be applied for all kinds of applications both in large and small scale.

Another issue associated with chamber flows and those in cone viscosimeters is the problem of artificial pressure near the interface, which affects directly the sediment-water fluxes. This issue has to be given special attention when the substrate is a porous sediment.

For the sake of completeness, the pressure field has been calculated and plotted in Figure 10.4. The analysis demonstrates an almost contact pressure for the entire radius. The pressure profile generated by the microcosm has two distinct gradients, leading to larger differential pressure. Hence, as far

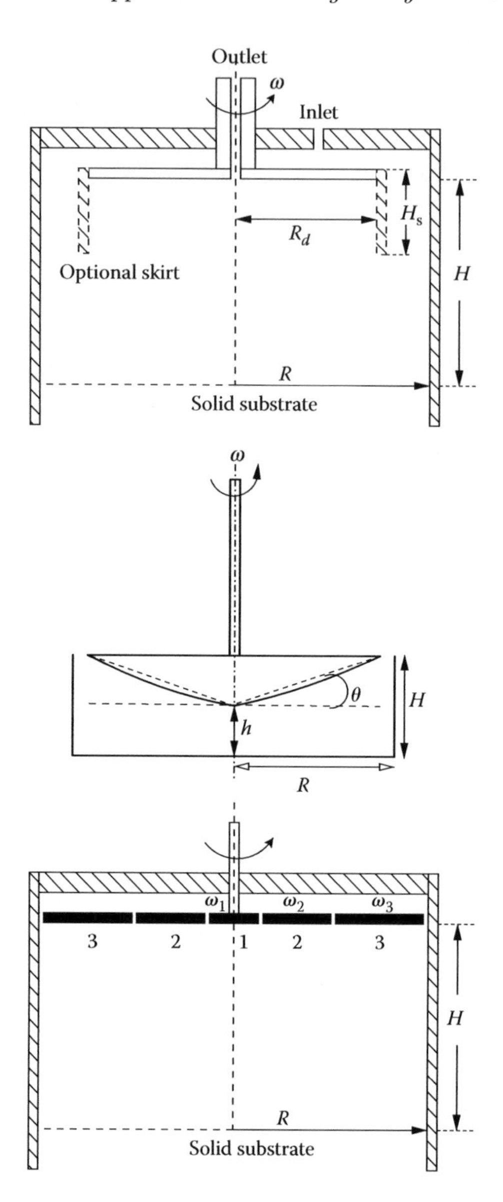

FIGURE 10.2
Geometries of shear stress uniformity devices. Top: microcosm with suction-injection after Gust (1990); middle: rotating cone with flat inclined sides (dashed: after Sun and Lee [2005]) and curved inclined cones (solid lines: after Ting and Chen [2008]) and bottom: multiring device of Khalili and Javadi (2009). Note that here only two rotating rings have been shown. Practically, four or more rings can be implemented to enhance the shear stress uniformity.

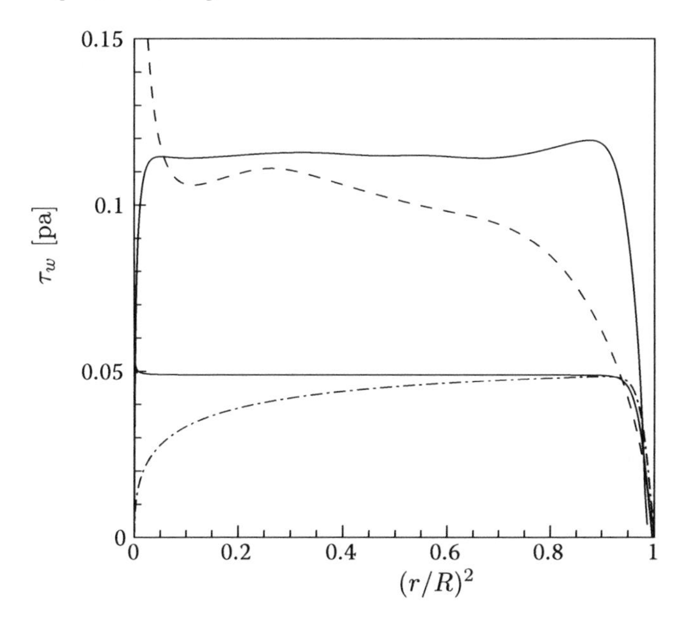

FIGURE 10.3
Shear stress uniformity obtained for three different devices. Dotted line: microcosm of Gust (1990) (Reynolds number is 68672); dashed line: shear inducer device with flat conical sides (Sun and Lee 2005) (cone-substrate distance $h = 100\,\mu\text{m}$, cone angle $\alpha = 1°$, angular velocity $\omega = 10$ rpm; solid line: multiring device of (Khalili and Javadi 2009) compared with the microcosm; bold solid line: multiring device miniaturized version.

as the differential pressure is concerned, the multiring device has a better performance.

10.3.2 Tortuosity of Marine Sediments

Permeability and porosity are two important physical properties of any seabed sediment or, in general, any porous media. Beside these two properties, there exists a third quantity known as tortuosity, which significantly influences the ongoing exchange processes in the field of marine geochemistry and geophysics. From hydrodynamic point of view, the tortuosity can be defined as follows: If a fluid particle located in the upstream can migrate on a purely horizontal path to a point downstream within a flow domain, then the tortuosity of the path would be $T = l/L = 1$ with l and L being the path-line length and the geometry length in the flow direction. Hence, the more tortuous the path-line of the fluid particle within a porous medium (because of the existence of solid obstacles) becomes, the larger is T (see Figure 10.5).

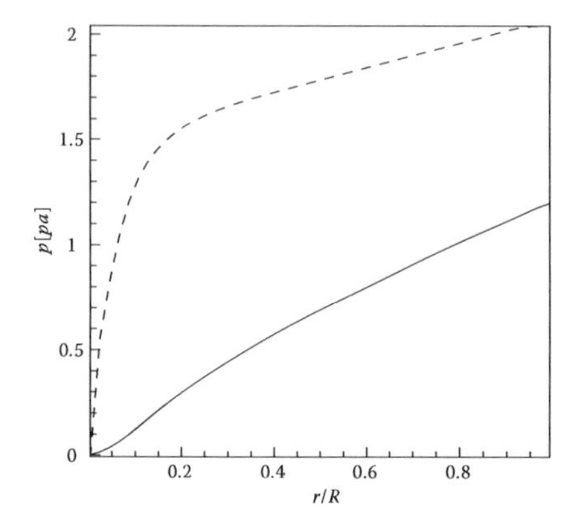

FIGURE 10.4
Distribution of pressure as a function of radius (Reynolds number = 70,000). Dashed line: microcosm of Gust (1990), solid line: multiring device of Khalili and Javadi (2009). The latter device produces an almost constant gradient and lower pressures compared to that of Gust.

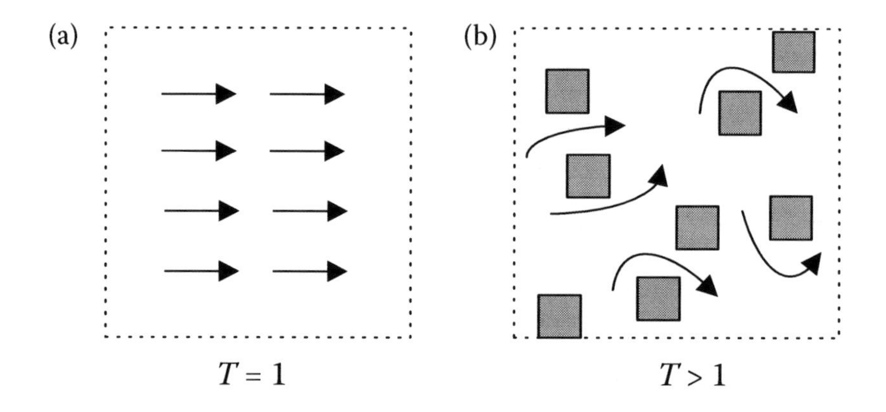

FIGURE 10.5
Comparison of free flow (a) to tortuous flow through a porous structure.

Unfortunately a direct measurement of tortuosity is not possible. This fact has led to diffusive (Nakashima and Yamaguchi 2004), electrical (Lorenz 1961), and acoustic (Johnson et al. 1982) tortuosity definitions. There were also further theoretical attempts by Koponen et al. (1996) to define tortuosity. However, all these tortuosities, in general, differ from each other. Except for some very simple models (Clennell 1997; Knackstedt 1994; Zhang and Knackstedt 1995), there is no clear consensus on its definition. In the literature, so far four different models for tortuosity have been provided given by

$$T(\phi) = \phi^{-p} \quad (10.31)$$

$$T(\phi) = 1 - p\ln\phi \quad (10.32)$$

$$T(\phi) = 1 + p(1-\phi) \quad (10.33)$$

$$T(\phi) = [1 + p(1-\phi)]^2 \quad (10.34)$$

with p as a constant factor and ϕ as porosity. The first, second, and third model are theoretical models whereas the fourth one is an empirical model. In sequence, the above equations go back to studies of Archie (1942); Weissberg (1963); Iversen and Jørgensen (1993); and Boudreau and Meysman (2006), respectively.

In a recent study, Matyka et al. (2008) developed an LBM (see Section 10.2) and studied the tortuosity problem from a mathematical perspective. For that purpose, they considered a rectangular flow domain with randomly distributed solid squares as solid obstacles with fixed locations (see Figure 10.6a). By calculating the velocity field and the streamlines (Figure 10.6b) the tortuosity could be calculated, and compared with the models discussed earlier. The comparison shows that the hydrodynamic-based tortuosity calculation of Matyka et al. (2008) matches well with the Weissberg relation (see Figure 10.6c).

For the mathematical modeling presented above, one may ask a question: what is the minimum size of the model system that is able to predict the behavior of the particles in the real world? The underlying basic assumption is that the porous material has to be homogeneous. Large model systems demand high-computational power. This is the main reason why in simulations, system sizes are kept as small as possible. To check this, computational analysis of the path of two particles traveling through a porous medium was performed. Two different alignments with the gravitational field was depicted (see Figure 10.7).

The model system should be homogeneous and should have similar properties in all directions (isotropy). Anisotropy is used to describe the variations of properties depending on the directions. As shown here the model system is too small. Therefore, the traveling particles do not follow direction determined by gravity vector. It was shown by Koza et al. (2009) that the model system has to be at least 100 times larger than the characteristic grain size.

10.3.3 Oscillating Flows over a Permeable Rippled Seabed

The sediment–water interface constitutes a dynamic and significant biologically active region in marine sediments. Within this region, sediments and porewater contact with the overlying water, and exchange between these reservoirs regulates oxygen or nutrient distributions. The importance of the solute transport across this zone has been long recognized as a key factor for accurate determination of sediment oxygen demand in marine environments (Berner 1976).

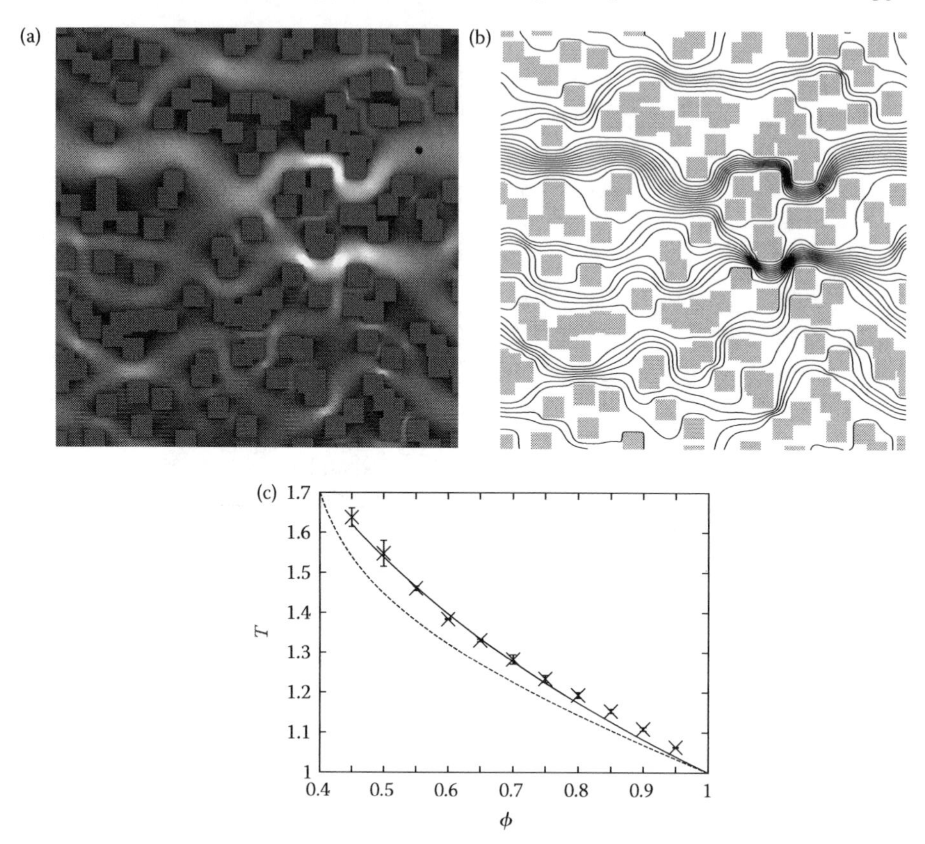

FIGURE 10.6
Calculation of tortuosity in sediments based on path-lines. (a) Velocity magnitudes squared ($u^2 = u_x^2 + v_y^2$). Light gray boxes show randomly placed fixed solid matrices (porosity is 0.7). (b) Streamlines calculated from the velocity field when the flow is induced by the action of gravity. (c) Comparison of tortuosity values calculated from our model (cross symbols with error bars), calculated after Weissberg relation (solid line) and after Koponen et al.

Coastal sediments are often sandy with uneven surfaces, above which the flows is induced into oscillatory motion under surface gravity waves. Gundersen and Jørgensen (1990) measured the vertical distribution of oxygen at the sediment–water interface. They found that the time-averaged concentration was indeed nearly constant except in a thin layer immediately above the sediment surface, where the mean concentration decreased linearly with decreasing elevation. Within this diffusive boundary layer (DBL), which was about 0.6 mm thick, molecular diffusion model was used to estimate the vertical flux of solutes. However, comparisons with actual flux measurements suggested that such an "empirical diffusion coefficient" would have to be a

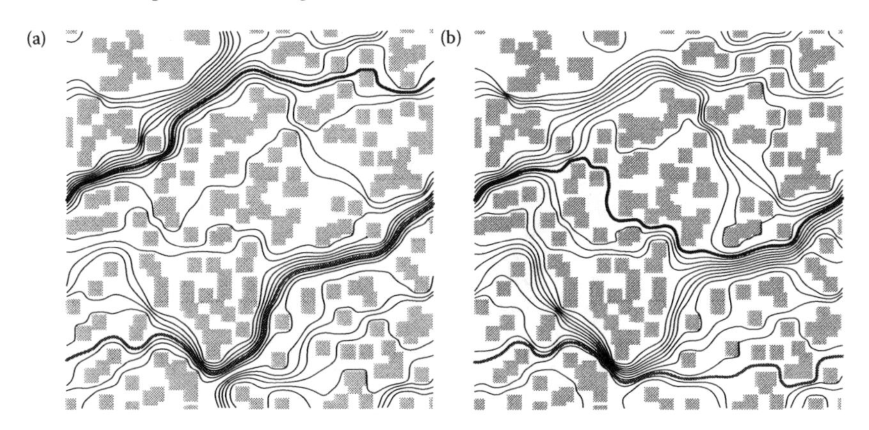

FIGURE 10.7
Computational analysis of the path of two particles traveling through a porous medium. Gravity vector parallel to the x axis (a) and rotated by 20°(b).

few times higher than that of molecular diffusivity. The oxygen concentration in the middle of this DBL oscillated in time with a magnitude of more than 10% of its mean value and at the frequency of the prevalent surface gravity wave. Gundersen and Jørgensen (1990) attributed the oscillations to the "numerous eddies which approach the sediment surface from the bulk of the following sea water and hit the viscous and diffusive sublayers," but details of the physical mechanism involved have yet not been explored.

However, mathematical models for quantifying fluxes across permeable seabeds in the presence of oscillatory flows are, in comparison, not numerous, and limited to the studies of Shum (1992a,b; 1995) and Hara et al. (1992). Although these models provide a good insight into solute distribution below the sediment–water interface, all of them are based on assuming linearized potential flows, and hence, of limited applications. To gain a better understanding of the solute transport in a wave-induced oscillating ambient flow, the LBM model was used to account for both advective and diffusive transport, allowing a clear identification and comparison of fluxes arising from diffusive as well as advective transport (Liu and Khalili 2010a).

In the study, an oscillatory flow has been generated on the surface of the water layer to follow $U = U_0 \sin(\omega t)$ (Figure 10.8) with ω and t being the oscillation frequency and time. The interfacial solute exchange depends on a number of different parameters. The first important parameter is the steepness factor, $s = 2a/L$, which characterizes the sinusoidal ripple. Next, the flow intensity is decided by Reynolds number, $Re = U_0 L/\nu$. Furthermore, the Strouhal number, $St = \pi L/U_0 T$), describes the oscillating intensity while the Schmidt number, $Sc = \nu/D$, describes the momentum and mass diffusion intensity. In the above relations, $2a, L, U_0, \nu, T$, and D are, respectively, the wave amplitude, the wave length, constant velocity, fluid viscosity, the

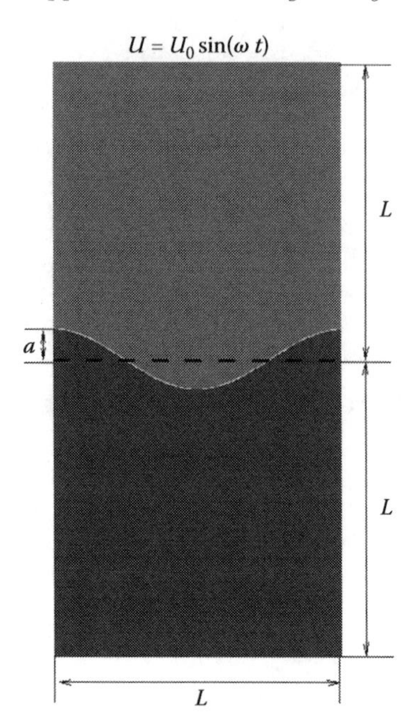

FIGURE 10.8
Illustration of the geometry and flow condition.

oscillation period, and the diffusion coefficient. Finally, the properties of the sediment, the porosity (ϕ) and permeability (k) are also important parameters for the interfacial solute exchange. As the results show, each of the above parameters (Re, St, Sc, s, ϕ, k) are important factors (Liu and Khalili 2010a), however, for the sake of brevity only two cases are shown. These are streamlines at one-quarter and three-quarter period of time, shown in Figure 10.9.

Here we only gave two examples. In Figures 10.10 and 10.11, it has been demonstrated that an increase in the steepness of the ripple and Reynolds number enhanced the advective transport of the solute at the water-sediment interface.

10.3.4 Nutrient Release from Sinking Marine Aggregates

Marine aggregates appear in different forms such as discarded feeding structures, fecal pellets, dead organisms, and other organic debris that sink from the ocean surface down the water depth to the seabed. Depending on their density and diameter, aggregates reach terminal velocities ranging from 15 to 30 m/d, and release/adsorb nutrient into/from the ambient seawater. A typical marine aggregate from the Atlantic with a diameter of 4 mm is shown in

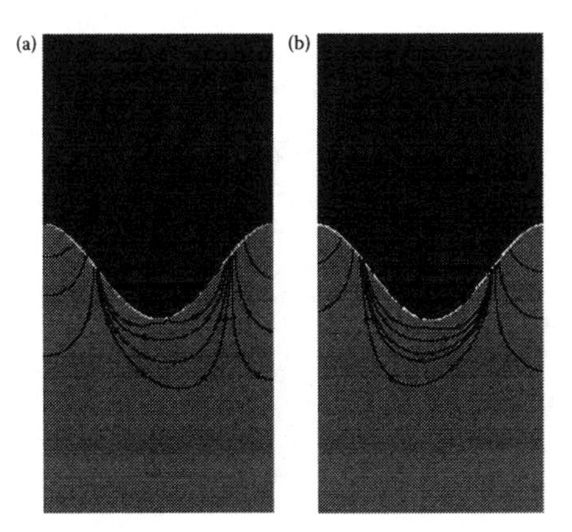

FIGURE 10.9
Streamlines at $t = \mathrm{T}/4$ (a) and $t = 3\mathrm{T}/4$ (b).

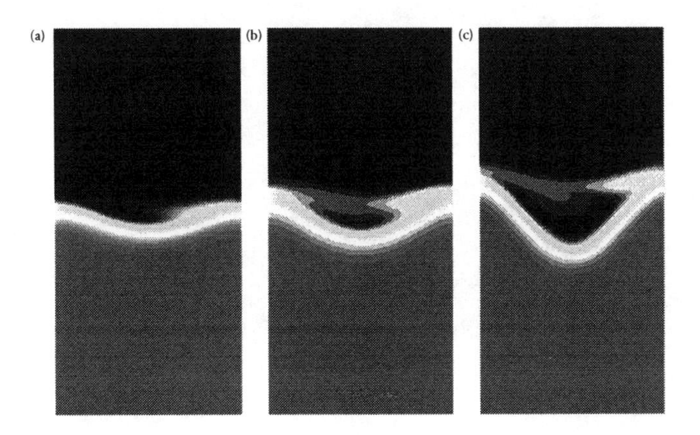

FIGURE 10.10
Solute release in different ripple steepness $s = 0.1$ (a), $s = 0.2$ (b), and $s = 0.4$ (c).

Figure 10.12. This specimen, in common with most from the Atlantic, comprises dead and decaying phytoplankton, zooplankton fecal matter, and their exoskeletons. They sink at rates from a few tens of meters per day to several hundred meters per day in contrast to phytoplankton cells that individually sink at no more than 1 m/d and typically 0.1 m/d.

Owing to this rapid sinking, aggregates are known as a vehicle for vertical flux of organic matter but also hotspots of microbial respiration responsible for a rapid and efficient turnover of particulate organic carbon in the sea (Logan and Wilkinson 1990).

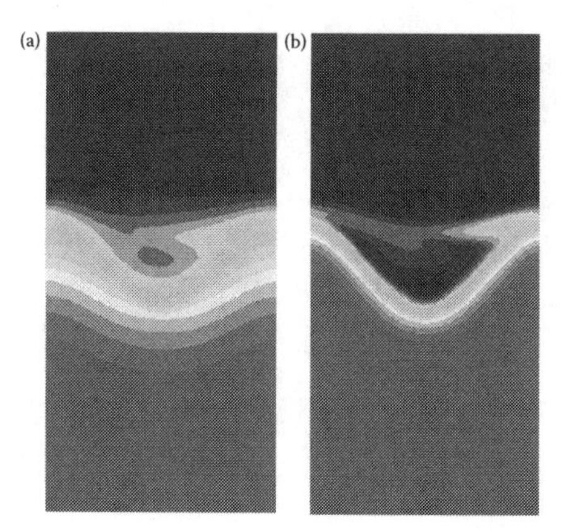

FIGURE 10.11
Solute release at different Reynolds number $Re = 125$ (a) and $Re = 1250$ (b).

FIGURE 10.12
A marine snow particle of diameter 4 mm. (Courtesy of R. Lampitt.)

This is the reason for the increased interest of the marine scientists in understanding the sinking and exchange mechanisms generated by the aggregates. Owing to the fact that *in-situ* and laboratory-experiments on living aggregates are not an easy task, attempts have been made to simulate their sinking procedure with mathematical techniques.

Until recently, the only model available in the literature was that of Kiørboe et al. (2001), in which aggregates were considered as a solid sphere. However, as mentioned earlier, transmission electron microscopy images have shown that aggregates have a porous structure (Leppard et al. 2004). Hence, there can be two flow scenarios (Figure 10.13). In the solid case, flow can only bypass the aggregates, whereas, in the porous case, a partial throughflow also exists.

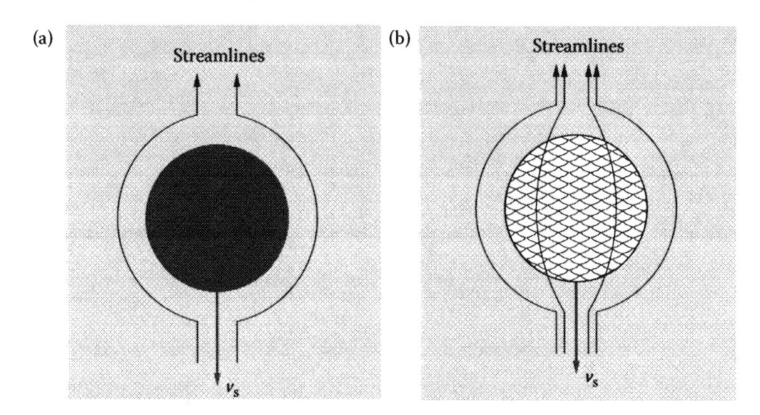

FIGURE 10.13
Schematics of the flow around a solid sphere (a) versus a porous sphere (b). In the latter case a partial throughflow also exits.

More recently, Bhattacharyya et al. (2006) have considered a circular cylindrical porous structure, and showed that porosity and permeability of the aggregate drastically alter the patterns of streamlines, vorticity, and nutrient transfer.

Because the aggregates have, in general, a complex shape, an efficient LBM code has been developed (Liu and Khalili 2008, 2009), which has made possible to treat not only spherical, but also arbitrarily-shaped porous domains (Liu and Khalili 2010b). A comparison between the nutrient release from a solid aggregate versus that of a porous aggregate solved by the model of Liu and Khalili (2010b) can be seen in Figure 10.14. Furthermore, while current calculations have assumed constant porosity, the LBM code can easily account for spatially heterogeneous porosity.

For comparison, a complex geometry is given in Figure 10.15b which consists of four different porous subdomains, a square, a rectangle, an oval, and a circle, which all lie within the same viscous ambient fluid.

As shown in the figure, the flow past the aggregate partially passes through the porous bodies and partially bypasses them. In the example shown, all subdomains have the same fixed porosity of $\phi = 0.993$.

The literature discussed so far invariably assumed a homogeneous fluid density. However, in lakes, oceans, and estuaries, vertical density gradients within the water column are ubiquitous. In freshwater systems, density gradients are caused by a decrease of temperature with depth, while in the ocean it is often salinity that increases with depth. The strength of the stratification is quantified by the Brunt-Väisälä frequency, $N = \sqrt{-(g/\rho_0)(\partial\rho/\partial z)}$, which measures the natural frequency of a fluid parcel in a stable density gradient,

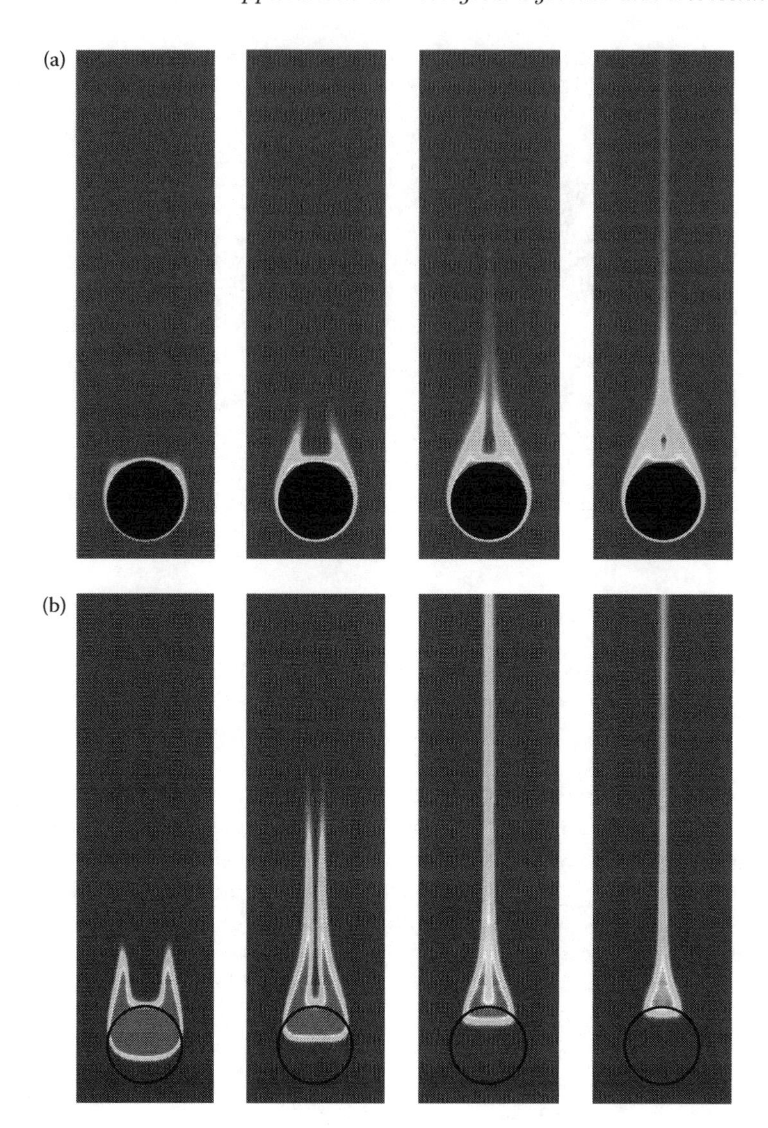

FIGURE 10.14
Release of nutrient from a solid aggregate (a) versus the same from a porous one (b) with a porosity of 0.993 and $Re = 10$. In (a) image, initially a maximum concentration is given at the aggregates surface, which is redistributed with time. In (b), however, the initial maximum concentration covers the entire aggregates interior (the entire sphere is fully red at time $t = 0$, having a maximum concentration). From left to the right, the nondimensional times plotted are 400, 1,000, 2,000, and 3,000, respectively. As can be seen from the figure, the mechanism of concentration release is entirely different in both cases (Liu and Khalili 2008, 2010b).

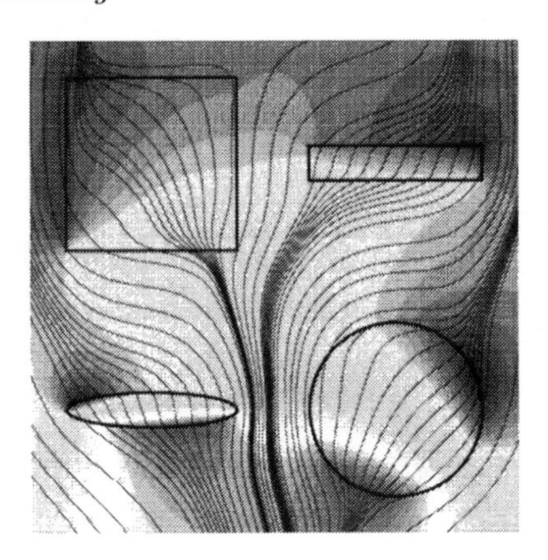

FIGURE 10.15
Streamlines (solid black lines) through and around a complex rectangular cell containing four different porous geometries (square, rectangle, ellipse, and circle). The gray contours represent the pressure distribution (high pressure below the circle and ellipse, low pressure at upper sides of the all geometries).

where g and ρ_0 denote acceleration due to gravity and a reference density, respectively. Naturally occurring stratifications range from $N \approx 0.01\,\mathrm{s}^{-1}$ in the ocean to $N \approx 0.2\,\mathrm{s}^{-1}$ in estuaries or fjords (Farmer and Ami 1999).

Stratification can have a significant impact on an important aspect of the marine carbon cycle, since the sedimentation of particulate organic matter is the main vector of carbon export from surface waters to the deep sea. Particles also affect marine ecology by providing an important resource for planktonic microorganisms. Marine particles of size $a \geq 0.5\,\mathrm{mm}$ are commonly referred to as marine snow. Marine snow typically consist of vestiges of phyto- and zooplankton, together with gel-like transparent exopolymer particles (TEP). Marine snow has high porosity, $\epsilon \geq 0.99$, and small excess density with respect to the ambient seawater, $\Delta\rho_p = \rho_p - \rho = O(10\,\mathrm{kg/m3})$ (Turner 2002). Because of the latter, the sinking of marine snow is characterized by low Reynolds numbers, $Re = aU/\nu = O(0.1-1)$, where U is the settling velocity and ν is the kinematic viscosity.

Marine snow is known to accumulate at pycnoclines in the ocean, forming thin layers that can persist for days and have highly elevated particle concentrations (McIntyre et al. 1995; Alldredge et al. 2002; McManus et al. 2003). It has been speculated that the retention of particles at pycnoclines is caused by the slow, diffusion-driven exchange between interstitial and ambient fluid at the pycnocline (Alldredge and Crocker 1995; Alldredge 1999). Diatom aggregates are nearly impermeable to flow (Ploug et al. 2002), hence the hydrodynamic properties of the aggregates are defined primarily by TEP. By reference to comparable gels, the permeability of TEP is estimated as $k \lesssim O(10^{-17}\,\mathrm{m}^{-2})$ (Jackson and James 1986). The diffusivity D of small molecules like sodium chloride is very nearly the same in TEP and in seawater (Ploug and Passow 2007).

Previous investigations of solid spheres settling through step-like stratifications at comparable Reynolds numbers, $Re=O(1)$, reported a reduction in settling speed at the pycnocline associated with an increase in drag. This excess drag resulted from the buoyancy of a wake of lighter fluid attached to and dragged downward by the particle (Srdic-Mitrovic et al. 1999). The magnitude of this "tailing" effect is governed by the relative importance of inertial and buoyancy forces, measured by the Froude number $Fr=U/(aN)$. A related effect has been observed for solid particles settling in linear stratifications, where a shell of lighter fluid is entrained by the particle, exerting a buoyancy force on the particle that retards its descent (Yick et al. 2007).

To study the effect of porosity on the settling process at a pycnocline, Kindler et al. (2010) recently conducted experiments with hydrogel spheres settling through a thin density interface with $N=7.2\,\mathrm{s}^{-1}$, using salt as the stratifying agent (Figure 10.16). The Reynolds number Re_0, based on the terminal velocity in the upper (lighter) phase U_0, was $O(0.1-1)$. The porosity of the spheres was $\phi=0.955$ and the permeability $k=10^{-15}\,\mathrm{m}^{-2}$, in general agreement with those of marine snow. The settling of porous, impermeable spheres through a pycnocline is based on two processes: the entrainment of lighter fluid from above, and the relaxation of the interstitial fluid (Figure 10.16). Depending on the initial particle excess density with respect to the lower (denser) layer, $\Delta\rho_p=\rho_1-\rho_{p_0}$, two limiting scenarios were identified. First,

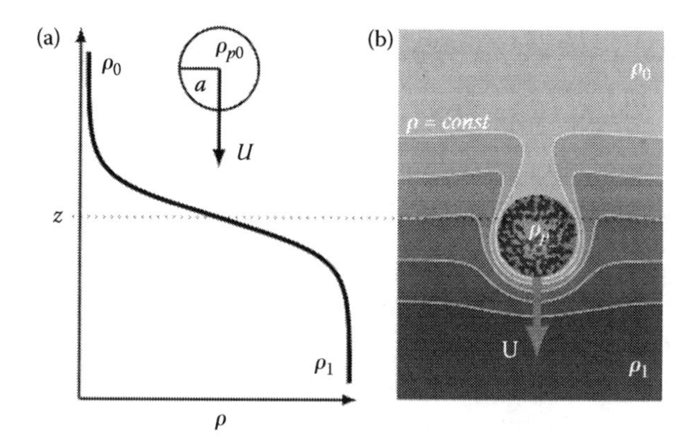

FIGURE 10.16
(a) Schematics of the density stratification. (b) Schematic illustration of the behavior of porous particles settling through a density interface, depending on the initial particle excess density ρ_{p_0} with respect to the lower phase density ρ_1. (From Kindler et al. (2010)).

when $\Delta\rho_p \geq 0$ the sphere decelerates as it enters the heavier layer, due to the decreased specific gravity. The sphere settles in response to the diffusional exchange of fluid from the pore space. In this case, the retention time at the pycnocline scales with the diffusional relaxation time, a^2/D. Second, if $\Delta\rho_p < 0$, particles can sink into the lower layer even in the absence of diffusional exchange of interstitial fluid. For large negative $\Delta\rho_p$, also implying large Re_0, the buoyancy of the interstitial fluid becomes negligible with respect to the effect of entrainment and the particle motion resembles that of a solid sphere in the lower layer.

In conclusion, Kindler et al. (2010) identified and verified a mechanism that can account for porous particle accumulation. However, the coupling of entrainment and diffusive effects for intermediate excess densities has to be clarified to consider more widely occurring, weaker stratifications. A better understanding of marine particle transport and retention within the water column will provide the basis for carbon transport modeling at the basin scale.

10.3.5 Enhanced Nutrient Exchange by Burrowing Macrozoobenthos Species

Chironomid larvae, known also as bloodworms, live on the river bed or lakes in u-tube-like burrows made from detritus (Figure 10.17a). The pupae of midges drift to the surface, where they rest before the adult fly emerges (Figure 10.17b). What makes these species interesting is that they enhance the exchange of dissolved substances between pore water and the overlying water body by their body motion while being in their burrows, and cause the so-called bioirrigation activity.

Microbial consequences and biogeochemical impacts of bioirrigation in benthic sediments have been long recognized and described in studies such as those related to filter feeding (Walshe 1947; Osovitz and Julian 2002), sediment biogeochemistry (Aller 1994; Stief and de Beer 2002; Lewandowski and Hupfer 2005) metabolic demand for oxygen (Polerecky et al. 2006; Timmermann et al. 2006), and the solute exchange between sediment and water (Meysman et al. 2006, 2007).

However, despite their high abundances ($\leq 4{,}000/\text{m}^2$) and their significant ecological role for processes both within and above the sediment, *Chironomus plumosus* provide challenging unsolved questions. Specifically, it was not clear until recently, how to quantify the flow rate pumped into the burrow. Using particle image velocimetry, Morad et al. (2010) and Roskosch et al. (2010) studied three different experimental setups to mimic the natural flow generated by the larvae. For this purpose, a setup was made allowing larvae to burrow their natural tubes in the sediment. A schematics of the burrow has been shown in Figure 10.18a. On the basis of velocity measurements, the volumetric flow rates could be calculated by integrating the velocities obtained by PIV (Figure 10.18b). Rigorous experiments performed showed

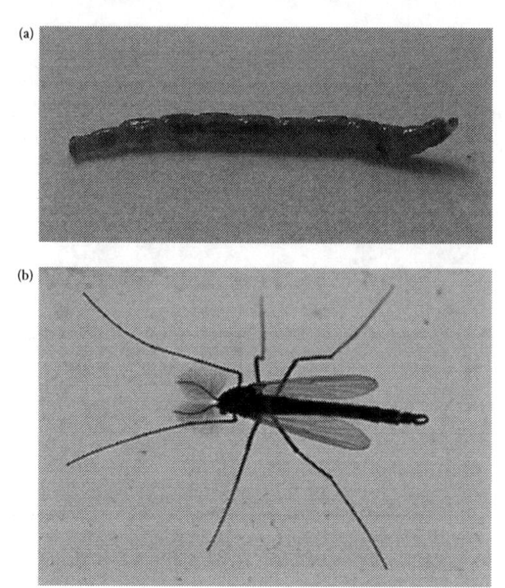

FIGURE 10.17
Chironomus plumosus larva (a) and after adult fly has emerged (b).

that the volumetric flow rates moved by the larvae was between 54.6 and 61.7 mm^3/s.

An early modeling of the effect of tube-dwelling animals was presented by Aller (1980). He defined a microenvironment in marine sediments as a single, tube-dwelling animal together with its surrounding sediment represented by a finite hollow cylinder. Ignoring advection, the transport of solutes within the bioturbated zone was then modeled within a microenvironment given by the diffusion-reaction equation:

$$\frac{\partial C}{\partial t} = D_s \frac{\partial^2 C}{\partial x^2} + \frac{D_s}{r} \frac{\partial}{\partial r}\left(r \frac{\partial C}{\partial r}\right) + R \qquad (10.35)$$

where x is depth in sediment relative to the sediment–water interface, r is the radial distance from the center of the tube/burrow, and t is time. Furthermore, the parameters C, D_s, and R are concentration of the dissolved solute, solute diffusion coefficient in bulk sediment, and reaction function, respectively. Equation (10.36) was solved subject to the initial and boundary conditions, such as constant concentration within the burrow by bioirrigation, or continuity of solute flux between the bioturbated and underlying sediment zones. The effect of sediment permeability was taken into account by correcting the diffusion coefficient via tortuosity.

Boudreau and Marinellli (1994) introduced modifications to the cylinder model allowing for periodic bioirrigation because the majority of infaunal

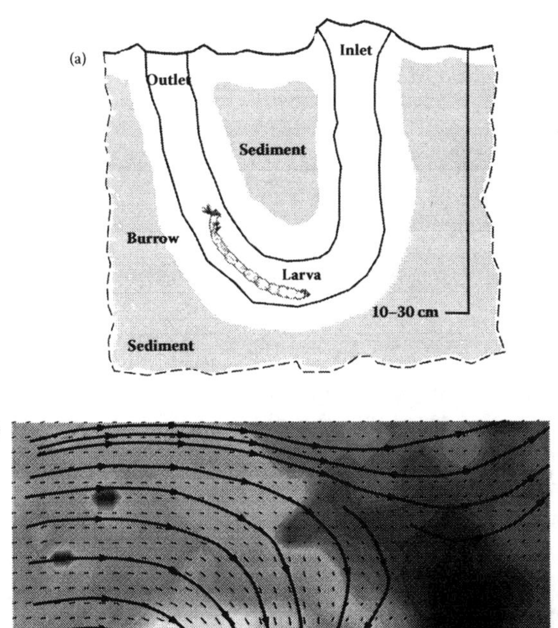

FIGURE 10.18
Schematics of a typical burrow made by *C. plumosus* (a) and the flow visualization near the burrow inlet (b).

burrow-dwelling organisms bioirrigate periodically. This model was further improved by Furukawa et al. (2001) who considered a more realistic depth-dependent distribution of burrows and burrow tilt angles rather than Aller's cylinder model with constant diameter and vertical direction for burrows.

Later, Timmermann et al. (2002) took the effect of advection into account. They injected water from overlying water column into a sediment depth they named Zone 2 (expected to be the feeding depth) to simulate bioturbation by *Arenicola marina*, a bioturbator which irrigates in J-shaped burrows. Zone 1 referred to the sediment above Zone 2, and Zone 3 was below the bioturbated zone where solutes are affected only by diffusion. Timmermann et al. (2002) considered the advection term and solved equation (10.37) considering the effect of sediment porosity in the set of advection–diffusion equations given by

$$
\begin{aligned}
&\text{Zone 1:} \quad \frac{\partial \phi C}{\partial t} = \frac{\partial}{\partial x}\left(D\frac{\partial \phi C}{\partial x}\right) + \frac{\partial v \phi C}{\partial x} \\
&\text{Zone 2:} \quad \frac{\partial \phi C}{\partial t} = \frac{\partial}{\partial x}\left(D\frac{\partial \phi C}{\partial x}\right) + \frac{\partial v \phi C}{\partial x} + S(x,t) \\
&\text{Zone 3:} \quad \frac{\partial \phi C}{\partial t} = \frac{\partial}{\partial x}\left(D\frac{\partial \phi C}{\partial x}\right)
\end{aligned}
\tag{10.36}
$$

where C is the solute concentration in a volume of pore water, ϕ is the porosity, $S(t,x)$ represents the source of solute because of injection of overlying water at feeding depth, $v(t,x)$ is the velocity of the advectively recirculating water and $D(x)$ represent the apparent diffusion coefficient in the sediment. Dividing pumping rate by the area of advection column and sediment porosity, they estimated the advective velocity used in equation (10.36).

A more recent model is presented by Meysman et al. (2006) for the same bioturbator, called the two-dimensional pocket injection model, which was regarded as the advective counterpart of Aller's "diffusive" two-dimensional tube bioirrigation model. They started from Darcy–Brinkman–Forchheimer equation as a general equation to model pore flow and neglected Brinkman-Forchheimer effects because of the low-pore velocity and large-length scales compared to the Brinkman layer involved in the problem, and finally employed the momentum balance reduced to Darcy's law (10.37).

$$
v_d = -\frac{k}{\mu}(\nabla p - \rho g \nabla x) \tag{10.37}
$$

In the above equation, k is the permeability, μ is the dynamic viscosity of the pore water, ρ is the pore water density, g is the gravitational acceleration, and x is the vertical coordinate. The Darcy velocity v_d is related to the actual velocity of pore water as $v_d = \phi v$, where ϕ is the porosity. A commercially available code (Comsol Multiphysics) was employed to solve (10.38) and the results were used to solve the reactive transport equations for concentrations.

To date, no exclusive modeling is performed on bioturbation because of U-shape burrows, nor is there any model to contribute the animal's motion characteristics to the flow generated along the burrows and in the sediment.

Beside this issue, there exists still a good number of challenging questions. One of the most prominent one is whether or not the pumping strength of *C. plumosus* is sufficient to mediate an additional advective flow through the burrow walls they construct. Some recent simple models are available for this problem (Shull et al. 1995), however, a real evidence in the form of a two-dimensional flow in a porous-fluid-burrow domain has not yet been provided. Such a model is under development (Morad et al. 2010b) for the geometry shown in Figure 10.19 (a) with a porous sediment having a fixed tube with permeable walls underneath a fluid layer. Using high speed cameras, the equation for the larva's motion has been obtained by digital analysis, and inserted as an input to the model. By solving the momentum equations

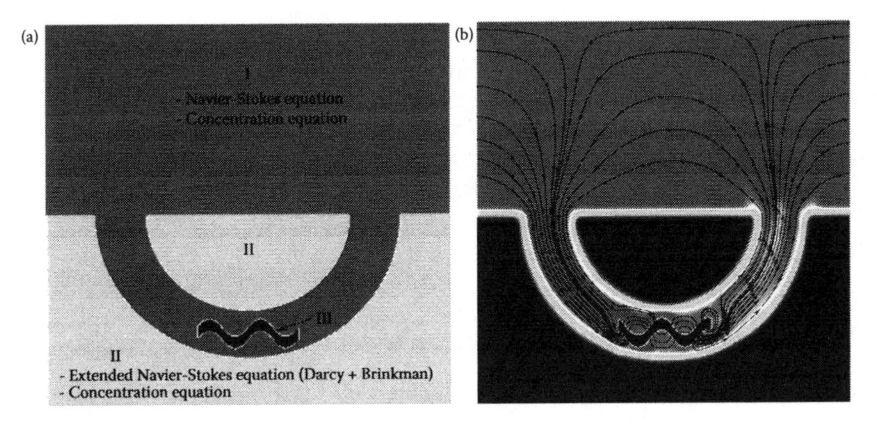

FIGURE 10.19
Geometry of the burrow and the overlying water column with the larva motion simulated (a) and the streamlines and the concentration distribution (gray scale) due to the larva's pumping (b).

as well as the concentration equation, the velocity and concentration field (Figure 10.19) (b) could be obtained precisely. The comparison of the outlet or inlet velocity field obtained in the simulation with the same from the PIV measurements resulted in a good agreement (Morad et al. 2010b).

10.4 Future Prospectives

Porous media applications are ubiquitous not only in technical but also in marine and biological studies, from which some current examples were given in this chapter. However, the real challenge is still to come, namely, in combining visualization and modeling in micro- and nano-scale biology, which, to a larger extent, depends on the progress in multiscale transport processes in complex porous media. Progress in porous media can help characterizing microbial communities to a small scale. Systems approaches require precise analyzes of the spatio-temporal properties of multiple existing microbial environments available in diverse microbiological applications. Exactly, this can be done by the lattice Boltzmann models that are capable of accounting for both the complex structures and the microbial growth and activity on changes to surface wetting induced by surfactants (O'Donnell et al. 2007).

10.5 References

Alldredge, A. L. (1999). The potential role of particulate diatom exudates in forming nuisance mucilaginous scums. *Annali Dell Istituto Superiore di Sanita*, **35**(3):397–400.

Alldredge, A. L. and Crocker, K. M. (1995). Why do sinking mucilage aggregates accumulate in the water column? *Science of the Total Environment*, **165**:15–22.

Alldredge, A. L., Cowles, T. J., McIntyre, S., Rins, J. E. B., Donaghay, P. L., Greenlaw, C. F., Holliday, D. V., Dekshenieks, M. M., Sullivan, J. M., and Zaneveld, J. R. V. (2002). Occurence and mechanisms of formation of a dramatic thin layer of marine snow in shallow Pacific fjord. *Marine Ecology Progress Series*, **233**:1.

Aller, R. C. (1980). Quantifying solute distributions in the bioturbated zone of marine sediments by defining an average microenvironment. *Geochimica Cosmochimica Acta*, **44**:1955–1965.

Aller, R. C. (1994). Bioturbation and remineralization of sedimentary organic matter: effects of redox oscillation. *Chemical Geology*, **114**:331–345.

Archie, G. (1942). The electrical resistivity log as an aid in determining some reservoir characteristics. *Transactions of the American Institute of Mining and Metallurgical Engineers*, **146**:54–62.

Azam, F. and Long, R. A. (2001). Sea snow microcosms. *Nature*, **414**:495–498.

Berelson, W. M., Hammond, D. E., and Fuller, C. (1982). Radon-222 as a tracer for mixing in the water column and benthic exchange in the southern california borderland. *Earth and Planetary Science Letters*, **61**:41–54.

Berner, R. A. (1976). The bentic boundary layer from the viewpoint of a geochemist. In I. N., Mc Cave, ed. *The benthic boundary layer*. Plenum Press, New York.

Bhattacharyya, S., Dhinakaran, S., and Khalili, A. (2006). Fluid motion around and through a porous cylinder. *Chemical Engineering Science*, **61**:4451–4461.

Boudreau, B. P. and Jørgensen, B. B., ed. (2001). *The Benthic Boundary Layer: Transport Processes and Biogeochemistry*. Oxford University Press, Oxford.

Boudreau, B. P. and Marinelli, R. L. (1994). A modelling study of discontinuous biological irrigation. *Journal of Magnetic Resonance*, **52**:947–968.

Boudreau, B. P. and Meysman, F. J. R. (2006). Predicted tortuosity of muds. *Geology*, **34**:693–696.

Chen, D., Mioshi, H., Akai, T., and Yazawa, T. (2005). Colorless transparent fluorescence material: Sintered porous glass containing rare-earth and transition-metal ions. *Applied Physics Letters*, **86**:231908–1–231908–3.

Chisholm, A. W. (2000). Oceanography: stirring times in the southern ocean. *Nature*, **407**:685–687.

Chopard, B. and Droz, M. (1998). *Cellular Automata Modeling of Physical Systems.* Cambridge University Press, Collection Aléa.

Clennell, M. B. (1997). Tortuosity: a guide through the maze. *Geological Society, London, Special Publications*, **122**:299–344.

Darcy, H. P. G. (1856). *Les fontaines publiques de la ville de Dijon.* Victor-Dalmont, Paris.

Dellar, P. J. (2003). Incompressible limits of lattice boltzmann equations using multiple relaxation times. *Journal of Computational Physics*, **190**:351–370.

Emerson, S., Jahnke, R., Bender, M., Froelich, P., Klinkhammer, G., Bowser, C., and Setlock, G. (1980). Early diagenesis in sediments from the eastern equatorial pacific, i. pore water nutrient and carbonate results. *Earth and Planetary Science Letters*, **49**:57–80.

Emery, K. O. (1968). Relict sediments on continental shelves of the world. *American AAPG Bulletin*, **52**:52.

Farmer, D. and Ami, L. (1999). The generation and trapping of solitary waves over topography. *Science*, **283**:188.

Furukawa, Y., Samuel, S. J., and Lavoie, D. L. (2001). Bioirrigation modeling in experimental benthic mesocosms. *Journal of Magnetic Resonance*, **59**:417–452.

Glud, R., Forster, S., and Huettel, M. (1996). Influence of radial pressure gradients on solute exchange in stirred benthic chambers. *Marine Ecology Progress Series*, **141**:303–311.

Glud, R. N., Berg, P., Fossing, H., and Jørgensen, B. B. (2007). Effect of diffusive boundary layer on benthic mineralization and O_2 distribution: a theoretical model analysis. *Journal of Limnology and Oceanography*, **52**: 547–557.

Gundersen, J. K. and Jørgensen, B. B. (1990). Microstructure of diffusive boundary layers and the oxygen uptake of the sea floor. *Nature*, **345**: 604–607.

Guo, Z., Zheng, C., and Shi, B. (2002). Discrete lattice effects on the forcing term in the lattice boltzmann method. *Physics Review E*, **65**:046308.

Gust, G. (1990). Patent US-4,973,165.

Hara, T., Mei, C. C., and Shum, K. T. (1992). Oscillating flows over periodic ripples of finite slope. *Physics of Fluids*, **4**:1373–1384.

He, X. and Luo, L.-S. (1997). Lattice boltzmann model for the incompressible Navier–Stokes equation. *Journal of Statistical Physics*, **88**: 927–944.

Huettel, M. and Gust, G. (1992a). Solute release mechanisms from confined sediment cores in stirred benthic chambers and flume flows. *Marine Ecology Progress Series*, **82**:187–197.

Huettel, M. and Gust, G. (1992b). Impact of bioroughness on interfacial solute exchange in permeable sediments. *Marine Ecology Progress Series*, **89**:253–267.

Iversen, N. and Jørgensen, B. B. (1993). Diffusion coefficients of sulfate and methane in marine sediments: Influence of porosity. *Geochimica Cosmochimica*, **57**:571–578.

Ivey, G. N., Winters, K., and Silva, I. P. D. S. D. (2000). Turbulent mixing in a sloping benthic boundary layer energized by internal waves. *Journal of Fluid Mechanics*, **48**:59–76.

Jackson, G. W. and James, D. F. (1986). The permeability of fibrous media. *Canadian Journal of Chemical Engineering*, **64**:364–374.

Javadi, K. and Khalili, A. (2009). On generating uniform bottom shear stress. Part II: shear stress inducing devices. *Recent Patents in Chemical Engineering*, **2**(3):223–229.

Johnson, D. L., Plona, T. J., Scala, C., Pasierb, F., and Kojima, H. (1982). Tortuosity and acoustic slow waves. *Physics Review Letters*, **49**:1840–1844.

Jue, T.-C. (2003). Numerical analysis of vortex shedding behind a porous square cylinder. *International Journal of Numerical Methods for Heat and Fluid Flow*, **14**:649–663.

Khalili, A. and Javadi, K. (2009). How to produce uniform shear stress? In preparation.

Khalili, A., Javadi, K., Saidi, A., Goharzadeh, A., Huettel, M., and Jørgensen, B. B. (2008). On generating uniform bottom shear stress, Part I: a quantitative study of microcosm chamber. *Recent Patents in Chemical Engineering*, **1**:174–191.

Kindler, K., Khalili, A., and Stocker, R. (2010). Accumulation of porous particles settling through pycnoclines. Submitted.

Kiørboe, T., Ploug, H., and Thygesen, U. H. (2001). Fluid motion and solute distribution around sinking aggregates. I. Small-scale fluxes and heterogeneity of nutrients in the pelagic environment. *Marine Ecology Progress Series*, **211**:1–13.

Knackstedt, M. A. (1994). Direct evaluation of length scales and structural parameters associated with flow in porous media. *Physics Review E*, **50**:2134–2138.

Koponen, A., Kataja, M., and Timonen, J. (1996). Tortuous flow in porous media. *Physics Review E*, **54**:406–410.

Koza, Z., Matyka, M., and Khalili, A. (2009). Finite-size anisotropy in statistically uniform porous media. *Physics Review E*, **79**:066306–1–066306–7.

Kroner, K. and Vogel, J. (2001). US20016193883B1.

Leppard, G. G., Mavrocordatos, D., and Perret, D. (2004). Electron-optical characterization of nano- and micro-particles in raw and treated waters: an overview. *Water Science and Technology*, **50**:1–8.

Lewandowski, J. and Hupfer, M. (2005). Effect of macrozoobenthos on two-dimensional smallscale heterogeneity of pore water phosphorus concentrations in lake sediments: a laboratory study. *Limnology Oceanography*, **50**:1106–1118.

Liu, B. and Khalili, A. (2008). Acceleration of steady-state Lattice Boltzmann simulation for exterior flow. *Physics Review E*, **78**:056701–1–056701–9.

Liu, B. and Khalili, A. (2009). Lattice Boltzmann model for exterior flows with an annealing preconditioning method. *Physics Review E*, **79**:066701–1–066701–7.

Liu, B. and Khalili, A. (2010a). Oscillatory flow over permeable beds, In preparation.

Liu, B. and Khalili, A. (2010b). Concentration release from sinking aggregates of an arbitrary shape, in preparation.

Logan, B. E. and Wilkinson, D. B. (1990). Fractal geometry of marine snow and other biological aggregates. *Limnology Oceanography*, **35**:130–136.

Lorenz, P. B. (1961). Tortuosity in porous meida. *Nature*, **189**:386–387.

Matyka, M., Khalili, A., and Koza, Z. (2008). Tortuosity–porosity relation in porous media flow. *Physics Review E*, **78**:026306–1–026306–8.

McIntyre, S., Alldredge, A. L., and Gotschalk, C. C. (1995). Accumulation of marine snow at density discontinuities in the water column. *Limnology and Oceanography*, **40**(3):449–468.

McManus, M. A., Alldredge, A. L., Barnard, A., Boss, E., Case, J. F., Cowles, T. J., Donaghay, P. L., Eisner, L. B., Gifford, D. J., Greenlaw, C. F., Herren, C. M., Holliday, D. V., Johnson, D., McIntyre, S., McGehee, D. M., Osborne,

T. R., Perry, M. J., Pieper, R. E., Rines, J. E. B., Smith, D. C., Sullivan, J. M., Talbot, M. K., Twardowski, M. S., Weidmann, A., and Zaneveld, J. (2003). Characteristics, distribution and persistence of thin layers over a 48 hour period. *Marine Ecology Progress Series*, **261**:1.

Meysman, F. J. R., Galaktionov, O. S., Cook, P. L. M., Janssen, F., Huettel, M., and Middelburg, J. J. (2007). Quantifying biologically and physically induced flow and tracer dynamics in permeable sediments. *Biogeosciences*, **4**:627–646.

Meysman, F. J. R., Galaktionov, O. S., Gribsholt, B., and Middelbur, J. J. (2006). Bioirrigation in permeable sediments: advective pore-water transport induced by burrow ventilation. *Journal of Limnology and Oceanography*, **51**:142–156.

Morad, M. R., Khalili, A., Roskosch, A., and Lewandowski, J. (2010). Quantification of pumping rate by chironomus plumosus larvae in real burrows. *Aquatic Ecology*, **44**(1):143–153.

Morad, M. R., Liu, B., and Khalili, A. (2010). Hydrodynamics generated by the *C. Plumosus* larva: an experimental and mathematical study, in preparation.

Nakashima, Y. and Yamaguchi, T. (2004). DMAP.m: a mathematica program for three-dimensional mapping of tortuosity and porosity of porous media. *Bulletin of the Geological Survey of Japan*, **55**:93–103.

Nikora, V., Goring, D., and Ross, A. (2002). The structure and dynamics of the thin near-bed layer in a complex marine environment: A case study in Beatrix Bay, New Zealand. *Estuarine, Coastal and Shelf Science*, **54**: 915–926.

O'Donnell, A. G., Young, I. M., Rushton, S. P., Shirley, M. D., and Crawford, J. W. (2007). Visualization, modelling and prediction in soil microbiology. *Nature Reviews Microbiology*, **5**:689–699.

Oldham, C. E., Ivey, G. N., and Pullin, C. (2004). Estimation of a characteristic friction velocity in stirred benthic chambers. *Marine Ecology Progress Series*, **279**:291–295.

Osovitz, C. J. and Julian, D. (2002). Burrow irrigation behaviour of *Urechis caupo*, a filter-feeding marine invertebrate, in its natural habitat. *Marine Ecology Progress Series*, **473**:149–155.

Pamatmat, M. M. and Banse, K. (1969). Oxygen consumption by the seabed. ii. in situ measurements to a depth of 180 m. *Journal of Limnology and Oceanography*, **14**:250–259.

Ploug, H. and Passow, U. (2007). Direct measurements of diffusivity within diatom aggregates containing transparent exopolymer particles. *Limnology and Oceanography*, **52**:1–6.

Polerecky, L., Volkenborn, N., and Stief, P. (2006). High temporal resolution oxygen imaging in bio-irrigated sediments. *Environmental Science Technology*, **40**:5763–5769.

Rao, A. M. F., Mccarthy, M. J., Gardner, W. S., and Jahnke, R. A. (2007). Respiration and denitrification in permeable continental shelf deposits on the south atlantic bight: Rates of carbon and nitrogen cycling from sediment column experiments. *Continental Shelf Research*, **27**:1801–1819.

Riisgård, H. U. and Larsen, P. S. (2005). Water pumping and analysis of flow in burrowing zoobenthos—an overview. *Journal of Experimental Biology*, **198**:283–294.

Roskosch, A., Morad, M. R., Khalili, A., and Lewandowski, J. Bioirrigation by Chironomus plumosus: advective flow investigated by particle image velocimetry. Accepted.

Røy, H. and Lee, J. S. (2008). Tide-driven deep pore-water flow in intertidal sands flats. *Limnology Oceanography*, **53**:1521–1530.

Sayles, F. L. (1979). The composition and diagenesis of interstitial solutions. fluxes across the seawater–sediment interface in the atlantic ocean. *Geochimica Cosmochimica Acta*, **43**:527–545.

Shull, D., Benoit, J., Wojcik, C., and Senning, J. (1995). Infaunal burrow ventilation and pore-water transport in muddy sediments. *Estuarine, Coastal, and Shelf Science*, **83**:277–286.

Shum, K. T. (1992a). Wave-induced advective transport below a rippled water-sediment interface. *Journal of Geophysics Research*, **97**(C1):789–808.

Shum, K. T. (1992b). The effects of wave-induced pore water circulation on the transport of reactive solutes below a rippled sediment bed. *Journal of Geophysics Research*, **98**(C6):10289–10301.

Shum, K. T. (1995). A numerical study of the wave-induced solute transport above a rippled bed. *Journal of Fluid Mechanics*, **299**:267–288.

Smith Jr., K. L. and Teal, J. M. (1973). Deep-sea benthic community respiration: an in situ study at 1850 meters. *Science*, **179**:282–283.

Srdic-Mitrovic, A. N., Mohamed, N. A., and Fernando, H. J. S. (1999). Gravitational settling of particles through density interfaces. *Journal of Fluid Mechanics*, **381**:175–198.

Stief, P. and de Beer, D. (2002). Bioturbation effects of chironomus riparius on the benthic n-cycle as measured using microsensors and microbiological assays. *Aquatic Microbial Ecology*, **2**:175–185.

Succi, S. (2001). *The Lattice Boltzmann Equation for Fluid Dynamics and Beyond.* Oxford University Press, Oxford.

Sun, M. and Lee, S. (2005). US20050032200.

Timmermann, K., Banta, G. T., and Glud, R. N. (2006). Linking arenicola marina irrigation behavior to oxygen transport and dynamics in sandy sediments. *Journal of Marine Research*, **64**:915–938.

Timmermann, K., Christensen, J. H., and Banta, G. T. (2002). Modeling of advective solute transport in sandy sediments inhabited by the lugworm arenicola marina. *Journal of Marine Research*, **60**:151–169.

Ting, T. and Chen, Y. (2008). US20080038816A1.

Vance-Harris, C. and Ingall, E. (2005). Denitrification pathways and rates in the sandy sediments of the georgia continental shelf, USA. *Geochemical Transactions*, **6**(1):12–18.

Vogel, H., Anspach, B., Kroner, K., Piret, J., and Haynes, C. (2002). Controlled shear affinity filtration (csaf): A new technology for integration of the cell separation and protein isolation from mammalian cell cultures. *Biotechnology Bioengineering*, **78**:806–814.

Walshe, B. M. (1947). Feeding mechanism of chironomus larvae. *Nature*, **160**:474.

Weissberg, J. (1963). Effective diffusion coefficient in porous media. *Journal of Applied Physics*, **34**:2636–2639.

Wu, H. R., He, Y. L., Tang, G. H., and Tao, W. Q. (2005). Lattice Boltzmann simulation of flow in porous media on non-uniform grids. *Progressive Computational Fluid Dynamics*, **5**(1/2):97–103.

Zhang, X. and Knackstedt, M. A. (1995). Direct simulation of electrical and hydraulic tortuosity in porous solids. *Geophysics Research Letters*, **22**:2333–2338.

Zou, Q., Hou, S., and Doolen, G. D. (1995). Analytical solutions of the lattice boltzmann bgk model. *Journal of Statistical Physics*, **81**(1/2):319–334.

11

The Transport of Insulin-Like Growth Factor through Cartilage

Lihai Zhang

Department of Civil and Environmental Engineering, The University of Melbourne, Melbourne, Australia

Bruce S. Gardiner, David W. Smith, Peter Pivonka

Faculty of Engineering, Computing and Mathematics, The University of Western Australia, Western Australia, Australia

Alan J. Grodzinsky

Center for Biomedical Engineering, Department of Electrical Engineering and Computer Science, Department of Mechanical Engineering, Massachusetts Institute of Technology, Cambridge, MA

CONTENTS

11.1 Overview

Articular cartilage is the smooth glistening white tissue, slippery to feel, that covers the surface of the diarthrodial joints. The functions of articular cartilage in joints are to reduce load per unit area and to provide a smooth, low-friction, and wear resistant bearing [1]. The composition and structural properties of cartilage allow it to achieve and maintain proper biomechanical function over the majority of a human lifespan [2]. Structurally, cartilage is a porous, fluid-filled, deformable material that is composed of cells (called chondrocytes), a fluid (predominantly water), and an ECM [3]. The ECM largely consists of collagen, proteoglycan, and small amounts of other molecules [4].

The mechanical properties of cartilage are determined, directly and indirectly, primarily by the collagen and proteoglycan components [5]. Chondrocytes also determine cartilage mechanical properties indirectly by regulating the cartilage composition through the synthesis and degradation of matrix components [5]. Chondrocytes do this by responding to their chemical and mechanical "microenvironment" [6]. Proteoglycans produced by chondrocytes carry a strong negative charge, and the repulsion between these molecules (electrical and osmotic pressures act in various proportions), giving cartilage a tendency to swell and so resist compressive forces [7–9]. A collagen network within the cartilage (which is anchored in the underlying bone) provides cartilage with tensile strength, and constrains the swelling of proteoglycans, and slows the movement of the proteoglycans in the tissue. In other words, equilibrium is achieved with the collagen in a state of tensile prestress that is induced by the proteoglycans in a state of compressive prestress [5]. If the collagen matrix is compromised (e.g., perhaps due to excessive mechanical loads or inflammatory processes), the cartilage can swell and proteoglycans can escape from the cartilage tissue [10]. These structural changes in the cartilage in turn lead to excessive cartilage deformation and so an abnormal chemical and mechanical microenvironment for the chondrocytes. The cartilage is then prone to increased rates of wear and degradation. When this occurs the joint is said to be diseased [11,12]. Recently, it was shown that this process of degradation can be very rapid following high-impact loads [13].

Unlike neighboring bone, articular cartilage lacks nerves to sense potentially damaging loads, and also lacks blood vessels to deliver nutrients and growth factors to the chondrocytes. As a consequence the cartilage has limited ability to repair itself. In many disease processes arising from mechanical damage (e.g., trauma) or biochemical injury (e.g., inflammatory processes) or both, there is a loss of proteoglycans and the cartilage degrades. When the cartilage wears away sufficiently, movement at an articular joint becomes painful, and now a days many diseased joints are replaced with artificial prostheses [14]. However, these prostheses require an expensive operation and all too frequently fail (revision rates are currently about 15%) [15]. Alternate techniques to treat cartilage defects include chondrocyte seeding (e.g., so-called autologous grafting [16]) and surgical damage to the subchondral bone to promote tissue repair [17], but fibrous rather than hyaline cartilage develops in about 25% of these cases [18]. Another approach, entering clinical trials in some countries, is growing neo-cartilage using cell-seeded scaffolds [19] though no such "tissue engineered" solutions are yet widely accepted [5]. While some approaches are very promising, all of them are basically a "let us try it and see" strategy. This approach stems from an absence of a fundamental understanding of the physical and chemical conditions that promote and maintain healthy cartilage tissue. It is quite conceivable that cartilage behavior is too complex to yield to intuition alone. Instead, computational modeling, which integrates mechanical, fluid, and biochemical processes, is essential to develop rational methodologies for developing new treatments. Understanding quantitatively the relative importance of these factors would substantially increase

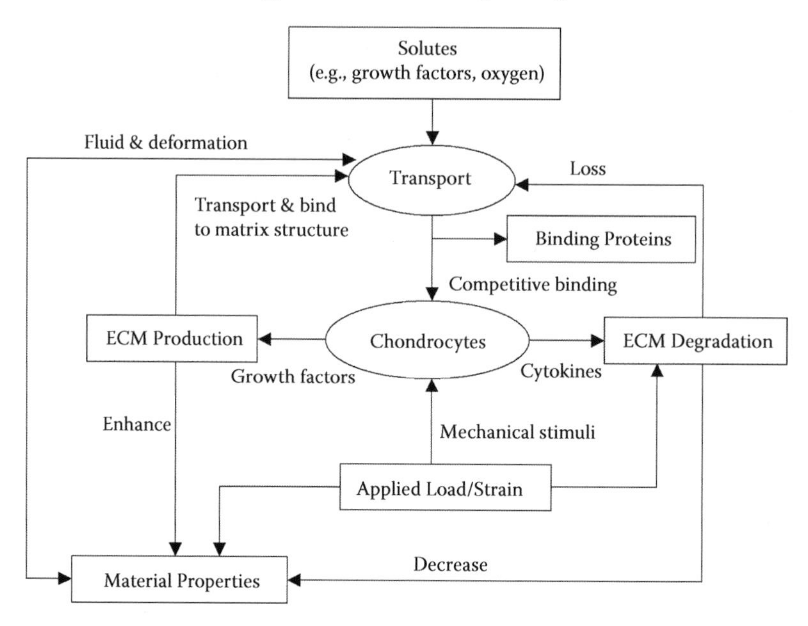

FIGURE 11.1
Schematic of feedback loops that control the mechanical and chemical environment sensed by chondrocytes embedded within the cartilage matrix; these feedback loops are at the "tissue level," but there exists many more feedback loops at the intercellular and intracellular levels, which enable cartilage to perform its physiological function.

our ability to coax cartilage toward ongoing health, or to optimize the conditions for cartilage repair once it is damaged.

As indicated earlier the current limitation in our understanding of cartilage homeostasis is largely due to the large number of complex processes occurring simultaneously. To illustrate this point diagrammatically, some of the key tissue level processes are shown in Figure 11.1. The transport of solutes (e.g., oxygen, growth factors, and cytokines) to the chondrocytes (which forms part of the chemical microenvironment "seen" by the cell), as well as the transport of newly synthesized matrix components [20] are coupled to the mechanical microenvironment, both directly through advective transport and indirectly through changes in the matrix physical properties (e.g., mechanical stiffness, diffusion properties, and hydraulic conductivity) due to changes in biosynthesis of proteoglycans. The system is further complicated by the presence of families of binding proteins, competitor growth factors, and cell surface receptors.

Approximately, 60%–80% of proteoglycans within the cartilage are believed to be in stable forms (i.e., effectively immobilized within the ECM) [20]. Newly synthesized proteoglycan monomers, secreted by the chondrocytes, are initially mobile until the majority become bound to hyaluronan in the form of supramolecular aggregates; other newly synthesized aggrecan monomers may be lost from the tissue by diffusion along a chemical potential gradient [20]. Finally under normal turnover conditions, a small population of newly synthesized and/or preexisting aggrecan are degraded, predominantly by one of the aggrecanases [21]. These aggrecan molecules are soluble and mobile in interstitial fluid, and may be transported through the ECM [22–24]. The newly synthesized aggrecans need to be transported away from chondrocytes after secretion and ultimately incorporated into the tissue matrix, while degraded aggrecans may find their way out of the cartilage ECM and are discharged into the synovial fluid. The turnover rate of proteoglycan in normal rabbit cartilage is around 21 days [25], however, the half-life of proteoglycan *in vivo* is much longer in human cartilage (estimated to be approximately 3.4 years) [26]. Additional complexity arises because cell responses are mediated via intracellular signaling pathways [27], which then lead to protein production. To "top it all off" many of these key processes are linked via feedback loops at a variety of length and time scales. The list of possible components and processes involved is truly daunting. Tackling problems of this complexity require multiscale hierarchical modeling approaches, and multiple models that suit the questions being investigated. Here we have begun the process of building a continuum reactive-transport poroelastic model at the tissue level. Clearly, this can be extended in the future.

We specifically focus on only two questions. First, what processes are involved in the transport of large molecules like growth factors into cartilage? The absence of blood and lymph vessels in cartilage implies that diffusive transport must play an important role in delivering nutrients and growth factors to chondrocytes. However, proteins such as growth factors are usually large molecules that do not diffuse easily. How might cartilage regulate its exposure to growth factors, and can we devise strategies to increase chondrocytes exposure to these molecules? Second, we build a model to investigate the interplay between two key stimulators of aggrecan production: IGF-I and cyclic mechanical loading. In doing so, we will describe a model framework that could be expanded naturally in a variety of ways.

In the next section a coupled solute transport and mechanical deformation model for articular cartilage [28] is presented, and this model is employed to examine the effect of advection on growth factor transport. This initial model will form the basis of a series of model "extensions" encompassing an increasing array of complex processes. Wherever possible, currently available experimental data [29–32] is used to validate the models.

More specifically in Section 11.3.2, the model is extended to incorporate reversible time-dependent binding of IGF to an IGFBP [33]. In Section 11.4.1.1, the model is extended to include a family of six IGFBPs in two functional groups to understand the effect of competitive binding on IGF-I transport in both normal and diseased cartilage [34]. In Section 11.4.2 a competitor growth factor is introduced into the transport model [34]. The models developed in Sections 11.4–11.5 are then used to describe the chemical microenvironment of chondrocytes in an aggrecan biosynthesis model presented in Section 11.5.2.3. This aggrecan biosynthesis model also includes fluid–shear-induced biosynthesis and aggrecan production, transport, and degradation.

11.2 Basic Solute Transport Model in a Deforming Articular Cartilage

11.2.1 Introduction

The absence of blood and lymph vessels in cartilage implies that diffusive transport must play an important role in delivering nutrients and growth factors to chondrocytes. However, important proteins such as growth factors are large molecules that do not diffuse easily. How then, might cartilage maximize its exposure to growth factors? Potentially, growth factor transport could be enhanced by advective transport. When cartilage is released from a compressive deformation, interstitial fluid can move into the cartilage, taking growth factors (and other molecules) with it.

It has been speculated that physiological relevant loading enhances growth factor transport (and subsequent matrix biosynthesis) via advection. That is, perhaps a walking pace induces cartilage deformation and interstitial fluid flow at a frequency and amplitude conducive to advective transport of growth factors. This speculation arose from an observed synergy between cyclic loading and growth factors on aggrecan production in cartilage explants and chondrocyte-seeded gels [35–37]. Further, several experiments directly aimed at measuring solute uptake or release from cartilage undergoing cyclic loading have also indicated enhanced solute transport [38,39]. Subsequently, several models have been constructed to examine the effect of cyclic loading on solute transport into cartilage [28,36,40]. These models are usually based on the theory of porous media.

To better understand this process and to introduce the underlying equations needed to describe coupled solute transport in a deforming tissue, a basic continuum coupled solute transport; poroelastic model is first presented. The use of porous media theory to model cartilage mechanical behavior (and other biological tissues) is not a recent idea [41–43], however, using a coupled reactive-transport poroelastic model to understand the transport of nutrients through a cyclically loaded cartilage is relatively new [36,40,44]. The basic model described in the next section is further developed in later sections, as new behaviors are added to the model.

11.2.1.1 Modeling Cartilage Using the Theory of Porous Media

In the framework of the theory of porous media, cartilage is modeled as an intrinsically incompressible solid phase, representing the collagen–proteoglycan matrix, and an incompressible fluid phase. This approach has demonstrated its capability in describing the experimentally observed mechanical response of cartilage [43] and other biological tissues [41–43]. Like many other biological tissues, articular cartilage exhibits significant viscoelastic-like behavior (i.e., time-dependent deformation processes under external loads), and this viscoelastic-like behavior is well described by the continuum porous media approach [45].

Mow et al. (1980) [46] proposed a so-called biphasic mixture model independent of the phenomenological approach to the theory of porous media. However, the biphasic **mixture theory of porous media** (based on Bowen's approach to mixture theory [47]) and the **phenomenological theory of porous media** approach lead to the same basic governing equations [42]. The biphasic model has found considerable use to describe cartilage mechanics [42,48–50]. Lai et al. [7] developed a triphasic model by adding ions into the biphasic model to describe the deformation and stress fields of cartilage under chemical potentials and mechanical loading. The total stress tensor within the cartilage can be treated as the sum of the solid matrix elastic stress, interstitial fluid pressure, and chemical-expansion pressure. The triphasic model has been extended by Gu et al. [51] to include multielectrolytes.

We begin by presenting a basic fully coupled reactive-transport poroelastic model for cartilage based on the phenomenological theory of porous media. Later this model is extended to include a wide-range of biological processes occurring with cartilage involving IGF-I transport, IGF interaction with its binding proteins (IGFBPs), IGF-I and interstitial fluid flow induced matrix biosynthesis, and the transport and degradation of matrix molecules.

11.2.2 Basic Solute Transport Model in Cyclically Loaded Cartilage

In this section, a basic model will be introduced by treating cartilage as a three phase mixture, a solid phase representing ECM, a fluid phase representing interstitial fluid, and a solute phase representing a dissolved solute.

The volume fraction of each phase is

$$\phi^\alpha = \frac{V^\alpha}{V} \tag{11.1}$$

where V is the overall volume of cartilage representative volume element (RVE) and V^α is the volume of α phase. The superscripts s, f, and w indicate the solid, fluid, and solute phase in the RVE, respectively. As the volume of solute phase is relatively small compared with the solid and fluid phases, it can be assumed that $\phi^s + \phi^f \approx 1$. With the volume fractions defined above the concentration of solute relative to the α phase (c^α) can be expressed as

$$c^\alpha = \frac{\bar{c}^\alpha}{\phi^\alpha} \tag{11.2}$$

where $\bar{c}^{\alpha}$ is the RVE volume-based solute concentration. Likewise, the density of the solid and fluid phase may now be expressed in terms of the phase volume fraction, that is,

$$\rho^{\alpha} = \frac{m^{\alpha}}{V^{\alpha}} = \frac{m^{\alpha}}{\phi^{\alpha} V} \tag{11.3}$$

where m^{α} is the mass of each phase.

11.2.2.1 Conservation of Mass

Conservation of mass of the solid phase and fluid phase can be expressed as

$$\frac{\partial(\phi^{s}\rho^{s})}{\partial t} + \nabla \bullet (\phi^{s}\rho^{s}\mathbf{v}^{\mathrm{s}}) = 0 \tag{11.4}$$

$$\frac{\partial(\phi^{f}\rho^{f})}{\partial t} + \nabla \bullet (\phi^{f}\rho^{f}\mathbf{v}^{\mathrm{f}}) = 0 \tag{11.5}$$

where ρ^{s} and ρ^{f} are the density of the solid and fluid phase, respectively; $\mathbf{v}^{\mathrm{s}}$ and $\mathbf{v}^{\mathrm{f}}$ are the velocity of solid and fluid phase, respectively, $\nabla\bullet$ denotes the divergence operator, and ∇ denotes the gradient operator.

For incompressible solid and fluid phases (a reasonable approximation for cartilage), equations (11.4) and (11.5) may be rewritten as

$$-\frac{\partial \phi^{f}}{\partial t} + \nabla \bullet \left[\left(1-\phi^{f}\right)\mathbf{v}^{\mathrm{s}}\right] = 0 \tag{11.6}$$

$$\frac{\partial \phi^{f}}{\partial t} + \nabla \bullet \left(\phi^{f}\mathbf{v}^{\mathrm{f}}\right) = 0 \tag{11.7}$$

Summing equations (11.6) and (11.7) leads to

$$\nabla \bullet \left[\left(1-\phi^{f}\right)\mathbf{v}^{\mathrm{s}} + \phi^{f}\mathbf{v}^{\mathrm{f}}\right] = 0 \tag{11.8}$$

The fluid phase velocity within an RVE can be related to the solid phase velocity using Darcy's law. The Darcy velocity ($\mathbf{v}_{\mathbf{d}}$) is the velocity of the fluid phase relative to the solid phase. Darcy velocity can be assumed to be proportional to the fluid pressure gradient (i.e., following Darcy's law). That is,

$$\mathbf{v}_{\mathbf{d}} = \phi^{f}\left(\mathbf{v}^{\mathrm{f}} - \mathbf{v}^{\mathrm{s}}\right) = -\boldsymbol{\kappa}\nabla p \tag{11.9}$$

such that

$$\mathbf{v}^{\mathrm{f}} = \mathbf{v}^{\mathrm{s}} - \frac{\kappa}{\phi^{f}}\nabla p \tag{11.10}$$

where $\boldsymbol{\kappa}$ is the hydraulic permeability tensor and p represents the interstitial fluid pressure.

Substituting equation (11.10) into (11.7) and (11.8), respectively, we obtain

$$\frac{\partial \phi^f}{\partial t} + \nabla \bullet \left(\phi^f \mathbf{v}^{\mathrm{s}} - \boldsymbol{\kappa} \nabla p\right) = 0 \tag{11.11}$$

$$\nabla \bullet \left(\mathbf{v}^{\mathrm{s}} - \boldsymbol{\kappa} \nabla p\right) = 0 \tag{11.12}$$

Equation (11.12) is a governing equation that links the cartilage matrix deformation to the interstitial fluid motion. Equation (11.11) is an intermediate result that can simplify the (soon to be introduced) solute transport governing equation.

In case of no mass sink, conservation of a solute phase (e.g., IGF) in a fluid phase can be expressed as

$$\frac{\partial \left(\phi^f c_I^f\right)}{\partial t} + \nabla \bullet J_{\mathrm{I}}^{\mathrm{f}} = 0 \tag{11.13}$$

where

$$J_{\mathrm{I}}^{\mathrm{f}} = -\phi^f D_I \nabla c_I^{\mathbf{f}} + \phi^f \mathbf{v}^{\mathbf{f}} c_I^f \tag{11.14}$$

Here $J_{\mathrm{I}}^{\mathrm{f}}$ is the mass flux of solute in the fluid phase and D_I is the effective diffusion coefficient of the solute in the cartilage including the tortuosity factor for the cartilage matrix.

The cartilage solid matrix volumetric strain, ε_v, is given by the divergence of the solid phase displacement

$$\varepsilon_v = \nabla \bullet \mathrm{u}^{\mathrm{s}} \tag{11.15}$$

For small strains,

$$\phi^f = \phi_0^f + \varepsilon_v \tag{11.16}$$

where ϕ_0^f represents the initial fluid volume faction.

Thus, the solute transport equation (11.13) can be simplified by using equation (11.11)

$$\phi^f \frac{\partial c_I^f}{\partial t} - \phi^f D_I \nabla^2 c_I^f + \left[\phi^f \mathbf{v}^{\mathrm{s}} - \boldsymbol{\kappa} \nabla p - \nabla \left(\phi^f D\right)\right] \bullet \nabla c_I^f = 0 \tag{11.17}$$

Equation (11.17) is the governing equation for the transport of solute in a deformable porous medium (in this case cartilage).

11.2.2.2 Conservation of Linear Momentum

The incremental total stress tensor σ inside the tissue is the sum of the incremental interstitial fluid pressure p, and the incremental elastic stress σ^e, resulting from deformation of the solid matrix [52]

$$\sigma = -p\mathbf{I} + \sigma^e \tag{11.18}$$

where $\mathbf{I}$ is the identity tensor.

Due to its charged nature, cartilage also exhibits complex electrochemical phenomena (e.g., streaming, diffusion potential, and Donnan osmotic pressure) [5]. For example, the inhomogeneous distribution of proteoglycans may lead to depth-dependent fixed charge density distribution [53]. In turn, inhomogeneities in the fixed charge density can potentially affect the mobile ion concentration, internal electrical potential and osmotic pressure [5]. Because of the composite structure of cartilage, the response of cartilage can be significantly different under tension, compression, and shear or throughout the cartilage thickness [53]. To account for different cartilage properties in tension and compression, Soltz and Ateshian [54] proposed that the mechanical behavior of cartilage's solid phase can be described by employing the "orthotropic octant-wise linear elasticity" model of Curnier et al. [55]. The model can be simplified to the more specialized case of cubic symmetry to reduce the number of material constants. The elastic stress σ^e resulting from deformation of the solid matrix can then be defined as

$$\sigma^e = \sum_{a=1}^{3} \left\{ \lambda_1 \left[\mathbf{A}_a : \mathbf{E}\right] \mathrm{tr}\left(\mathbf{A}_a \mathbf{E}\right) \mathbf{A}_a + \sum_{\substack{b=1 \\ b \neq a}}^{3} \lambda_2 \mathrm{tr}\left(\mathbf{A}_a \mathbf{E}\right) \mathbf{A}_b \right\} + 2\mu \mathbf{E} \qquad (11.19)$$

where $\mathbf{E}$ is the infinitesimal strain tensor related to the solid-phase displacement $\mathbf{u}$ defined as $\mathbf{E} = \frac{1}{2}\left[\nabla \mathbf{u} + (\nabla \mathbf{u})^{\mathrm{T}}\right]$, λ_1 and μ are the Lamé constants, $\mathbf{A}_a$ is a texture tensor ($\mathbf{A}_a = \mathbf{a}_a \otimes \mathbf{a}_a$) corresponding to each of the three preferred material directions defined by unit vector $\mathbf{a}_a$ ($\mathbf{a}_a \bullet \mathbf{a}_a = 1$, no sum over a) with $\mathbf{a}_1$ parallel to the split line direction; $\mathbf{a}_2$ perpendicular to the split line direction; and $\mathbf{a}_3$ normal to the articular cartilage surface (Figure 11.2). The term $\lambda_1 \left[\mathbf{A}_a : \mathbf{E}\right]$ denotes that λ_1 is a function of $\mathbf{A}_a : \mathbf{E}$

$$\lambda_1 \left[\mathbf{A}_a : \mathbf{E}\right] = \begin{cases} \lambda_{-1}, & \mathbf{A}_a : \mathbf{E} < \mathbf{0} \\ \lambda_{+1}, & \mathbf{A}_a : \mathbf{E} > \mathbf{0} \end{cases} \qquad (11.20)$$

which suggests that the material properties, λ_1, differ whether the normal strain component along the a_a is compressive or tensile, and so can incorporate the bimodulus response of the cartilage into the solute transport model. Equation (11.19) is adopted throughout this study.

The momentum equation for the mixture, under quasistatic conditions (in the absence of body forces) is given by

$$\nabla \bullet \sigma = 0 \qquad (11.21)$$

Equations (11.12), (11.17), and (11.21) form the basic set of governing equations describing the fully coupled mechanical and solute transport behavior in a three-dimensional porous media system with solute diffusion only in the fluid phase. This basic model can be further extended to explore, for example, the influence of solute binding, IGF-I, and mechanical stimuli mediated biosynthesis and transport of matrix molecules (e.g., aggrecan).

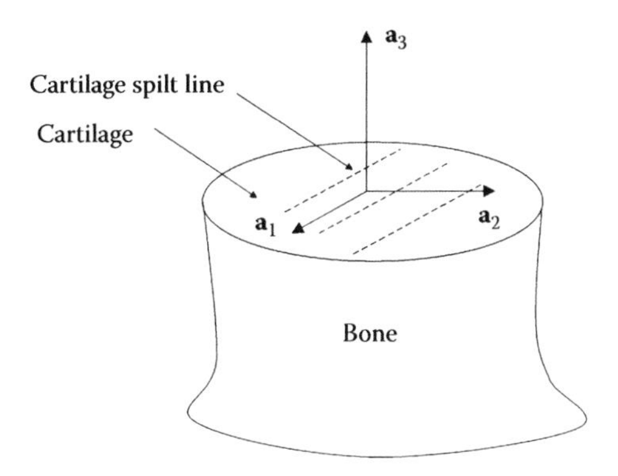

FIGURE 11.2
Application of orthotropic Conewise Linear Elasticity model [55] in the solid phase of cartilage. Three preferred directions of material symmetry: $\mathbf{a}_1$ parallel to the split line direction; $\mathbf{a}_2$ perpendicular to the split line direction; and $\mathbf{a}_3$ normal to the articular cartilage surface.

11.2.2.3 Model Geometry for Radial Solute Transport in Cartilage under Unconfined Cyclic Compression

One common experimental method for testing cartilage is the unconfined compression test [9,54,56,57] (as shown in Figure 11.3). Typically, a thin, cylindrical cartilage disc explant is obtained (to minimize inhomogeneities in cartilage properties) [57]. For experiments investigating the role of cyclic loading on nutrient uptake, the cartilage disc is loaded between two impermeable platens in a bathing solution containing IGF-I [58]. Under this configuration, the radial expansion (r) caused by compression applied along the z direction is unconstrained and fluid can either be exuded or imbibed across the outer surface of the cartilage disc [57]. As the axial normal strain is homogeneous, due to effectively frictionless platens the system is symmetrical about the z-axis, as demonstrated below, the problem can be reduced to a one-dimensional problem in the radial coordinate.

Due to symmetry, it can be assumed that $\partial()/\partial\theta = 0$ and $u_\theta = 0$. Thus,

$$u_r^s = u_r(r,t), \quad v_r = \frac{\partial u_r^s}{\partial t}, \quad p = p(r,t), \quad \varepsilon_z(t) = \frac{\partial u_z^s}{\partial z}, \quad c_I^f = c_I^f(r,t) \tag{11.22}$$

where r and z are radial and axial coordinates, respectively, and t is time. ε_z is the axial strain due to the applied axial load, and is therefore assumed to take a sinusoidal form,

$$\varepsilon_z = \frac{\varepsilon_0}{2}\left[1 - \cos(2\pi f t)\right] \tag{11.23}$$

where ε_0 is the peak-to-peak strain amplitude and f is the frequency of the axial strain.

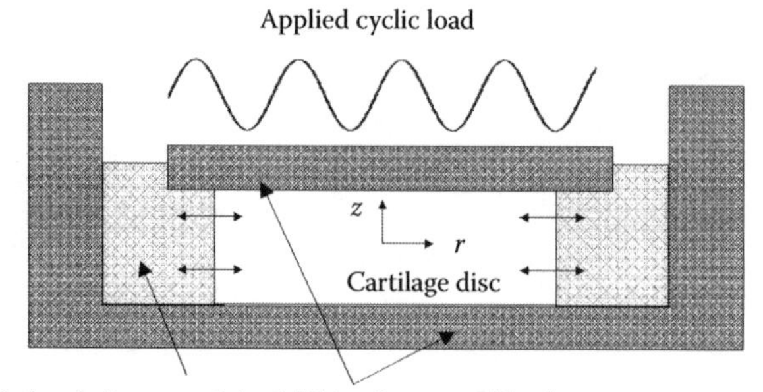

FIGURE 11.3
A schematic diagram of an axisymmetrical cylindrical cartilage disc undergoing cyclic deformation between frictionless impermeable platens. The cartilage is within a bath containing IGF-I.

With consideration of earlier assumptions, the governing equations (11.12) and (11.17) can be rewritten in radial coordinates as

$$\frac{v_r}{r}+\frac{\partial v_r}{\partial r}+\frac{\partial \varepsilon_z}{\partial t}-\kappa_r\left(\frac{\partial^2 p}{\partial r^2}+\frac{1}{r}\frac{\partial p}{\partial r}\right)=0 \tag{11.24}$$

$$\phi^f\frac{\partial c_I^f}{\partial t}-\phi^f D_I\left(\frac{\partial^2 c_I^f}{\partial r^2}+\frac{1}{r}\frac{\partial c_I^f}{\partial r}\right)+\left(\phi^f v_r-\kappa_r\frac{\partial p}{\partial r}\right)\frac{\partial c_I^f}{\partial r}=0 \tag{11.25}$$

In the radial IGF-I transport equation (11.25), the first term represents the change of concentration of IGF-I with respect to time; the second term represents the IGF-I transported by diffusion, and the third and fourth terms describe the contribution of mechanical loading and advection in the deforming porous media.

Furthermore, under this unconfined, axisymmetric loading geometry, the elastic stress σ^e in equation (11.19) reduces to

$$[\sigma^e]=\begin{bmatrix} H_{+A}\frac{\partial u_r}{\partial r}+\lambda_2\left(\frac{u_r}{r}+\frac{\partial u_z}{\partial z}\right) & 0 & 0 \\ 0 & H_{+A}\frac{u_r}{r}+\lambda_2\left(\frac{\partial u_r}{\partial r}+\frac{\partial u_z}{\partial z}\right) & 0 \\ 0 & 0 & H_{-A}\frac{\partial u_z}{\partial z}+\lambda_2\left(\frac{u_r}{r}+\frac{\partial u_r}{\partial r}\right) \end{bmatrix} \tag{11.26}$$

where $H_{+A}=\lambda_{+1}+2\mu$ and $H_{-A}=\lambda_{-1}+2\mu$ are the tensile and compressive aggregate modulus, respectively. Thus, the radial component of the balance of linear momentum equation (11.21) can be simply written as

$$-\frac{\partial p}{\partial r}+H_{+A}\left(-\frac{u_r}{r^2}+\frac{1}{r}\frac{\partial u_r}{\partial r}+\frac{\partial^2 u_r}{\partial r^2}\right)=0 \tag{11.27}$$

That is, under this loading condition, the governing equation (11.27) depends only on the tensile aggregate modulus H_{+A}, which is largely determined by the collagen matrix.

11.2.2.4 Boundary Conditions

At the center of the cartilage (i.e., at $r = 0$),

$$u_r(0,t) = 0, \quad \left(\frac{\partial p}{\partial r}\right)_{r=0} = 0, \quad \left(\frac{\partial c_I^f}{\partial r}\right)_{r=0} = 0 \tag{11.28}$$

At the outer edge of the cartilage (i.e., at $r = r_0$), the following boundary conditions apply:

$$p(r_0,t) = 0, \quad c_I^f(r_0,t) = c_{I0}^f \tag{11.29}$$

Furthermore, at this cartilage–solute bath interface, the quantities ϕ^f and ϕ^s each exhibit a discontinuity. The traction condition across the interface implies that

$$\left(\frac{\partial u_r}{\partial r}\right)_{r=r_0} = -\frac{\upsilon}{1-\upsilon}\left(\frac{u_r|_{r=r_0}}{r_0} + \varepsilon_z\right) \tag{11.30}$$

where υ is Poisson's ratio defined as $\upsilon = \dfrac{\lambda_2}{H_{+A} + \lambda_2}$.

11.2.2.5 Initial Conditions

For simplicity, the following initial conditions can be assumed

$$u_r(r,0) = 0, \quad p(r,0) = 0, \quad c_I^f(r,0) = 0 \tag{11.31}$$

That is, both, the radial displacement of the solid matrix and the fluid pressure are initially set to zero throughout the tissue and we have assumed that there is no growth factor within the tissue.

11.2.2.6 Numerical Method

Numerical results obtained throughout this chapter were obtained by solving the derived governing equations using the commercial finite element software COMSOL MULTIPHYSICS [59]. The applied strain protocol was discretized into time steps. At each time step, the solute concentration, c_I^f, solid phase displacement, u_r, solid phase velocity, v_r, and the interstitial fluid pressure, p, were determined. A one-dimensional domain using 300 quadratic Galerkin elements was used for all calculations, and a relative tolerance 10^{-10} and an absolute tolerance 10^{-11} were adopted. The Finite Element (FEM) discretization of the time-dependent partial differential equation (PDE) was solved using an implicit solver of COMSOL MULTIPHYSICS so that oscillations in

TABLE 11.1
Material Parameters Used in the Model

Parameter	Value	References
IGF-I diffusion coefficient (D_I)	$4.1 \times 10^{-11}\,\mathrm{m^2/sec}$	[29]
Hydraulic permeability (κ)	$1.6 \times 10^{-15}\,\mathrm{m^4/Ns}$	[9]
Aggregate elastic modulus (H_{+A})	13.2 MPa	[30]
Fluid phase volumetric fraction (ϕ^f)	0.8	[30,60]

the solution for any size of incremental time step were avoided. The material parameters used in the model prediction are shown in Table 11.1. These parameters are typical of the bovine calf cartilage typically used in explant experiments [9,29,30,60].

11.3 The Effect of Cyclic Loading and IGF-I Binding on IGF-I Transport in Cartilage

11.3.1 Introduction

IGFs are key regulators of cellular proliferation, survival, and differentiation in mammals [61] and are implicated in many human diseases (including osteoarthritis [OA] and atherosclerosis) [62]. As chondrocytes synthesize little or no IGF-I under normal physiological conditions [63], the primary source of IGF-I is the liver. IGF-I must penetrate into cartilage from synovial fluid to reach chondrocytes, where it exerts its biological effects. However, because of its avascular nature and the thickness of cartilage at diarthroidal joints, IGF-I is unable to diffuse into the cartilage easily.

Cyclic loading has been shown experimentally to enhance the transport of solutes depending on the size of the solute. Bonassar et al. [37] showed experimentally that fluid flow induced by dynamic compression may alter the pericellular concentrations of growth factors, cytokines, newly synthesized matrix macromolecules, or other nutrients. This observation may potentially explain some of the increase in matrix synthesis observed in cartilage undergoing dynamic loading compared with cartilage not undergoing dynamic loading. From a transport experiment involving radiolabeled bovine serum albumin, IGF-I, urea, and sodium. O'Hara et al. [39] concluded that dynamic compression enhanced the desorption of large solutes much more than that of small solutes. This might be expected based on increase in the Peclet number for larger molecules.

Recent theoretical studies generally support the idea that cyclic loading enhances large molecule transport through advective transport. Mauck et al.

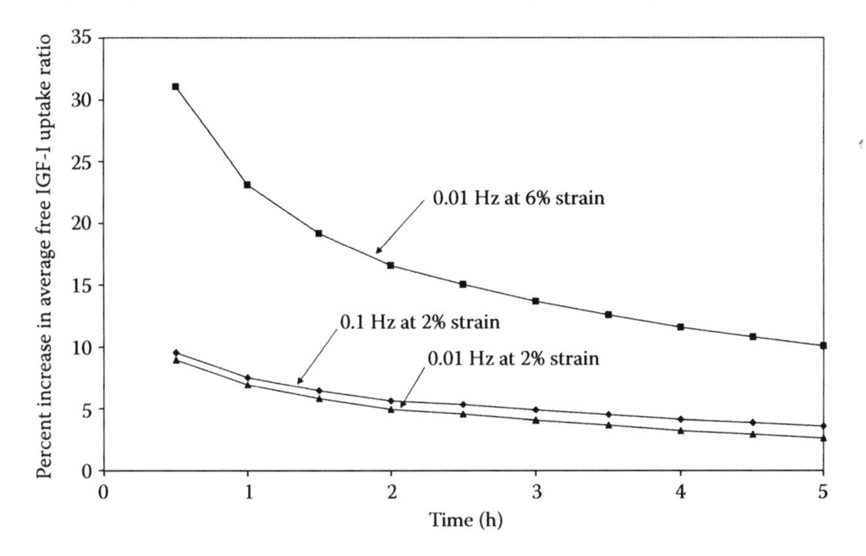

FIGURE 11.4
Predictions of the percent increase (in comparison to case of free diffusion) in the spatially averaged IGF-I uptake due to cyclic loading. Prediction based on model presented in Section 11.2.2 using data contained in Table 11.1.

[36] developed a theoretical model based on the theory of porous media to quantify the effect of cyclic loading on nutrient transport in articular cartilage. Dynamic loading with 0.01–1 Hz and 0%–20% strain were investigated. The numerical outcomes provocatively suggested that cyclic loading can actually concentrate solute inside a cartilage in a variety of cases.

By incorporating strain-dependent diffusion coefficient and hydraulic permeability into a porous media transport model, Zhang and Szeri [40] showed that dynamic loading can enhance solute transport in the surface layer more than in deeper layers, and the beneficial effect is more obvious for large molecules, even in deeper layers.

However, a recent computational study by Gardiner et al. [28], demonstrated that any enhancement of solute uptake is strongly time dependent (see Figure 11.4). More specifically, enhancement only occurs when concentration gradients are large and are colocalized to high Darcy velocities. As high Darcy velocities occur only for high-frequency deformations or high-strain amplitudes and near the cartilage–solute bath interface (see Figure 11.5 Darcy velocity), colocalization of large Darcy velocity and high-solute gradients only occurs in the initial stages of solute uptake into a solute-depleted cartilage. When these conditions are no longer met, that is, longer timescales of several hours or more, dynamic loading has negligible effect on IGF-I transport.

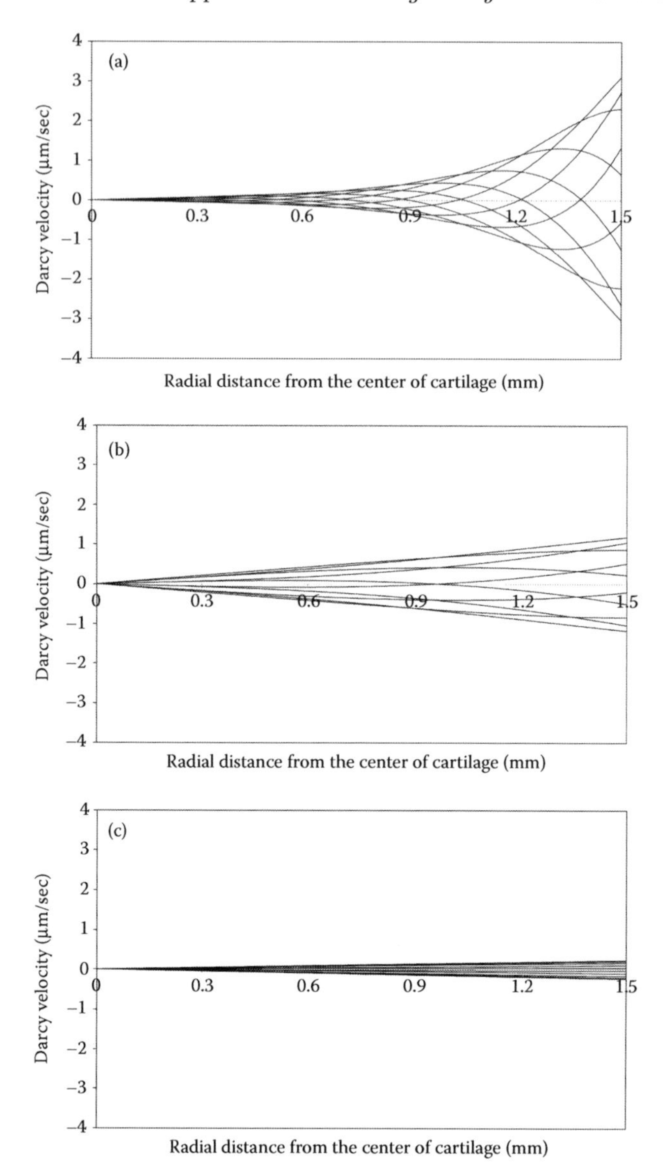

FIGURE 11.5
The Darcy velocity at multiple stages within the first loading cycle as a function of the radical distance from the center of the cartilage for (a) 0.1 Hz, (b) 0.01 Hz, and (c) 0.001 Hz at 10% strain, respectively. It can be seen that a high-frequency dynamic loading (i.e., 0.1 Hz) induces higher Darcy velocities in the periphery region, intermediate-frequency loading (i.e., 0.01 Hz) produces lower Darcy's velocities but over a wider spatial distance, whereas low-frequency loading (i.e., 0.001 Hz) results in little fluid movement. Any potential enhancement of transport due to advection is constrained to the regions of nonzero Darcy velocities. Once these regions become saturated by solute, the transport through the remaining cartilage is equivalent to free diffusion.

Gardiner et al. [28] went on to state that before cyclic loading can significantly enhance solute transport by advection, there needs to be a so-called "symmetry breaking" mechanism, otherwise solutes are just advectively transported into and out of the cartilage within each loading cycle with equal facility. That is, there needs to be a directional bias imposed on the solute transport before cyclic fluid motion can enhance transport. Diffusion provides one symmetry breaking mechanism, as concentration gradient provides a bias to the transport. Solutes can then be transported into the cartilage under advection and diffusion in one half of the loading cycle, but in the other half of the loading cycle advection and diffusion are in opposite directions. This cyclic advection plus diffusive transport still provides enhanced solute uptake compared to diffusion only transport, as advection allows solute to reach cartilage at deeper layers sooner (although this may potentially flatten the concentration gradient). Hence for diffusion to act as a symmetry breaking mechanism, to significantly enhance transport, large concentration gradients are required. Using their model Gardiner et al. [28] showed that enhanced transport was only observed at early times near the surface of the cartilage when the solute concentration gradient was the steepest and colocalized with maximum Darcy velocities.

One of the more promising other potential symmetry breaking mechanisms is the "capture" of solute to the cartilage matrix (by binding to the solid phase), so that they are not released when fluid moves back out of the cartilage. To work effectively, this requires a matching of timescales of cyclic loading to the association/dissociation rates of the capture process. If the timescale of capture is too slow or if release is too quick, relative to the loading cycle frequency, solute capture cannot operate as an effective symmetry breaking mechanism. In the case of IGF-I there is a range of IGFBPs contained within the cartilage matrix which may perform this symmetry breaking function. Here we will develop a model of this process by extending the model equations developed in the previous section.

11.3.1.1 The Effect of IGF Binding on IGF Transport in Cartilage

A family of at least six IGFBPs regulates the bioavailability of IGFs in cartilage [64]. It is suggested that the major functions of these binding proteins are to prolong the half-life of IGFs in cartilage and to regulate the bioavailability of IGFs in their interaction with cell surface receptors [65–67]. In addition, IGFBPs may also exert IGF-independent effects by transcriptional activation of genes by IGFBPs transported into the nucleus via their nuclear localization signal [64]. However, the actual mechanisms are far from clear.

There are relatively few studies that experimentally and theoretically investigate the effect of IGF binding to its binding proteins on IGF transport in cartilage. Bhakta et al. [29] conducted a series of experiments to study the effect of graded levels of unlabeled IGFs competing with radio-labeled

^{125}I-IGFs (~0.003 nM) in adult bovine articular cartilage discs (3 mm diameter × 400 μm depth), which were equilibrated in solutions for 48 hours containing both radiolabeled ^{125}I-IGFs and unlabeled IGFs. Their results indicate that the binding of IGFs to cartilage, through the IGFBPs, regulates the transport of IGFs in cartilage, probably contributing to the control of their bioavailability.

Garcia et al. [58] experimentally and theoretically studied the role of IGFBPs on the transport and binding of IGF-I within the cartilage tissue. ^{125}I-IGF-I was allowed to diffuse across a 400 μm-thick cartilage disk. It was observed that the transport of ^{125}I-IGF-I was dramatically slowed by binding of IGF-I to its IGFBPs. Garcia et al. [58] also presented a theoretical model based on the Langmuir isotherm to predict the equilibrium binding behavior of IGF-I. Together, the studies of Bhakta et al. [29] and Garcia et al. [58], provide strong evidence that IGFBPs can potentially influence the transport of IGFs in cartilage.

Recently, Barta and Maroudas [27] modeled the cartilage as a two-layer continuum, namely a thin surface layer exposed to synovial fluid and a deeper layer with impermeable bony endplate to study the diffusion of IGF-I, its interaction with soluble binding protein and fixed binding sites on ECM, as well as cell receptor binding and internalization. However, without the implementation of the porous mechanics theory coupled with reactive transport, Barta and Maroudas's study cannot fully explore the effect of physiologically relevant loading on IGF transport.

In this section, the basic solute transport presented in the previous section is extended by including solute binding, and then governing equations are particularized to the specific geometry of a cylindrical cartilage disc undergoing unconfined axial loading with radial transport of IGF-I. Finally, model predictions are compared with the experimental findings [29].

11.3.2 Interaction between IGF-I and Its IGFBPs

The total solute concentration ($\bar{c}_I$) is comprised of both "free" and "bound" solute, that is,

$$\bar{c}_I = \bar{c}_I^f + \bar{c}_I^b \tag{11.32}$$

where $\bar{c}_I^f$ is the concentration of free (or unbound) solute and $\bar{c}_I^b$ represents the concentration of solute bound onto the solid phase (e.g., via binding proteins). Both concentrations are given with respect to the overall RVE volume.

11.3.2.1 Law of Mass Action

Assuming that IGF-I can combine with IGFBP to form an IGF/IGFBP complex and that the complex can also break into its original constituents (i.e., the reaction is reversible), we can express the interaction as.

$$\text{IGF-I} + \text{IGFBP} \underset{k_{-1}}{\overset{k_{+1}}{\leftrightarrow}} \text{IGF/IGFBP} \tag{11.33}$$

where k_{+1} is the rate constant for the association reaction (binding) and k_{-1} is the rate constant for the dissociation reaction. Ignoring the possibility of diffusion and advection of fixed IGFBPs and complexes, the law of mass action [68] can be used to quantitatively describe the IGF-I and IGFBP interaction and leads to the following set of differential equations.

$$\frac{d\bar{c}_I^b}{dt} = k_{+1}\bar{c}_I^f \bar{c}_{BP}^b - k_{-1}\bar{c}_I^b \tag{11.34a}$$

$$\frac{d\bar{c}_{BP}^b}{dt} = -k_{+1}\bar{c}_I^f \bar{c}_{BP}^b + k_{-1}\bar{c}_I^b \tag{11.34b}$$

Summing equations (11.34a) and (11.43b) it is easily confirmed that when there is no net production/degradation of binding proteins

$$\frac{d\bar{c}_{BP}^b}{dt} + \frac{d\bar{c}_I^b}{dt} = 0 \tag{11.35}$$

Thus, $\bar{c}_{BP}^b(t) + \bar{c}_I^b(t) = m$. The integration constant, m, can be obtained using the initial condition, and leads to

$$\bar{c}_{BP}^b(t) + \bar{c}_I^b(t) = \bar{c}_{BP0}^b + \bar{c}_{I0}^b, \quad \text{where } \bar{c}_{BP}^b(t=0) = \bar{c}_{BP0}^b,\ \bar{c}_I^b(t=0) = \bar{c}_{I0}^b \tag{11.36}$$

Finally, equation (11.36) is substituted into (11.34a) to obtain

$$\frac{d\bar{c}_I^b}{dt} = k_{+1}\left(\bar{c}_{BP0}^b + \bar{c}_{I0}^b - \bar{c}_I^b\right)\,\bar{c}_I^f - k_{-1}\bar{c}_I^b \tag{11.37}$$

Equation (11.37) provides a relationship linking c_I^f and c_I^b.

11.3.2.2 Model of Solute Transport and Binding in a Deformable Cartilage

Now that IGF-I is able to bind to IGFBPs on the solid phase, the solute conservation law derived previously needs to be modified and an additional expression needs to be introduced for solute conservation on the solid phase. Specifically the conservation of mass of solute in the free and bound state can be described by

$$\frac{\partial\left(\phi^f c_I^f\right)}{\partial t} + \nabla \bullet \left(-\phi^f D \nabla c_I^f + \phi^f \mathbf{v}^{\mathbf{f}} c_I^f\right) = -s \tag{11.38}$$

$$\frac{\partial\left[\left(1-\phi^f\right) c_I^b\right]}{\partial t} - \nabla \bullet \left[\left(1-\phi^f\right)\mathbf{v}^{\mathbf{s}} c_I^b\right] = s \tag{11.39}$$

TABLE 11.2
IGF-I, IGFBP Association/Dissociation Rates Parameters Used in the Model

Parameter	Value	References
Association rate constant (k_{+1})	$3.67 \times 10^5\,\mathrm{M}^{-1}\mathrm{sec}^{-1}$	[69]
Dissociation rate constant (k_{-1})	$0.001\,\mathrm{sec}^{-1}$	[69]

where s is the solute sink due to the binding of free solute to binding proteins attached to the solid phase.

The addition of equations (11.38) and (11.39) leads to

$$\frac{\partial\left(\phi^f c_I^f\right)}{\partial t}+\frac{\partial\left[\left(1-\phi^f\right)c_I^b\right]}{\partial t}+\nabla\bullet\left(-\phi^f D\nabla c_I^f+\phi^f \mathbf{v}^{\mathbf{f}} c_I^f\right)$$
$$-\nabla\bullet\left[\left(1-\phi^f\right)\mathbf{v}^{\mathbf{s}} c_I^b\right]=0 \tag{11.40}$$

Using equation (11.10) and performing some algebraic manipulations [33], we obtain the general transport governing equation for the solute in the deformable cartilage with consideration of binding of solute to the solid matrix.

$$\phi^f\frac{\partial c_I^f}{\partial t}+\left(1-\phi^f\right)\frac{\partial c_I^b}{\partial t}-\phi^f D_I\nabla^2 c_I^f+\left[\phi^f\mathbf{v}^{\mathbf{s}}-\kappa\nabla p-\nabla\left(\phi^f D\right)\right]\bullet\nabla c_I^f$$
$$+\left[\left(1-\phi^f\right)\mathbf{v}^{\mathbf{s}}\right]\bullet\nabla c_I^b=0 \tag{11.41}$$

With consideration of above assumptions, the governing equations (11.12) and (11.41) can be rewritten in radial coordinates as

$$\frac{v_r}{r}+\frac{\partial v_r}{\partial r}+\frac{\partial \varepsilon_z}{\partial t}-\kappa_r\left(\frac{\partial^2 p}{\partial r^2}+\frac{1}{r}\frac{\partial p}{\partial r}\right)=0 \tag{11.42}$$

$$\phi^f\frac{\partial c_I^f}{\partial t}+\left(1-\phi^f\right)\frac{\partial c_I^b}{\partial t}-\phi^f D_I\left(\frac{\partial^2 c_I^f}{\partial r^2}+\frac{1}{r}\frac{\partial c_I^f}{\partial r}\right)$$
$$+\left(\phi^f v_r-\kappa_r\frac{\partial p}{\partial r}\right)\frac{\partial c_I^f}{\partial r}+\left(1-\phi^f\right)v_r\frac{\partial c_I^b}{\partial r}=0 \tag{11.43}$$

In the radial IGF-I transport equation (11.43), the first term represents the change of concentration of IGF-I with respect to time; the second term represents the IGF-I transported by diffusion, and the third and fourth terms describe the contribution of mechanical loading and advection in the deforming porous media.

11.3.2.3 Boundary and Initial Conditions

In addition to the boundary conditions provided in Section 11.2.2.4, at the center of the cartilage (i.e., at $r = 0$),

$$\left(\frac{\partial c_I^b}{\partial r}\right)_{r=0} = 0 \tag{11.44}$$

Likewise the following additional initial condition is assumed:

$$c_I^b(r,0) = 0 \tag{11.45}$$

That is, we have assumed that initially there is no bound growth factor within the tissue.

11.3.3 Results and Discussion

Since the solute concentration is generally nonuniform along the radial direction, it is useful to define an average total ($\overline{R}_U$) IGF-I uptake ratio, which is the sum of the average free uptake ratio ($\overline{R}^f$) and the average bound uptake ratio ($\overline{R}^b$),

$$\overline{R}_u = \overline{R}^f + \overline{R}^b \tag{11.46}$$

where

$$\overline{R}^f = \frac{\int_0^{r_0} 2\pi r \bar{c}_I^f dr}{\bar{c}_{I0}^f \int_0^{r_0} 2\pi r dr} \quad \text{and} \quad \overline{R}^b = \frac{\int_0^{r_0} 2\pi r \bar{c}_I^b dr}{\bar{c}_{I0}^b \int_0^{r_0} 2\pi r dr} \tag{11.47}$$

Here $\bar{c}_{I0}^f$ is the solute concentration at the outer edge of cartilage disc, and $\bar{c}_{I0}^b$ is the initial volume-based bound solute concentration. Simply stated, the uptake ratios describe the average IGF-I concentration within the cartilage (free, bound, or total) relative to the free IGF-I concentration in the surrounding IGF-I bath solution.

11.3.3.1 Free Diffusion

Figure 11.6 allows comparison between the numerical predictions of free diffusion (with and without consideration of binding) and the experimental data of Bhakta et al. [29]. It can be seen that the model is able to produce a reasonable fit to the steady-state experimental data when the model includes binding of IGF-I to IGFBPs with an IGFBP concentration $c_{BP0} = 45\,\text{nM}$. If the bath IGF-I concentration is low, ignoring binding results in significant underestimation of total IGF-I uptake; however, the influence of binding becomes negligible if the concentration is high due to the saturation of the finite number of binding sites.

Figure 11.7 shows the total solute uptake ratio profiles at time intervals of 5 h, with and without binding (40 nM bath concentration). It can be seen

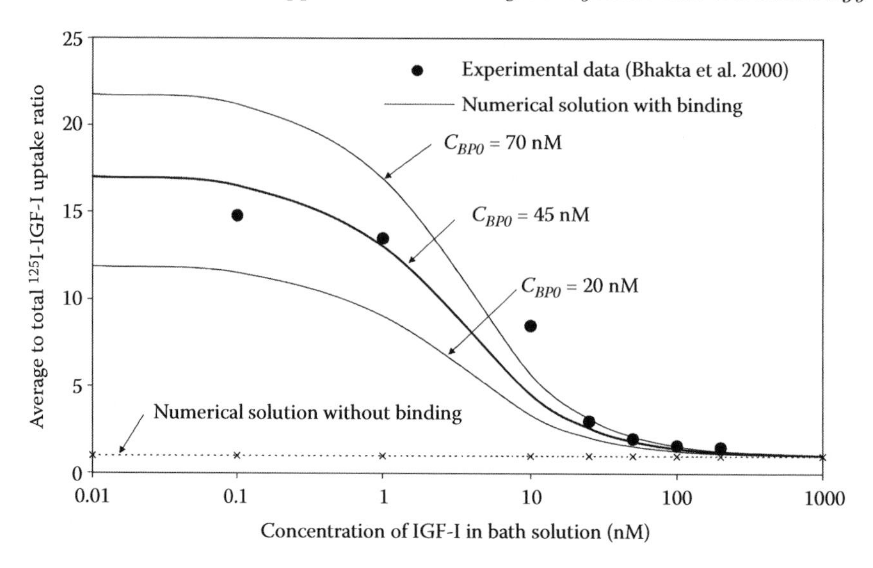

FIGURE 11.6
Comparison of numerical predictions with experimental results of Bhakta et al. [29].

that binding results in the total solute concentration throughout the cartilage exceeding the solute bath concentration, that is, $R_u > 1$ (as shown in Figure 11.7a). On the other hand, if binding is not included, the curves of diffusion profile are constrained by c^f_{I0} at every stage of solute transport (see Figure 11.7b). On the other hand, binding was seen to inhibit the rate of free IGF-I uptake as shown in Figure 11.8. This inhibition is due to removal of free IGF-I from the diffusion process by binding to the IGFBP. However, once the binding sites are saturated the free IGF-I concentration achieves a steady-state level equal to the surrounding bath concentration.

11.3.3.2 Diffusion with Cyclic Deformation and IGF-I, IGFBP Interaction

From the results and discussion presented in Section 11.3.1, we would expect that the optimal loading regime to enhance IGF-I transport should correspond to high-strain magnitudes and frequencies, so as to obtain large advective Darcy velocities. Indeed this expectation is confirmed by the numerical predictions shown in Figures 11.9 and 11.10 of the average percent increase in total IGF-I uptake ratio into a cyclically loaded cartilage in comparison to free diffusion. Specifically it can be seen that a loading regime with a combination of high-strain amplitudes (e.g., 6% strain) and high frequencies (e.g., 0.1 Hz) produces the most dramatic total IGF-I enhancement, especially at early time (i.e., time < 2 h). More critically to the current discussion is the

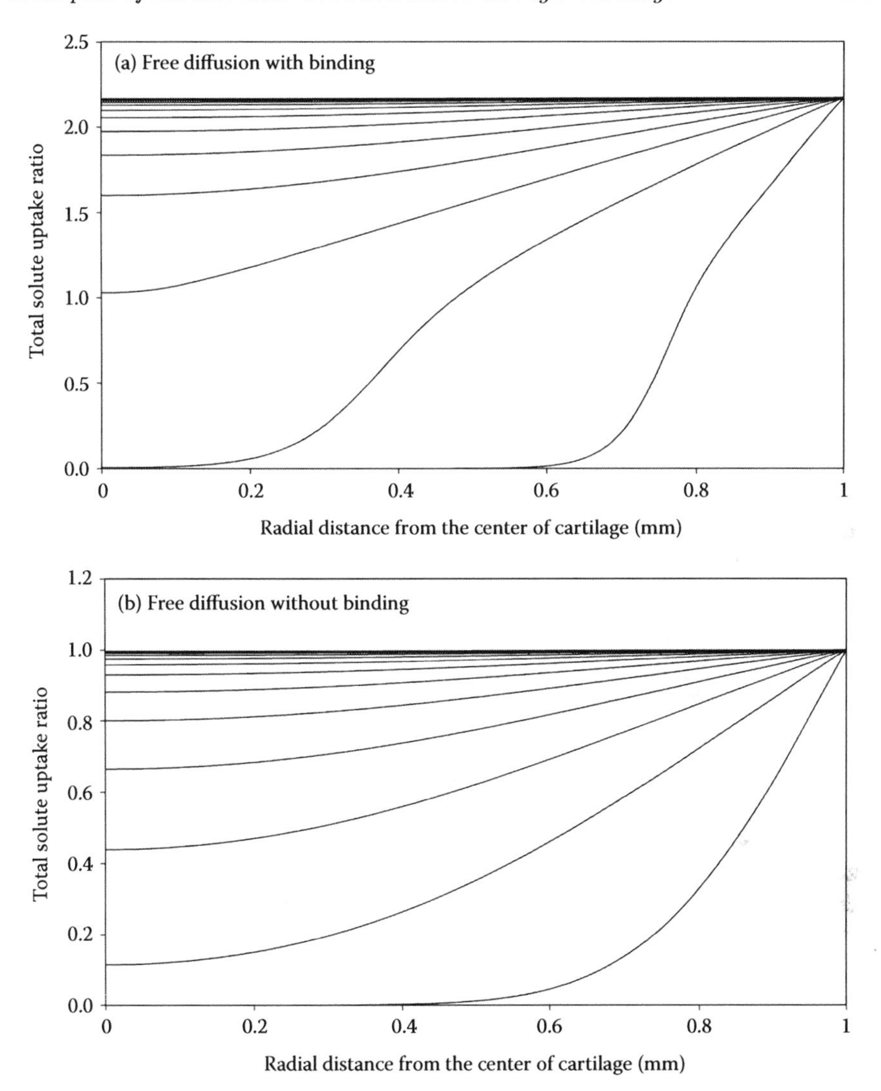

FIGURE 11.7
Total solute uptake ratio ($\overline{R}_u$) as a function of radial distance from center of cartilage undergoing free diffusion, with or without consideration of binding effects ($c_{I0}^f = 40$ nM). Time between each curve is 5 h.

role that binding proteins play in modifying this cyclic-loading enhanced transport. When IGF-I interaction with IGFBPs are included in the model we see that the enhancement of the free IGF-I is greater and for a longer time period than seen in the equivalent model which ignored IGFBPs (i.e., Figure 11.11). Therefore, IGFBPs appear to provide a mechanism to enhance the transport

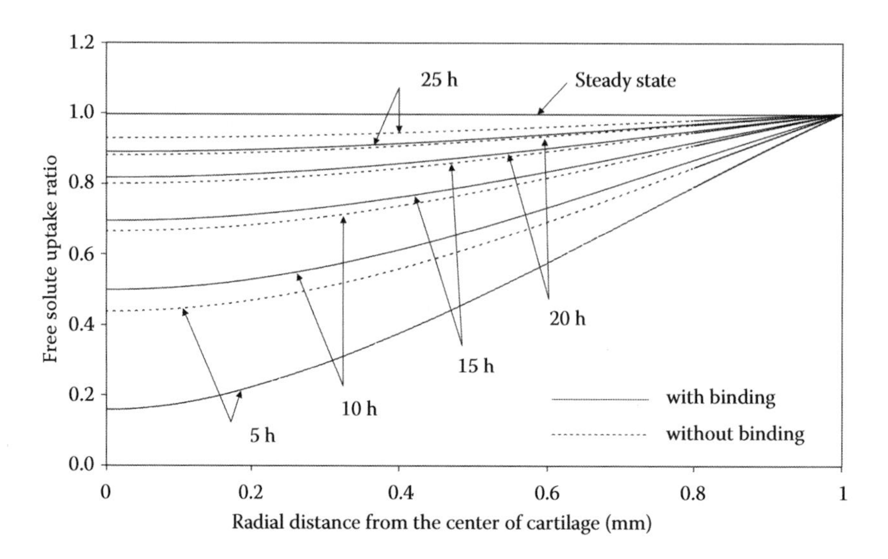

FIGURE 11.8
Free solute uptake ratio ($\overline{R}^f$) as a function of radial distance from center of cartilage undergoing free diffusion, with or without consideration of binding effects ($c_{I0}^f = 40$ nM). Time between each curve is 5 h.

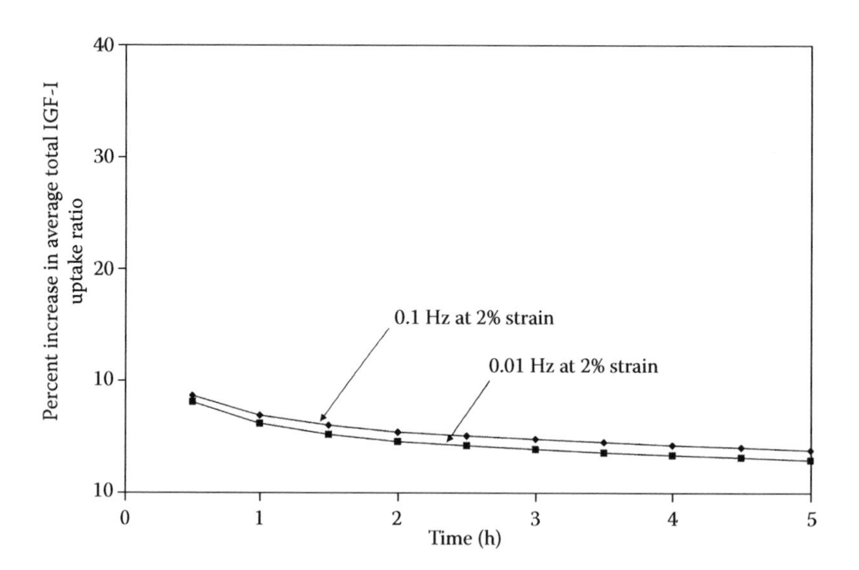

FIGURE 11.9
Percent increase in average total IGF-I uptake ratio ($\overline{R}_u$), as a function of time, at 2% peak-to-peak strain amplitude and various frequencies ($c_{I0}^f =$ 40 nM).

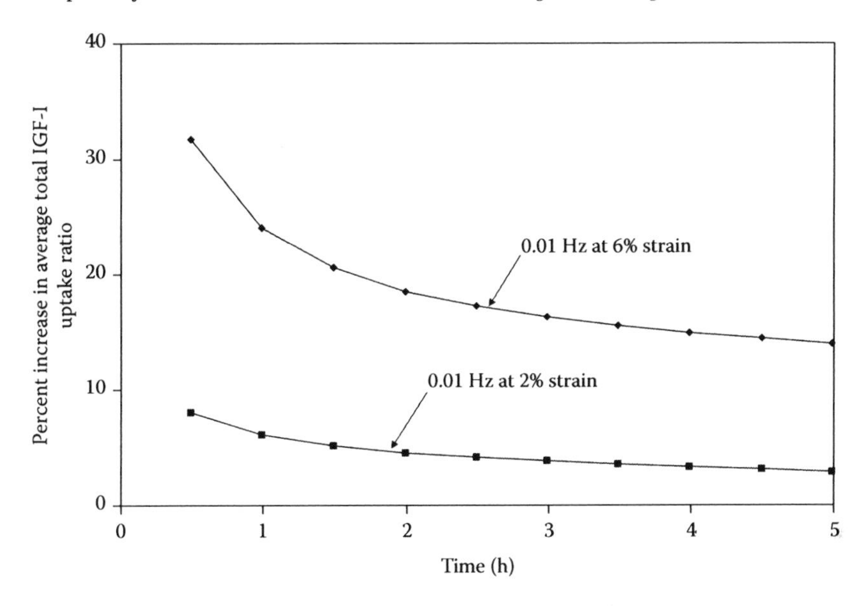

FIGURE 11.10
Percent increase in the average total IGF-I uptake ratio ($\overline{R}_u$), as a function of time, at 0.01 Hz and various peak-to-peak strain amplitudes ($c^f_{I0} = 40$ nM).

of IGF-I through capturing IGF-I, thereby breaking the symmetry of advective transport within a loading cycle.

11.4 IGF Transport with Competitive Binding in a Deforming Articular Cartilage

11.4.1 Introduction

In the previous section, we saw that the presence of IGFBPs attached to the cartilage matrix can lead to enhanced IGF-I transport under cyclic-loading conditions but can inhibit the transport under static (free diffusion) conditions. Of course, the interaction between IGF-I and IGFBPs is not as straightforward as depicted in the previous section. IGF-I is not the only growth factor that can bind with IGFBPs, and as stated earlier there are at least six different IGFBPs, presumably each with their own association/dissociation rates with IGF-I. To begin to delve into this complexity, here we extend the previous model to include an additional growth factor, that is, IGF-II and some of the interaction subtleties with the family of six IGFBPs.

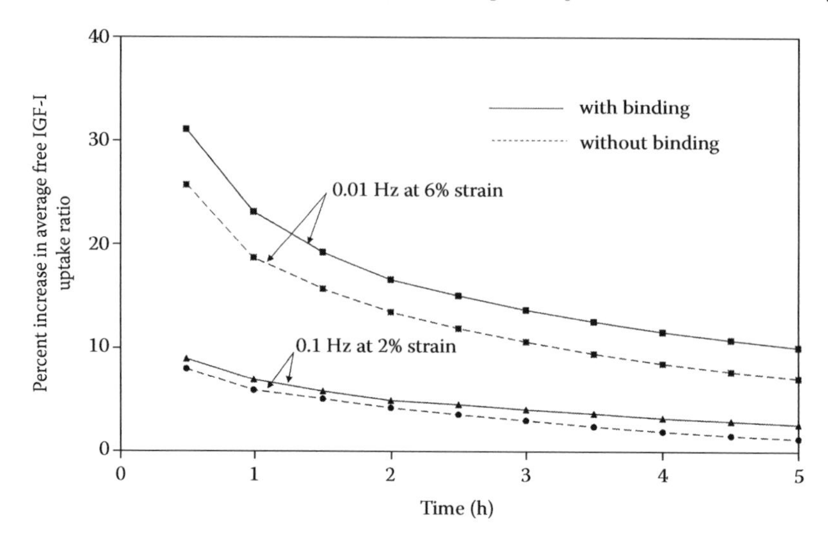

FIGURE 11.11
Comparison of the predicted percent increase in the average free IGF-I uptake ratio ($\overline{R}_u$), as a function of time for cases of no binding (interaction) with IGFBP and with a binding interaction. Note two loading regimes are shown. In each case ($c^f_{I0} = 40$ nM).

11.4.1.1 Competitive Binding of IGFs to Their IGFBPs in Cartilage

IGF-I and -II are important stimuli for cartilage ECM synthesis and assembly [63]. The content of IGFs and IGFBPs in cartilage may vary under various conditions. In diseased cartilage (e.g., OA and rheumatoid arthritis) the level of IGF-I in arthritic synovial fluid is found to be increased relative to normal cartilage, but no change is recorded in the level of IGF-II [62]. IGFBP-3 is the most abundant binding in human cartilage and significant elevation of its content was observed in OA cartilage [70]. IGFBPs in cartilage also vary between species. For example, high content of IGFBP-6 has been identified in bovine cartilage [29], while IGFBP-6 was too little to detect in human cartilage [70]. All these changes may influence IGF uptake and ultimately influence cartilage homeostasis. The effect of biological changes (e.g., IGF concentration in synovial fluid) can be investigated through parametric studies.

In vivo, IGF-I and -II competitively bind to a family of at least six IGF-BPs simultaneously with differential affinities [65, 71–73] (Figure 11.12), and therefore potentially provides a mechanism for modifying the bioavailability of IGFs to chondrocytes. Measurements of the interaction kinetics between IGFs and their binding proteins in solutions has revealed that IGFBPs 1–5 have a similar binding preference for IGF-I and -II (although IGFBP-2 has a slight IGF-II binding preference), whereas the IGFBP-6 differs from other

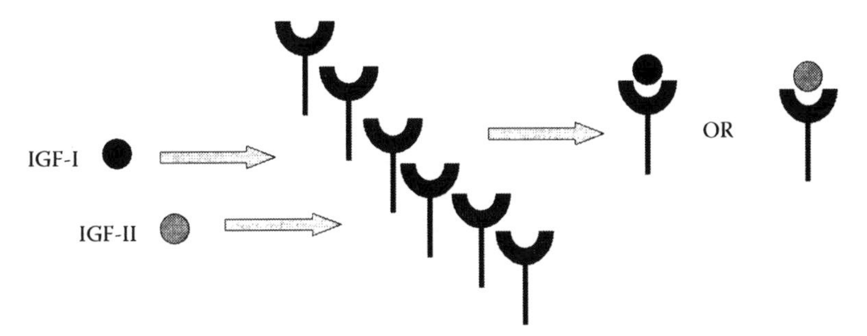

FIGURE 11.12
IGF-I and -II competitively bind to six IGFBPs to form corresponding complexes.

binding proteins in that it displays a 20- to 100-fold higher affinity for IGF-II than IGF-I [69, 72, 74–76]. We can then reasonably assume that the six IGFBPs in the cartilage can be described by two functional groups. That is, one group binds IGF-I and II with equal affinity (e.g., IGFBP 1-5), and the other group has high IGF-II binding preference but low IGF-I affinity (e.g., IGFBP-6) [34].

11.4.2 Model Development for a Competitor Growth Factor

The reactive-transport model shown in the previous sections will now be further extended to include competitive binding of IGF-I and -II to a family of six IGFBPs attached to the ECM.

Let IGF_j be used to represent the two growth factors (i.e., IGF-I and -II), IGFBP_i represent the two functional groups of IGFBPs; k_{+ji} represents the association rate constant of each growth factor with each of the two IGFBP functional groups; and k_{-ji} the dissociation rate constant of each growth factor with each IGFBP functional group. With the nomenclature the chemical reaction between IGFs and IGFBPs can be efficiently described as

$$\text{IGF}_j + \text{IGFBP}_i \underset{k_{-ji}}{\overset{k_{+ji}}{\rightleftarrows}} \text{IGF/IGFBP}_{ji}, \quad j = 1,2; \quad i = 1,2 \tag{11.48}$$

Thus, with analogy to equation (11.31), the total concentration of IGF-I and -II ($\bar{c}_j$) in cartilage is assumed to be composed of free IGF ($\bar{c}_j^f$) and its immobile complexes ($\bar{c}_{ji}^b$) attached to two functional groups of IGFBPs. That is,

$$\bar{c}_j = \bar{c}_j^f + \sum_{i=1}^{6} \bar{c}_{ji}^b, \quad j = 1,2 \tag{11.49}$$

where the subscript i refers to each of the six IGFBPs, and the subscript $j=1$ refers to IGF-I and $j=2$ to IGF-II.

11.4.2.1 Law of Mass Action with Competitive Binding

The chemical reactions between two growth factors (i.e., IGF-I and -II) and two IGFBP functional groups can be described by

$$\frac{d\bar{c}^s_{BPi}}{dt}=\sum_{j=1}^{2}\left(k_{-ji}\bar{c}^s_{ji}-k_{+ji}\bar{c}^s_{BPi}\bar{c}^f_j\right),\quad i=1,2 \tag{11.50a}$$

$$\frac{d\bar{c}^s_{1i}}{dt}=k_{+1i}\bar{c}^f_1\bar{c}^s_{BPi}-k_{-1i}\bar{c}^s_{1i},\qquad i=1,2 \tag{11.50b}$$

$$\frac{d\bar{c}^s_{2i}}{dt}=k_{+2i}\bar{c}^f_2\bar{c}^s_{BPi}-k_{-2i}\bar{c}^s_{2i},\qquad i=1,2 \tag{11.50c}$$

where $\bar{c}^f_1$ and $\bar{c}^f_2$ are the volume-based mobile (free) IGF-I and IGF-II concentration, respectively; $\bar{c}^s_{1i}$ and $\bar{c}^s_{2i}$ are their immobile complexes attached to two functional groups (i.e., $\bar{c}^s_{BP1}$ and $\bar{c}^s_{BP2}$); k_{+ji} is the association rate constant of IGF-I and -II with each two IGFBP functional group; and k_{-ji} is the dissociation rate constant of IGF-I and -II with each two IGFBP functional group.

Summing equations (11.50a–c) leads to

$$\frac{d\left(\bar{c}^s_{BPi}+\bar{c}^s_{1i}+\bar{c}^s_{2i}\right)}{dt}=0,\quad i=1,2 \tag{11.51}$$

Thus, $\bar{c}^s_{BPi}(t)+\bar{c}^s_{1i}(t)+\bar{c}^s_{2i}(t)=m_i$. Similar to equation (11.34), the integration constants m_i can again be obtained from the initial condition, such that,

$$\bar{c}^s_{BPi}(t)+\bar{c}^s_{1i}(t)+\bar{c}^s_{2i}(t)=\bar{c}^s_{BPi}(0)+\bar{c}^s_{1i}(0)+\bar{c}^s_{2i}(0),\quad i=1,2$$

where

$$\bar{c}^s_{BPi}(t=0)=\bar{c}^s_{BPi0},\bar{c}^s_{1i}(t=0)=\bar{c}^s_{1i0}\text{ and }\bar{c}^s_{2i}(t=0)=\bar{c}^s_{2i0} \tag{11.52}$$

Substituting equation (11.52) into (11.50a) leads to

$$\frac{1}{k_{+1i}}\frac{d\bar{c}^s_{1i}}{dt}=\bar{c}^f_1\left(\bar{c}^s_{BPi0}+\bar{c}^s_{1i0}+\bar{c}^s_{2i0}-\bar{c}^s_{2i}\right)-\left(K_{D1i}+\bar{c}^f_1\right)\bar{c}^s_{1i},\quad i=1,2 \tag{11.53a}$$

$$\frac{1}{k_{+2i}}\frac{d\bar{c}^s_{2i}}{dt}=\bar{c}^f_2\left(\bar{c}^s_{BPi0}+\bar{c}^s_{1i0}+\bar{c}^s_{2i0}-\bar{c}^s_{1i}\right)-\left(K_{D2i}+\bar{c}^f_2\right)\bar{c}^s_{2i},\quad i=1,2 \tag{11.53b}$$

where

$$K_{D1i}=\frac{k_{-1i}}{k_{+1i}},\quad K_{D2i}=\frac{k_{-2i}}{k_{+2i}},\quad i=1,2 \tag{11.54}$$

11.4.2.2 Steady-State Growth Factor Uptake

At steady state, equations (11.53a–b) reduce to

$$\bar{c}_1^f\left(\bar{c}_{BPi0}^s+\bar{c}_{1i0}^s+\bar{c}_{2i0}^s-\bar{c}_{2i}^s\right)-\left(K_{D1i}+\bar{c}_1^f\right)\bar{c}_{1i}^s=0, \quad i=1,2 \tag{11.55a}$$

$$\bar{c}_2^f\left(\bar{c}_{BPi0}^s+\bar{c}_{1i0}^s+\bar{c}_{2i0}^s-\bar{c}_{1i}^s\right)-\left(K_{D2i}+\bar{c}_2^f\right)\bar{c}_{2i}^s=0, \quad i=1,2 \tag{11.55b}$$

Solving equations (11.55a) and (11.55b) for $\bar{c}_{1i}^s$ and $\bar{c}_{2i}^s$ we obtain

$$\bar{c}_{1i}^s=\frac{\bar{c}_{BPi0}^s K_{D2i}\bar{c}_1^f}{K_{D11}\left(K_{D21}+\bar{c}_2^f\right)+K_{D21}\bar{c}_1^f}, \quad i=1,2 \tag{11.56a}$$

$$\bar{c}_{2i}^s=\frac{\bar{c}_{BPi0}^s K_{D1i}\bar{c}_2^f}{K_{D11}\left(K_{D21}+\bar{c}_2^f\right)+K_{D21}\bar{c}_1^f}, \quad i=1,2 \tag{11.56b}$$

The results are a pair of competitive Langmuir sorption isotherms [77].

The total growth factor uptake ratio (R_{ui}) (analogous to definitions in equation (11.47)) can be expressed as

$$R_{ui}=\frac{1}{\bar{c}_{i0}^f}\left(\bar{c}_i^f+\sum_{j=1}^{2}\bar{c}_{ij}^b\right), \quad i=1,2 \tag{11.57}$$

where $\bar{c}_{i0}^f$ are IGF-I and -II concentration in synovial fluid at the outer surface of cartilage, respectively.

At steady state, the IGF concentration within cartilage is equal to that in synovial fluid (i.e., $\bar{c}_i^f=\bar{c}_{i0}^f$). Further, substituting equations (11.56a,b) into (11.57) leads to

$$R_{u1}=1+\sum_{i=1}^{2}\frac{\bar{c}_{BPi0}^b K_{D2i}}{K_{D11}\left(K_{D21}+\bar{c}_2^f\right)+K_{D21}\bar{c}_1^f} \tag{11.58a}$$

$$R_{u2}=1+\sum_{i=1}^{2}\frac{\bar{c}_{BPi0}^b K_{D1i}}{K_{D11}\left(K_{D21}+\bar{c}_2^f\right)+K_{D21}\bar{c}_1^f} \tag{11.58b}$$

11.4.2.3 Model Calibration

The six unknown parameters in Equations (11.58a,b), namely, K_{D11}, K_{D12}, K_{D21}, K_{D22}, c_{BP10}^b, and c_{BP20}^b, can be determined by fitting the experimental data of Bhakta et al. (2000) (shown in Figure 11.13) using a nonlinear least

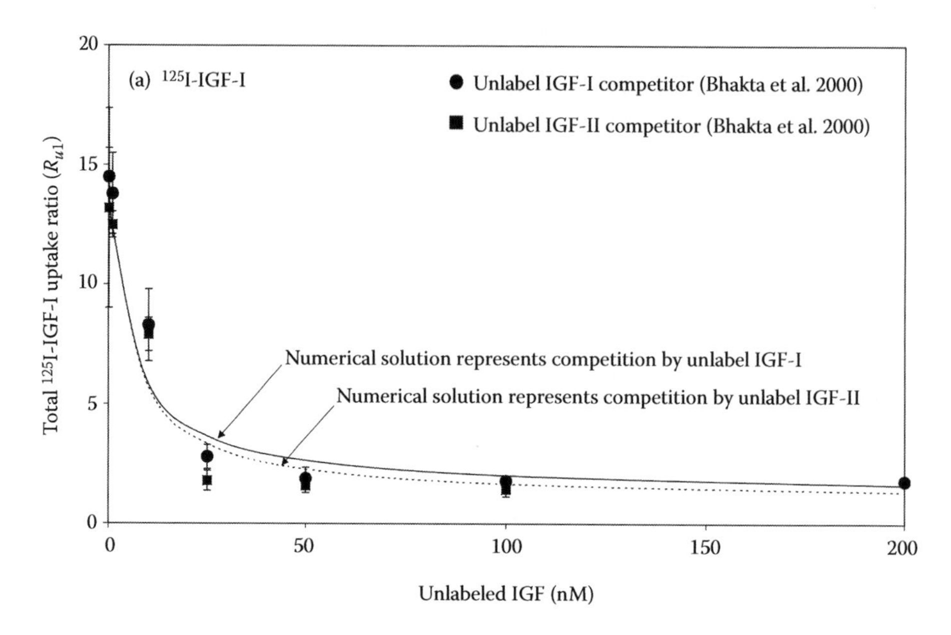

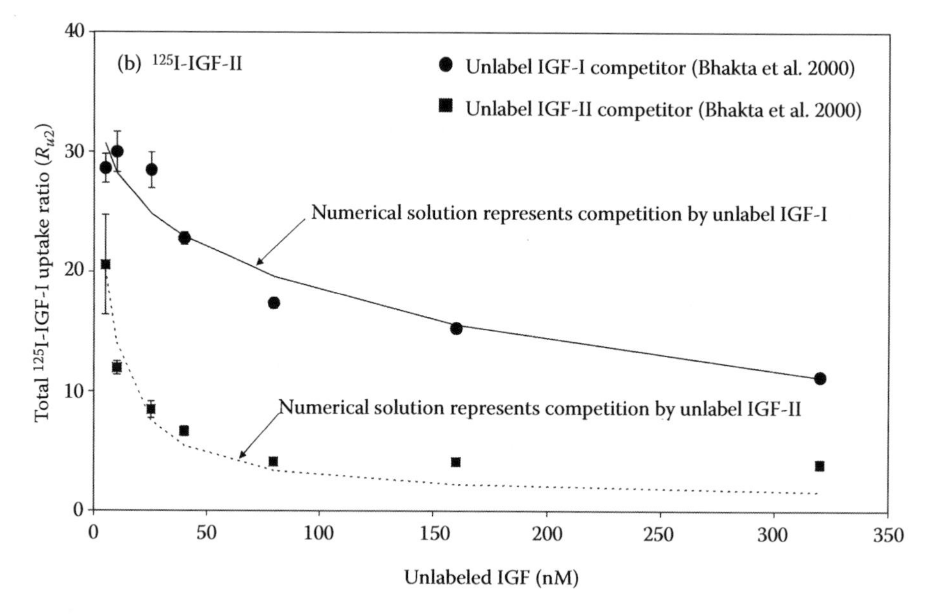

FIGURE 11.13
The determination of K_{D11}, K_{D12}, K_{D21}, K_{D22}, c^b_{BP10}, and c^b_{BP20} using the experimental data of Bhakta et al. [29], (a) Competition of unlabeled IGF-I and -II with ^{125}I-IGF-I for binding to bovine articular cartilage; (b) Competition of unlabeled IGF-I and -II with ^{125}I-IGF-II for binding to bovine articular cartilage.

squares and curve-fitting feature in MATLAB's optimization toolbox [78]. The fitting results through the optimization process are shown in Table 11.3. Details of this calibration procedure can be found in the work by Zang et al. [34]. The first thing to note is that the model is able to be fitted to the experimental results of Bhakta et al. [29]. This adds support to the notion that the differential uptake observed by Bhakta et al. [29] when the radiolabeled IGF-I or IGF-II was used was due to differences in binding affinities of these two growth factors for the local IGFBPs [59]. Second, the estimates of IGFBP concentrations and the various binding affinities derived from the calibration are consistent with previous experimental findings [69,72,74–76] showing IGFBP-6 has 20- to 100-fold higher binding affinity for IGF-II than IGF-I and IGFBP-6 is the major IGFBP species in bovine cartilage with a range of 30–150 nM. Therefore, the competitive interaction model presented here appears to be consistent with known experimental data and adds support to our current understanding of this system. Now that we have gained confidence in this competitor model we can begin to extend our knowledge by combining it with the transport model in a cartilage undergoing cycle deforming. That is, in the following section, we will extend our growth factor transport model to see how the presence of a competitor growth factor influences IGF-I uptake in a cyclically deforming cartilage.

11.4.2.4 Competitive Binding in a Deforming Cartilage

Assuming the contribution of chondrocyte consumption of IGFs to the IGF sink are not considered in the model due to the relatively low total-receptor concentration and binding affinities in comparison with that of IGFBPs [27, 58], conservation of mass of free IGF-I and -II can be expressed as

$$\frac{\partial\left(\phi^f c_j^f\right)}{\partial t} + \nabla \bullet \left(-\phi^f D_j \nabla c_j^f + \phi^f \mathbf{v}^{\mathrm{f}} c_j^f\right) = -\sum_{i=1}^{2} s_{ji}, \quad j = 1,2 \tag{11.59}$$

where s_{ji} is source/sink term for IGF-I and -II, respectively, representing their interaction with each functional group of IGFBPs attached to the solid phase. D_1 and D_2 are the effective diffusion coefficients for IGF-I and -II in the uniform cartilage, respectively.

The conservation of mass of the IGF-I and -II attached to the two functional groups on the solid phase is described by

$$\frac{\partial\left[\left(1-\phi^f\right) c_{ji}^b\right]}{\partial t} + \nabla \bullet \left[\left(1-\phi^f\right) \mathbf{v}^{\mathrm{s}} c_{ji}^b\right] = s_{ji}, \quad j = 1,2 \tag{11.60}$$

TABLE 11.3
Values of Parameters Estimated by the Optimization Process [34]

Parameters	Values (nM)
K_{D11}	4.8
K_{D21}	5.2
K_{D12}	222.1
K_{D22}	5.7
$\bar{c}^s_{BP10}$	45
$\bar{c}^s_{BP20}$	101

Adding equations (11.59) and (11.60) leads to

$$\frac{\partial\left(\phi^f c_j^f\right)}{\partial t} + \nabla \bullet \left(-\phi^f D_j \nabla c_j^f + \phi^f \mathbf{v}^{\mathbf{f}} c_j^f\right) + \frac{\partial}{\partial t}\left[\left(1-\phi^f\right)\sum_{i=1}^{2} c_{ji}^b\right] + \nabla \bullet \left[\sum_{i=1}^{2}\left(1-\phi^f\right)\mathbf{v}^{\mathbf{s}} c_{ji}^b\right] = 0, \; j = 1,2 \quad (11.61)$$

which describes IGF-I and -II transported by diffusion and advection in a deforming porous media with competition for binding sites.

Equation (11.61) can be further simplified to

$$\phi^f \frac{\partial c_j^f}{\partial t} + \left(1-\phi^f\right)\sum_{i=1}^{2}\frac{\partial c_{ji}^s}{\partial t} - \phi^f D_j \nabla^2 c_j^f + \left[\phi^f \mathbf{v}^{\mathbf{s}} - \kappa \nabla p - \nabla\left(\phi^f D_j\right)\right] \bullet \nabla c_j^f + \left[\left(1-\phi^f\right) v^{\mathrm{s}}\right] \bullet \nabla\left(\sum_{i=1}^{2} c_{ji}^b\right) = 0, \quad j = 1,2 \quad (11.62)$$

Equation (11.62) is the governing transport equation for IGF-I and -II. The mechanical quantities, such as fluid pressure, displacement, and velocity can be determined using methods described in the previous sections.

11.4.2.5 Radial IGF-I and -II Transport in Cartilage under Unconfined Dynamic Compression

To allow direct comparison with the studies presented in previous sections, unconfined compressive loading between two impermeable platens will again be considered. Thus, equation (11.62) can be rewritten as

$$\phi^f \frac{\partial c_j^f}{\partial t} + (1-\phi^f) \sum_{i=1}^{2} \frac{\partial c_{ji}^s}{\partial t} - \phi^f D_j \left(\frac{\partial^2 c_j^f}{\partial r^2} + \frac{1}{r} \frac{\partial c_j^f}{\partial r} \right)$$
$$+ \left(\phi^f v_r - \kappa_r \frac{\partial p}{\partial r} \right) \frac{\partial c_j^f}{\partial r} + (1-\phi^f)\, v_r \sum_{i=1}^{2} \frac{\partial c_{ji}^s}{\partial r} = 0, \quad j = 1,2 \qquad (11.63)$$

The average free and bound (complexed) IGF uptake ratio become

$$\text{Free IGF uptake ratio}: \quad \overline{R}_{ui}^f = \frac{1}{\bar{c}_{i0}^f} \left(\frac{\int_0^{ro} 2\pi r \bar{c}_i^f \, dr}{\int_0^{ro} 2\pi r dr} \right) \qquad (11.64)$$

$$\text{Bound IGF uptake ratio}: \quad \overline{R}_{ui}^s = \frac{1}{\bar{c}_{i0}^f} \left[\sum_{j=1}^{2} \left(\frac{\int_0^{ro} 2\pi r \bar{c}_{ij}^b \, dr}{\int_0^{ro} 2\pi r dr} \right) \right] \qquad (11.65)$$

11.4.2.6 Free Diffusion with Competitor

Experiments showed a significant increase in the free IGF-I level in human synovial fluid in both OA and rheumatoid arthritis patients in comparison with normal cartilage (e.g., increasing from 20 to 80 ng/mL), but no change in the level of IGF-II [62]. Shown in Figure 11.14 are results of a "mathematical experiment" to understand the bioavailability of free IGF-I under various cartilage conditions. The IGF-I concentration ($\bar{c}_{10}^f$) in the solute bath (synovial fluid) is varied to explore the possible IGF-I transport behavior for "OA" (i.e., "diseased") conditions. The results show that the rate of free IGF-I uptake is relatively high in the diseased condition (Figure 11.14[a]), whereas a decrease of bound IGF-I uptake is observed.

11.4.2.7 Growth Factor Transport with Competitor and Cyclic Deformation

In cartilage synovial fluid, free IGF-I concentration is low (around 20 ng/mL) in comparison to that of free IGF-II (around 200 ng/mL) [62]. In previous sections it was shown that cyclic loading enhances IGF-I transport. However, the combination of competitive binding and cyclic loading might further modify the transport behavior of IGF-I. Figure 11.15 shows the percent increased in time-dependent free and bound IGF-I uptake in normal cartilage condition, in the absence (i.e., $\bar{c}_{10}^f = 20$ ng/mL, $\bar{c}_{20}^f = 0$ ng/mL) or in the presence of IGF-II competition (i.e., $\bar{c}_{10}^f = 20$ ng/mL, $\bar{c}_{20}^f = 200$ ng/mL), after 5 h 0.1 Hz at 6% strain cyclic deformation. The results indicate that cyclic loading still enhances IGF-I uptake in both cases; however, the enhancement is reduced when a competitor for binding sites is introduced. The competitor effectively

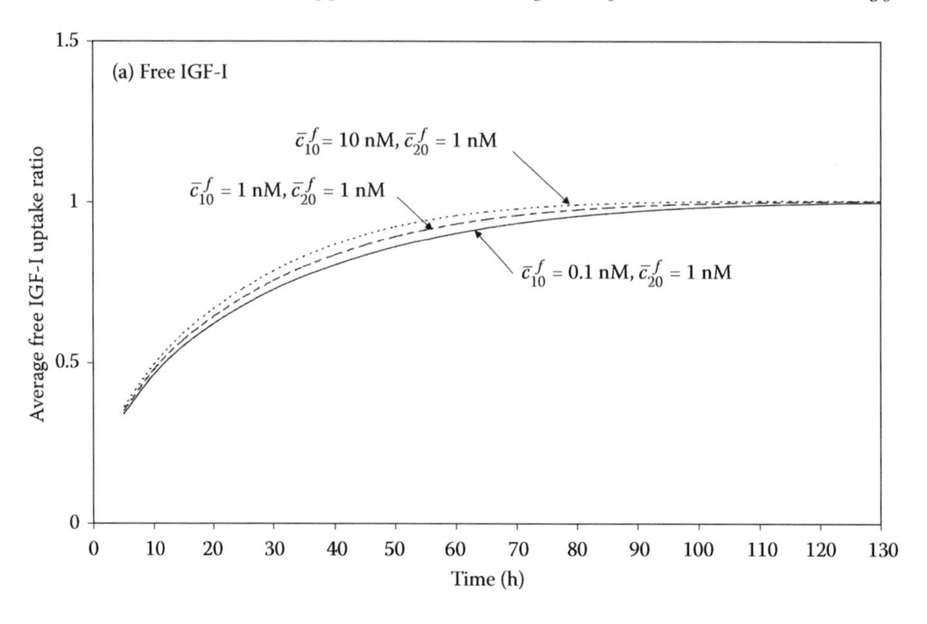

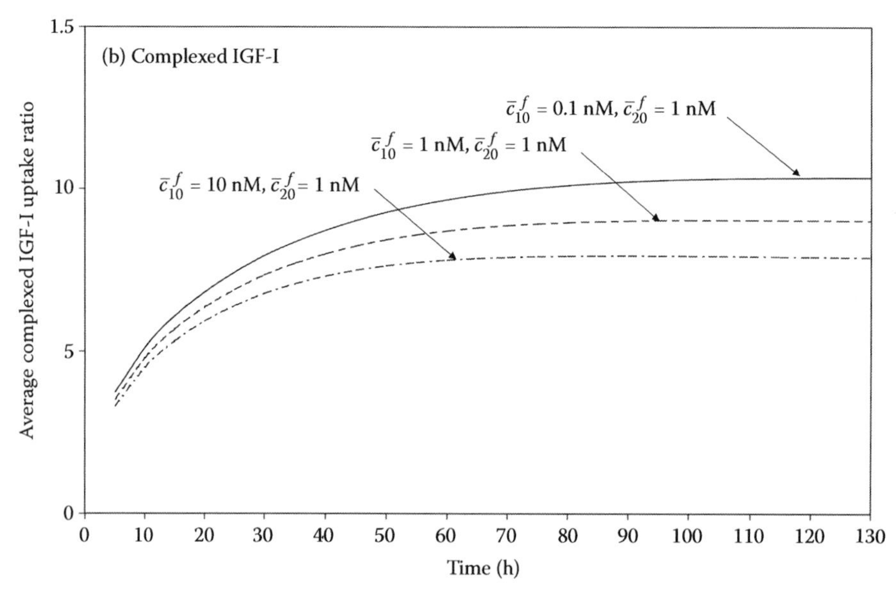

FIGURE 11.14
Comparison of time-dependent average IGF-I uptake ratio in the cartilage under various boundary IGF-I concentrations ($c^b_{BP1} = 45$ nM; $c^b_{BP2} = 101$ nM). (a) Free IGF-I; (b) Bound IGF-I.

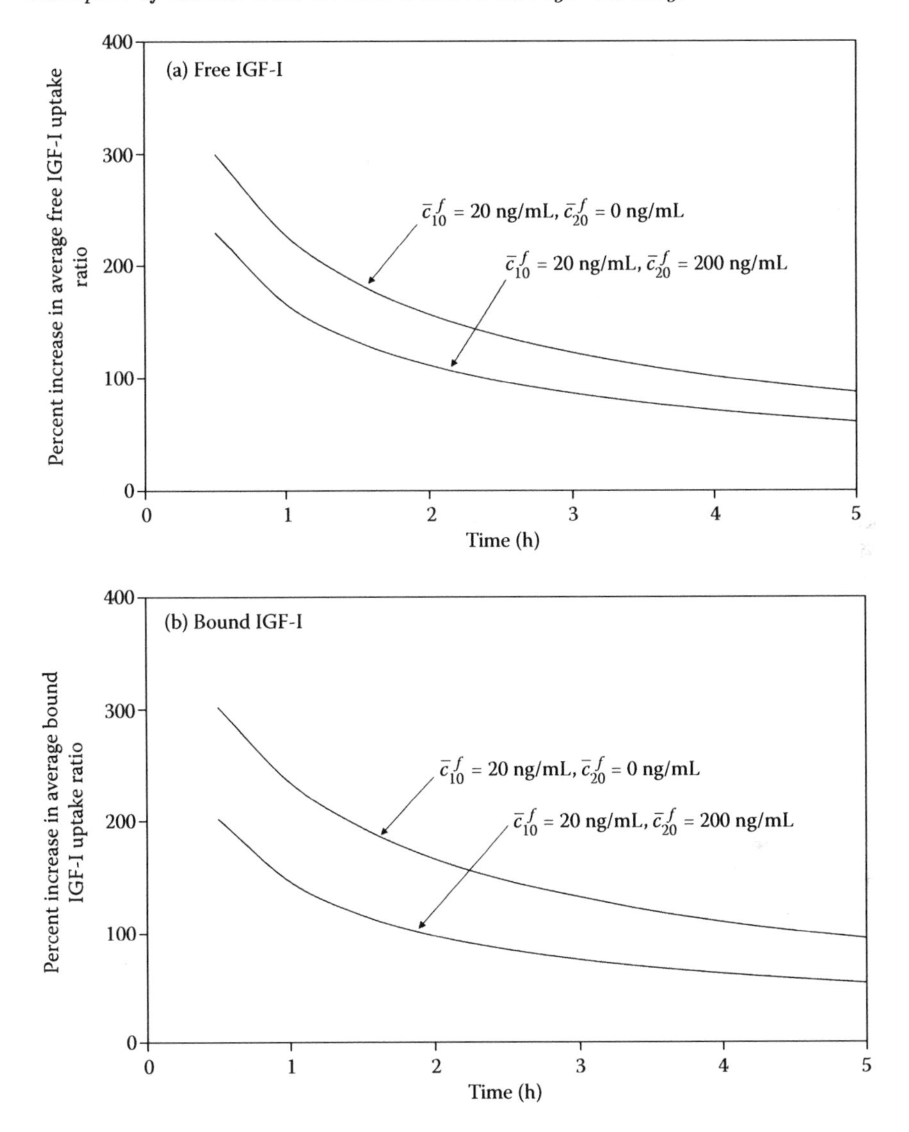

FIGURE 11.15
Percent increase in time-dependent, free, and bound IGF-I uptake in normal cartilage condition, in the absence (i.e., $\bar{c}^f_{10} = 20$ ng/mL, $\bar{c}^f_{20} = 0$ ng/mL) or in the presence of IGF-II competition (i.e., $\bar{c}^f_{10} = 20$ ng/mL, $\bar{c}^f_{20} = 200$ ng/mL), after 5 h 0.1 Hz at 6% strain cyclic deformation.

reduces the number of binding sites which then reduces the influence of binding sites in enhancing IGF-I transport under cyclic loading. Therefore, it is important to incorporate the competitive binding in the theoretical model.

11.5 An Integrated Model of IGF-I and Mechanical–Loading-Mediated Biosynthesis in a Deformed Articular Cartilage

11.5.1 Introduction

In the previous sections a series of models of IGF-I transport into cartilage was presented. Ultimately the motivation for this work was to understand how the microenvironment of chondrocytes might be shaped, so that we could begin to understand how chondrocytes might respond to its chemical and mechanical environment, and to change the local ECM composition. It would seem natural then to couple the previous model with a model for matrix biosynthesis in response to these stimuli. This is our goal of the final section of this chapter, though only selected aspects of matrix biosynthesis are dealt with here.

A numbers of studies have theoretically modeled the turnover of ECM macromolecules. Sengers et al. [79] proposed a solute transport and biosynthesis model for cartilage constructs by using the finite element approach. The matrix synthesis rate was assumed to linearly depend on solute and cell concentration. However, the model ignored the influence of IGFBPs and the consumption of solutes by chondrocytes.

Recently, Klein and Sah [80] extended a continuum model of matrix metabolism and transport developed by DiMicco and Sah [20] to understand the time and spatially dependent accumulation of proteoglycan in engineered cartilage constructs. Most importantly, the model considered the matrix macromolecules as soluble, bound, and degraded components and investigated the synthesis, binding, and transport behavior of each component, and their interactions. However, without the implementation of the theory of porous media, the model cannot describe the mechanical behavior of cartilage and the influence of mechanical stimuli on matrix biosynthesis.

To construct or biosynthesis model first we need to briefly review some experimental findings on IGF-I and load induced biosynthesis in cartilage. We will see that very little is really known about the signaling pathways between stimulus (e.g., IGF-I, mechanical loading) and the production rate of matrix molecules. We will therefore tend to use Hill functions to connect these two processes. Once these matrix molecules are synthesized they need to move away from the chondrocytes and into the surrounding cartilage. Our model will also be extended to include these processes as well. We will show that the equations for this process are very similar to what we have already seen for IGF-I transport.

11.5.1.1 IGF-I and Mechanical–Loading-Mediated Cartilage Biosynthesis

There are two key biosynthesis experimental results that we will use to construct the biosynthesis model. An important requirement of these experimental studies is that they are all performed on similar cartilage explants, so as to avoid variation in chondrocyte behavior due to species or animal age.

The first key experimental results were obtained by Bonassar et al. [30]. They showed that IGF-I increased both proline and sulfate incorporation within cartilage tissue (a measure of collagen and proteoglycan synthesis) in a dose-dependent manner. However, it is unclear exactly how IGF-I and the presence of its complexes stimulate ECM synthesis. IGF-I may stimulate synthesis of cartilage matrix components, such as GAG (a component of aggrecan) and collagen, through binding to chondrocyte membrane receptors [63]. The binding of IGF-I to receptors is a relatively specific and reversible process depending on time, pH, temperature, and the concentration of the components involved [81].

It is known that in the microenvironment of the chondrocytes the mechanical stimuli alone can significantly affect the synthesis and degradation rate of matrix macromolecules [5]. Clinical research on human knee joints indicated that moderate exercise helped to improve GAG content in the knee cartilage of patients who were at high risk of developing OA [82]. Experimental studies of Davisson et al. [83] on tissue engineered cartilage constructs found that static compression diminished aggrecan synthesis whereas dynamic loading enhanced aggrecan production. Buschmann et al. [32] measured spatially localized changes of aggrecan synthesis in response to a range of cyclic mechanical loading in bovine cartilage explants. Their results suggested that the mechanical stimulation was dependent on the flow of interstitial fluid induced by mechanical loading and a certain threshold of interstitial fluid velocity. This is the second key biosynthesis experimental result upon which the model is based.

11.5.2 Biosynthesis Model Construction

The previous sections have introduced a reactive-transport poroelastic porous media model to investigate the coupled processes of growth factor transport through cartilage undergoing mechanical deformation. Here the model is further extended to include biosynthesis and degradation of matrix molecules. The model is validated using three independent experimental data sets. It is found that a single set of parameters can describe the experimental results. The model is then employed to make predictions about changes in proteoglycan content under a variety of conditions including free, bound, and degraded conditions. This model may prove useful in predicting the behavior of tissue engineering constructs, or predicting the outcome of repair processes in cartilage.

11.5.2.1 IGF-I Transport and Interaction with IGFBPs and Receptors

It is reasonable to assume that IGF-I may bind to IGFBPs, or cell surface receptors (R1). Binding of IGF-I to receptors forms the IGF-I/R1 complex (R1I) and initiates an intracellular signaling cascade, ultimately leading to the production of cartilage ECM constituents (e.g., aggrecan). IGF-I is may be internalized during this initiation process [27], and a new, unoccupied receptor is returned to the cell's surface [81]. The reversible reaction involving IGF-I, receptors and IGFBPs can be described by

$$\text{IGF-I} + \text{IGFBP} \underset{k_{-1}}{\overset{k_{+1}}{\longleftrightarrow}} \text{Complex} \tag{11.66}$$

$$\text{IGF-I} + \text{R1} \underset{k_{-2}}{\overset{k_{+2}}{\longleftrightarrow}} \text{R1I} \overset{k_0}{\leftrightarrow} \text{R1} \tag{11.67}$$

Equations (11.66) and (11.67) can be included in the reactive-transport equations for IGF-I and IGFBPs as follows:

$$\frac{d\bar{c}_I^f}{dt} = -\nabla \bullet \left(-D_I \nabla \bar{c}_I^f + \mathbf{v^f} \bar{c}_I^f\right) - k_{+1}\bar{c}_I^f \bar{c}_{BP} + k_{-1}\bar{c}_I^b - k_{+2}\bar{c}_I^f \bar{c}_{R1} + k_{-2}\bar{c}_{R1I} \tag{11.68}$$

$$\frac{d\bar{c}_I^b}{dt} = -\nabla \bullet \left(\mathbf{v^s} \bar{c}_I^b\right) + k_{+1}\bar{c}_I^f \bar{c}_{BP} - k_{-1}\bar{c}_I^b \tag{11.69}$$

$$\frac{d\bar{c}_{BP}}{dt} = -k_{+1}\bar{c}_I^f \bar{c}_{BP} + k_{-1}\bar{c}_I^b \tag{11.70}$$

$$\frac{d\bar{c}_{R1}}{dt} = -k_{+2}\bar{c}_I^f \bar{c}_{R1} + (k_{-2} + k_0)\,\bar{c}_{R1I} \tag{11.71}$$

$$\frac{d\bar{c}_{R1I}}{dt} = k_{+2}\bar{c}_I^f \bar{c}_{R1} - (k_{-2} + k_0)\,\bar{c}_{R1I} \tag{11.72}$$

where $\bar{c}_I^f$, $\bar{c}_{BP}$, $\bar{c}_I^b$, $\bar{c}_{R1}$, and $\bar{c}_{R1I}$ are volume-based IGF-I, IGFBP, IGF-I/IGFBP complex, receptor, and IGF-I/receptor complex concentrations, respectively. k_{+1}, k_{-1}, k_{+2}, k_{-2}, and k_0 are the respective reaction rate constants.

In unconfined compression, equations (11.68)–(11.72) can be written in radial coordinate as

$$\begin{aligned} &\phi^f \frac{\partial c_I^f}{\partial t} - \phi^f D_I \left(\frac{\partial^2 c_I^f}{\partial r^2} + \frac{1}{r}\frac{\partial c_I^f}{\partial r}\right) + \left(\phi^f v_r - \kappa \frac{\partial p}{\partial r}\right) \frac{\partial c_I^f}{\partial r} \\ &= -k_{+1}\phi^f \left(1-\phi^f\right) c_I^f c_{BP} + k_{-1}\left(1-\phi^f\right) c_I^b - k_{+2}\phi^f \left(1-\phi^f\right) c_I^f c_{R1} \\ &\quad + k_{-2}\left(1-\phi^f\right) c_{R1I} \end{aligned} \tag{11.73}$$

$$\frac{\partial c_I^b}{\partial t} + v_r \frac{\partial c_I^b}{\partial r} = k_{+1}\phi^f c_I^f c_{BP} - k_{-1} c_I^b \tag{11.74}$$

$$\frac{dc_{BP}}{dt} = -k_{+1}\phi^f c_I^f c_{BP} + k_{-1} c_I^b \tag{11.75}$$

$$\frac{dc_{R1}}{dt} = -k_{+2}\phi^f c_I^f c_{R1} + (k_{-2} + k_0) c_{R1I} \tag{11.76}$$

$$\frac{dc_{R1I}}{dt} = k_{+2}\phi^f c_I^f c_{R1} - (k_{-2} + k_0) c_{R1I} \tag{11.77}$$

11.5.2.2 Cartilage ECM Biosynthesis

As a first estimate, the total aggrecan biosynthesis rate is assumed to be a sum of the basal production rate s_{p0} [80], the matrix production rate induced by IGF-I (s_{pI}) and the mechanical–stimuli-induced production rate (s_{pm}), respectively. Further experiments have suggested that aggrecan biosynthesis is limited by the possible maximum allowable aggrecan concentration (i.e., $\bar{c}_{a\,\max}$) in cartilage tissue [30,84,85]. Thus, the total aggrecan biosynthesis rate s_p may be expressed as

$$s_p = (s_{p0} + s_{pI} + s_{pm}) \left(1 - \frac{\bar{c}_a^f + \bar{c}_a^b}{\bar{c}_{a\,\max}}\right) \tag{11.78}$$

where $\bar{c}_{a\,\max}$ is the maximum allowable maximum aggrecan content in cartilage.

Following the experimental result of Sah et al. [86] the basal rate of aggrecan is assumed to be 1.9×10^{-8}.

11.5.2.3 IGF-I Mediated Aggrecan Biosynthesis

IGF-I mediated production rate (s_{pI}) depends on the concentration of complex IGF-I/R1 (i.e., $\bar{c}_{R1I}$) and as a first approximation the link between s_{pI} and IGF-I/R1 complex concentration can be described by an activator Hill function [87]. That is, s_{pI} may be expressed as

$$s_{pI} = \left(\frac{\beta \bar{c}_{R1I}^n}{K^n + \bar{c}_{R1I}^n}\right) \tag{11.79}$$

where parameter K is the "activation coefficient" and determines the threshold IGF-I/R1 complex concentration at which aggrecan biosynthesis is "switched on." The smaller the K, the lower the IGF-I/R1 concentration required to initiate biosynthesis. K can be directly related to receptor occupancy. The index n defines the steepness of the Hill function when $K = \bar{c}_{R1I}$ and usually takes between 1 and 5. Hence when $n = 1$ aggrecan production is largely dose dependent, whereas $n = 5$ corresponds to threshold or "on-off" dependency of aggrecan production with an increase in $\bar{c}_{R1I}$. Finally, β is the IGF-I mediated maximum production rate.

11.5.2.4 Mechanical–Stimuli-Mediated Aggrecan Biosynthesis

On the basis of experimental observations, cellular response to mechanical stimuli exhibits a threshold behavior, with aggrecan production only triggered when the interstitial fluid velocity (i.e., the Darcy velocity $\mathbf{v_d}$) exceeds a certain threshold (i.e., $v_0 = 0.25\,\mu\text{m/s}$). Further increases in the fluid velocity were not observed to increase the aggrecan production rate [32]. These observations can be described mathematically as

$$s_{pm} = \begin{cases} \lambda, & (|\mathbf{v_d}| \geq \mathbf{v_0}) \\ 0, & (|\mathbf{v_d}| < \mathbf{v_0}) \end{cases} \tag{11.80}$$

Recall that the maximum Darcy velocity, in the unconfined compression tests we have focused on previously, occurs near the periphery of the cartilage discs. It is expected then that the mechanical–loading-induced aggrecan production will be mainly confined to the outer edge of the cartilage disc and for the larger strain and loading frequencies cases. This expectation is borne out in Figure 11.16, in which typical variations in the Darcy velocity profiles as a function of radial distance from center of the cartilage under various loading conditions (i.e., 0.1 Hz at 10% strain, 0.01 Hz at 10% strain, and 0.001Hz at 10% strain) are shown. It can be seen that a loading regime of high frequency

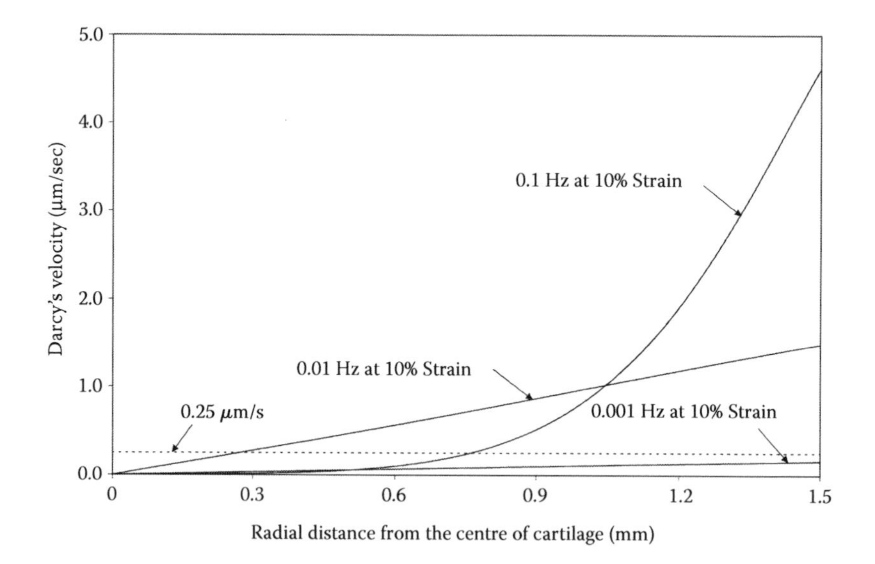

FIGURE 11.16
Profiles of Darcy velocity as a function of radial distance from center of the cartilage under various loading conditions. The aggrecan synthesis is stimulated in the region where the Darcy velocity $\mathbf{v_d} \geq v_0$ (i.e., 0.25 [32]) and is indicated by the dotted line.

0.1 Hz results in the greatest fluid velocity and the mechanical stimulation is localized to the periphery domain. For the intermediate frequency of 0.01 Hz, the region experiencing mechanical stimulation extends into the inner region of the domain, whereas a low frequency at 0.001 Hz there was no stimulatory effect.

11.5.2.5 Aggrecan Molecule Transport in Cartilage

Newly synthesized matrix molecules are initially mobile after secretion by chondrocytes. They are then transported into the surrounding matrix, and can become either bound to hyaluronan as immobile proteoglycan aggregates, be degraded, or potentially be released from the cartilage matrix [20] into the synovial fluid. The total matrix aggrecan concentration in cartilage ($\bar{c}_a$) is the sum of chondrocyte-derived mobile component transported in fluid phase ($\bar{c}_a^f$); "immobilized" proteoglycan component effectively "bound" within ECM ($\bar{c}_a^b$), and finally the mobile component degraded from the ECM structure ($\bar{c}_a^d$) (which may escape from the cartilage altogether). Therefore the total aggrecan concentration is given by

$$\bar{c}_a = \bar{c}_a^f + \bar{c}_a^b + \bar{c}_a^d \tag{11.81}$$

The transport equation of each component is represented as

$$\frac{d\bar{c}_a^f}{dt} + \nabla \bullet \left(-K_d D_a \nabla \bar{c}_a^f + K_a \mathbf{v}^{\mathbf{f}} \bar{c}_a^f\right) = s_p - s_b \bar{c}_a^f \tag{11.82}$$

$$\frac{d\bar{c}_a^b}{dt} = s_b \bar{c}_a^f - s_d \bar{c}_a^b \tag{11.83}$$

$$\frac{d\bar{c}_a^d}{\partial t} + \nabla \bullet \left(-K_d D_a \nabla \bar{c}_a^d + K_a \mathbf{v}^{\mathbf{f}} \bar{c}_a^d\right) = s_d \bar{c}_a^b \tag{11.84}$$

where D_a is diffusion coefficient of free aggrecan and s_b and s_d are aggrecan binding and degradation rates, respectively, which are obtained from experimental studies [20,80]. K_d and K_a are the dimensionless tortuosity and hindrance coefficient, respectively, and their values depend on molecular size and shape of the molecule being transported through the ECM. Newly synthesized mobile aggrecans are large molecules with very small diffusion coefficients in the range of 10^{-10}–10^{-9} cm^2/s [88], and so cartilage ECM may decrease the mobility of macromolecules and limit their transport. For this reason hindered advective transport of these matrix molecules is adopted.

Equations (11.82)–(11.84) can be presented in radial coordinates as

$$\begin{aligned}\phi^f \frac{\partial c_a^f}{\partial t} - \phi^f K_d D_a \left(\frac{\partial^2 c_a^f}{\partial r^2} + \frac{1}{r}\frac{\partial c_a^f}{\partial r}\right) + K_a \left(\phi^f v_r - \kappa_r \frac{\partial p}{\partial r}\right) \frac{\partial c_a^f}{\partial r} \\ = s_{p0} + \left(\frac{\beta c_{R1I}^n}{K^n + c_{R1I}^n} + \lambda\right)\left[1 - \frac{\phi^f c_a^f + (1-\phi^f) c_a^b}{c_{a\max}}\right] - s_b \phi^f c_a^f\end{aligned} \tag{11.85}$$

$$\left(1-\phi^f\right)\frac{\partial c_a^b}{\partial t}=s_b\phi^f c_a^f-s_d\left(1-\phi^f\right)c_a^b \tag{11.86}$$

$$\phi^f\frac{\partial c_a^d}{\partial t}-\phi^f K_d D_a\left(\frac{\partial^2 c_a^d}{\partial r^2}+\frac{1}{r}\frac{\partial c_a^d}{\partial r}\right)+K_a\left(\phi^f v_r-\kappa_r\frac{\partial p}{\partial r}\right)\frac{\partial c_a^d}{\partial r}=s_d\left(1-\phi^f\right)c_a^b \tag{11.87}$$

The normalized average free, bound, and degraded aggrecan in the radial direction can be defined using the following equations as:

$$\bar{c}_{aavg}^f=\frac{1}{\bar{c}_{a0}}\left(\frac{\int_0^{r_0}2\pi r\bar{c}_a^f dr}{\int_0^{r_0}2\pi r dr}\right) \tag{11.88a}$$

$$\bar{c}_{aavg}^b=\frac{1}{\bar{c}_{a0}}\left(\frac{\int_0^{r_0}2\pi r\bar{c}_a^b dr}{\int_0^{r_0}2\pi r dr}\right) \tag{11.88b}$$

$$\bar{c}_{aavg}^d=\frac{1}{\bar{c}_{a0}}\left(\frac{\int_0^{r_0}2\pi r\bar{c}_a^d dr}{\int_0^{r_0}2\pi r dr}\right) \tag{11.88c}$$

11.5.3 Biosynthesis Model Validation and Predictions

The developed numerical model is validated by using the experimental data from Bonassar et al. [30], Jin et al. [31], and Buschmann et al. [32]. In experiments investigating the aggrecan synthesis when cartilage explants were exposed to a range of IGF-I concentrations under conditions of free diffusion, Bonassar et al. [30] treated cartilage discs with IGF-I ranging from 0 to 300 ng/mL for 48 h with the ^{35}S-sulfate only present in the final 24 h, while in Jin et al.'s [31] experiment, cartilage explants were exposed to IGF-I in the range of 0–300 ng/mL together with ^{35}S-sulfate for the first 24 h. The ^{35}S-sulfate incorporation was later used to assess the effect of IGF-I on aggrecan synthesis. Differing from the experiments of Bonassar et al. [30] and Jin et al. [31], the study of Buschmann et al. [32] focused on understanding the correlation between cyclic loading and the spatial distribution of biosynthesis in the cartilage explants. The rate of aggrecan synthesis was assessed by measuring the ^{35}S-sulfate incorporation during the last 8 h after cartilage explants were under cyclic mechanical loading at frequencies of 0.01–0.1 Hz at displacement of amplitude of 50 μm for 23 h. Thus, these experimental studies provide three sets of independent biosynthesis data on similar cartilage explants but under different experimental conditions. The model should be able to reproduce these experimental observations.

The rate of aggrecan formation is determined by currently unknown model parameters, namely the coefficient n, the activation coefficient K in equation (11.79), the IGF-I mediated maximum production rate β, and mechanical–stimuli-mediated production rate λ, as well as estimating the initial aggrecan concentration $\bar{c}_{a0}$. A key challenge here is to reproduce the experimental outcome using a single set of parameters.

As shown in Figure 11.17, with the adoption of the aforementioned set of parameters in the theoretical model (Table 11.4), the numerical solutions

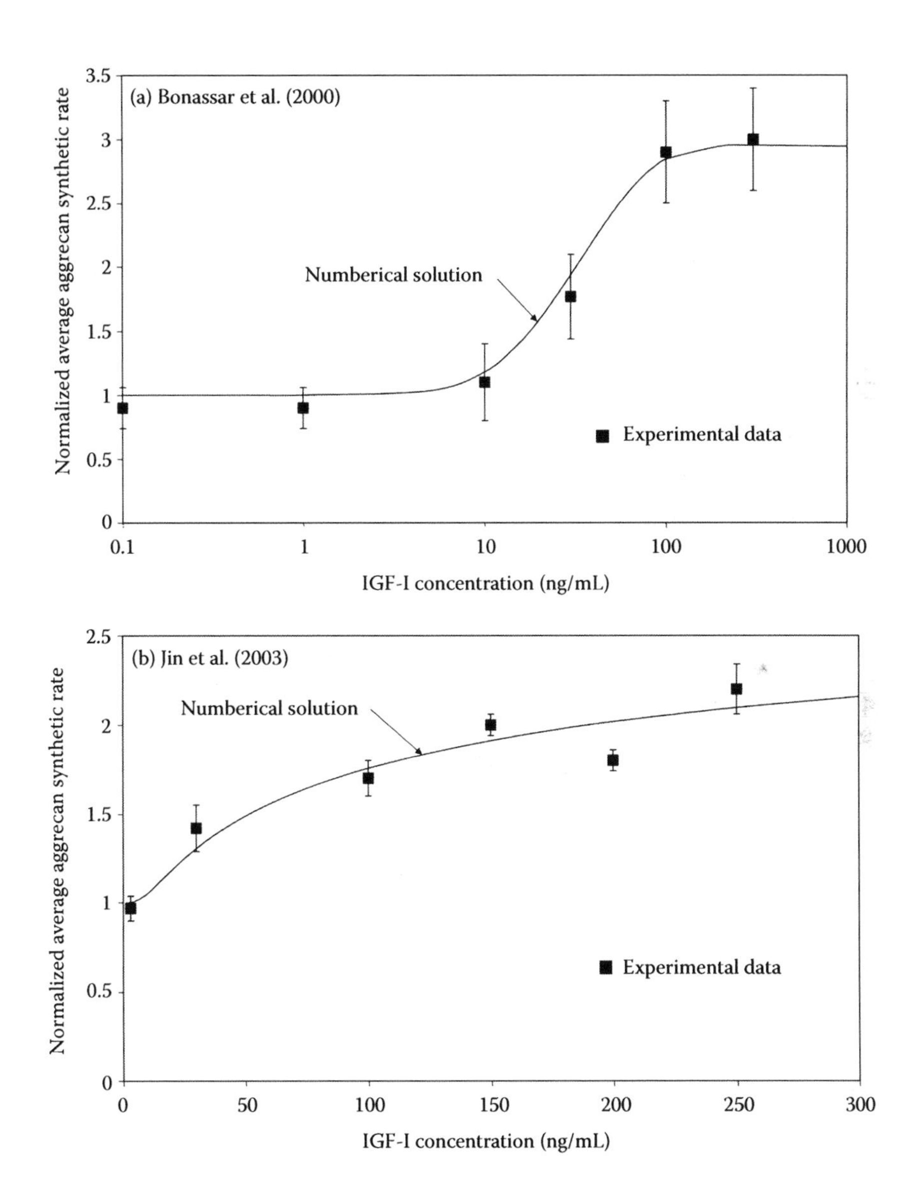

FIGURE 11.17
Comparison of model predictions with three independent experimental results [30–32]. Note the set of model parameters is identical in each case. (*Continued*)

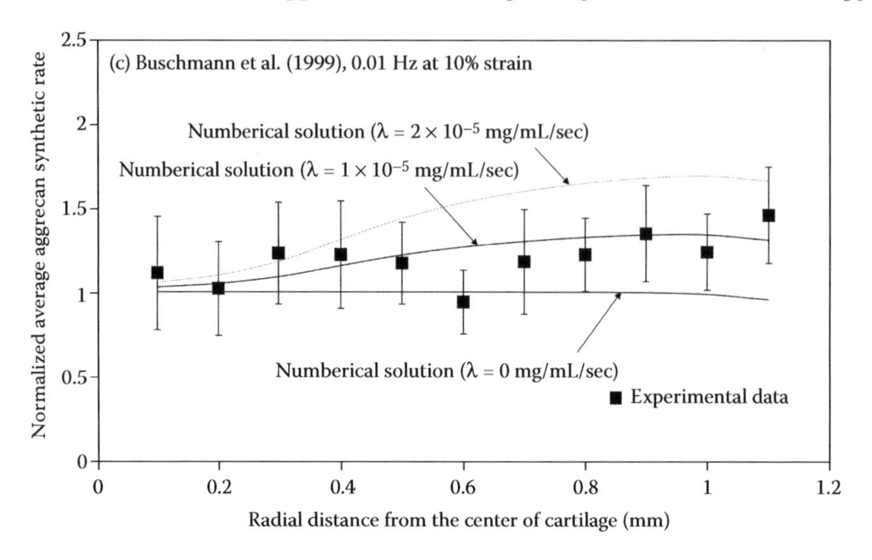

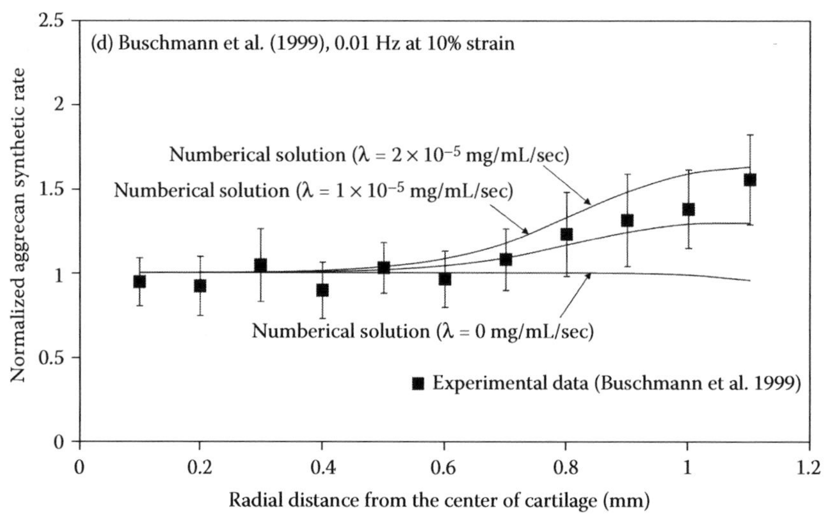

FIGURE 11.17
(Continued)

provide a remarkably good fit with all three sets of experimental data by using least squares optimization. Figure 11.17a shows that there is no obvious aggrecan production if IGF-I concentration is less than certain threshold (i.e., 10 ng/mL) but the production can be rapidly "switched on" once IGF-I concentration is over this threshold. It can be seen from Figure 11.17c–d that a loading regime of high frequency (i.e., 0.1 Hz) stimulates aggrecan biosynthesis only at the peripheral region, while the region experiencing mechanical

TABLE 11.4
Parameters Used in Model Validation [89]

Parameters	**Values**
N	3
K	30% total-receptor concentration
β	3.8×10^{-5} mg/mL/sec
Initial aggrecan content	16 mg/mL

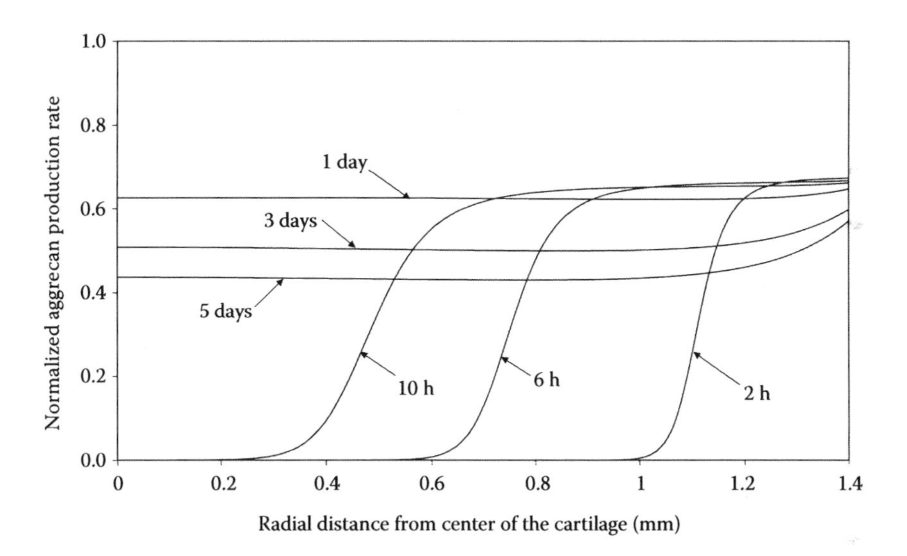

FIGURE 11.18
Normalized IGF-I mediated aggrecan production rate, as a function of radial distance from the center of cartilage undergoing free diffusion. The calculated aggrecan production rate is normalized to the value of parameter β. ($K = 0.3\bar{c}_{RT}$, $\beta = 3.8 \times 10^{-5}$ mg/mL/sec, $n = 3$, $\bar{c}_{a0} = 16$ mg/mL, and $c_{I0}^{f} = 40$ nM.)

stimulation extends into the inner region of the domain at an intermediate frequency (i.e., 0.01 Hz).

Figure 11.18 shows predictions of the spatial dependence of IGF-I induced aggrecan synthesis rate in the presence of 300 ng/mL bath IGF-I concentration undergoing free diffusion. It can be seen that a strong spatially dependent aggrecan production develops at early times (i.e., time $<$ 1 day) when IGF-I initially diffuses into the cartilage with a steep concentration gradient from the outer edge of the cartilage. Over time, as IGF-I gradually saturates the cartilage disc, aggrecan production rate becomes uniform throughout the disc and reaches its maximum level. Eventually aggrecan production is "switched off"

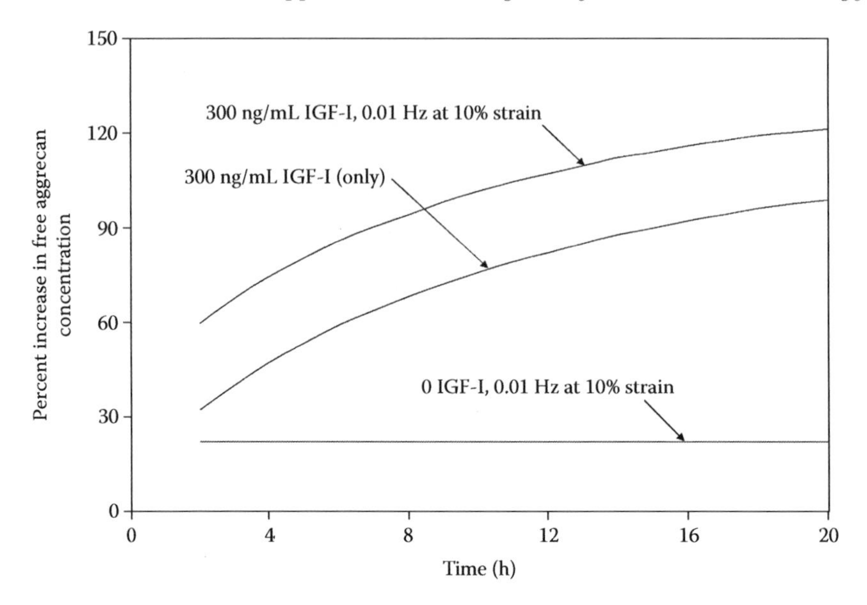

FIGURE 11.19
Percent increase in time-dependent free aggrecan concentration in cartilage disk response to individual and combination of IGF-I (300 ng/mL) and dynamic loading (0.01 Hz at 10% strain) in comparison to basal condition (i.e., 0 ng/mL IGF-I, no loading).

as a new dynamic balance of aggrecan production and degradation is established.

Figure 11.19 presents predictions of the aggrecan accumulation in cartilage after chondrocytes are exposed to either IGF-I (300 ng/mL) or mechanical stimuli (i.e., 0.01 Hz at 10% strain) or a combination of both. It shows that in comparison to basal condition (i.e., 0 ng/mL IGF-I, no loading), the component of free aggrecan synthesis stimulated by IGF-I increases with time as IGF-I concentration increases. The component of biosynthesis induced by mechanical loading is constant in time, that is, it is equivalent to an increased basal rate. It can be seen that IGF-I induced biosynthesis has a larger contribution than the mechanical loading regime investigated. When IGF-I and mechanical stimulus are applied simultaneously, the degree of aggrecan stimulation is greater than that achieved by either IGF-I or mechanical loading alone.

11.6 Summary

This chapter started with a basic but fully coupled reactive-transport poroelastic model within the theory of porous media. The model was extended to include a series of biological processes taking place within cartilage involving IGF transport, IGF interaction with its binding proteins (IGFBPs), IGF and interstitial fluid flow induced matrix biosynthesis, and the transport of matrix molecules. The model was validated using several independent experimental data sets. Of particular significance was that by using this model, all the available experimental data that is comparable, can be reproduced using only a single set of model parameters. The results of this new biosynthesis model show that IGF-I and mechanical loading are able to stimulate aggrecan synthesis independently, but when applied simultaneously, the degree of aggrecan stimulation is greater than that achieved by either IGF-I or mechanical loading alone. The dynamic balance of aggrecan metabolism and catabolism in cartilage depends on the IGF-I concentration in synovial fluid, and aggrecan production can be rapidly "switched on" when the concentration of IGF-I reaches a certain threshold. This model may prove useful in understanding normal homeostatic processes in cartilage, and in identifying optimal cell density distributions for efficient protein synthesis and more uniform cartilage distributions in tissue engineered constructs.

11.7 References

[1] Maroudas, A. and Bullough, P. (1968). Permeability of articular cartilage. *Nature*, **219**:1260–1261.

[2] Krishnan, R., Seonghun, F., Park, E., and Ateshian, G. A. (2003). Inhomogeneous cartilage properties enhance superficial interstital fluid support and frictional properties, but do not provide a homogenous state of stress. *Journal of Biomechanical Engineering*, **125**:569–577.

[3] Mow, V. C. and Lai, W. M. (1980). Recent developments in synovial joint biomechanics. *Siam Review*, **22**:275–317.

[4] Maroudas, A. and Kuettner, K. (1990). *Methods in Cartilage Research.* Academic Press, San Diego.

[5] Grodzinsky, A. J., Levenston, M. E., Jin, M., and Frank, E. H. (2000). Cartilage tissue remodeling in response to mechanical forces. *Annual Review of Biomedical Engineering*, **2**:691–713.

[6] Bachrach, N. M., Valhmu, W. B., Stazzone, E., Ratcliffe, A., Lai, W. M., and Mow, V. C. (1995). Changes in proteoglycan synthesis of chondrocytes in articular cartilage are associated with the time-dependent changes in their mechanical environment. *Journal of Biomechanics*, **28**: 1561–1569.

[7] Lai, W. M., Hou, J. S., and Mow, V. C. (1991). A triphasic theory for the swelling and deformation behaviors of articular cartilage. *Journal of Biomechanical Engineering*, **113**:245–258.

[8] Sun, D. N., Gu, W. Y., Guo, X. E., Lai, W. M., and Mow, V. C. (1999). A mixed finite element formulation of triphasic mechano-electrochemical theory for charged, hydrated biological soft tissues. *International Journal for Numerical Methods in Engineering*, **45**:1375–1402.

[9] Lu, X. L., Sun, D. D. N., Guo, X. E., Chen, F. H., Lai, W. M., and Mow, V. C. (2004). Indentation determined mechanoelectrochemical properties and fixed charge density of articular cartilage. *Annals of Biomedical Engineering*, **32**:370–379.

[10] Rowan, A. D. (2001) Cartilage catabolism in arthritis: factors that influence homeostasis. *Expert Reviews in Molecular Medicine*, **3**:1–20.

[11] Pelletier, J.-P., Martel-Pelletier, J., and Abramson, S. B. (2001). Osteoarthritis, an inflammatory disease: potential implication for the selection of new therapeutic targets. *Arthritis Rheumatism*, **44**: 1237–1247.

[12] Attur, M. G., Dave, M., Akamatsu, M., Katoh, M., and Amin, A. R. (2002). Osteoarthritis or osteoarthrosis: the definition of inflammation becomes a semantic issue in the genomic era of molecular medicine. *Osteoarthritis and Cartilage*, **10**:1–4.

[13] DiMicco, M. A., Patwari, P., Siparsky, P. N., Kumar, S., Pratta, M. A., Lark, M. W., Kim, Y.-J., and Grodzinsky, A. J. (2004). Mechanisms and kinetics of glycosaminoglycan release following in vitro cartilage injury. *Arthritis & Rheumatism*, **50**:840–848.

[14] Faulkner, A., Kennedy, L. G., Baxter, K., Donovan, J., Wilkinson, M., and Bevan, G. (1998). Effectiveness of hip prostheses in primary total hip replacement: a critical review of evidence and an economic model. *Health Technology Accessment*, **2**:1–133.

[15] Buckwalter, J. A. and Mankin, H. J. (1997). Instructional course lectures. The American Academy of Orthopaedic Surgeons—Articular Cartilage. Part II: degeneration and osteoarthrosis, repair, regeneration, and transplantation. *Journal of Bone & Joint Surgery (American Volume)*, **79**: 612–632.

[16] Grande, D. A., Pitman, M. I., Peterson, L., Menche, D., and Klein, M. (2005). The repair of experimentally produced defects in rabbit articular cartilage by autologous chondrocyte transplantation. *Journal of Orthopaedic Research*, **7**:208–218.

[17] Steadman, J. R., Rodkey, W. G., and Rodrigo, J. J. (2001). Microfracture: surgical technique and rehabilitation to treat chondral defects. *Clinical Orthopaedics and Related Research*, **391**:362–369.

[18] Mauck, R. L., Nicoll, S. B., Seyhan, S. L., Ateshian, G. A., and Hung, C. T. (2003). Synergistic action of growth factors and dynamic loading for articular cartilage tissue engineering. *Tissue Engineering*, **9**:597–611.

[19] Blunk, T., Sieminski, A. L., Gooch, K. J., Courter, D. L., Hollander, A. P., Nahir, A. M., Langer, R., Vunjak-Novakovic, G., and Freed, L. E. (2002). Differential effects of growth factors on tissue-engineered cartilage. *Tissue Engineering*, **8**:73–84.

[20] DiMicco, M. A. and Sah, R. L. (2003). Dependence of cartilage matrix composition on biosynthesis, diffusion and reaction. *Transport in Porous Media*, **50**:57–73.

[21] Dudhia, J. (2005). Aggrecan, aging and assembly in articular cartilage. *Celluar and Molecular Life Sciences*, **62**:2241–2256.

[22] Bolis, S., Handley, C. J., and Comper, W. D. (1989). Passive loss of proteoglycan from articular cartilage explants. *Biomedica Biochimica Acta*, **993**:157–167.

[23] Sandy, J. D., O'Neill, J. R., Ratzlaff, L. C. (1989). Acquisition of hyaluronate-binding affinity in vivo by newly synthesized cartilage proteoglycans. *Biochemical Journal*, **258**:875–880.

[24] Pottenger, L. A., Webb, J. E., and Lyon, N. B. (1985). Kinetics of extraction of proteoglycans from human cartilage. *Arthritis & Rheumatism*, **28**: 323–330.

[25] Sandy, J. D. and Plaas, A. H. K. (1986). Age-related changes in the kinetics of release of proteoglycans from normal rabbit cartilage explants. *Journal of Orthopaedic Research*, **4**:263–273.

[26] Maroudas, A., Baylissb, M. T., Uchitel-Kaushanskya, N., Schneidermana, R., and Gilav, E. (1998). Aggrecan turnover in human articular cartilage: use of aspartic acid racemization as a marker of molecular age. *Archives of Biochemistry and Biophysics*, **350**:61–71.

[27] Barta, E. and Maroudas, A. (2006). A theoretical study of the distribution of insulin-like growth factor in human articular cartilage. *Journal of Theoretical Biology*, **241**:628–638.

[28] Gardiner, B. S., Smith, D. W., Pivonka, P., Grodzinsky, A. J., Frank, E. H., and Zhang, L. (2007). Solute transport in cartilage undergoing cyclic deformation. *Computer Methods in Biomechanics and Biomedical Engineering*, **10**:265–278.

[29] Bhakta, N. R., Garcia, A. M., Frank, E. H., Grodzinsky, A. J., and Morales, T. I. (2000). The insulin-like growth factors (IGFs) I and II bind to articular cartilage via the IGF-binding proteins. *The Journal of Biological Chemistry*, **275**:5860–5866.

[30] Bonassar, L. J., Grodzinsky, A. J., Srinivasan, A., Davila, S. G., and Trippel, S. B. (2000). Mechanical and physiochemical regulation of the action of insulin-like growth factor-ion articular cartilage. *Archives of Biochemistry and Biophysics*, **379**:57–63.

[31] Jin, M., Emkey, G. R., Siparsky, P., Trippel, S. B., and Grodzinsky, A. J. (2003). Combined effects of dynamic tissue shear deformation and insulin-like growth factor I on chondrocyte biosynthesis in cartilage explants. *Archives of Biochemistry and Biophysics*, **414**:223–231.

[32] Buschmann, M. D., Kim, Y.-J., Wong, M., Frank, E., Hunziker, E. B., and Rodzinsky, A. J. (1999). Stimulation of aggrecan synthesis in cartilage explants by cyclic loading is localized to regions of high interstitial fluid flow. *Archives of Biochemistry and Biophysics*, **366**:1–7.

[33] Zhang, L., Gardiner, B. S., Smith, D. W., Pivonka, P., and Grodzinsky, A. J. (2007). The effect of cyclic deformation and solute binding on solute transport in cartilage. *Archives of Biochemistry and Biophysics*, **457**:47–56.

[34] Zhang, L., Gardiner, B. S., Smith, D. W., Pivonka, P., and Grodzinsky, A. J. (2008). IGF uptake with competitive binding in articular cartilage. *Journal of Biological Systems*, **16**:175–195.

[35] Hung, C. T., Mauck, R. L., Wang, C. C.-B., Lima, E. G., and Ateshian, G. A. (2004). A paradigm for functional tissue engineering of articular cartilage via applied physiologic deformational loading. *Annals of Biomedical Engineering*, **32**:35–49.

[36] Mauck, R. L., Hung, C. T., and Ateshian, G. A. (2000). Modeling of neutral solute transport in a dynamically loaded porous permeable gel: implications for articular cartilage biosynthesis and tissue engineering. *Journal of Biomechanical Engineering*, **125**:602–614.

[37] Bonassar, L. J., Grodzinsky, A. J., Frank, E. H., Davila, S. G., Bhaktav, N. R., and Trippel, S. B. (2001). The effect of dynamic compression on the response of articular cartilage to insulin-like growth factor-I. *Journal of Orthopaedic Research*, **19**:11–17.

[38] Quinn, T. M., Grodzinsky, A. J., and Meister, J. J. (2002). Preservation and analysis of nonequilibrium solute concentration distributions within mechanically compressed cartilage explants. *Journal of Biochemical and Biophysical Methods*, **52**:83–95.

[39] O'Hara, B. P., Urban, J. P. G., and Maroudas, A. (1990). Influence of cyclic loading on the nutrition of articular cartilage. *Annals of Rheumatic Diseases*, **49**:536–539.

[40] Zhang, L. and Szeri, A. Z. (2005). Transport of neutral solute in articular cartilage: effects of loading and particle size. *Proceedings of the Royal Society*, **461**:2021–2042.

[41] Simon, B. R., Liable, J. P., Pflaster, D., Yuan, Y., and Krag, M. H. (1996). A poroelastic finite elment formulation including transport and swelling in soft tissue structures. *Journal of Biomechanical Engineering*, **118**:1–9.

[42] Ehlers, W. and Markert, B. (2001). A linear viscoelastic biphasic model for soft tissues based on the theory of porous media. *Journal of Biomechanical Engineering*, **123**:418–424.

[43] Oloyede, A. and Broom, N. D. (1994). The generalized consolidation of articular cartilage: an investigation of its near-physiological response to static load. *Connective Tissue Research*, **31**:75–86.

[44] Gardiner, B. S., Smith, D. W., Pivonka, P., Grodzinsky, A. J., Frank, E. H., and Zhang, L. (2006). Solute transport in cartilage undergoing cyclic deformation. *Computer Methods in Biomechanics and Biomedical Engineering*, accepted.

[45] Ehlers, W. and Markert, B. (2000). On the viscoelastic behaviour of fluid-saturated porous materials. *Granular Matter*, **2**:153–161.

[46] Mow, V. C., Kuei, S. C., Lai, W. M., and Armstrong, C. G. (1980). Biphasic creep and stress relaxation of articular cartilage in compression: theory and experiment. *Journal of Biomechanical Engineering*, **102**:73–84.

[47] Bowen, R. M. (1980). Incompressible porous media models by use of the theory of mixtures. *International Journal of Engineering Science*, **18**:1129–1148.

[48] Soulhat, J., Buschmann, M. D., and Shirazi-Adl, A. (1999). A fibril-network-reinforced biphasic model of cartilage in unconfined compression. *Journal of Biomechanical Engineering*, **121**:340.

[49] Disilvestro, M. R. and Suh, J.-K. F. (2002). Biphasic poroviscoelastic characteristics of proteoglycan-depleted articular cartilage: simulation of degeneration. *Annals of Biomedical Engineering*, **30**:792–800.

[50] Ateshian, G. A., Warden, W. H., Kim, J. J., Grelsamer, R. P., and Mow, V. C. (1997) Finite deformation biphasic material properties of bovine articular cartilage from confined compression experiments. *Journal of Biomechanics*, **30**:1157–1164.

[51] Gu, W. Y., Lai, W. M., and Mow, V. C. (1998). A mixture theory for charged-hydrated soft tissues containing multi-electrolytes: passive transport and swelling behaviors. *Journal of Biomechanical Engineering*, **120**:169–180.

[52] Lai, W. M., Rubin, D., and Krempl, E. (1993). *Introduction to Continuum Mechanics*. Pergamon Press, London.

[53] Mow, V. C. and Guo, X. E. (2002). Mechano-electrochemical properties of articular cartilage: their inhomogeneities and anisotropies. *Annual Review of Biomedical Engineering*, **4**:175–209.

[54] Soltz, M. A. and Ateshian, G. A. (2000). A Conewise linear elasticity mixture model for the analysis of tension-compression nonlinearity in articular cartilage. *Journal of Biomechanical Engineering*, **122**:576–586.

[55] Curnier, A., He, Q.-C., and Zysset, P. (1995). Conewise linear elastic materials. *Journal of Elasticity*, **37**:1–38.

[56] Burasc, P. M., Obitz, T. W., Eisenberg, S. R., and Stamenovic, D. (1999). Confined and unconfined stress relaxation of cartilage: appropriateness of a transversely isotropic analysis. *Journal of Biomechanics*, **32**:1125–1130.

[57] Armstrong, C. G., Lai, W. M., and Mow, V. C. (1984). An analysis of the unconfined compression of articular cartilage. *Journal of Biomechanical Engineering*, **106**:165–173.

[58] Garcia, A. M., Szasz, N., Trippel, S. B., Morales, T. I., Grodzinsky, A. J., and Frank., E. H. (2003). Transport and binding of insulin-like growth factor I through articular cartilage. *Archives of Biochemistry and Biophysics*, **415**:69–79.

[59] COMSOL-Multiphysics (2007). COMSOL, Inc., MA.

[60] Mow, V. C., Hou, J. S., Owens, J. M. and Ratcliffe, A. (1990). Biphasic and quasilinear viscoelastic theories for hydrated soft tissues. In *Biomechanics of Diarthrodial Joints*. Springer-Verlag, New York, pp. 215–260.

[61] Cohick, W. S. and Clemmons, D. R. (1993). The insulin-like growth factors. *Annual Review of Physiology*, **55**:131–153.

[62] Tavera, C., Abribat, T., Reboul, P., Dore, S., Brazeau, P., Pelletier, J.-P., and Martel-Pelletier, J. (1996). IGF and IGF-binding protein system in the synovial fluid of osteoarthritic and rheumatoid arthritic patients. *Osteoarthritis and Cartilage*, **4**:263–274.

[63] Luyten, F. P., Hascall, V. C., Nissley, S. P., Morales, T. I., and Reddi, A. H. (1988). Insulin-like growth factors maintain steady-state metabolism

of proteoglycans in bovine articular cartilage explants. *Archives of Biochemistry and Biophysics*, **267**:416–425.

[64] Mohan, S. and Baylink, D. J. (2002). IGF-binding proteins are multifunctional and act via IGF-dependent and -independent mechanisms. *Journal of Endocrinology*, **175**:19–31.

[65] Jones, J. I. and Clemmons, D. R. (1995). Insulin-like Growth Factors and their binding proteins: biological actions. *Endocrine Reviews*, **16**:3–34.

[66] Collett-Solberg, P. F. and Cohen, P. (2000). Genetics, chemistry, and function of the IGF/IGFBP system. *Endocrine*, **12**:121–136.

[67] Arai, T., Parker, A., Walker Busby, J., and Clemmons, D. R. (1994). Heparin, heparan sulfate, and dermatan sulfate regulate formation of the insulin-like growth factor-I and insulin-like growth factor-binding protein complexes. *The Journal of Biological Chemistry*, **269**:20388–20393.

[68] Lee, A. S. (1984). *Modeling Dynamic Phenomena in Molecular and Cellular Biology*. Cambridge University Press, Oxford.

[69] Cassino, T. R. (2002). Quantification of the binding of Insulin-like growth factor-I (IGF-I) and IGF binding protein-3 (IGFBP-3) using surface plasmon resonance. Master of Science thesis. Department of Chemical Engineering, Virginia Polytechnic Institute and Sate University.

[70] Morales, T. I. (2002). The insulin-like growth factor binding proteins in uncultured human cartilage. *Arthrits & Rheumatism*, **46**:2358–2367.

[71] Blum, W. F., Reppin, W. J. E, F., Kietzmann, K., Ranke, M. B., and Bierich, J. R. (1989). Insulin-like growth factor I (IGF - I)-binding protein complex is a better mitogen than free IGF-I. *Endorinology*, **125**:766–772.

[72] Wong, M.-S., Fong, C.-C., and Yang, M. (1999). Biosensor measurement of the interaction kinetics between insulin-like growth factors and their binding proteins. *Biochimica et Biophysica Acta*, **1432**:293–301.

[73] Dubaquie, Y. and Lowman, H. B. (1999). Total alanine-scanning mutagenesis of insulin-like growth factor I (IGF-I) identifies differential binding epitopes for IGFBP-1 and IGFBP-3. *Biochemistry*, **38**:6386–6396.

[74] Heding, A., Gill, R., Ogawa, Y., Meyts, P. D., and Shymko, R. M. (1996). Biosensor measurement of the binding of insulin-like growth factors I and II and their analogues to the insulin-like growth factor-binding protein-3. *The Journal of Biological Chemistry*, **271**:13948–13952.

[75] Vorwerk, P., Hohmann, B., Oh, Y., Rosenfeld, R. G., and Shymko, R. M. (2002). Binding properties of insulin-like growth factor binding protein-3 (IGFBP-3), IGFBP-3 N- and C-terminal fragments, and structurally

related proteins mac25 and connective tissue growth factor measured using a biosensor. *Endocrinology*, **143**:1677–1685.

[76] Headey, S. J., Leeding, K. S., Norton, R. S., and Bach, L. A. (2004). Contributions of the N- and C-terminal domains of IGF binding protein-6 to IGF binding. *Journal of Molecular Endocrinology*, **33**: 377–386.

[77] Stumm, S. and James, J. M. (2004). *Aquatic Chemistry.* John Wiley & Sons, NY.

[78] MATLAB (2004). *MATLAB.* The MathWorks, Inc.

[79] Sengers, B. G., Oomens, C. W. J., and Baaijens, F. P. T. (2004). An integrated finite-element approach to mechanics, transport and biosynthesis in tissue engineering. *Journal of Biomechanical Engineering*, **126**:82–91.

[80] Klein, T. J. and Sah, R. L. (2007). Modulation of depth-dependent properties in tissue-engineered cartilage with a semi-permeable membrane and perfusion: a continuum model of matrix metabolism and transport. *Biomechanics and Modeling in Mechanobiology*, **6**:21–23.

[81] Schalch, D. S., Sessions, C. M., Farley, A. C., Masakawa, A., Emler, C. A., and Dills, D. G. (1986). Interaction of insulin-like growth factor I/somatomedin-C with cultured rat chondrocytes: receptor binding and internalization. *Endocrinology*, **118**:1590–1597.

[82] Roos, E. M. and Dahlberg, L. (2005). Positive effects of moderate exercise on glycosaminoglycan content in knee cartilage: a four-month, randomized, controlled trial in patients at risk of osteoarthritis. *Arthritis & Rheumatism*, **52**:3507–3514.

[83] Davisson, T., Kunig, S., Chen, A., Sah, R., and Ratcliffe, A. (2002). Static and dynamic compression modulate matrix metabolism in tissue engineered cartilage. *Journal of Orthopaedic Research*, **20**(4):842–848.

[84] Wu, J. Z., Herzog, W., and Epstein, M. (1999). Modelling of location- and time-dependent deformation of chondrocytes during cartilage loading. *Journal of Biomechanics*, **32**:563–572.

[85] Obradovic, B., Meldon, J. H., Freed, L. E., and Vunjak-Novakovic, G. (2000). Glycosaminoglycan deposition in engineered cartilage: Experiments and mathematical model. *AIChE Journal*, **46**:1860–1871.

[86] Sah, R. L., Chen, A. C., Grodzinsky, A. J., and Trippel, S. B. (1994). Differential effects of bGFG and IGF-I on matrix metabolism in calf and adult bovine cartilage explants. *Archives of Biochemistry and Biophysics*, **308**:137–147.

[87] Alon, U. (2006). *An Introduction to Systems Biology*. Chapman & Hall/CRC, Taylor & Francis Group, London.

[88] Comper, W. D. and Williams, R. (1987). Hydrodynamics of concentrated proteoglycan solutions. *Journal of Biological Chemistry*, **262**:13464–13471.

[89] Zhang, L., Gardiner, B. S., Smith, D. W., Pivonka, P., and Grodzinsky, A. J. (2008). A fully coupled poroelastic reactive-transport model of cartilage. *Molecular & Cellular Biomechanics*, **5**:133–153.

12

Biotechnological and Biomedical Applications of Magnetically Stabilized and Fluidized Beds

Teresa Castelo-Grande
Laboratory of Environmental Process Engineering and Energy (LEPAE), Department of Chemical Engineering, Faculty of Engineering, Universidade do Porto, Porto, Portugal
Faculty of Natural Sciences, Engineering Technology (FCNET), Universidade Lusófona do Porto, Porto, Portugal
Department of Chemical Engineering and Textile, Faculty of Sciences, Chemistry, Universidad de Salamanca, Salamanca, Spain

Paulo A. Augusto
Department of Chemical Engineering and Textile, Faculty of Sciences, Chemistry, Universidad de Salamanca, Salamanca, Spain
Department of Chemical Engineering, Faculty of Engineering, Universidade do Porto, Porto, Portugal

Angel M. Estevéz
Department of Chemical Engineering, Faculty of Engineering, Universidade do Porto, Porto, Portugal

Domingos Barbosa
Laboratory of Environmental Process Engineering and Energy (LEPAE), Department of Chemical Engineering, Faculty of Engineering, Universidade do Porto, Porto, Portugal

Jesus Mª. Rodríguez, Audelino Álvaro
Department of Chemical Engineering and Textile, Faculty of Sciences, Chemistry, Universidad de Salamanca, Salamanca, Spain

Acknowledgment
Fundação para a Ciência e a Tecnologia is acknowledged for its Grant SFRH/BD/29893/2006

CONTENTS

12.1 Introduction

The processing of cells, biomedical substances, and/or high-valued biotechnological products has always been a major issue in bioengineering and biomedicine. In fact, from approximately 5% (in the case of wholesale products) up to 90% (in the case of pharmaceuticals) of the production costs are due to downstream processing (Böhm and Pittermann 2000). Isolation, separation, and purification of various types of proteins, peptides, and other specific molecules are required in almost all branches of biosciences and biotechnologies (Safarik and Safarikova 2004). Therefore, separation technologists are under constant pressure to develop more efficient separation processes, both at laboratorial and large scale levels, capable to separate and/or purify the target biosubstances, even if present at very small concentrations, and also even when they are in the presence of particulate matter.

The mostly used technologies to achieve the separation and purification of peptides and proteins are based on chromatographic, electrophoretic, ultrafiltration, and precipitation methods. Actually, the most commonly used technique is the affinity chromatography due to its good performance (both in terms of selectivity and in terms of recovery). However, among other disadvantages, in fact, technique is not able to handle samples containing particulate material (typically present in early stages of the isolation/purification of biological materials when suspended solid and fouling components exist), and therefore, alternative separation processes, such as magnetic affinity, ion exchange, hydrophobic or adsorption batch separations, magnetically stabilized fluidized beds (MSFBs), and magnetically modified two-phase systems, have shown their usefulness (Safarik and Safarikova 2004).

Magnetic methods are very promising for the purification and isolation of substances, especially when allied with the new magnetic tagging and carrying technologies, because they have a low cost, and are very effective. This is even more the case when every day new methods to create and effectively manipulate magnetic tags and carriers are being developed.

In the case of biomedical and biotechnological applications the principle of magnetic separation is very simple (Safarik and Safarikova 2004): magnetic carriers with some sort of affinity, hydrophobic ligands, or specially chosen ion-exchange groups are mixed with a sample containing target compound(s) (e.g., crude cell lysates, whole blood, plasma, ascites fluid, milk, whey, urine, cultivation media, wastes from food and fermentation industry); after the necessary incubation period for the target compound(s) to attach to the magnetic particles, the resulting magnetic complex is easily and rapidly removed from the sample by using a magnetic separator. Then, after washing out the contaminants, the treated sample can be eluted and the isolated target compound(s) further processed.

Details on the main advantages of application of magnetic methods can be found in this chapter and also in Safarik and Safarikova (2004).

Magnetically stabilized fluidized beds exhibit a unique combination of the properties of packed and fluidized beds (Hausmann et al. 2000). They have been one of the key magnetic separation techniques used during the last decade to achieve the purification of biological materials. In MSFBs the magnetic particles (magnetic carriers or other type of tags) are placed under the influence of an external magnetic field that is capable of sustaining them against flow velocities higher than those verified in packed beds. These magnetic particles have some artificially manipulated affinity that enables them to collect/adsorb/attach the target biological substance from the media that flows through them.

In this chapter, the applications of MSFBs to biotechnology and biomedicine will be addressed. First it will be given an historical overview of the developments and applications of MSFBs in the biotechnological and biomedical areas. Next, the main principles behind MSFB technology will be reviewed, including the behavior of MSFB's as porous media. Then, the theory that supports MSFB applications, including the supporting theory for applications such as biorreactions and bioadsorptions will be presented. The main current and past applications of MSFB in the areas of biotechnology and biomedicine are described in detail in a subsequent section. Detail on the type of particles used to form these beds, are also given. The main advantages and disadvantages of this technique, as compared to other existing techniques, will be also addressed through the chapter. At the end, some conclusions are presented.

12.2 Historical Overview of Magnetically Stabilized and Fluidized Beds

12.2.1 General

The first paper published on magnetically fluidized beds (MFBs) as an all, was by Kirko and Filipov (1960). Filipov further developed this subject by investigating the behavior of a fluidized bed of iron particles under the influence of a magnetic field and of a water stream (Filipov et al. 1961a,b; Filipov 1962), which complemented by the works of Nekrassov and Chekin (Nekrassov and Chekin 1956, 1961), initiated the investigation and development of the MFBs and MSBs beds. Figure 12.1 represents the device studied and developed by Filipov. Many publications on the subject have followed this breakthrough, and, as examples, we may stand out the following works: Rosensweig (1979) presented the first basic description of MSB and claimed that Filipov and the other previous researchers apparently did not realized the importance of the orientation and uniformity of the applied field, besides having neglected the existence of a distinct range of superficial gas velocities between the minimum fluidization velocity and the transition velocity (Liu et al. 1991). Ivanov, Zrunchev, Popova, and Grozev (Ivanov and Grozev 1970; Zrunchev 1975;

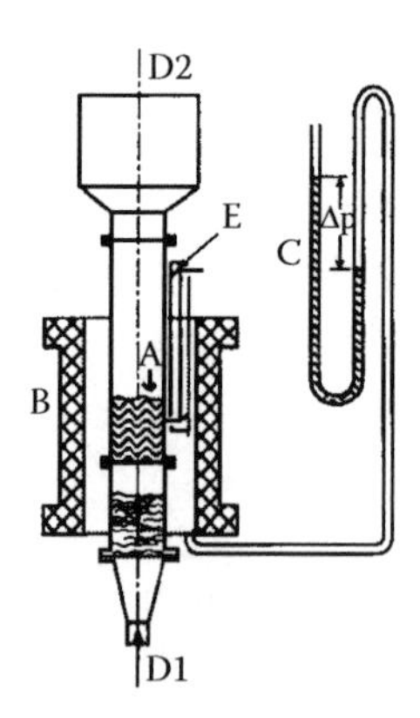

FIGURE 12.1
The first setup of a magnetically fluidized bed for the treatment of gases (based on Filipov [1961a] and Hristov [2002]). A—magnetically fluidized bed; B—magnetic field generator; C—measure of pressure drop in the bed; D—air flow: 1- entrance, 2—exit; E—Periscope.

Zrunchev and Popova 1975) were the first to apply this technique to ammonia synthesis. Some important developments were also reported by Herschler (1969), which consisted of the application of magnetic fields to liquid metals and to fluidized beds formed by magnetizable particles. Herschler also showed that the magnetic field would induce a good mixture if it was created by an alternating current. Later, some works were published on the hydrodynamics of MSB and MFB (Penchev et al. 1990; Conta et al. 1998), and even on some less common techniques like countercurrent contacting gas–solid magnetic valves (Jaraiz 1983; Jaraiz and Estevez 1987). Mass transfer in MSFB has always been an important research topic because of the important and singular mass transfer characteristics that MSFB possesses. Arnaldos et al. (1985) and Neff and Rubinsky (1983) studied the way heat is transferred in MSFB, and Arnaldos and Casal (1987) and Terranova and Burns (1991) described the observed heat transfer in liquid and gas fluidized beds. In what concerns applications, we may refer works on aerosol filtration (Albert and Tien 1985; Geuzens and Thoens 1988a,b), biotechnology (Terranova and Burns 1991), bioreactors (Moffat et al. 1994; Hristov and Ivanova 1999), adsorption (e.g., waste removal (Nuñez and Kaminski 1998), removal of metal ions (Anacleto and Carvalho 1996; Sedzimir 2002; Zouboulis and Katsoyiannis 2002)), analytical separations (Siegell 1988; Graves 1992), and bioseparations (Graves 1992; Evans and Burns 1995). This topic continues to deserve the attention of many researchers, and it is easy to find recent publications on the fundamentals of MSBs and MFBs for coal separation (Fan et al. 2003), biological and medical applications (Özkara et al. 2004; Uzun and Denizli 2006; Al-Qodah and Al-Shannag 2007)—these applications will be further detailed in this chapter—environmental applications (Paice and Jurasek 1984; Nyens 1995; Rodriguez et al. 1999; Estevez et al. 2008), hydrodynamics (Britton et al. 2005), mass transfer studies (Hausmann et al. 2000), and applications to catalysis (Graham et al. 2006; Gui and Zhang 2006; Dong et al. 2008).

Good reviews on MSFB have been published (Bologa and Syutkin 1977; Colver 1979; Siegell 1989; Sonolikar 1989; Liu et al. 1991; Saxena et al. 1994), and more recently Hristov presented a series of reviewing papers detailing the major developments in this field (Hristov 2002, 2003a,b, 2004, 2006b, 2007, 2009).

The goal of this chapter is to present a basic understanding of MFB and MSB, its principles and background theory, and complement the above cited reviews by considering the applications of this technique to biotechnology and biomedicine, thus giving a concise review of the field and its current applications.

12.2.2 Biotechnology and Biomedicine

Several applications, at laboratorial and industrial scale, of the MSFBs in several areas of science have been developed through the years. In the last decades a significant effort has been put into the development of biotechnological and medical applications of MSFBs; examples of this are as follows: separation of proteins (Cocker et al. 1997; Seibert et al. 1998; Tanyolac and Özdural 2000; Tong and Sun 2003; Özkara et al. 2004; Ding and Sun 2005; Safarik and Safarikova 2004; Türkmen et al. 2006; Yao et al. 2006), antibody removal/processing and antibiotics production (Webb et al. 1996; Al-Qodah 2000), fermentation and cell/enzyme immobilization (Ivanova et al. 1996; Al-Qodah and Al-Hassan 2000; Bahar and Celebar 2000; Böhm and Pittermann 2000; Al-Qodah and Al-Shannag 2006; Qiu et al. 2006; Van Hee et al. 2006; Betancor and Luckarift 2008), processing of high-value substances (e.g., plant cell culture processing) (Al-Qodah and Al-Shannag 2006), bilirubin removal (Uzun and Denizli 2006), cell separation (Qiu et al. 2006; Al-Qodah and Al-Shannag 2007; Karatas et al. 2007), immunoglobulin G depletion and erythrocyte fractionation (Özkara et al. 2004; Karatas et al. 2007; Özturk et al. 2007), bioanalytical techniques (e.g., continuous affinity chromatography) (Cocker et al. 1997; Türkmen et al. 2006; Yao et al. 2006; Özturk et al. 2007), water treatment by enzymes (Arica 2000), and nucleic acids purification (Hultman et al. 1989; Yamaura et al. 2004).

All these applications will be detailed in Section 12.5.

12.3 MSBs and MFBs

12.3.1 Principles of MSBs and MFBs

A MFB is a device in which a bed of particles suffers the influence of an external magnetic field. This magnetic field may act axially or transversely to the flow (the former is more usual, although some authors defend the latter as being the most effective configuration (Hristov 2006a; Augusto et al. 2008)). The interparticle forces between the particles forming the bed are increased by the action of this magnetic field because it stimulates the appearance of a magnetic force between them (as the bed is formed by ferromagnetic or strong paramagnetic particles or by agglomerates presenting strong magnetic characteristics), leading to an enhancement of the properties and efficiency of the fluidized beds. Distinguishing between MFBs and MSBs is not an easy task and, indeed, this distinction is not accepted by all the scientific community. However, the majority of researchers consider that the MFBs include the stabilized beds, establishing that MSBs are fluidized beds in which the bubbling effect, usually present in ordinary fluidized beds, is suppressed (this suppression is due to the influence of the magnetic field and thus of the interparticle magnetic forces), thus "stabilizing" the bed. This bubbling effect is not always present in MFBs, and therefore the definition of MFBs includes the cases of MSBs, the cases where the field is not uniform, and even uncommon applications such as the magnetic valve for solids (MVS) (Zhang et al. 1984) and the

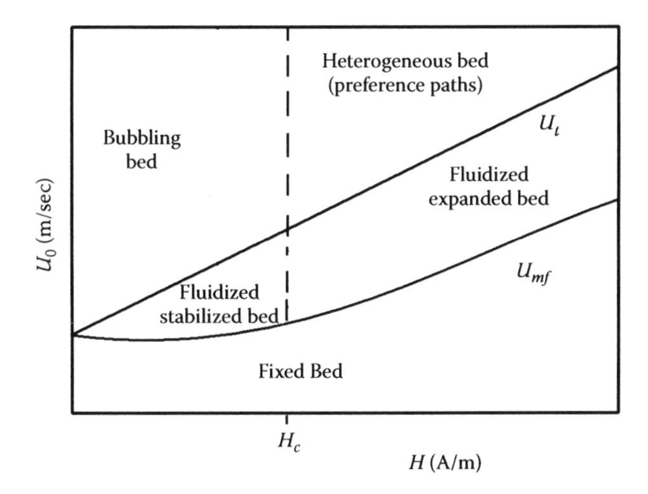

FIGURE 12.2
Regions of behavior of a MSB depending on the velocity and magnetic field values. (Adapted from Augusto et al. 2008.)

magnetic distribution downcomer (MDD) (Zhang et al. 1984). Many authors prefer to ally both definitions to avoid controversy, speaking of MSFBs or of magnetic field assisted fluidization (Hristov 2002, 2003a,b, 2004, 2006b). In this chapter we will adopt the former denomination of MSFB to indicate that we are generally speaking about both forms of magnetic beds.

The magnetic field in MSFB is usually fixed (i.e., it does not usually vary with time) and its magnitude is high enough to strengthen the immobilization of the particles in the magnetic bed and to avoid the appearance of bubbles (if the velocity is below a given value). In fact, for these systems, a graphical depiction of the velocity of the gas (U_0) versus the magnitude of the magnetic field, like the one presented in Figure 12.2, shows five distinct regions: (1) *Fixed Bed*—in this region, for a certain magnitude of the magnetic field acting externally, the bed behaves as a fixed bed, as long as the velocity is maintained below a given value—the velocity of fluidization U_{mf}; (2) *Fluidized Stabilized Bed*—this is the region where the system behaves as a fluidized bed, but stabilized (i.e., with no creation of bubbles), as long as the magnetic field is maintained below a certain magnitude and for velocities between the velocity of fluidization, U_{mf}, and the velocity of transition, U_t; (3) *Fluidized Expanded Bed*—in this region the system behaves as an expanded stabilized bed (without formation of bubbles), and expands as a piston for increasing velocities, as long as the magnetic field is maintained above a certain magnitude and for velocities between the velocity of fluidization, U_{mf}, and the velocity of transition, U_t; (4) *Bubbling Bed*—this is the region where the system behaves as an ordinary fluidized bed containing bubbles, and exists as long as the magnetic field is maintained below a certain magnitude and for

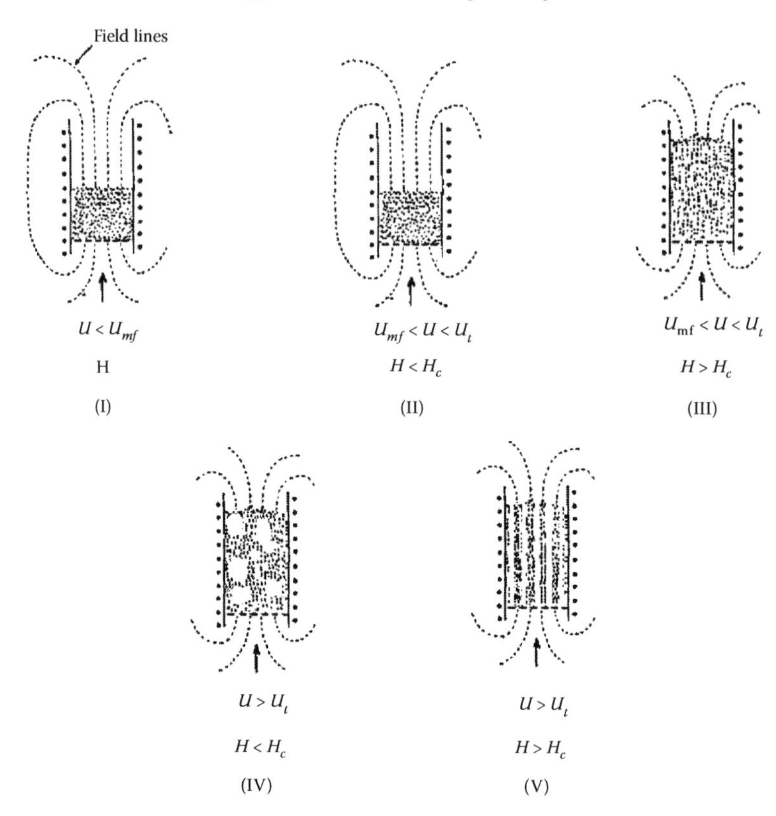

FIGURE 12.3
Different behaviors of a magnetic stabilized and fluidized bed: (I) fixed bed; (II) fluidized stabilized bed; (III) fluidized expanded bed; (IV) bubbling bed; (V) heterogeneous bed.

velocities above the velocity of transition, U_t; (5) *Heterogeneous Bed*—in this region it is observed that the system starts creating preference paths for the fluid to flow, as the particles tend to align with the field. This phenomenon appears when the magnetic field is maintained above a certain magnitude and for velocities above the velocity of transition, U_t.

In Figure 12.3 we present a graphical depiction of the bed behavior in each of these regions. Increasing the velocity too much will lead to the collapse of the bed and the particles will be dragged by the flow—this is called *elutriation* and the corresponding velocity is known as elutriation velocity. On the other hand, if the magnetic field is increased too much, regardless of the velocity value, the magnetic particles in the bed will start to align with the field creating the above cited preference paths, which allow a tunneling effect of the flow, and the pressure drops abruptly.

There are several important characteristics of the MSFB that make them the best option for several biotechnological and biomedical applications, and many of them are referred in several sections of this chapter. Nonetheless, we may point out as a general characteristic of the MSFB, its capability to maintain the stabilization of the bed (suppressing bubbling and slugging) for a much larger range of velocities than that of conventional fluidized beds (i.e., maintaining the fixed bed behavior much longer) while keeping a low pressure drop at all times (a characteristic which is typical of fluidized beds). MSFB also enables to keep the mixing of the solids while processing solutions containing suspended solids. Also, very important in biomedical and biotechnological applications is the ability of MSFB to process biological cultures in a gentle and harmless way.

12.3.2 MSBs and MFBs as Porous Media

Particles used in MSBs and MFBs may be porous or nonporous. Beds of nonporous magnetic particles seem to be more resistant to diffusional limitations, attrition, and fouling than beds of porous particles (Halling and Dunnill 1980). Several nonporous particles have been used, as, for example, iron, cobalt, and their oxides; and there are four main methods for their preparation: direct use of a coupling agent, adsorption, encapsulation, and formation of a layer of a thin polymer film. However, porous particles are the most interesting media for biotechnological and biomedical applications, and in many cases such applications would be impossible to perform with nonporous magnetic particles. The main feature of the magnetic porous particles, when used as media for MSFB, is their adsorption capabilities, which, when properly controlled, have a direct action and influence in the separation/purification of important biological substances. The protein adsorption is one of the examples of their spreading use. MSFBs are replacing existing packed-bed operating systems, as they have the advantage of a low and constant pressure drop, which is independent of the flow rate. Besides the ability to process feed streams with suspended solids (as previously discussed), magnetic stabilization of fluidized beds also reduces axial dispersion and enhances separation (Lucchesi et al. 1979; Rosensweig 1979; Rosensweig et al. 1981a,b; Siegell et al. 1984; Burns and Graves 1985; Siegell 1987, 1989; Geuzens and Thoens 1988a,b; Sajc et al. 1994; Rosensweig and Ciprios 1991; Wallace et al. 1991).

The porous particles are usually polymeric (generally having a magnetite core and a polymeric shell) and contain biologically active centers that help in the adsorption/separation of biological materials. Details of these beads will be discussed later in Section 12.5. Magnetic porous particles have a high-magnetic susceptibility, a large reactive polymer surface area, and pores on which functional molecules, such as enzymes, antibodies, antigens, cells, drugs, and biotins, among many others, can be immobilized (Ugelstad et al. 1992; Ding et al. 1998; Qiu et al. 2006). These magnetic particles have many advantages as supports for cell immobilization due to their controlled surface area and porous structure, good thermal resistance, reasonable chemical inertness, and the possibility of being reused (Qiu et al. 2006). These polymer beads can be formed either by suspension polymerization and specific treatment with acid and base or by seed swelling polymerization in the presence of the magnetite core (Bahar and Celebar 2000).

12.4 General Supporting Theory

12.4.1 MSBs and MFBs

The acting forces that cannot be neglected in a MSFB may be divided into four main categories: gravitational forces, buoyancy forces, drag forces, and interparticle forces. These four main categories of forces interact in such a way that they reach a balance in a magnetic particle of the bed, until the elutriation velocity is reached.

The Drag forces, F_d, are expressed by Stoke's equation

$$F_d = 3\pi\mu d_p(U_f - U_p) \tag{12.1}$$

for laminar flows, and by Newton's equation,

$$F_d = C_d A \frac{\rho_f (U_p - U_f)^2}{2} \tag{12.2}$$

for turbulent flow, where μ is the viscosity of the fluid, d_p is the diameter of the bed particle, U_f is the velocity of the fluid, U_p is the velocity of the particle (which is zero in the case of MSBs), C_d is the drag coefficient for the particle (Shames 1982), A the cross-sectional area of the particle in the direction of the flow, and ρ_f the fluid density.

The Gravitational forces, F_g, are expressed by the well-known equation,

$$F_g = -\rho_p V_p g \tag{12.3}$$

where ρ_p is the density of the particle, V_p is the volume of the particle, and g is the acceleration of gravity.

The Buoyancy force, F_b, may be expressed by

$$F_b = \rho_f V_p g \tag{12.4}$$

The Interparticle forces, F_{ip}, represent a class of important forces in the MSFBs. However, it is necessary to distinguish between their different natures. The most important interparticle force for the particles of a MSFB is the magnetic force (Fan and Zhu 1998), but other, sometimes, nonnegligible forces may also be considered, as in the case of the short-range interparticle forces, like van der Waals, electrostatic and even collisional forces.

12.4.1.1 Magnetic Forces

The stabilization of the bed is accomplished by the magnetic forces acting on it due to the presence of a magnetic field generated by Helmholtz current-bearing coils. In a homogeneous magnetic field these forces are a function of the magnetic susceptibility of the particles forming the bed and the field gradient.

From Rosensweig (1985), we know that the magnetic force, F_m, is expressed by

$$F_m = \frac{1}{4\pi} M \nabla(\mu_0 H) \tag{12.5}$$

where μ_0 is the magnetic permeability of the vacuum, H is the magnetic field intensity and M is the magnetization. If all particles are oriented according to the direction of the magnetic field, for the x direction, equation (12.5) reduces to

$$F_m = \frac{1}{4\pi} \chi_m \mu_0 H \frac{dH}{dx} \tag{12.6}$$

where χ_m is the mass magnetic susceptibility of the particles, and dH/dx is the magnetic field gradient over the x direction.

12.4.1.2 Van der Waals Forces

These forces are not only present for atoms and molecules, but also for solid particles. If we assume the particles to be spherical, the Van der Waals forces, F_{vw},may be expressed by (Hamaker 1937)

$$F_{vw} = \frac{A_{12} a}{6 l^2} \tag{12.7}$$

where A_{12} is the Hamaker's constant, l is the distance between the surfaces of the two solid particles, and a is the relative radius defined as

$$\frac{1}{a} = \frac{1}{a_1} + \frac{1}{a_2} \tag{12.8}$$

where a_1 and a_2 are the radii of the spherical solid particles.

12.4.1.3 Electrostatic Forces

Electrostatic effects are always present, but they are more significant when working with a gas–solid two-phase flow. The simple contact or collision of the particles with each other, or with other surfaces, is enough to trigger the electrostatic charging of the particles. This force, F_e, acts along a straight line from one charged object to the other, and is expressed by

$$F_e = \frac{q_1 q_2}{4\pi\varepsilon\, r^2} \tag{12.9}$$

where q_1 and q_2 are the charges of the solid particles, ε is the permittivity of the surrounding medium, and r is the distance between the particles.

12.4.1.4 Collisional Forces

Collisions between the particles in the bed exist when they move, even if they quickly stabilize. For elastic spheres, the maximum collisional force in a collinear impact between them is expressed by

$$F_{\text{col}} = \frac{4}{3} E\sqrt{a} \left(\frac{15 m U^2}{16 E \sqrt{a}} \right)^{\frac{3}{5}} \tag{12.10}$$

where U is the relative velocity of the two particles, a is the relative radius as defined by equation (12.8), E is the contact module, and m is the relative mass of the particles, defined as

$$\frac{1}{m} = \frac{1}{m_1} + \frac{1}{m_2} \tag{12.11}$$

12.4.1.5 Force Balances and Parameters Computation

The equilibrium in a MSFB is reached (i.e., the bed particles do not move, and therefore the bed "gets stabilized") if the resultant force acting on the particles is zero:

$$F_b + F_d + F_{ip} - F_g = 0 \tag{12.12}$$

with

$$F_{ip} = F_m + F_{vw} + F_e + F_{col} \tag{12.13}$$

Examples of the application of this balance may be found in literature, and we may point out as an example (Augusto et al. submitted).

In the majority of the cases the magnetic force in Equation 12.13 is much higher than other interparticle forces and, in these cases we obtain the following equation from the force balance (Estevez et al. 1995):

$$H_g = S\frac{u^2}{\varepsilon^{2n}} - R \tag{12.14}$$

where $H_g = H\nabla H$, ε is the void fraction (bed porosity), n is the Richardson and Zaki index, u is the superficial velocity, and S and R are constants defined as $S = \frac{3f_{D\infty}\rho_f}{4d_p\chi\mu_0}$ and $R = \frac{g(\rho_p - \rho_f)}{\chi\mu_0}$, and f_{D_∞} is the individual drag coefficient for immersed bodies. This equation means that if we represent H_g versus u^2 we obtain a straight line in the area where the magnetic bed behaves as fluidized, as may be seen in the practical experiments depicted in Figure 12.4.

The minimum fluidization velocity, U_{mf}, may be computed by using the general equations like (Hristov 2002):

$$U_{mf} = \mathrm{Re}_{mfo}\frac{\nu}{d_p}e^{aB-c} \tag{12.15}$$

where Re_{mfo} is the Reynolds number of the minimum fluidization velocity (and a function of the Arquimedes number), a and c are constants, υ is the kinematic viscosity of the fluid, and B is the acting magnetic field inside the particles.

Elutriation velocity may be computed using the formula (Rodriguez et al. 2000):

$$\frac{K^*_{e,i}d^2_{pi}g}{\mu(u_o - u_{ti})^2} = 1.5\times10^{-9}\,Re_{t0.6} + 2.5\times10^{-5}\,Re_{t1.2} \tag{12.16}$$

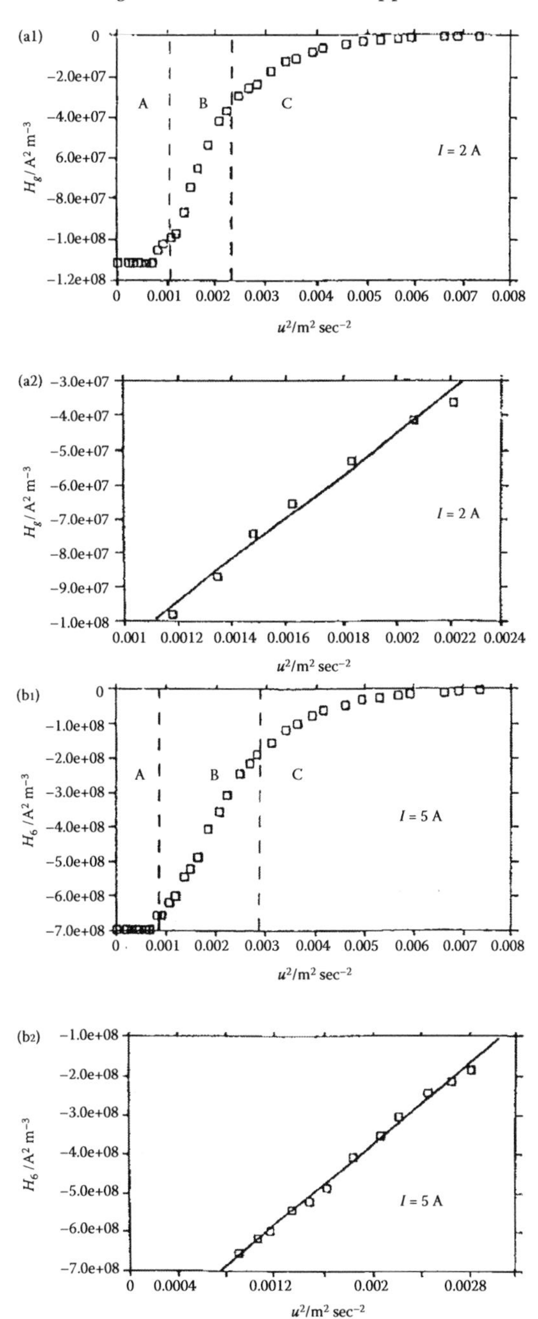

FIGURE 12.4
H_g versus the square of the water superficial velocity at a current intensity of: (a) 2 A; (b) 5 A; (c) 9 A. A is the fixed bed zone, B is the fluidized bed zone, and C is also a fluidized zone but where the velocity is very high and the top of the bed o particles is well above the coil, and thus, the magnetic traction force is very low. (From Estevez et al. 1995.) (*Continued*)

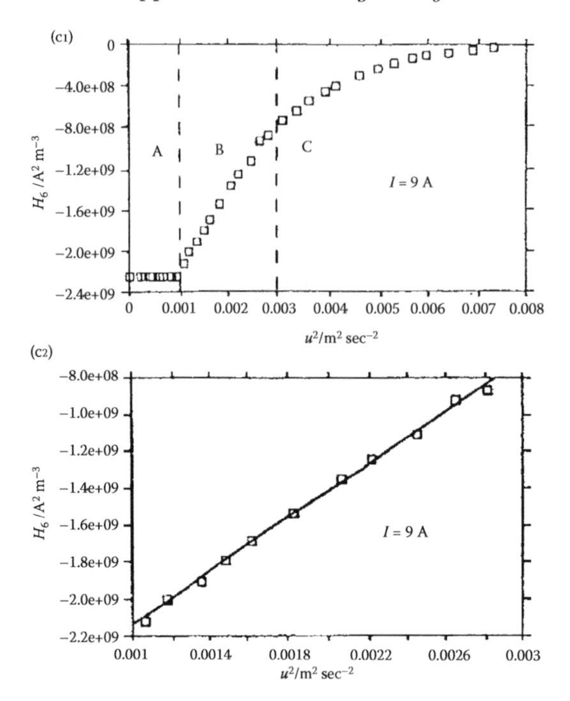

FIGURE 12.4
(Continued)

where $K^*_{e,i}$ is the elutriation rate constant for particles of size i, d_{pi} is the diameter of particles of size i, μ is the air viscosity, u_o is the superficial air velocity, u_{ti} is the terminal velocity of particles of size i, and Re_{ti} is the terminal Reynolds number for particles of size i. Experimental data is shown in Figure 12.5.

Rosensweig and coworkers (Rosensweig et al. 1981b) have studied the arrangement of the particles in a MSFB and concluded that it can be quantitatively correlated by a number that they denominated E_G, which is an adimensional number defined as

$$E_G = \frac{24\rho_S g d_p}{\mu_0 H_p^2} \tag{12.17}$$

where H_p is the apparent applied magnetic field acting on the particles due to their magnetization.

When $E_G < 1$, the particles align with the field lines and form chain-like structures that create the tunneling effect (due to the strong magnetization). For $E_G > 10$, the structure looks random and compact with no preferred orientation. When $1 < E_G < 10$, the bed is partially structured.

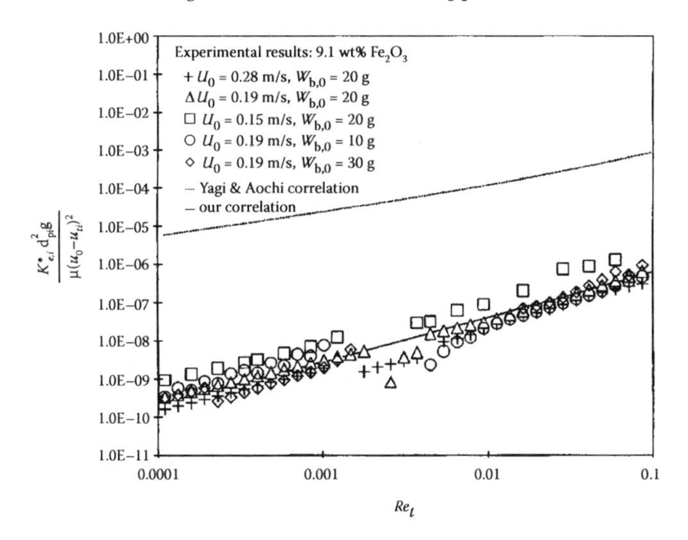

FIGURE 12.5
Comparison of the new correlation presented as equation (12.16) with that of Yagi and Aochi and with experimental results. (From Alvaro et al. 2007.)

12.4.2 Extra Forces or Equations Usually Required When MSFBs Are Applied in Biotechnology and Medicine

Overall equations as described in the previous sections apply in general terms to every application of MSFBs. However, for each specific application they must be complemented by other equations and also by other complementing support theory.

This is the case of the applications of MSFBs in the areas of biotechnology and biomedicine, and we will see some of these complementing equations and theoretical support, in what follows.

In these kind of applications mass transfer is always involved and thus, equations like the one that regulates the diffusion of a certain compound into and within a porous magnetic compound forming the bed, must be taken into account. We may use, for example, the dispersion model, given by

$$\frac{\partial c_i}{\partial t} + u\frac{\partial c_i}{\partial z} - D_{ax}\frac{\partial^2 c_i}{\partial z^2} = \nu\left(c_i\right) \tag{12.18}$$

which, when applied, justifies mass transfer behavior like the one depicted in Figure 12.6.

Quite often we must also include in our theoretical analyses complementing theories and equations that describe chemical or biological reactions and/or adsorptions, as many applications of MSFBs in the biotechnological and biomedical areas use a chemical or biological reaction as the main support

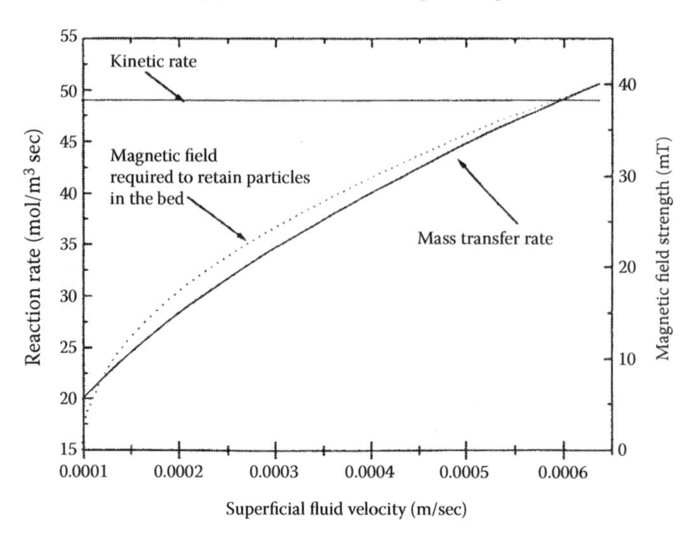

FIGURE 12.6
Model prediction of MSFB reaction mass transfer rates at different flow rates showing also the magnetic field strength required to enable operation at the necessary superficial fluid velocities. (From Webb et al. 1996.)

for their action. As these processes are variable the reaction and/or adsorption equations will also vary accordingly. We may point out as an example the adsorption of *Saccharomyces cerevisiae* in activated carbon or zeolites governed by the sorption isotherm,

$$Y = \frac{AX}{B + X} \tag{12.19}$$

where X is the free cell concentration in 10^7 cells/g support, Y is the adsorbed cell concentration in 10^6 cells/g support, and A and B are the coefficients that describe the maximum possible cell loading and the sorptive immobilization constant of yeast cells on these magnetic particles (Al-Hassan et al. 1991).

First-order rate equations such as

$$\ln A = \ln A_0 - K_i t \tag{12.20}$$

are very often observed for the description of reactions (as e.g., in the immobilization of horse radish peroxidase [HRP] for phenols degradation [Lai and Lin 2005; Bayramoglu and Arica 2008]), where K_i is the inactivation rate constant, and A_0 and A are the initial activity and the activity after time t (min).

12.5 Main Biotechnological and Biomedical Applications

As referred in the introduction, several applications, at laboratorial and industrial scale, of the MFBs and MSBs in the areas of biotechnology and biomedicine have been developed through the years (mainly cell, antibody, and protein removal/ processing, fermentation, and cell/enzyme immobilization). The major applications will be detailed in this chapter. We start by describing generically the type of magnetic particles that may form the MSBs and MFBs in this type of processes.

12.5.1 Particles (Beads)

Magnetic stabilized and fluidized beds require the use of particles/substances possessing sufficiently high values of magnetic susceptibility and/or permeability. Two main types of systems have been developed: those that use 100% magnetic particles and those that use a mixture of magnetic and nonmagnetic particles. The systems having only magnetic particles proved to be superior in recently developed applications, when the fulfillment of all the goals of separation and reaction are considered (Seibert et al. 1998; Hristov 1999).

The magnetic beads used in MSFBs may assume many forms and compositions, but they usually consist of a central core made of magnetite, maghemite, or any other kind of strong paramagnetic or ferromagnetic material, which is then covered by some kind of polymeric adsorbent or resin that is usually later functionalized and surface modified to have higher adsorption and active sites, and that will be responsible for the interaction with the biological media. This interaction may be directly on the substances we are interested in removing, modifying, or purifying, or it may be indirect by the attachment of biological material to the surface of the beads (e.g., enzymes, zeolites, etc.), which will then interact with the desired biological substances.

Some reviews on magnetic beads used in MSFBs may be found in literature (e.g., Hristov 1999; Tanyolac and Özdural 2000).

As examples of magnetic beads we may point out: Fe_3O_4/(alginate or activated carbon) particles (see Figure 12.7) (Böhm and Pittermann 2000; Al-Qodah and Al-Shannag 2006); polymeric adsorbent magnetic particles (Seibert et al. 1998) like the highly used mPHEMA beads (magnetic polyhydroxyethylmethacrylate) (Basar et al. 2007; Özturk et al. 2007), mPEGDMA beads (magnetic polyethyleneglycoldimethylcrylate) (Özturk et al. 2007), PAM beads (polyacrylamide/magnetite) (Cocker et al. 1997), mPVB beads (polyvinylbutyral) (Tanyolac and Özdural 2000), P(GMA-MMA) beads (poly glycidylmethacrylate-methylmethacrylate) (Bayramoglu and Arica 2008); and ferromagnetic beads with immobilized enzymes or catalysts (Al-Hassan et al. 1991; Ivanova 1996 Bayramoglu and Arica 2008). These magnetic beads are

FIGURE 12.7
Fe_3O_4/alginate particles with a sphericity of 0.87. (From Böhm and Pittermann 2000.)

usually prepared by deposition of a polymer/epoxy resin/alginate layer on the magnetic core, and subsequent application of some kind of chemical reagent or thermal technique that modifies the surface of the beads for it to be fully functionalized, according to our specific goals.

12.5.2 Applications

In the last couple of years, the number of applications of MSFBs in biotechnology and biomedicine has had a steep increase. In fact, until very recently the most known applications of MSFBs in these areas were all directed toward bioreactions (by the use of immobilized enzymes and magnetic ion-exchange systems for protein purification), analytical procedures (such as chromatographic methods) and protein adsorption.

In this chapter, we will review these well-established processes and also point out some recently developed important applications.

12.5.2.1 Enzyme or Cell Immobilization/Bioreactions

Biocatalysis uses enzymes or cells to catalyze a remarkable spectrum of reactions. However, enzymes usually present a problem of solubility, and so, the recovering process at the end of the reaction step is always difficult. Therefore, enzyme immobilization is very often the best solution to avoid enzymes to dissolve in the effluent stream, and thus prevent product contamination, poor stability, and limited reuse (Webb et al. 1996; Betancor and Luckarift 2008). However, many of the methods used for enzyme immobilization are not suitable as they lead to reduction of its activity by causing structural damages (Betancor and Luckarift 2008).

Immobilization of enzymes or cells in MSFBs has been a very well-accepted method and its high potential known for some time (Webb et al. 1996). Although the majority of enzyme and cell supports for these applications are nonporous, there are some cases were porous media is used.

The operating principles of enzyme or cell immobilization in MSFBs are quite simple. The first step is concerned with the selection of the appropriate enzyme/cell and reaction process. As examples we may refer the use of *glucoamylase* for the hydrolysis of maltodextrin (Bahar and Celebar 2000), *S. cerevisiae* for ethanol fermentation (Ivanova 1996; Liu et al. 2009), *lipase* to catalyze the hydrolysis of triacylglycerol to glycerol and fatty acids (Lei et al. 2009), *Thiobacillus thioparus* for biodesulfurization of industrial water (Qiu et al. 2006), *horseradish peroxidase* for the treatment of phenolic wastewater in continuous systems (Bayramoglu and Arica 2008), and *Penicillium chrysogenum* for antibiotics production (Al-Qodah 2000).

The following step is the immobilization of the enzymes/cells in a magnetic support. These magnetic supports may present many different compositions, but usually are based on a ferromagnetic core, covered with a layer of a certain polymer and/or some sort of epoxy resin that attaches the enzymes/cells. Besides the examples of magnetic supports referred in Section 12.5.1, the reader can find more detailed information in the literature (Al-Hassan et al. 1991; Liu et al. 1991, 2009; Ivanova 1996; Hristov 1999; Al-Qodah 2000; Al-Qodah and Al-Hassan 2000; Bahar and Celebar 2000; Qiu et al. 2006; Safarik and Safarikova 2004; Van Hee et al. 2006; Webb et al. 1996; Hristov and Fachikov 2007; Bayramoglu and Arıca 2008; Betancor and Luckarift 2008; Lei et al. 2009). Betancor and Luckarift (2008) refer several cases concerning also their encapsulation in silica supports.

The encapsulated enzymes and their magnetic support matrices are then inserted in a MSBs and MFB where they are able to sustain higher flow velocities and present all the advantages inherent to these kinds of beds, which were referred previously.

A typical flow process using a MSB and MFBs with an immobilized enzyme is depicted in Figure 12.8.

Many examples of immobilized enzyme/cells applications may be found in literature, and it may be referred as typical examples: ethanol and other fermentations (Ivanova 1996; Al-Qodah and Al-Hassan 2000; Liu et al. 2009), antibiotics production (Al-Qodah 2000), phenol and p-chlorophenol removal from wastewaters (Bayramoglu and Arica 2008) or its biodesulfurization (Qiu et al. 2006), synthesis of hexyl acetate (Lei et al. 2009), glucose production (Bahar and Celebar 2000), starch hydrolysis (Webb et al. 1996), and waste-treatment (Karam and Nicell 1997).

12.5.2.2 Protein Purification/Adsorption

Protein purification by adsorption is one of the few "classical" applications of MSFBs in biological and health sciences. Its use has been only moderate

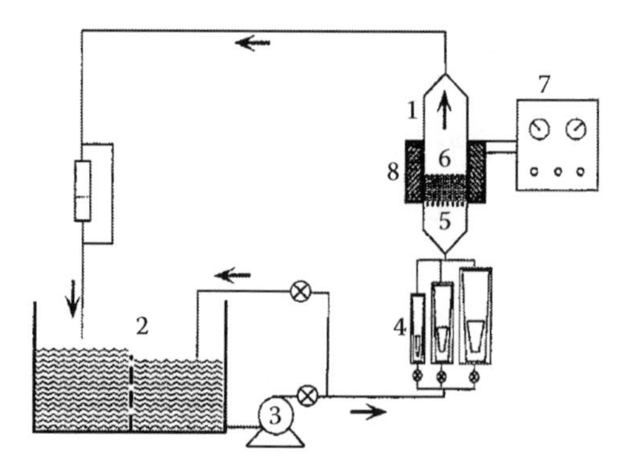

FIGURE 12.8
Experimental setup of a enzyme immobilized MSFB: 1—MSFB column, 2—Recirculation tank, 3—Pump, 4—Rotameters, 5—Distributor, 6—Enzyme immobilized magnetic particle bed, 7—Power supply, 8—Magnetic coil. (From Webb et al. 1996.)

mainly due to the great efficiency of another well-known competitive method: affinity chromatography. However, in situations where particulate solids are present (mainly in the early stages of the separation/purification processes) MSFB is the most effective existing method.

Protein purification/adsorption by MSFB depends on the possibility to create porous nanomagnetic beads that are capable to adsorb the target protein. These nanomagnetic beads may have a large spectrum of different compositions, and more details of these types of particles, and how they are manufactured and functionalized are given by Safarik and Safarikova (2004). As previously stated, these nanomagnetic beads are formed by a ferromagnetic core covered with a polymeric layer, which is then functionalized (polymers possess a large variety of surface functional groups, which can be tailored to be used in different kinds of applications [Türkmen et al. 2006]). Polymer chains are then attached to the surface of the polymeric layer of the porous nanomagnetic beads to form tentacle-type supports that enhance the protein adsorption capability (Türkmen et al. 2006). These magnetic adsorbents are then inserted in a MSFB setup, as in the example shown in Figure 12.9.

Several proteins are and have been purified by this method, as detailed by Safarik and Safarikova (2004). As typical examples we may refer: cytochrome-*c* adsorption (Türkmen et al. 2006), lysozyme adsorption (with the kind of beads described above, or with dye-affinity mPHEMA beads) (Tong and Sun 2003; Basar et al. 2007), bovine serum albumin adsorption (Tanyolac and Özdural 2000; Ding and Sun 2005), and horsehearth myoglobin (Seibert et al. 1998).

12.5.2.3 MSFB Chromatography

A technique called MSFB chromatography was initiated several decades ago and it has recently reached a high efficiency level. More recently, a novel

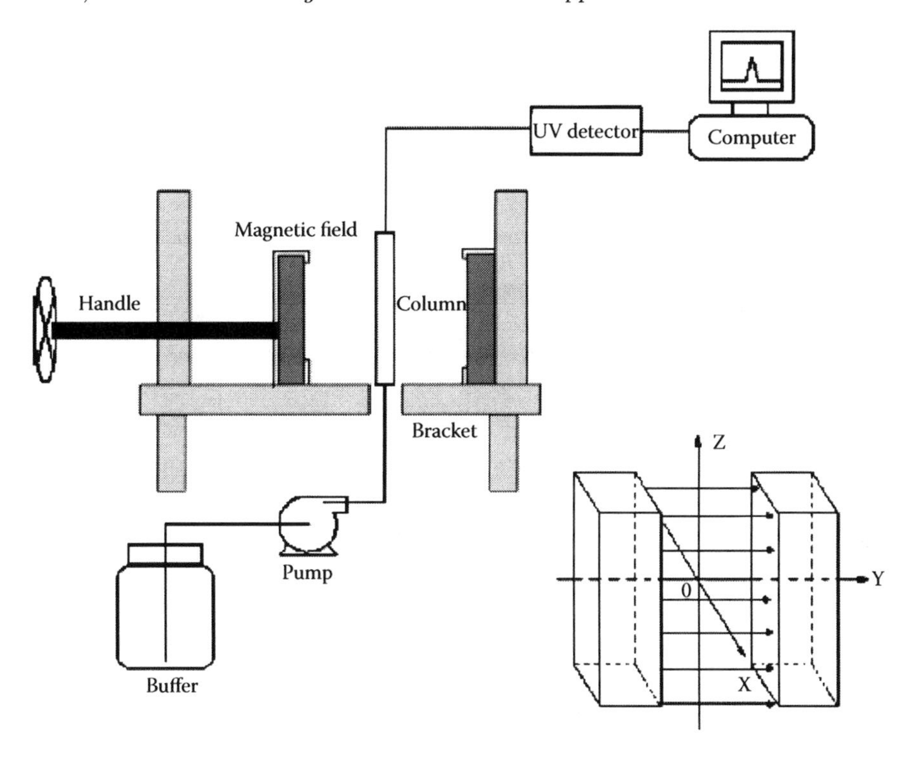

FIGURE 12.9
Experimental setup of a MSFB for protein adsorption. (From Tong and Sun 2003.)

chromatographic technique using continuous supermacroporous monolith has been used for the direct purification of products.

This chromatographic technique is useful for the selective separation of proteins and enzymes. There are two typical operation setups: one based on the use of nanomagnetic particles covered by a functionalized porous polymeric layer that is assembled in a magnetic bed structure (Cocker et al. 1997), and the other being composed by a supermacroporous cryogel column containing embedded ferromagnetic nanoparticles that are acted by the external magnetic field, a setup that allows to attain flows similar to those obtained in conventional chromatographic columns, having, at the same time, several pores that selectively delay the target substances (Yao et al. 2006).

Typical application examples of MSFB chromatography in biotechnology are the chromatographic separation of bovine serum albumin (Yao et al. 2006) and the affinity separation of trypsin from chymotrypsin (Cocker et al. 1997).

12.5.2.4 Novel Separations

In this section, some details will be given on recent applications of MSFBs in our areas of concern. Some of them could have been considered as being part

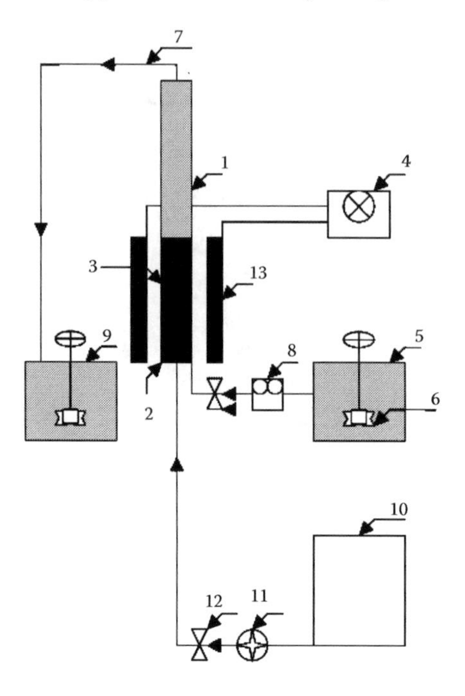

FIGURE 12.10
Schematic of the experimental setup of the application of MSFBs to the removal of yeast cells: 1—column; 2—supporting grid; 3—magnetic particles; 4—power supply; 5—feed tank; 6—mixer; 7—effluent stream; 8—peristaltic pump; 9—effluent receiver; 10—distilled water; 11—centrifugal pump; 12—valve; 13—magnetic system. (From Al-Qodah and Al-Shannag 2006.)

of the applications previously described, however, due to their importance, they are referred in this separate section.

Immunoglobulin G separation (Özkara et al. 2004; Karatas et al. 2007; Özturk et al. 2007), and *antibody removal* (Odabas et al. 2005) are recent applications of MSFBs. They represent an important step concerning clinical analysis and the treatment of diseases like lupus and cancer. They are based in the same process described for protein purification and, while in immunoglobulin G, functionalized mag-poly (EGDMA-MAH) beads are used, in the antibody removal functionalized mPHEMA beads are the usual choice.

Separation of cells and nucleic acids are also important applications of MSFBs, and, as before, the process is very similar to the one previously described for protein purification (Al-Qodah and Al-Shannag 2006, 2007). A schematic drawing of such a process setup is depicted in Figure 12.10, for the case of yeast cells, where magnetic beads are composed by a magnetite (Fe_3O_4) core, and covered by a stable layer of activated carbon whose function is to adsorb the cells from the suspension.

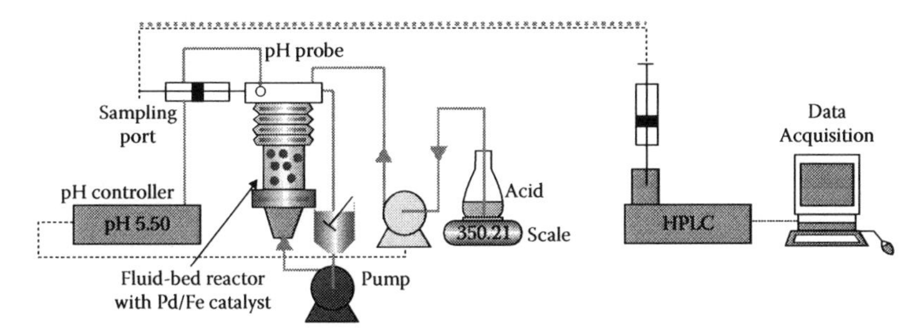

FIGURE 12.11
Schematic of the experimental setup of the application of MSFBs with Fe_3O_4/alginate complexes to the chlorophenol dehalogenation. (From Graham et al. 2006.)

Unusual applications like the *purification of caprolactam* (Menga et al. 2003) or *bilirubin removal* (Uzun and Denizli 2006) may also be found in recent literature, and while in the former, simple ferromagnetic beads of a novel amorphous nickel alloy catalyst (SRNA-4) are used, in the latter typical mPHEMA particles are used with human serum albumin immobilized on them. The process setup for the latter application is similar to the ones used for protein purification, opposite to the experimental setup of the former, which is constituted by a MSFB reactor with some previous pretreatment of the process flow.

Applications exist were *Fe_3O_4/alginate complexes* (as the one shown in Figure 12.7) are used (Böhm and Pittermann 2000). An example is the *chlorophenol dehalogenation* (Graham et al. 2006). In this case the process setup is not the conventional one, as may be seen in Figure 12.11.

12.6 Conclusion and Future Perspectives

MSBs and MFBs have been subject of study and development for more than six decades. However, biotechnological and biomedical applications of MSFBs appeared only more recently. In this chapter, the main features of MSFBs were reviewed, including the background theory and main principles of operation. The main biotechnological and biomedical applications of MSFBs are described, including enzyme immobilization for bioreactions, protein separation, magnetic chromatography, immunoglobulin G separation, antibody removal, separation of cells and nucleic acids, purification of caprolactam, bilirubin removal, and environmental improvements. The magnetic particles (beads) used in these applications were also reviewed and described.

The spectrum of future biotechnological and biomedical applications of MSFBs is vast as they depend mainly upon the development of appropriated magnetic beads, which are nowadays cheaper and easier to fabricate and functionalize.

12.7 References

Albert, R. V. and Tien, C. (1985). Particle collection in magnetically stabilized fluidized filters. *AIChE Journal,* **31**:288–295.

Álvaro, A., Rodriguez, J. M., Augusto, P. A., and Estévez, A. M. (2007). Magnetic filtration of an iron oxide aerosol by means of magnetizable grates. *China Particuology,* **5**:140–144.

Al-Hassan, Ivanova, V., Dobreva, E., Penchev, R., Hristov, J., Rachev, R., and Petrov, R. (1991). Non-porous magnetic supports for cell immobilization. *Journal of Fermentation and Bioengineering,* **71**:1–6.

Al-Qodah, Z. (2000). Antibiotics production in a fluidized bed reactor utilizing a transverse magnetic field. *Bioprocess Engineering,* **22**:299–308.

Al-Qodah, Z. and Al-Hassan, M. (2000). Phase holdup and gas to liquid mass transfer coefficient in magneto G-L-S airlift fermenter. *Chemical Enginering Journal,* **79**: 41–52.

Al-Qodah, Z. and Al-Shannag, M. (2006). Separation of yeast cells from aqueous solutions using magnetically stabilized fluidized beds. *Letters in Applied Microbiology,* **43**:652–658.

Al-Qodah, Z. and Al-Shannag, M. (2007). Application of magnetically stabilized fluidized beds for cell suspension filtration from aqueous solutions. *Semiconductor Science and Technology,* **42**:421–438.

Anacleto, A. L. and Carvalho, J. R. (1996). Mercury cementation from chloride solutions using iron, zinc and aluminium. *Mineral Engineering,* **9**(4): 385–397.

Arica, M. Y. (2000). Immobilization of polyphenol oxidase on carboxymethylcellulose hydrogels beads: preparation and characterization. *Polymer International,* **49**:775–781.

Arnaldos, J. and Casal, J. (1987). Study and modelling of mass transfer in magnetically stabilized beds. *International Journal of Heat and Mass Transfer,* **30**(7): 1525–1529.

Arnaldos, J., Casal, J., Lucas, A., and Puigjaner, L. (1985). Magnetically stabilized fluidization: modelling and application to mixtures. *Powder Technology,* **44**:57–62.

Augusto, P. A., Castelo-Grande, T., Estevéz, A. M., Barbosa, D., Rodríguez, J. Mª.; Álvaro, A., and Sanchéz, J. (2008). Magnetically stabilized and fluidized beds: heat and mass transfer. *Defect and Diffusion Forum,* **273–276**:46–51.

Augusto, P. A., Castelo-Grande, T., Estevéz, Angel M., Rodríguez, J. Mª., Álvaro, A. Magnetic fluidized bed: background theory and experimental determination of susceptibilities, submitted.

Bahar, T. and Celebar, S. S. (2000). Performance of immobilized glucoamylase in a magnetically stabilized fluidized bed reactors. *Enzyme Microb Technol,* **26**:28–33.

Basar, N., Uzun, L., Güner, A., Denizli, A. (2007). Lysozyme purification with dye-affinity beads under magnetic field. *International Journal of Biological Macromolecules,* **41**:234–242.

Bayramoglu, G. and Arıca, M. Y. (2008). Enzymatic removal of phenol and p-chlorophenol in enzyme reactor: horseradish peroxidase immobilized on magnetic beads. *Journal of Hazardous Materials,* **156**: 148–155.

Betancor, L. and Luckarift, H. R. (2008). Bioinspired enzyme encapsulation for biocatalysis. *Trends in Biotechnology,* **26**(10):566–572.

Böhm, D. and Pittermann, B. (2000). Magnetically stabilized fluidized beds in biochemical engineering—investigations in hydrodynamics. *Chemical Engineering & Technology,* **23**(4):309–312.

Bologa, M. K. and Syutkin, S. V. (1977). Effect of an electromagnetic field on the structure and hydrodynamic characteristics of a fluidized bed. *Electroanna. Obrabotka Materialov* (Russia), **1**:37–42.

Britton, M. M., Sederman, A. J., Taylor, A. F., Scott, S. K., and Gladden, L. F. (2005). Magnetic resonance imaging of flow-distributed oscillations. *Journal of Physical Chemistry A,* **109**:8306–8313.

Burns, M. A. and Graves, D. J. (1985). Continuous affinity chromatography using a magnetically stabilised fluidised bed. *Biotechnology Progress,* **1**:95–103.

Cocker, T. M., Fee, C. J., and Evans, R. A. (1997). Preparation of magnetically susceptible polyacrylamide/magnetite beads for use in magnetically stabilized fluidized bed chromatography. *Biotechnology and Bioengineering,* **53**:79–87.

Colver, G. M. (1979). The influence of electric and magnetic fields on air-fluidized beds. In *Proceedings of the NSF Workshop on Fluidization and Fluid-Particle Systems: Research Needs and Priorities*, ed. H. Littman, 57. National Science Foundation, Washington, DC.

Conta, P., Gonthier, Y., Bernis, A., and Lacour, G. (1998). Etude et modélisation du comportement hydrodynamique d'un lit fluidisé gaz-solide stabilisé magnétiquemen. *Powder Technology,* **99**:201–209.

Ding, X. B., Sun, Z. H., Wan, G. X., and Jiang, Y. Y. (1998). Preparation of thermosensitive magnetic particles by dispersion polymerization. *Reactive & Functional Polymers,* **38**:11–5.

Ding, Y. and Sun, Y. (2005). Small-sized dense magnetic pellicular support for magnetically stabilized fluidized bed adsorption. *Chemical Engineering Science,* **60**:917–924.

Dong, M., Pan, Z., Peng, Y., Meng, X., Mu, X., Zong, B., and Zhang, J. (2008). Selective acetylene hydrogenation over core–shell magnetic Pd-supported catalysts in a magnetically stabilized bed. *AIChE Journal,* **54**(5): 1358–1364.

Estevez, A. M. et al. (1995). Behavior of magnetizable composites in a liquid-solid magnetofluidized bed. Potential use as a bioreactor. *Chemical and Biochemical Engineering Quarterly,* **9**(2):67–77.

Estevez, A. M., Rodriguez, J. M., Alvaro, A., Augusto, P. A., Jiménez, O., Castelo-Grande, T., and Barbosa, D. (2008). Preparation, characterization, and testing of magnetic carriers for arsenic removal from water. *IEEE Transactions on Magnetic,* **44**(11):4436–4439.

Evans, L. L. and Burns, M. A. (1995). Countercurrent gradient chromatography: A continuous focusing technique. *Biotechnology & Bioengineering,* **48**(5):461–475.

Fan, L. S. and Zhu, C. (1998). *Principles of Gas–Solid Fluidization.* Cambridge University Press, New York.

Fan, M., Chen, Q., Zhao, Y., Luo, Z., and Guan, Y. (2003). Fundamentals of a magnetically stabilized fluidized bed for coal separation. *Coal Preparation,* **23**:47–55.

Filipov, M. V. (1961a). The resistance and expansion of a magnetite fluidized bed in a magnetic field. *Izvestiya Akademii Nauk, Latv. SSR,* **12**(173):47–51.

Filipov, M. V. (1961b). A fluidized bed of ferromagnetic particles and the action of a magnetic field on it. *Trouddi Instituta, Latv. SSR,* **12**:215–236.

Filipov, M. V. (1962). Some properties of a suspended bed of ferromagnetic particles in a magnetic field. In *Voprosii magn. Gdrodinamiki I dinamiki plasmii* (Riga). SSSR, 637–642.

Geuzens, P. and Thoenes, D. (1988a). Magnetically stabilized fluidization, part II: continuous gas filtration. *Chemical Engineering Communications,* **67**:229–242.

Geuzens, P. and Thoens, D. (1988b). Magnetically stabilized fluidization, Part I: Gas and solids flow. *Chemical Engineering Communications,* **67**:217.

Graham, L, J., Atwater, J. E., and Jovanovic, G. N. (2006). Chlorophenol dehalogenation in a magnetically stabilized fluidized bed reactor. *AIChE Journal,* **52**(3):1083–1093.

Graves, D. J. (1992). Bioseparations in magnetically stabilized fluidized bed. In *Preparative and Production Scale Chromatography.* eds. G. Ganetsos and P. E. Barker, Marcel Dekker, New York, pp. 187–207.

Gui, K. and Zhang, Q. (2006). Study on the mechanism of the flue gas desulphurization in magnetically fluidized bed. In *Proceedings of 11th Workshop on Transport Phenomena in Two-Phase Flow,* eds. C. Boyadiev and J. Hristov, Slanchev, Bryag, Bulgaria, pp. 61–68.

Halling, P. J. and Dunnill, P. (1980). Magnetic supports for immobilized enzymes and bioaffinity adsorbents. *Enzyme Microbiology Technology,* **2**: 1–10.

Hamaker, H. C. (1937). The London van der Waals attraction between spherical particles. *Physica IV,* **10**:1058.

Hausmann, R., Hofmann, C., Franzreb, M., and Holl, W. H. (2000). Mass transfer rates in a liquid magnetically stabilized fluidized bed of magnetic ion-Exchange particles. *Chemical Engineering Science,* **55**:1477–1482.

Herschler, A. (1969). Method for the production and control of fluidized beds. *United States Patent nr.* 3439899

Hristov, J. Y. (1999). Remarks on the equipment and the opertaing modes of fluidized magnetizable particle bed bioreactors (liquid-solid systems). In *Proceedings of the 2nd Conference "Magnetic Separations in Biosciences and Biotechnologies,"* ed. I. Safarik and M. Safarikova, Ceske Budejovice, pp. 97–107.

Hristov, J Y. (2002). Magnetic field assisted fluidization—a unified approach. Part 1. Fundamentals and relevant hydrodynamics. *Reviews in Chemical Engineering,* **18**(4–5):295–509.

Hristov, J. Y. (2003a). Magnetic field assisted fluidization—a unified approach. Part 2. Solids batch gas-fluidized beds: versions and rheology. *Reviews in Chemical Engineering,* **19**(1):1–132.

Hristov, J. Y. (2003b). Magnetic field assisted fluidization—a unified approach. Part 3. Heat transfer- a critical re-evaluation of the results. *Reviews in Chemical Engineering,* **19**(3):229–355.

Hristov, J. Y. (2004). Magnetic field assisted fluidization—a unified approach. Part 4. Moving gas-fluidized beds. *Reviews in Chemical Engineering,* **20**(5–6):377–550.

Hristov, J. Y. (2006a). Separation techniques performed by magnetically stabilized beds. In *Proceedings of the Symposium on Magnetic Separation and Nanomagnetics.* eds. H. Nirschl, M. Franzreb, and H. Anlauf, University of Karlsruche, Karlsruhe, Germany.

Hristov, J. Y. (2006b). Magnetic field assisted fluidization—a unified approach. Part 5. A hydrodynamic treatise on liquid–solid fluidized beds. *Reviews in Chemical Engineering,* **22**(4–5):195–375.

Hristov, J. Y. (2007). Magnetic field assisted fluidization—a unified approach. Part 6. Topics of gas-liquid-solid fluidized bed hydrodynamics. *Reviews in Chemical Engineering,* **23**(6):373–526.

Hristov, J. Y. (2009). Magnetic field assisted fluidization—a unified approach. Part 7. Mass transfer: chemical reactors, basic studies and practical implementations there of. *Reviews in Chemical Engineering,* **25**(1–3):1–254.

Hristov, J. and Fachikov, L. (2007). An overview of separation by magnetically stabilized beds: State-of-the-art and potential applications. *China Particuology,* **5**:11–18.

Hristov, J. Y. and Ivanova, V. (1999). Magnetic field assisted bioreactors. In *Recent Research Developments in Fermentation and Bioengineering 2,* SignPost Research, Trivandrum, pp. 41–95.

Hultman, T., Stal, S., Hornes, E., and Uhlen, M. (1989). Direct phase sequencing of genomic and plasmid DNA using magnetic beads as solid support. *Nucleic Acid Research,* **17**:4937–4946.

Ivanov, D. G. and Grozev, G. T. (1970). Determination of the critical velocity of fluidization of ferrochrome catalyst in a magnetic field. *Zhizn i Priklyucheniya Khimiya* (Russia), **43**:2200–2204.

Ivanova, V., Hristov, J., Dobreva, E., Al-Hassan, Z., and Penchev, I. (1996). Performance of a magnetically stabilized bed reactor with immobilized yeast cells. *Applied Biochemistry and Biotechnology,* **59**:187–198.

Jaraíz, E M. 1983. Nuevas tecnologías para el procesamiento de sólidos utilizando campos magnéticios. *Ingeniería Química* (Spain), Julio:149–154.

Jaraíz, E. M. and Estevez, A. M. (1987). Design concepts for a collar-type magnetic valve for small-size solids (MVS): theory and experiments. *Powder Technology,* **53**:1–9.

Karam, J. and Nicell, J. A. (1997). Potential applications of enzymes in waste treatment. *Journal of Chemical Technology and Biotechnology,* **69**: 141–148.

Karataş, M., Akgöl, S., Yavuz, H., Say, R., and Denizli, A. (2007). Immunoglobulin G depletion from human serum with metal-chelated beads under magnetic field. *International Journal of Biological Macromolecules,* **40**:254–260.

Kirko, I. M. and Filipov, M. V. (1960). The special features of a fluidized bed of ferromagnetic particles in a magnetic field. *Journal of Technological Physics* (in Russian), **30**(9):1081–1084.

Lai, Y. C. and Lin, S. C. (2005). Application of immobilized horseradish peroxidase for the removal of p-chlorophenol from aqueous solution. *Process Biochemistry,* **40**:1167–1174.

Lei, L., Bai, Y., Li, Y., Yi, L., Yang, Y., Xia, C. (2009). Study on immobilization of lipase onto magnetic microspheres with epoxy groups. *Journal of Magnetism and Magnetic Materials,* **321**:252–258.

Liu, C.-Z., Wang, F., Ou-Yang, F. (2009). Ethanol fermentation in a magnetically fluidized bed reactor with immobilized Saccharomyces cerevisiae in magnetic particles. *Bioresource Technology,* **100**:878–882.

Liu, Y. A., Hamby, R. K., and Colberg, R. D. (1991). Fundamental and practical. Developments of magnetofluidized beds: a review. *Powder Technology,* **64**:3–41.

Lucchesi, P. J., Hatch, W. H., Mayer, F. X., and Rosensweig, R. E. (1979). Magnetically stabilized beds – new gas-solids contacting technology. In *Proceedings of the 10th World Petroleum Congress*, 4:419–425. Bucharest, Heyden, Philadelphia, PA.

Menga, X., Mua, X., Zonga, B., Mina, E., Zhub, Z., Fub, S., and Luob, Y. (2003). Purification of caprolactam in magnetically stabilized bed reactor. *Catalysis Today,* **79–80**:21–27.

Moffat, G., Williams, R. A., Webb, C., and Stirling, R. (1994). Selctive separations in environmental and industrial-processes using magnetic carrier technology. *Minerals Engineering,* **7**:1039–1056.

Neff, J. and Rubinsky, B. (1983). The effect of a magnetic field on the heat transfer characteristics of an air fluidized bed of ferromagnetic particles. *International Journal of Heat and Mass Transfer*, **16**:1885–1889.

Nekrassov, Z. I. and Chekin, V. V. (1956). The effective viscosity of a fluidized bed of polydisperse ferromagnetic solids in alternating magnetic field. *Izv. Acad. Nauk* (Russia)—*Metall and Fuels.*, **1**:56–59.

Nekrassov, Z. I. and Chekin, V. V. (1961). The effect of an alternating magnetic field on a fluidized bed of ferromagnetic particles. *Izv. Acad. Nauk.* (Russia)—*Metall and Fuels,* **6**:25–29.

Nuñez, L. and Kaminski, M. D. (1998). Magnetically assisted chemical separation process. *Filtration & Separation,* **35**:349–352.

Nyns, E. J. (1995). High-rate continuous biodegradation of concentrated chlorinated aliphatics by a durable enrichment of methanogenic origin undercarrier-dependent conditions. *Biotechnology and Bioengineering,* **47**:298–307.

Odabas¸ M., Özkayar, N., Özkara, S., Ünal, S., and Denizli, A. (2005). Pathogenic antibody removal using magnetically stabilized fluidized bed. *Journal of Chromatography B,* **826**:50–57.

Özkara, S., Akgöl, S., Çanak, Y., and Denizli, A. (2004). A novel magnetic adsorbent for immunoglobulin-G purification in a magnetically stabilized fluidized bed. *Biotechnology Progress,* **20**:1169–1175.

Özturk, N., Gu1nay, M. E., Akgo1l, S., and Denizli, A. (2007). Silane-modified magnetic beads: application to immunoglobulin G separation. *Biotechnology Progress,* **23**:1149–1156.

Paice, M. G. and Jurasek, L. (1984). Peroxidase-catalyzed colour removal from bleach plant effluent. *Biotechnology and Bioengineering,* **26**:477–480.

Penchev, I. P. and Hristov, J. Y. (1990). Behaviour of fluidized beds of ferromagnetic particles in an axial magnetic field. *Powder Technology,* **61**: 103–118.

Qiu, G.-L., Li, Y.-L., and Zhao, K. (2006). Thiobacillus thioparus immobilized by magnetic porous beads: preparation and characteristic. *Enzyme and Microbial Technology,* **39**:770–777.

Rodríguez, J. M., Macias-Machin, A., Alvaro, A., Sánchez, J. R., and Estevez, A. M. (1999). Removal of iron oxide particles in a gas stream by means of a magnetically stabilized granular filter (MSF). *Industrial & Engineering Chemistry Research,* **38**:276–283.

Rodríguez, J. M., Sánchez, J. R., Alvaro, A., Florea, D. F., and Estévez, A. M. (2000). Fluidization and elutriation of iron oxide particles. A study of attrition and agglomeration processes in fluidized beds. *Powder Technology,* **111**(3): 218–230.

Rosensweig, R. E. (1979). Fluidization: hydrodynamic stabilization with a magnetic field. *Science,* **204**:57.

Rosensweig, R. E. (1985). *Ferrohydrodynamics.* Cambridge University Press, Cambridge.

Rosensweig, R. E. and Ciprios, G. (1991). Magnetic liquid stabilization of fluidization in a bed of nonmagnetic sphere. *Powder Technology,* **64**:115.

Rosensweig, R. E., Jerauld, G. R., and Zahn, M. (1981a). Structure of magnetically stabilized fluidized solids. In *Continuum Models of Discrete Systems*, eds. O. Brulin and R. K. T. Hsieh, North-Holland, Amsterdam, pp. 137–144.

Rosensweig, R. E., Siegell, J. H., Lee, W. K., and Mikus, T. (1981b). Magnetically stabilized fluidized solids. *AIChE Symposium Series*, **77**(205):8–16.

Safarik, I. and Safarikova, M. (2004). Magnetic techniques for the isolation and purification of proteins and peptides. *BioMagnetic Research and Technology,* **2**(7):17.

Sajc, L. M., Jovanovic, Z. R., Vunjak-Novakovic, G., Jovanovic, G. N., Pesic, R. D., and Vukovic, D. V. (1994). Liquid dispersion in a magnetically stabilized fluidized bed (MSFB). *Transactions of the Institution of Chemical Engineers*, **72**:236–240.

Saxena, S. C., Ganzha, V. L., Rahman, S. H., and Dolidovich, A. F. (1994). Heat transfer and relevant characteristics of magnetofluidized beds. *Advanced Heat Transfer*, **25**:151–249.

Sedzimir, J. A. (2002). Precipitation of metals by metals (cementation)–Kinetics, equilibria. *Hydrometallurgy,* **64**(3):161–167.

Seibert, K. D. and Burns, M. A. (1998). Effect of hydrodynamic and magnetic stabilization on fluidized-bed adsorption. *Biotechnology Progress,* **14**: 749–755.

Shames, I. (1982). *Mechanics of Fluids.* McGraw-Hill, New York.

Siegell, J. H. (1987). Liquid fluidized magnetically stabilized beds. *Powder Technology,* **52**:139–148.

Siegell, J. H. (1988). Applications of crossflow magnetically stabilized fluidized beds. *Chemical Engineering Communications,* **67**(1):43–54.

Siegell, J. H. (1989). Early studies of magnetized-fluidized beds. *Powder Technology,* **57**:213–220.

Siegell, J. H., Pirkle Jr., J. C., and Dupre, G. D. (1984). Crossflow magnetically stabilized bed chromatography. *Separation Science and Technology,* **19**:213.

Sonolikar, R. L. (1989). Magneto-fluidized beds. In *Transport in Fluidized Particle Systems.* Elsevier, New York, pp. 359–422.

Tanyolaç, D. and Özdural, A. R. (2000). A new low cost magnetic material: magnetic polyvinylbutyral microbeads. *Reactive & Functional Polymers,* **43**:279–286.

Terranova, B. E. and Burns, M. A. (1991). Continous cell suspension processing using magnetically stabilized fluidized beds. *Biotechnology and Bioengineering,* **37**:110–120.

Tong, X.-D. and Sun, Y. (2003). Application of magnetic agarose support in liquid magnetically stabilized fluidized bed for protein adsorption. *Biotechnology Progress,* **19**:1721–1727.

Türkmen, D., Yavuz, H., and Denizli, A. (2006). Synthesis of tentacle type magnetic beads as immobilized metal chelate affinity support for cytochrome c adsorption. *International Journal of Biological Macromolecules,* **38**:126–133.

Ugelstad, J., Berge, A., and Ellingsen, T. (1992). Preparation and application of new monosized polymer particle. *Progress in Polymer Science,* **17**: 526–529.

Uzun, L. and Denizli, A. (2006). Bilirubin removal performance of immobilized albumin in a magnetically stabilized fluidized bed. *Journal of Biomaterials Science, Polymer Edition,* **17**(7):791–806.

van Hee, P., Hoeben, M. A., van der Lans, R. G. J. M., and van der Wielen, L. A. M. (2006). Strategy for selection of methods for separation of bioparticles from particle mixtures. *Biotechnology and Bioengineering,* **94**(4):689–709.

Wallace, A. K., Ranawake, U. A. and Levien, K., (1991). Experimental feasibility study of a magnetic elevator for particles. *Powder Technology,* **64**:125.

Webb, C., Kang, H.-K., Moffat, G., Williams, R. A., Estévez, A.-M., Cuéllar, J., Jaraíz, E., and Galán, M.-A. (1996). *The Chemical Engineering Journal,* **61**:241–246.

Yamaura, M., Camilo, R. L., Sampaio, L. C., Macedo, M. A., Nakamura, M., and Toma, H. E. (2004). Preparation and characterization of (3-aminopropyl)trieth-oxysilane-coated magnetite nanoparticles. *Journal of Magnetism and Magnetic Materials,* **279**:210–217.

Yao, K., Yun, J., Shen, S., Wang, L., He, X., Yu, X. (2006). Characterization of a novel continuous supermacroporous monolithic cryogel embedded with nanoparticles for protein chromatography. *Journal of Chromatography A,* **1109**:103–110.

Zhang, G. T., Jaraiz-M. E., Wang, Y., and Levenspiel, O. (1984). Theory and operational characteristics of the magnetic valve for solids. Part II: collar design. *AIChE Journal,* **30**:951–959.

Zouboulis, A. I. and Katsoyiannis, I. A. (2002). Arsenic removal using iron oxide loaded alginate beads. *Industrial Engineering Chemistry Research,* **41**(24):6149–6155.

Zrunchev, I. A. (1975). Ammonia synthesis in a fluidized bed with and without a magnetic field. *Ann. Reports of UCTM* (Sofia), **22**(4):171–181.

Zrunchev, I. A. and Popova, T. F. (1975). Ammonia synthesis in a fluid bed of magnetized catalyst. *Ann. Reports of UCTM* (Sofia), **22**(3): 105–110.

13

In Situ Characterizations of Porous Media for Applications in Biofuel Cells: Issues and Challenges

Bor Yann Liaw
Hawaii Natural Energy Institute, SOEST, University of Hawaii at Manoa, Honolulu, HI

CONTENTS

13.1 Introduction

Porous media in either electrode configurations or membranes used in biofuel cell applications are often required to have specific functionalities for use in aqueous solutions at nearly neutral pH (Atanassov et al. 2007). These unique requirements and operating conditions can only be better understood by *in situ* characterizations that combine both physical and chemical analyses in a noninvasive manner (Minteer et al. 2007; Cooney and Liaw 2008). The functionalities of these porous media also need to be studied in small physical dimensions from biological species on the order of a few nanometers to pore structures covering the entire range of micropores to macropores; that is, a wide range of volume, surface area, or thickness shall be involved. Some of the characterizations (*in situ* in solutions) can hardly be achieved by conventional physical or chemical characterization techniques (e.g., porosimetry or nanoscale imaging), nor can they be accomplished easily with sufficient accuracy. Furthermore, additional issues with ex situ characterizations can also emerge. For instance, misleading results associated with exposure to the ex situ environment during sample preparation or characterization may inhibit interpretations of the underlying surface or interface phenomena. Facing these problems, one shall find that *in situ* characterizations are thus more desirable, critical, and necessary for defining the properties of the porous media used in such applications.

These porous media can be categorized into inorganic, organic (most likely, polymeric) materials, and hybrid composites. Porous silicon oxides and metal oxides are examples of the inorganic materials. Conductive (electronic and/or ionic) polymers are usually referred to as the organic materials. Polymer-carbon nanotube composites are a good example for the hybrids. These materials are usually used as the backbone for immobilization of biocatalytic species, including microbial, enzymatic catalysts, cofactors, or mediators. They can also act as the current collector for mediators or microbial cells to enable electron transfer or as membranes to separate reactants to prevent interference in electrode reaction or fuel crossover (and thus efficiency loss). To characterize these materials for their feasibility in the biofuel cell applications, engaging *in situ* observations can be advantageous to help us monitor and understand their behavior under operating conditions. Such noninvasive, *in situ* observations remain challenging yet rewarding with great payoffs. In this chapter, some recent advancement in the *in situ* characterization techniques, primarily spectroscopic in nature, are reviewed to provide some introduction of how they can be utilized in the understanding of porous media applications in biofuel cells.

A variety of *in situ* characterization techniques can be utilized particularly for studying small quantities or sizes of porous materials or conductive membranes, usually in thin film forms. Applying such techniques, often in a liquid environment, does require some nonconventional approaches to characterize the pore structure and related (either physical or chemical) properties. Particularly interested are approaches incorporating various techniques including spectroscopic imaging ellipsometry (SIE), quartz crystal microbalance (QCM), x-ray diffraction, fluorescence, and reflectometry (XRD, XRF, and XRR), or laser scanning confocal microscopy (LSCM), in combination with electrochemical techniques to yield insightful information of porous media's unique properties in solutions and at interfaces. These techniques are either surface or bulk sensitive, or both; and, they can provide either spatial or temporal information simultaneously or separately to reveal the dynamic nature of the processes in a biofuel cell, or on an electrode surface, or in the membrane. Combining the information obtained from these *in situ* techniques carefully, a wide spectrum of data can be obtained to achieve sufficient understanding of the electrode or cell behavior to allow us to clearly identify cell operating limitation, understand limiting mechanism, and improve cell performance.

13.2 Biofuel Cell Applications

Biofuel cells are attractive as promising micropower generation devices particularly for low-power niche applications (e.g., for enzymatic biofuel cells, some recent efforts have been reported by Palmore et al. 1998; Ikeda and Kano 2001; Katz et al. 2003; Akers et al. 2005; De Lacey et al. 2000; Qian et al. 2002; Karyakin et al. 2002; Atanassov et al. 2007; Minteer et al. 2007; Cooney and Liaw 2008; and, for microbial biofuel cells, Delaney et al. 1984; Allen and Bennetto 1993; Kim et al. 1999; Bond et al. 2002; Chaudhuri and Lovley 2003; Bond and Lovley 2003; Rabaey et al. 2003; Liu and Logan 2004; Cheng et al. 2006; Ringeisen et al. 2006; Weber et al. 2006; Richter et al. 2008). A recent review by Calabrese-Barton et al. (2004) summarized the work up to early 2000 for implantable microdevices. An abiotic glucose biofuel cell configuration recently reported by Kerzenmacher et al. is shown in Figure 13.1 to illustrate a potential power source for medical implant applications (Kerzenmacher et al. 2008).

The seminal work by A. Heller and his coworkers has shown the viability of using miniature biofuel cells in implants (Mano et al. 2003; Heller 2004), as illustrated in Figure 13.2. Recently a landmark effort in commercializing glucose biofuel cells was demonstrated by Sony Corp. (Sakai et al. 2009) in its launch of a glucose battery to power small electronic devices such as MP3 players (Figure 13.3). Another interesting glucose battery without the use of precious metal or biological catalysts was reported by Scott and Liaw lately, as shown in Figure 13.4 (Scott and Liaw 2009). Some earlier work of microbial biofuel cell in mediatorless configurations using *Shewanella putrefaciens* was reported by Kim et al. as illustrated in Figure 13.5 (Kim et al. 2002).

Similar concepts are also being investigated for biosensor and other bioelectrocatalytic applications. For instance, Bianco discussed protein modified and membrane electrodes for biomolecular sensor applications up to early 2000 (Bianco 2002). More recently, Katz and Willner promote nanoparticle (NP)-enzyme hybrid systems for nanobiotechnology applications (Katz and Willner 2004; Willner et al. 2007). Figure 13.6 illustrates an earlier example of how a surface-modified gold (Au) electrode can be used in biosensor applications (Xiao et al. 2003; Willner et al. 2007). In the example, the reconstitution of apo-flavoenzyme and apo-glucose oxidase (GOx) is illustrated by the immobilization of pyrroloquinoline quinone (PQQ)-flavin adenine dinucleotide (FAD) (Figure 13.6 [a] and [b]) on gold electrode, which was used as the platforms to immobilize GOx for glucose biosensor applications. The Au is functionalized with the cofactor flavin adenine dinucleotide (FAD) and PQQ to incorporate the enzymes to perform bioelectrocatalysis of glucose oxidation. The amperometric responses of the sensor with respect to the glucose concentrations were illustrated in Figure 13.6(c). Under a potentiostatic operation, the sensor will produce current in proportion to the glucose concentration in the ambient environment.

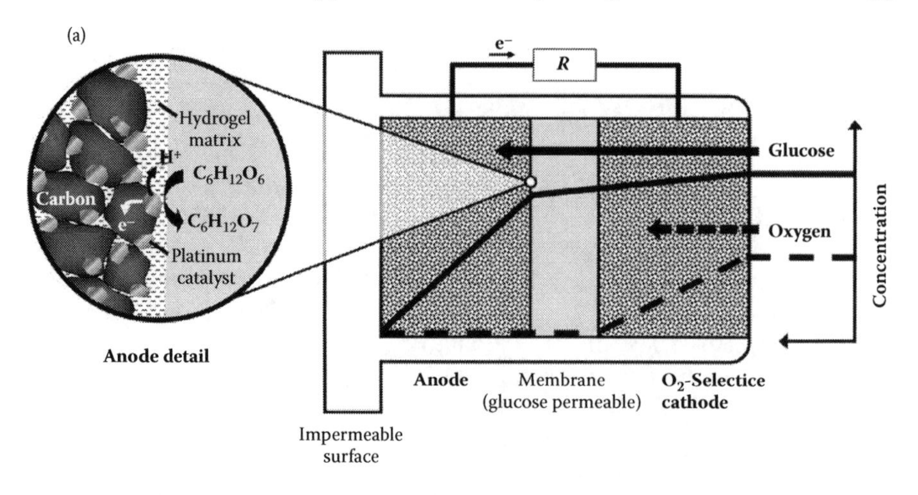

(b)

Anode:	$C_6H_{12}O_6 + H_2O \rightarrow C_6H_{12}O_7 + 2\,H^+ + 2\,e^-$
Cathode:	$1/2\,O_2 + 2H^+ + 2\,e^- \rightarrow H_2O$
Overall:	$C_6H_{12}O_6 + 1/2\,O_2 \rightarrow C_6H_{12}O_7$

FIGURE 13.1
(a) A schematic of an abiotic biofuel cell powered by glucose using an oxygen-selective air electrode as a cathode (positive electrode) and a hydrogel matrix embedded with carbon-supported Pt catalyst to oxidize glucose to gluconic acid to harness electricity and power a small implant, (b) the electrochemical reactions occur in the cell. (Reprinted from Kerzenmacher, S., Ducrée, J., Zengerle, R., and von Stetten, F., *J. Power Sour.*, 182, 2008. With permission from Elsevier.)

Potential merits of these enzymatic or microbial systems may include (anticipated low) cost and almost unlimited supply of biocatalytic species that can be regenerated through reproduction, less requirements for fuel purity due to better selectivity but also avoidance to intermediate poisoning (such as the CO poisoning on Pt-based catalysts), fuel flexibility (from hydrogen to carbohydrates), adaptability to improved performance by bioengineering, and possibility of self-assembly to allow *in situ* fabrication (on unique electrode surfaces). The foundation of many biosensors, bioreactors, and biofuel cells is established on bioelectrocatalysis, an important concept that couples enzymatic reactions with electrochemical ones. A recent review by Ikeda and

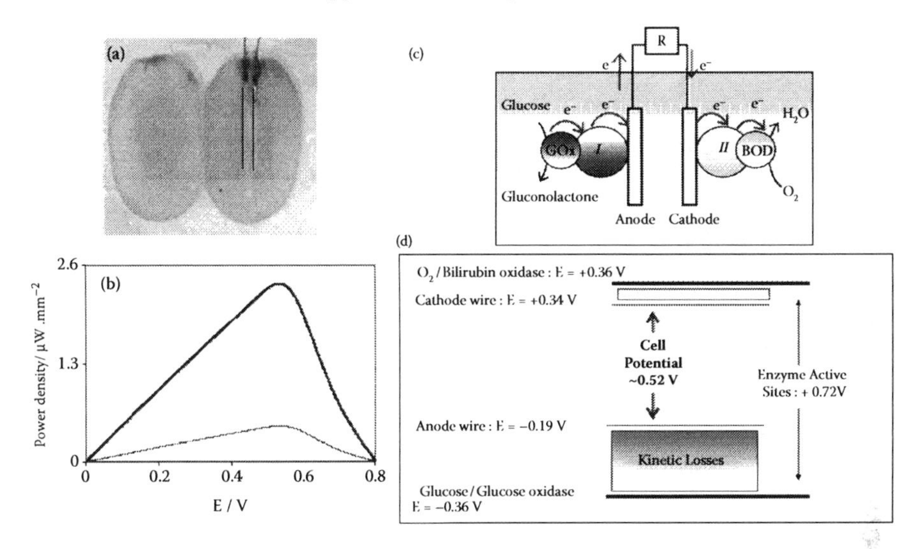

FIGURE 13.2
(a) Photo of a sliced grape with an implanted glucose biofuel cell, (b) power density output versus cell voltage when the air electrode is close to skin (bold) or the inner part of the grape (thin), (c) schematic of the cell, (d) illustration of the cell voltage as comprised by the electrode potentials. (Reprinted with permission from Mano, N., Mao, F., and Heller, A., *J. Am. Chem. Soc.*, 125, 2003. Copyright 2008 American Chemical Society.)

Kano (2001) provides a good discussion and introduction of the subject. In the bioelectrocatalysis, the oxidation or reduction of the substrates by the enzymatic redox reactions require the use of organic dyes or metal complexes as electron acceptors or donors (mediators) to transfer the electrons to the conductive electrode surface and be regenerated at such a surface. Figure 13.7 illustrates this important concept and shows how the enzymatic species and mediators are regenerated in the bioelectrocatalysis.

Although the concept and a working model of the first fuel cell were available in the mid-1800s, it was 100 years later that the hydrogen fuel cell was developed for application in space explorations and other niche applications. Since then the hydrogen fuel cell has been promised to be a clean power source. Its potential, however, has been hampered by the high costs of its components and fuel-related production, storage, and distribution (Wu et al. 2008). The growing demand for cleaner power in conjunction with the advent of biotechnology and the ability to produce a significant amount of biocatalysts, spurred investigations into the biofuel cells (Ikeda and Kano 2001; Katz et al. 2003; Chaudhuri and Lovley 2003; Liu

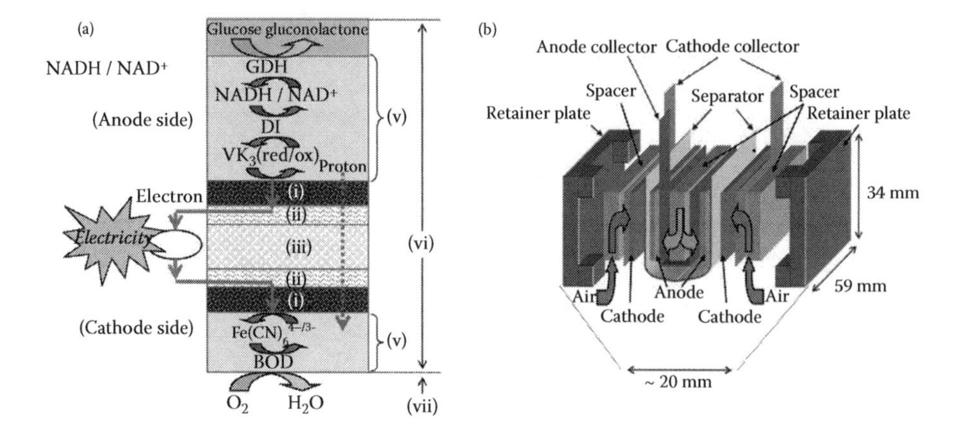

FIGURE 13.3
(a) A schematic showing a glucose-oxygen biofuel cell that can be operated at room temperature and quiescent conditions, (b) schematic of a two-cell stack that can generate 100 mW in 80 cm^3 and weighted about 40 g as revealed by Sony Corp. in 2007. (Sakai, H., Nakagawa, T., Tokita, Y., Hatazawa, T., Ikeda, T., Tsujimura, S., and Kano, K., *Energy Env. Sci.*, 2, 2009. Reproduced by permission of the Royal Society of Chemistry.)

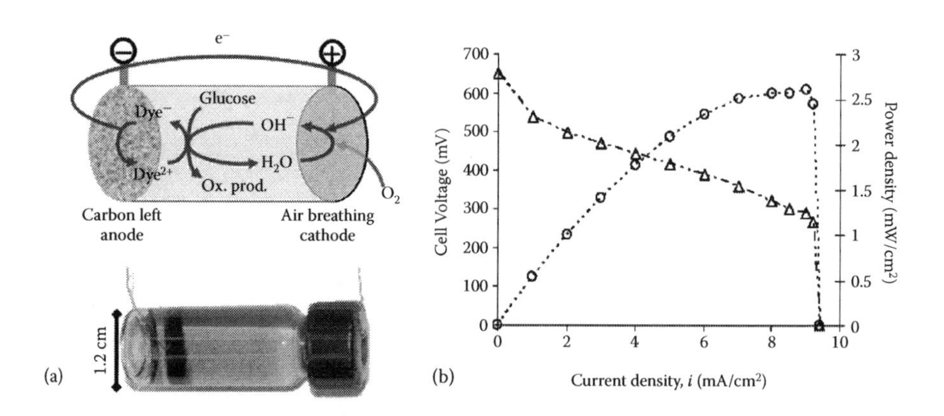

FIGURE 13.4
(a) A schematic and photo of a simple glucose-air battery without use of precious metal or biological catalysts in the operation at room temperature and ambient pressure, (b) the polarization curve and the power density profile versus current density for such a battery with 2 M glucose in 3 M KOH and 28 mM methyl viologen. (Scott, D. and Liaw, B.Y., *Energy Env. Sci.*, 2, 2009. Reproduced by permission of the Royal Society of Chemistry.)

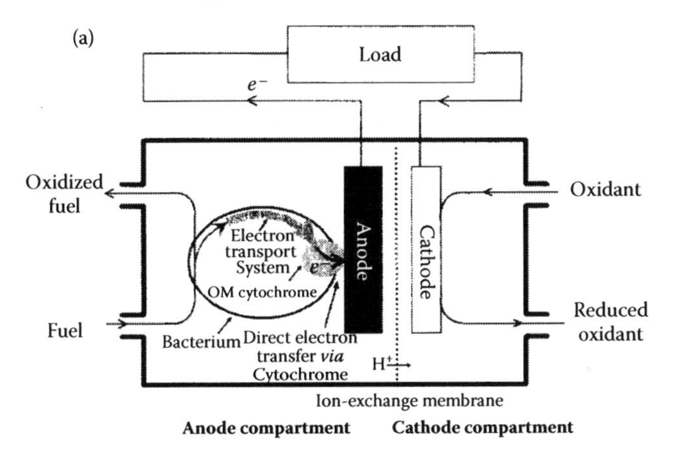

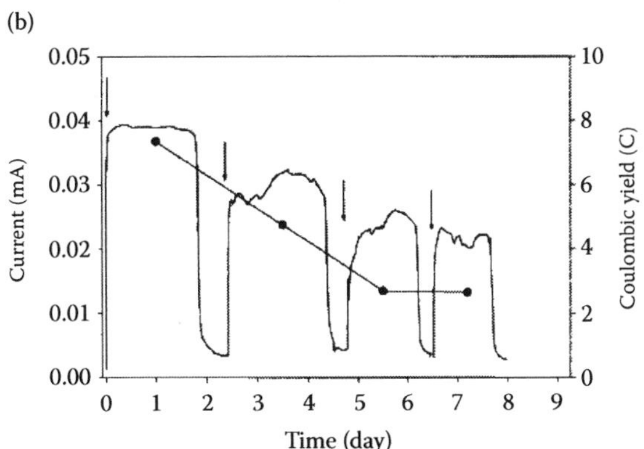

FIGURE 13.5
(a) Schematic diagram of a mediatorless microbial fuel cell, (b) current and coulombic yield of a cell (apparent electrode surface area: 50 cm^2) containing *Shewanella putrefaciens* IR-1 (0.2 ± 0.02 g dry cell weight/L) with a 1 kΩ resistor and intermittent additions of 10 mM lactate (shown by arrows). (Reprinted from Kim, H.J., Park, H.S., Hyun, M.S., Chang, I.S., Kim, M., and Kim, B.H., *Enzyme Micro. Tech.*, 30, 2002. With permission from Elsevier.)

and Logan 2004; Willner et al. 2007) as the next generation of power sources. Such biofuel cells extend the hydrogen fuel cell concept for replacing the expensive inorganic catalyst and fuel with renewable biocatalysts and fuels; for example, alcohol dehydrogenase for an enzymatic biofuel cell and alcohol (Akers et al. 2005) or Shewanella-based microbial biofuel cells (Kim et al. 1999; Ringeisen et al. 2006) with accompanying substrates (e.g., ethanol or glucose).

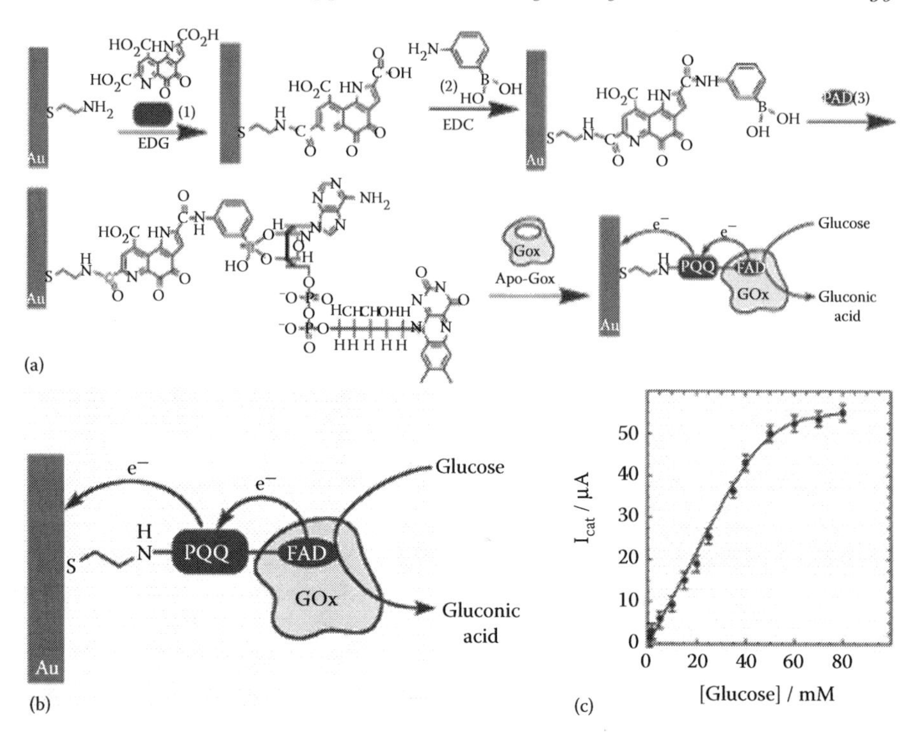

FIGURE 13.6
(a) Assembly of reconstituted GOx bioelectrode, which consists of electroactive relay units, based on functionalized PQQ-FAD and immobilized on Au electrode. (b) The functionalized GOx electrode that can conduct partial oxidation of glucose to gluconic acid. (c) A calibration plot of the electrocatalytic current at 0.2 V versus SCE as a function of glucose concentrations.

The advancement of the fuel cells, and more so with biofuel cells, has increased the need for additional characterization techniques that were not essential in the case of previous power generating systems. As opposed to an internal combustion engine in which reactants are mixed before the reaction and thermal energy is generated to create mechanical displacement for power, a fuel cell generates power when a reactant reacts with a catalyst in a compartment to realize half of the cell reaction (a reduction or oxidation as a half-cell reaction) while the other half-cell reaction (in complement to the reduction or oxidation) occurs at the other compartment, whereas the electrolyte keeps the electrons flowing in an external circuit to power the load. Although, this electrochemical route allows electron transfer to achieve a more efficient chemical energy conversion directly to an electrical one than

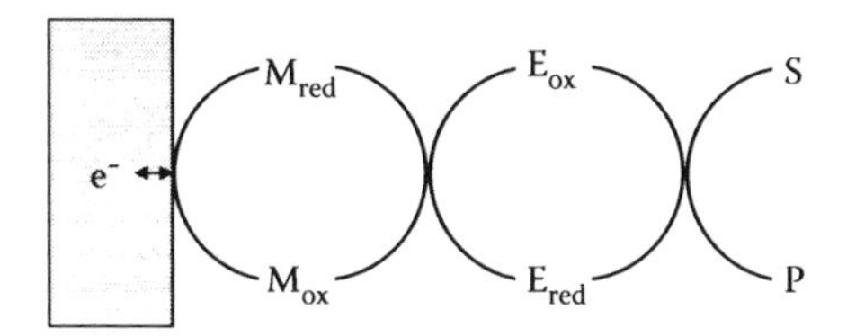

FIGURE 13.7
Schematic illustrating the regeneration of redox species (enzymes/coenzymes and mediators) in the bioelectrocatalysis. (S: substrate, P: product, E_{ox} and E_{red}: oxidized and reduced form of the enzyme/coenzyme, M_{red} and M_{ox}: reduced and oxidized form of mediator.) (Modified from Figure 1 in Ikeda, T. and Kano, K., *J. Biosci. Bioeng.*, 92, 2001.)

a thermal mechanical pathway, it also requires an extremely complex activation and separation of the reactants to enable an efficient energy conversion process.

Such a "balance-of-the-plant" must allow a variety of tasks to accomplish efficiently. For a biofuel cell to fulfill its promise, unlimited flow of reactants and products to reach and remove, respectively, from the stationary catalyst layer must be conducive to fuel delivery and product removal (Cooney and Liaw 2008). If direct electron transfer is desired (the more promising and efficient form of a biofuel cell operation) biocatalyst needs to be further supported on a conductive surface such that electrons can transfer from the reaction center to the catalyst support (Katz et al. 2003). Because catalysts in an enzymatic biofuel cell are inherently unstable the desired immobilization approach must also provide conditions that will stabilize and protect the enzyme from gradually denaturing (Minteer et al. 2007). For microbial biofuel cells, more recent studies on nanowired electron transfer from the microbial species to electrode surface indeed are very intriguing (Reguera et al. 2005; Gorby et al. 2006), which also illustrate the complexity involved in the charge transfer processes in a biofuel cell system.

13.3 Desirable Properties and Functionalities

Conductive materials (for electrons or ions, or both) with high porosity (i.e., high surface area and proper pore size distribution [PSD]) and suitable chemical backbone are more likely to fulfill the requirements for efficient energy transfer. Ideally the porosity needs to be on the tens of nanometer scale such that enzymes can be immobilized in a well-controlled environment and individually accessible to the substrate which is the fuel, and on the other hand, on the scale of hundreds of microns such that transport of the fuel in its liquid phase will not become the limiting factor in the power producing process. The balance between the considerations of optimal enzyme immobilization and fuel transport will determine the ultimate pore configuration for the power generation. In other words, macropores in the size of hundreds of microns with network of channels are required for facile fuel transport, whereas the pore structure needs to accommodate sublayered mesopores on the surface to enable effective enzyme immobilization. In the microbial electrode design, the pore structure of the transport media is less stringent, but PSD from submicron to several hundred microns remains critical in the design. Besides pore structure, prolonging the lifetime of the biocatalytic species in frequently strained fuel cell operating conditions to sustain the power generation is also a critical requirement. It usually requires an appropriate and delicate immobilization technique to retain enzymatic catalyst's activity or facilitate biofilm development. Furthermore, to achieve an effective immobilization of the biocatalytic species, it is also desirable to increase its loading (density of biocatalytic species per electrode surface area) to enhance power generation. A high loading is to create a large number of reactive sites per unit surface area. Complicated with substrate transport to allow effective biocatalytic reaction, it is also attractive to design electrodes with optimal three-dimensional architecture and optimized pore structure to allow the best combination of effective surface loading and mass transport to increase the loading density per unit volume for power generation (Cooney and Liaw 2008).

To achieve these objectives, an interesting possibility that has been proposed by many is to use conductive polymer network (such as those made of polypyrrole [PPy] and its derivatives) as an immobilization platform. Such a conductive backbone remains a promising possibility to enable favorable three-dimensional architecture for power generation in batteries or fuel cells (Scott et al. 2008) and as proposed for its use in biosensors (Ramanavicius et al. 1999). For instance, PPy is known for the ease in fabrication using electrochemical deposition techniques and in manipulation of its property through the deposition conditions and electrolyte compositions (which affect growth mechanism and dopant level) (e.g., Diaz et al. 1979; Salmon et al. 1982; Yang et al. 1991; Sutton and Vaughan 1995; Li et al. 1995; Kaynak 1997; Miles et al. 2000). In many literature reports, the film morphology and pore structure indeed depend on a wide range of fabrication conditions. For instance, PPy films with significant differences in density and morphological microstructure can be developed using control of pyrrole monomer concentration and growth rate under various potentiostatic and galvanostatic conditions (Sutton and Vaughan 1995; Kaynak 1997). Another interesting example is the fabrication of biocompatible chitosan or its derivatives into scaffolds that can be used as immobilization media (Cooney et al. 2008).

13.4 Needs for *in situ* Characterization: Issues and Challenges

As many porous media in various forms and structures are being used as electrode supports, immobilization matrices, or separators, and having unique functionalities operable in aqueous solution at nearly neutral pH, their performance and behavior should be better understood by *in situ* characterization techniques in a noninvasive manner. In the quest of finding useful *in situ* characterization techniques that can be applied to studying the behavior of these porous media, one often found that many commonly known *in situ* techniques might not be useful or feasible for such studies in aqueous solution. For instance, although atomic force microscopy (AFM) can be a useful tool for studying the surface morphology of a solid, with high spatial resolutions, it would be awkward to study a porous medium, where open pores are part of the morphological features. AFM is also difficult for studies in a dynamic situation where its temporal resolution is poor for a large surface area. It is therefore interesting to realize that there are only few *in situ* characterization techniques available for studying solid–liquid interfaces in solution to extract physical and chemical information with temporal and spatial resolutions tentative for kinetic and dynamic studies.

There is further complication for *in situ* characterizations due to a sample's small quantity or physical dimensions. In light of more recent interests in nanomaterials and nanotechnologies, such *in situ* characterizations may become problematic for porous media in small quantities or thin film forms. Traditional porosimetry (gas adsorption Brunauer-Emmett-Teller (BET) or mercury porosimetry) usually require a large amount of sample. These techniques are also ex situ, thus the sample properties could be altered by the preparation or characterization process. For *in situ* characterizations of porous films (in solution), it is more desirable to determine pore structure in solution or operating environment, in which factors that can affect the bulk (such as swelling due to hydration) and interfacial (such as wettability due to hydrophobic and hydrophilic interaction) properties can be investigated. One should anticipate that the properties determined in the solution and operating environment could be substantially different from those under dry and/or rigid conditions, especially the properties associated with a solid/gas or solid/liquid interface. Such a detailed solid/liquid interface characterization is a critical point of discussions in this chapter.

13.5 Applicable *in situ* Techniques

13.5.1 Spectroscopic Imaging Ellipsometry

Ellipsometry is an optical technique that is known for a century, versatile and powerful for the investigation of a material on its dielectric properties (complex refractive index or dielectric function) especially in a thin film form. Due to the interaction between the incident polarized light and the dielectric property of the material in the study, we can use an optical model based on Fresnel's equations to characterize such an interaction and derive useful information regarding the material's physical properties. This utility makes ellipsometry a useful tool in many different fields, from semiconductor physics to microelectronics and biology, and from basic research to industrial application. Ellipsometry is a very sensitive technique with superior capabilities for thin film metrology. As an optical technique, spectroscopic ellipsometry is noninvasive and contactless, therefore useful for *in situ* studies. There are several review articles in the literature (Tompkins 1993; Tompkins and McGahan 1999; Arwin 2005), which provide excellent overviews of this technique and its limitation.

The utility of ellipsometry is based upon the analysis of the change in polarization of light as it is reflected off a sample to yield information about layers that are thinner than the wavelength of the probing light itself, down to an atomic layer. Ellipsometry can probe the complex refractive index or dielectric function tensor, provide fundamental physical parameters, thus reveal such quantities in relevance to a variety of sample properties, including morphology, crystal quality, chemical composition, or electrical conductivity. Therefore, ellipsometry is commonly used to characterize film thickness for a single layer or complex multilayer stacks ranging from a few angstroms (or tenths of a nanometer) to several micrometers with excellent accuracy.

The name "ellipsometry" comes from the origin due to light polarization that is often elliptic and so used in this technique. The ellipsometry is becoming more interesting to researchers in many disciplines including biology and medicine. These new applications pose new challenges to the technique since *in situ* measurements on unstable liquid surfaces and microscopic imaging are frequently desired and required.

Ellipsometers are commonly operated in a nulling or a photometric mode in their measurements. A null ellipsometer is traditionally more common than the photometric configuration in the polarizer-compensator-sample-analyzer (PCSA) setup due to the merits of inherent stability, ease of operation, high resolution, simple data acquisition and analysis, and cost. Possible drawbacks include limitations to single wavelength and low speed in measurements, which prompt to the utilization of photometric mode for high speed and multiwavelength spectroscopic applications. For imaging ellipsometry, off-nulling design to enable imaging capability is typically used. To allow *in situ* characterization, particularly in solution, a liquid test cell to accommodate sample, solution, and ellipsometric measurement has to be designed and set up with the ellipsometer. A few relevant considerations in the design may include sample size, solution volume, stirring mechanism, temperature control, optical window property, angle of incidence (AOI, and bear in mind that the incident beam needs to be perpendicular to the window), sample mounting, electrode configuration (in an electrochemical cell), flow control (in a flow cell design), and so on. A recent application of wave guides (Benjamins et al. 2002) to permit a range of AOI through liquid/solid interface can be an interesting approach for use in such *in situ* study. If measurements have to be performed in an opaque solution or medium, an internal reflection approach can be applied as attenuated total reflection (ATR) (Tiwald et al. 1998) or total internal reflection ellipsometry (TIRE) (Rekveld 1997; Poksinski et al. 2000). In these measurements the optical path is critical and the probe beam needs to be reflected at its second interface to achieve sensitivity for layers on the surface. It is also of paramount importance to maintain a "clean" surface without contamination. It is known that when the sample surface passes through a liquid–air interface, a contamination layer can be easily formed by the Langmuir–Blodgett (LB) effect. One should also note that several surface layer formations and phenomena such as the double layer formation and the presence of a stagnant diffusion layer (even under mechanical stirring) could vary the optical function in these interface layers before a steady quasiequilibrium condition is attained in dynamic or transient measurements. Although metallic surfaces such as those with Au, Ag, Ti, Fe, Cu, Cr, Hg, and their oxides when available, are used in the ellipsometric studies, most of the studies are often carried out on Au or oxidized silicon surfaces. Oxidized silicon is of particular interest for macromolecules and porous media, because of its reasonable cost, availability with flat surface, high purity, well-known optical properties, high refractive index, thus high contrast, sensitivity, and resolution. The oxidized silicon surface can be modified chemically to become more hydrophobic or hydrophilic (Arkles 1977; Pluddeman 1980; Welin-Klintström et al. 1993a; Welin-Klintström et al. 1993b) to facilitate the study of molecular interaction on such surfaces. Some interesting studies of protein adsorption in porous oxidized silicon have been demonstrated (Zangooie et al. 1998; Arwin et al. 2000). Another example by using infrared spectroscopic ellipsometry (IRSE) to detect residual water in annealed sol-gel thin films is illustrated in Figure 13.8 (Bruynooghe et al. 1998). By FTIR spectrophotometry, ATR configuration with spectroscopic guided wave absorption, SIMS, and IRSE measurements, residual amounts of water remains in the thermally annealed sol-gel silica samples were shown, and IRSE is the most sensitive technique among all.

The application of ellipsometry to life science has been reviewed by Arwin (2005) recently. The approach used in the study of macromolecular adsorption could be applicable to the situations with porous media. For future application in nanotechnology, ellipsometry with nanoscale sensitivity to follow kinetic measurements at the solid/liquid interface is particularly interesting. Due to the size of microbial species (in the vicinity of micron range), ellipsometric techniques might have limited value for *in situ* studies. However, applications to biofilms remain interesting.

Although the high sensitivity in the nanoscale and *in situ* capability with temporal and spatial resolutions make ellipsometry very attractive as a characterization tool, there are pitfalls that need to be mentioned. One of the drawbacks in the ellipsometry is the prerequisite of very inert, smooth, and flat surfaces that needs to be used as the model surface to permit quantitative

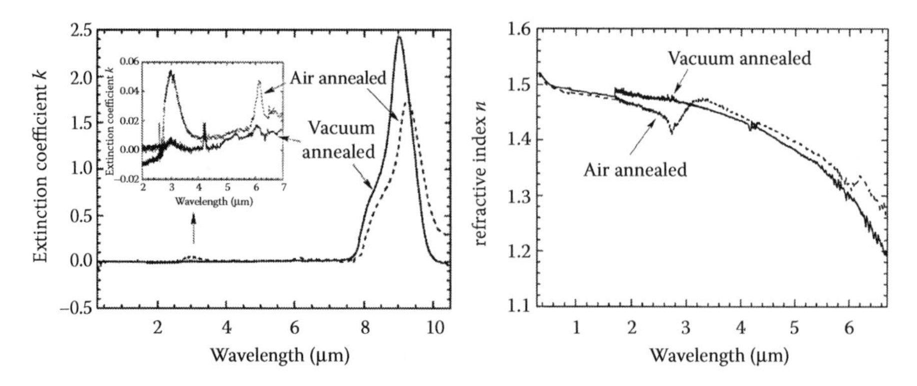

FIGURE 13.8
The extinction coefficient, k, and refractive index, n, versus wavelength for 300 nm thick sol-gel silica films on silicon substrate. One of them was heat treated at 1,000°C for 1 min in air (dashed line), while the other in vacuum (solid line). (Reprinted from Bruynooghe, S., Bertin, F., Chabli, A., Gay, J-C., Blanchard, B., and Couchaud, M., *Thin Solid Film.*, 313–314, 1998. With permission from Elsevier.)

analysis of the data. Only materials that can provide such surface properties are usable, including Si, Cr, Ti (and their respective oxides), and Au. Another challenging issue is the vulnerability to contamination at the surface/interface of interest. Since the information of the contact surface layer is often lacking in most experiments, care should be exercised in interpreting the data in relevance to the chemical composition of the layer on the surface or the interface. In some cases, intended or unintentional surface modification could change the surface property significantly. Reproducibility of the ellipsometric measurements is often a challenging aspect in actual practice. The initial characterization of the "clean" surface may take up a significant effort before engaging the intended study of the surface phenomena. It is particularly challenging for transient or dynamic studies of the kinetic property of such phenomena due to ambiguity in the baseline condition. When dealing with measurements in a liquid cell, there are other limitations in ellipsometry. For instance, the liquid phase and the optical window have to possess a high degree of transparency to the wavelength range used in the experiment to enable the measurements. The optical property of the liquid and the lens also constrain the range of AOI, which in most cases is fixed to simplify the experimental set up. If an electrochemical interface is involved, the space charge region near the surface could become a challenging issue in the data analysis due to the variation in the optics of the interface. Hopefully, these effects might reside as secondary, so the premise of the data interpretation remains valid. The solution optics may depend on solute concentration. It is prudent to check such concentration effects in the initial measurement. Despite

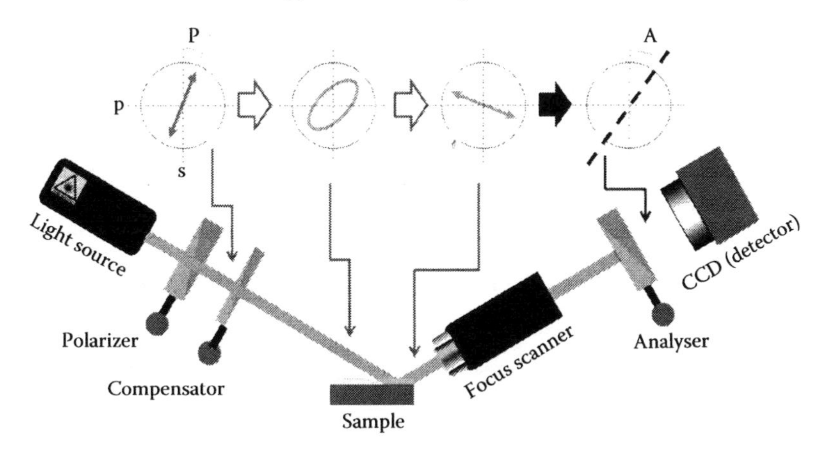

FIGURE 13.9
Schematic of an imaging ellipsometer, showing the polarized laser source and the detection methods. When operating under null ellipsometric condition, the ellipsometric angles (Ψ and Δ) can be measured with a range of AOI to derive d and N. (Courtesy of Accurion, Inc. 2009.)

these potential difficulties, ellipsometry remains as an attractive approach to study interfacial properties and thin porous media, due to the molecular-level sensitivity, noninvasiveness, and *in situ* capability, in combination with well-established optical model, that allows monitoring the dynamic changes at the surface/interface, without labeling the species of interest.

Successful practice of ellipsometry relies on a proper interpretation of the measured results. Although the ellipsometric measurements directly determine the changes in the optics at the interface/surface, the understanding of the phenomenon however comes from the optical model interpretation. Independent verification of the interpreted results from other supplementary data is essential. Such an independent verification is, however, difficult due to limited availability of supplementary data from few other techniques. Fortunately, some earlier work (Benesch et al. 2000) showed that ellipsometry is at least as accurate as other state-of-the-art techniques.

Figure 13.9 shows an imaging ellipsometer schematic, which details the light source and detection mechanism. In the ellipsometric measurements, a pair of ellipsometric angles (Ψ and Δ) is often measured along with wavelength, AOI, time, and spatial references. A monochromatic linearly polarized beam is radiated on and reflected from the sample, while polarization and intensity of the incident and reflected beams determine the sample's Ψ and Δ. Delta (Δ) specifies the phase shift (δ) between the parallel-to-surface (s) and perpendicular (p) components of the incident (δ_i) and reflected (δ_r) beam.

$$\Delta = \delta_i - \delta_r$$

Reflectance is a ratio between the reflected and the incidence beam amplitude. A ratio between the perpendicular and parallel-to-surface reflectance components (R^p and R^s) specifies tangents of the ellipsometric angle Ψ:

$$\tan\Psi = \frac{|R^p|}{|R^s|}$$

In complex values, the relation of Δ and Ψ can be expressed as follows:

$$\tan\Psi e^{j\Delta} = \frac{R^p}{R^s}$$

Although both angles reflect thickness and material properties, practically Δ reflects primarily thickness and Ψ material's properties or processes occurred in the material.

Variations in the incident wavelength, AOI, time, and spatial references give the basis for the spectroscopic, variable angle, dynamic, and imaging ellipsometry, respectively. From Ψ and Δ and optical model analysis, the refractive index ($N = n + ik$), thickness (d), and coverage (θ) of the surface film can be derived. For porous media, a parameter of interest is surface mass density (Γ), which is reduced from thickness and refractive index into a comprehensive, simple parameter.

If the film is sufficiently transparent, so the extinction coefficient, k, is rather constant, $N = n$ can be assumed to make the observation of the film thickness variation much easier directly from the change of the ellipsometric angles Δ. If the k varies with film thickness, the interpretation of the film thickness and optical property becomes more complicated. A common practice to deal with this situation is to use a simple volume averaging approach (Equation 13.1) or the effective-medium theory and approximation (Equation 13.2) to assess the variations in the dielectric function. Thus,

$$\varepsilon = \theta\varepsilon_i + (1-\theta)\varepsilon_{\text{amb}} \tag{13.1}$$

$$1/\varepsilon = \theta/\varepsilon_i + (1-\theta)/\varepsilon_{\text{amb}} \tag{13.2}$$

where the dielectric function $\varepsilon = N^2$ is for the "equivalent" interface layer, while ε_i is for the intrinsic layer composition, and ε_{amb} is for the ambient, respectively; θ is the surface coverage for the thin film. For porous media, θ can be replaced with a volume fraction of the solid media ϕ. This simplified approach is only valid when a simple geometry is involved so the electric field can be treated fairly in a perpendicular and parallel fashion. With oblique incidence, the use of such approximations should be cautious, especially when the pore shape is irregular or the pore size variation is significant. The anisotropy and possible birefringence should be considered in due diligence. To apply effective-medium approximation, the Bruggeman theory (von Bruggeman 1935) is often used to give the following equation for the calculation of the surface coverage θ or the volume fraction of the solid media ϕ (the two are interchangeable in this case):

$$\theta[(\varepsilon_i - \varepsilon)/(\varepsilon_i + 2\varepsilon)] + (1-\theta)[(\varepsilon_{\text{amb}} - \varepsilon)/(\varepsilon_{\text{amb}} + 2\varepsilon)] = 0 \tag{13.3}$$

The porosity $(1-\theta)$ or $(1-\phi)$ can thus be obtained or estimated. One may often find that the derived surface mass density can be a useful and easy-to-comprehend parameter to represent the surface property of the film or porous medium as $\Gamma = d\rho_i\theta$ (or ϕ), where ρ_i is the intrinsic density of the material constitutes the mass of the film or the porous medium. Depending on the purpose of study and experimental conditions, there are variations in the data reduction to reach reliable interpretation (de Feijter et al. 1978; Cuypers et al. 1978; Stenberg et al. 1980; Cuypers et al. 1983; Stenberg and Nygren 1983). More detailed description of these variations can be found in the discussion by Arwin (2005).

A very useful application in the biofuel cell development is to understand the behavior of adsorption of bio- or macromolecules into a porous medium. Due to the shear size of the biomacromolecules, their interaction with a porous medium or thin film creates a complicate boundary that makes the ellipsometric optical model difficult to apply. Microscopically, this interaction between macromolecules and the pore surface makes no difference to the adsorption of such molecules on any model surface. Macroscopically, in the ellipsometric perspective, this interaction can be treated as adsorption into porous medium with volume filling in an effective-medium theory as proposed by Bruggeman (von Bruggeman 1935). Before adsorption, the porous medium needs to be characterized with spectroscopic techniques to obtain the porosity profile and relevant pore size information. There are three constituents involved in the system: the porous medium bulk, the ambient medium, and the macromolecules, of which their optical properties need to be determined. If the porous structure does not possess an iso-symmetric geometry, anisotropic effects due to form birefringence needs to be considered. There might be other anisotropic effects including those arising from ordering of the alignment of the molecules in the adsorption layer (den Engelsen 1976; Sano 1988; Tronin and Konstantinova 1989; Schubert 1996). Another important issue is the surface roughness, which often makes the optical model interpretation difficult due to light scattering that undermines the quantitative sensitivity. As a matter of fact, the presence of macromolecules itself creates an inherent roughness toward light scattering, and such a technical surface roughness is of the order of or larger than the macromolecule. In the case that the lateral roughness is larger than the wavelength of the incident light, the loss of the intensity due to scattering can be determined by the porous structure characterization as reflected in the optical property measurements of the porous medium. Depending on the severity and lateral size of the roughness, this rough boundary can be modeled as an equivalent thin effective-medium layer. Nevertheless, roughness remains a tough issue for quantitative analysis of the adsorption in porous medium with rough surface.

Recently, Murray et al. (2002) made a comparison of complementary techniques among laser-generated surface acoustic waves (LSAW), ellipsometric porosimetry (EP), Rutherford backscattering (RBS), and nanoindentation to study a range of mesoporous xerogel low-k dielectric films for their

TABLE 13.1
Summary by Murray et al. Illustrates the Ability for *in situ* Characterization of Porous Films among Four Complementary Techniques

	Film thickness	**Refractive index (film and skeleton)**	**Mean film density**	**Porosity**	**Pore size distri-bution**	**Surface area**	**Elastic modulus**	**Hardness**	**Atomic density**
LSAW			○	△			○		
EP	○	○	△	○	○	△			○
RBS/ERD			△	△					
Nanoindentation							□	□	

○ = Proven,
△ = Possible,
□ = Proven for dense films.
LSAW = laser-generated surface acoustic waves.
RBS/ERD = Rutherford backscattering/elastic recoil detection.
EP = ellipsometric porosimetry.
Source: Courtesy of Elsevier. Murray et al. (2002).

density, porosity, PSD, cumulative surface area, elastic modulus, and hardness (Table 13.1). They used multiangle single wavelength EP to measure the film refractive index and applying the known skeletal SiO_2 refractive index to calculate the "full" porosity using the Lorentz–Lorenz equation:

$$\pi = 1 - \{[(n_{\text{film}})^2 - 1]/[(n_{\text{film}})^2 + 2]\}/\{[(n_{\text{skeleton}})^2 - 1]/[(n_{\text{skeleton}})^2 + 2]\} \quad (13.4)$$

and in different media of known refractive index using the effective-medium theory. They also measured changes in the ellipsometric angles during desorption of toluene vapor and calculated the "open" porosity that is accessible to the vapor penetration. The comparison of the two porosities can indicate the interconnectivity of the pores. The PSD can also be calculated from the same set of data using the Kelvin and BET equations (Baklanov et al. 2000). By integrating the PSD data for a cylindrical pore model, they derived the cumulative surface area.

More porosimetry techniques have been reported recently by Gidley et al. (2007) regarding the experimental determination of porosity of low dielectric constant materials, particularly in thin film forms. Several important properties of such porous thin films such as density, stiffness, strength, thermal conductivity, and chemical reactivity depend on the pore structure; thus, the determination of the pore structure and porosity of these films are of significant importance to their applications. To achieve the control of pore structure in these films, it is critical to measure the pore structure quantitatively, including porosity, average pore size, PSD, and pore interconnectivity. However, conventional porosimetric techniques for bulk materials such as stereology analysis by microscopic methods, intrusive analysis by gas adsorption, mercury porosimetry, and nonintrusive methods by radiation scattering and wave propagation (Julbe and Ramsay 1996) are hardly applicable to thin films because of the shear size of the total pore volume and surface area, not to mention that these techniques are ex situ. There are several advanced techniques are thus being developed in the past decade, including EP (Dultsev and Baklanov 1999; Baklanov et al. 2000), x-ray porosimetry (XRP) (Lee et al. 2003), small-angle neutron and x-ray scattering (SANS and SAXS) (Wu et al. 2000; Huang et al. 2002; Omote et al. 2003), and positron annihilation lifetime spectroscopy (PALS) (Petkov et al. 1999; Gidley et al. 2000). Here we shall use examples of the deposition of thin polymer films to illustrate the utility available from a few of these techniques.

Controlling deposition of redox polymer films has been of interest in the past few decades (Hamnett and Hillman 1985, 1987; Greef et al. 1989; Redondo et al. 1988; Rishpon et al. 1990; Rubinstein et al. 1990; Rishpon and Gottesfeld 1991; Sabatani et al. 1993a,b; Tjaernhage and Sharp 1994; Gottesfeld et al. 1995; Severin and Lewis 2000; McMillan et al. 2005; Richter and Brisson 2005; Hillman and Mohamoud 2006; Wang et al. 2007), because of their potential in utilizing either passive (e.g., for corrosion protection) or conductive (e.g., for energy conversion and storage systems or electronics) properties for many practical applications. The recent rapid expansion of interests on nanomaterials and fabrications demands better temporal and spatial characterization techniques to facilitate research and development. However, only few nano- or meso-scale characterization techniques (such as atomic force microscope, or AFM; e.g., Picart et al. 2007; Xu and Siedlecki 2007; Zhitomirsky and Hashambhoy 2007) are available for *in situ* observations, particularly, in solution. In addition, few can easily allow any moderate to large area observation with respect to experimental conditions in transient. Thus, any *in situ*, nonintrusive technique to offer such utility appears very attractive, especially for surface or interface characterizations in liquid environments. Here we present some simple illustrations to show that imaging ellipsometric measurements could offer useful utility for studying conductive polymer film depositions.

The first example is the deposition of conductive PPy films on Pt or glassy carbon electrode surface. Depending on the surface condition, the PPy film could have different morphological appearances. It also depends on the deposition conditions, such as monomer concentration, electrolyte species and its concentration, current density or working potential. In Figure 13.10, an example is shown. By varying the current density (thus, with different deposition rates), films of different density and morphology can be developed (Liaw et al. 2005).

In another example (Figure 13.11), as shown by the deposition of polymeric methylene green, using the imaging ellipsometric capability, Svoboda et al. (Svoboda et al. 2007) show that a dense film of the polymer can be obtained, judging from the retention of the microstructure and roughness of the underlying Pt surface. Methylene green is an important mediator for NADH (the reduced form of nicotinamide adenine dinucleotide or NAD^+) reoxidation in biofuel cell operation, and the polymeric form of this azine has been suggested (Zhou et al. 1996; Akkermans et al 1999; Karyakin et al. 1999) more effective in performing such a function. Without the shuttling of electrons by the coenzyme NAD^+–NADH redox reaction and the poly-azine to transfer the electrons to the electrode surface, the energy harnessing from the

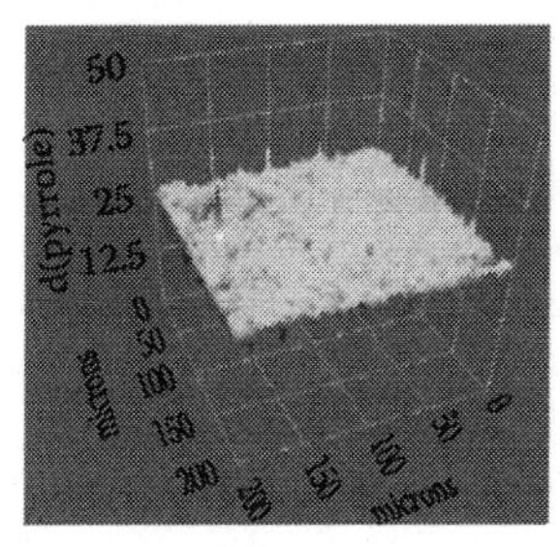

0.5 mA/cm^2

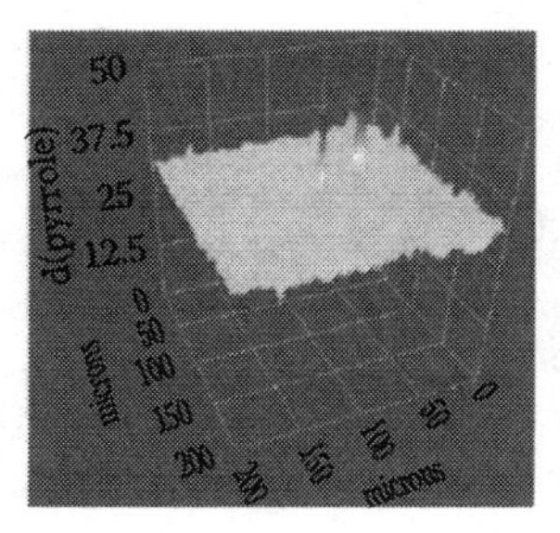

1 mA/cm^2

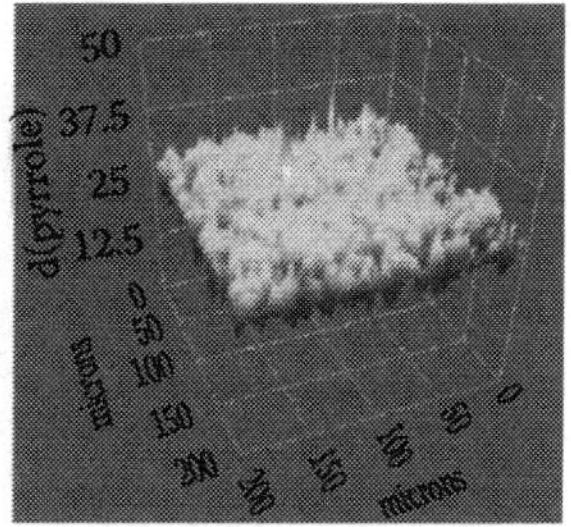

2 mA/cm^2

FIGURE 13.10
Polypyrrole deposition using galvanostatic technique at various current densities was shown to produce different morphologies and densities in the films.

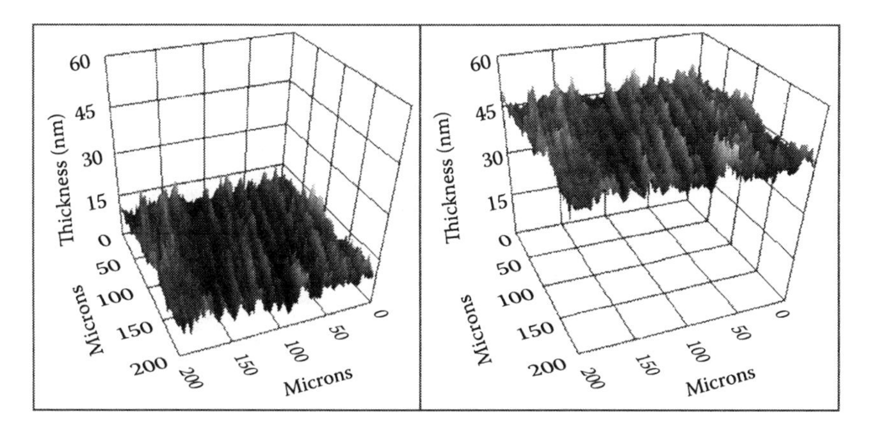

FIGURE 13.11
Surface morphology of a bare Pt electrode surface and the subsequently deposited poly-methylene green film about 36 nm thick. (Courtesy of the Svoboda et al., *Electrochem. Soc.*, 2007.)

oxidation of the biofuels (e.g., methanol, ethanol, lactate, glucose, etc.) cannot be facilitated. However, the actual deposition process of the poly-azine is much more complicated than the images show. When combining with microgravitational measurements using QCM during the cyclic voltammetric deposition of the film (Figure 13.12), the adsorption of a precursor methylene green layer on the electrode surface was observed (Figure 13.13).

This phenomenon complicated the derivation of the ellipsometric data significantly. The evidence of the adsorption was shown in Figure 13.14, which shows that the progression of the mass change is not in sync with that of the current. While the ellipsometric angle, Δ synchronizes with the mass change, Ψ shows with a different profile, reflecting the nature of the redox reaction involved in the cyclic voltammetry. More detailed discussion of this poly-methylene green deposition process has been reported by Svoboda and Liaw (2008). The mechanism of the deposition can be modeled by estimating the thickness of the solid poly-methylene green film and the methylene green adsorption layer on the surface with their respective optical parameters. Figure 13.15 shows the results of such model simulations at four distinct points of a cyclic voltammetric deposition to illustrate the changes in the optical property due to the thickness changes in the polymer and the adsorption layer, respectively. This illustration highlights the benefit of such *in situ* characterization and capability from the imaging spectroscopic ellipsometry in providing such a superior resolution and accuracy in a noninvasive approach.

13.5.2 Quartz Crystal Microbalance

Quartz crystal microbalance can measure nanograms of mass change on an electrode surface (Buttry 1991; O'Sullivan and Guilbault 1999). A QCM consists of a thin quartz disk with electrodes of defined area and property

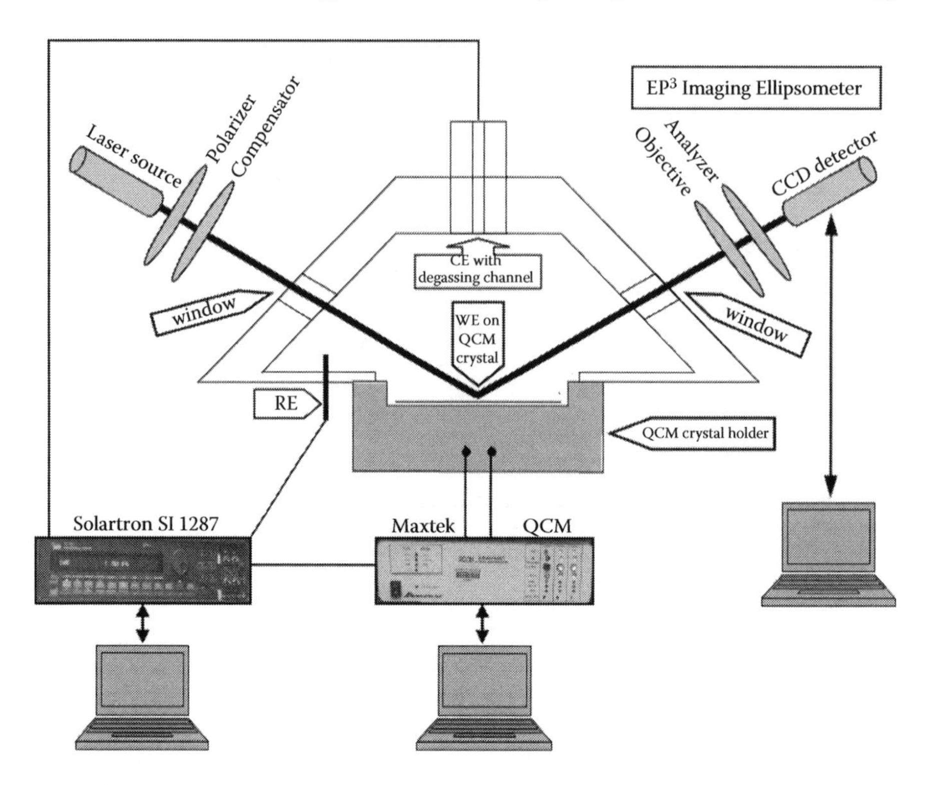

FIGURE 13.12
The experimental setup for the combined electrochemical microgravimetric imaging ellipsometric measurements that can correlate the changes in mass, electric charge, surface film thickness, optical property, and morphology. (Figure 1 in Svoboda, V. and Liaw, B.Y., *Pure App. Chem.*, 80, 2008. IUPAC. 2008.)

as a device for sensitive mass detection. The application of an external oscillating electrical field on the quartz disk, which is a piezoelectric material, will induce an acoustic wave that propagates through the disk with the electrode. When the thickness of the disk is a multiple of a half wavelength of the acoustic wave, the device will exhibit the minimum impedance and create an oscillation resonance with a standing shear wave. The ratio of frequency and bandwidth, called *quality factor* or *Q factor*, can be as high as 10^6, which provides the sensitivity in the gravimetric measurement. A common instrument of this type can offer resolution down to less than 1 Hz with a baseline resonant frequency on the order of 4–6 MHz. A typical setup for the QCM includes water cooling tubes, a retaining unit, a frequency sensing unit, an oscillation source, and a measurement and recording device. Since the shear-mode oscillation acoustic wave propagates in a direction perpendicular to the crystal surface, the quartz crystal has to be cut along a few commonly used orientations, so-called the

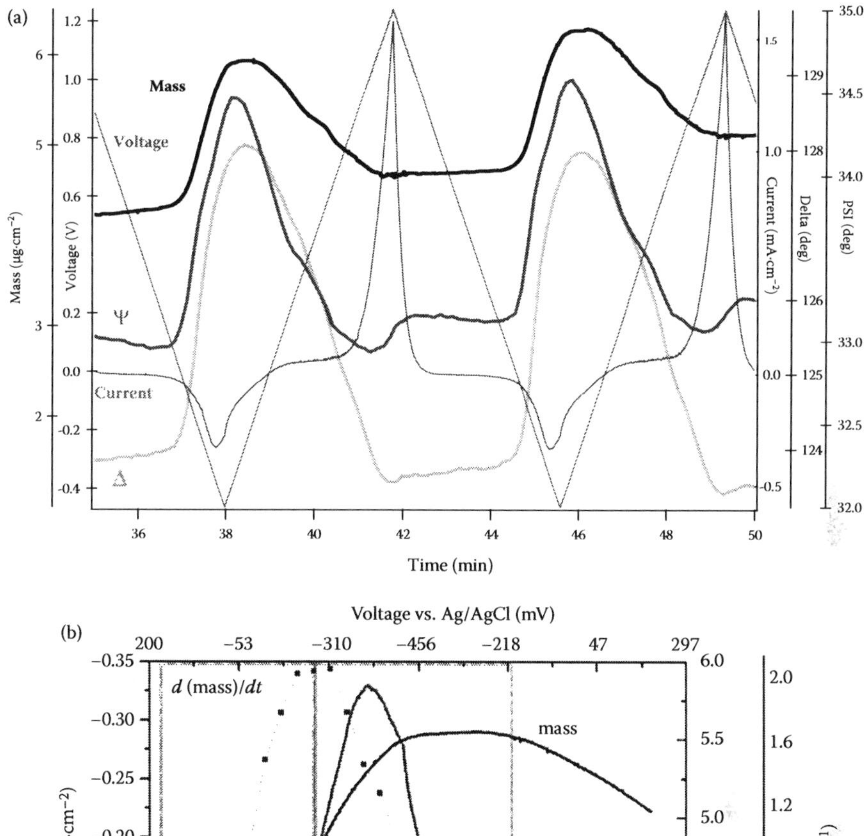

FIGURE 13.13
Combined ellipsometric, quartz crystal microbalance, and cyclic voltammetric measurements during the deposition of poly-methylene green film in a temporal correlation. (a) The mass change-voltage-current density-ellipsometric angles correlation in two consecutive cycles and (b) correlated distinct adsorption and reduction regimes in the redox cycle.

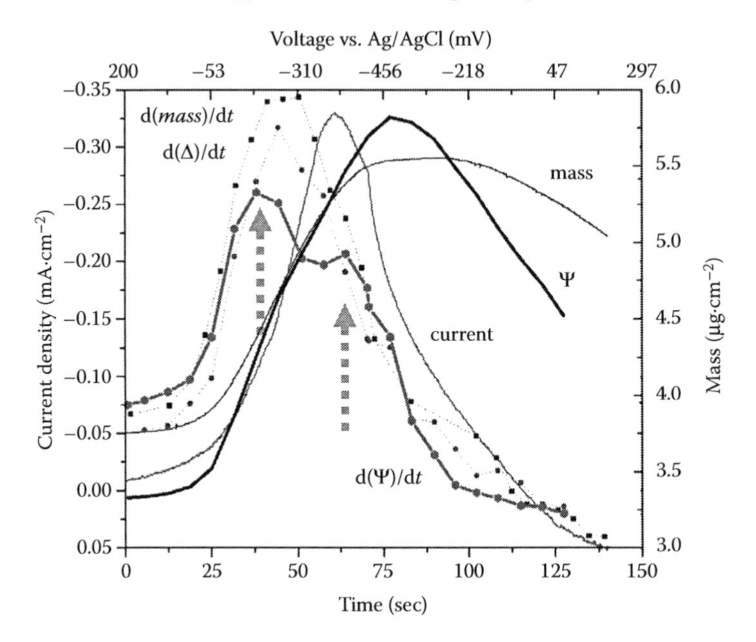

FIGURE 13.14
Correspondence of changes in ellipsometric angles, mass, and current, showing that the physical process (adsorption) and reduction-induced chemical changes can be detected in the combined electrochemical microgravimetric ellipsometric measurement which gives the temporal and spatial (thickness) resolution to reveal the physical and chemical processes involved in the poly-methylene green deposition.

AT-, BT-, and rotated Y-cuts, in the crystal structure. The oscillation frequency is related to the crystal thickness and the crystallographic orientation. Deposition of a thin layer of material on the crystal surface will affect the oscillation frequency of the crystal in proportion to the mass deposited, a relationship described by the Sauerbrey's equation (Sauerbrey 1959):

$$\Delta f = -2f_o^2 \cdot \Delta m/A \cdot \rho_q^{1/2} \cdot \mu_q^{1/2} \tag{13.5}$$

where Δf is the frequency change, f_o the resonant frequency of the quartz crystal, Δm the mass change, A the piezoelectrically active crystal surface area, ρ_q the density of quartz, μ_q the shear modulus of quartz for AT-cut crystal ($\mu_q = 2.947 \times 10^{11}$ g/cm·sec^2). It should be noted that the Sauerbrey's equation is strictly applied only to rigid mass on the surface. To perform QCM in a liquid, a viscosity-related frequency change will be observed (Kanazawa and Gordon 1985). In such a case, the frequency change should follow:

$$\Delta f = f_o^{3/2} \cdot (\eta_l \cdot \rho_l/\pi \cdot \rho_q \cdot \mu_q)^{1/2} \tag{13.6}$$

where ρ_l is the density and η_l the viscosity of the liquid.

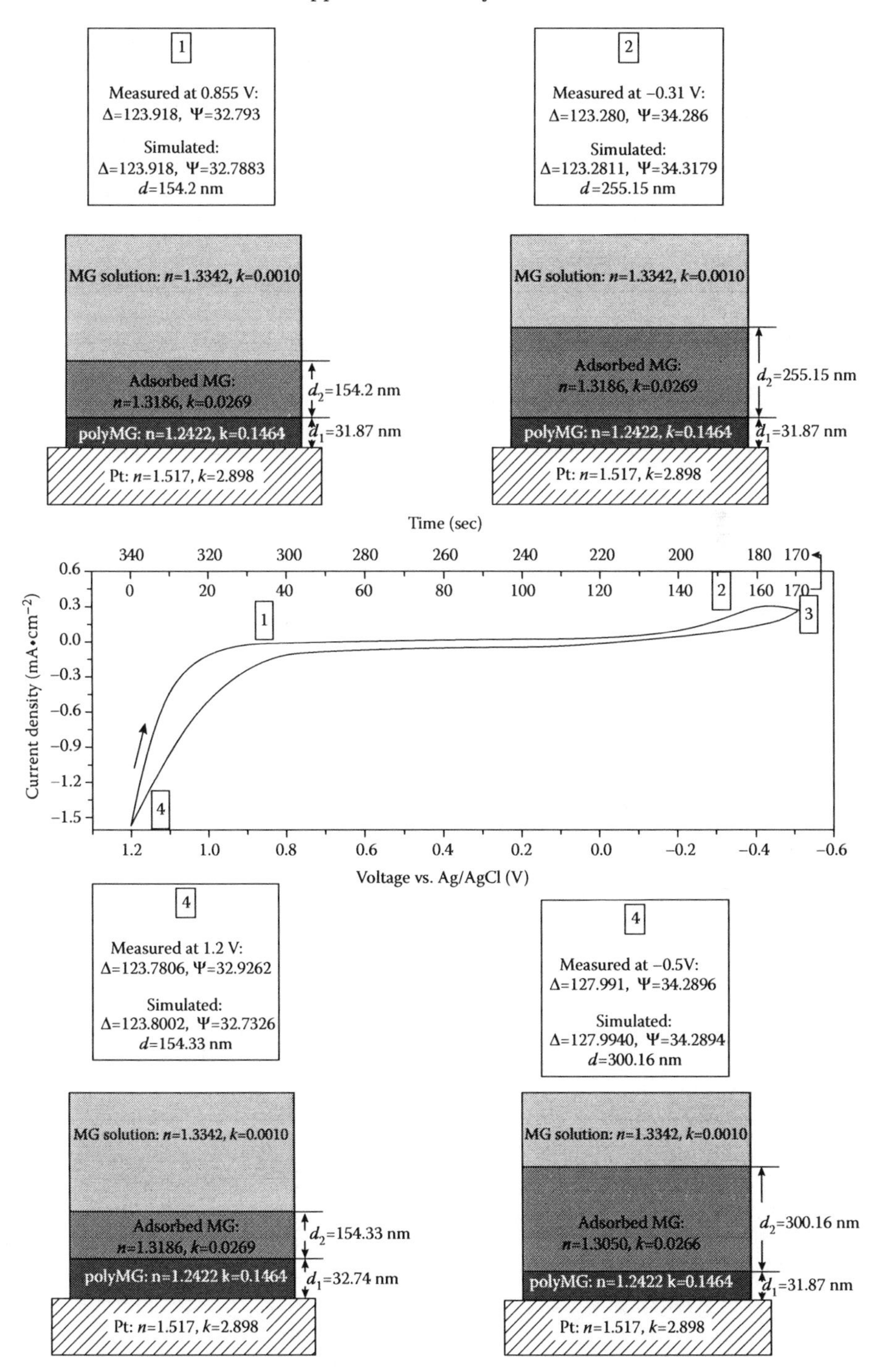

FIGURE 13.15
An optical model simulating the adsorption layer and polymer film deposition in progression in the solution using cyclic voltammetric method.

Nivens et al. (1993) have used this technique in the study of microbial biofilm growth. Since the observation was conducted over a long period of time, they have to remove the baseline drift due to the fluctuations of hydrostatic pressure and temperature in the liquid test cell. They thus generated a calibration curve for the frequency shift corresponding to the number of bacteria within the biofilm with a detection limit of 3×10^5 cells cm^{-2}. With this calibration curve, they can study the number of attached cells versus time and derive the rate of biofilm formation. Marx (2003) published a recent review on QCM for studying thin polymer films and complex biomolecular systems at the solution-surface interface. The wide detection range makes QCM attractive for studies ranging from detection of monolayer surface coverage to measuring much larger mass bound to surface in solution that contains either biopolymers or biomacromolecules. The QCM technique also provides information about the energy dissipating properties of the bound surface mass. In combination with electrochemical techniques, one can study the ion or solute transport in a film during changes in the film environment or (chemical and/or electrochemical) state. The molecular systems include micellar systems, self-assembling monolayers (and their phase transition behavior), molecularly imprinted polymers, chemical sensing systems, films made by layer-by-layer (LBL) assembly technique, and biopolymer films.

There are numerous attempts in the past to use various combinations of these *in situ* techniques (even with other spectroscopic techniques) to investigate film deposition and the resulting film properties. For example, Hamnett and Hillman used spectroscopic ellipsometry to study electrodeposition of poly-thionine and -thiophene in solutions (Hamnett and Hillman 1985, 1987). They were able to calculate film composition and thickness with optical parameters (n, k) accounted for solvent ingress in the film development, thus fully explored the utility of the spectroscopic ellipsometry with electrochemical techniques. They, however, did not use QCM to help them reap out the adsorption effect that should be accounted for in the optical model. Gottesfeld and his coworkers (Redondo et al. 1988; Rishpon et al. 1990; Rubinstein et al. 1990; Rishpon and Gottesfeld 1991; Sabatani et al. 1993a,b; Gottesfeld et al. 1995) and many others (e.g., Tjaernhage and Sharp 1994; Severin and Lewis 2000; McMillan et al. 2005; Richter and Brisson 2005) attempted to utilize various combinations of these techniques to study conjugated conductive polymers. In general, it is important to point out that the ellipsometric angles not only change with a material's thickness but also reflect its optical, electrical, and other physical property changes. Probing evolutions of ellipsometric angles simultaneously with imaging of surface morphological development and synchronously with nanoscale measurement of mass changes in relation to other experimental parameters and conditions (such as those imposed by electrochemical cyclic voltammetry [Rusling and Suib 1994]) to reduce the information into detailed temporal and spatial resolutions is a very powerful tool to probe reaction kinetics or surface phenomena at a (solid–liquid or solid–gas) interface.

13.5.3 X-Ray Spectroscopic Techniques

X-ray spectroscopic techniques (XRD, XAS, XRF, and XRR) offer a variety of utilities for studying functional materials related to crystal structure, fine structure for nearest-neighbor chemical bonding, and other relevant physical and chemical information from atomic to micrometer scale. X-ray spectrometry has been reviewed by Szalóki et al. (2000). For *in situ* investigations, XRD has been used to study crystal structure changes and phase transformations in topotactic electrochemical reactions in battery electrodes quite commonly to date (e.g., Yang, Sun, McBreen 2000; Dubarry et al. 2008). Russell and Rose (2004) provided a comprehensive review about XAS for the study of fuel cell catalysts is a good example of using *in situ* investigations to reveal the catalyst behavior in a working condition. The behavior of the catalyst can be studied as a function of particle size, composition, and morphology in porous structure. Analysis of the x-ray absorption near-edge structure (XANES) can provide the information of the oxidation state of the catalyst center, and the extended x-ray absorption fine structure (EXAFS) can reveal the short-range crystal structure information of the reaction site. This type of information can be valuable to the understanding of the reaction pathway and mechanism. These techniques are, however, not designed to study the porous structure regarding pore size, porosity, and pore distribution.

X-ray scattering at glancing angles known as grazing incidence x-ray reflectometry (GIXR) is a powerful tool for thin film analysis (Stoev and Sakurai 1997). Similar in principle to XRD and ellipsometry, x-ray reflectometry (XRR) is a nondestructive and noninvasive technique commonly for thin film thickness determination between a few to several hundreds nm with a precision of about 1–3 Å. XRR takes the advantage of surface reflectivity changes due to film coverage or roughness by measuring the intensity of X-rays reflected from a surface as a function of the AOI. Thin films on a surface can give rise to oscillations of the x-ray intensity with the AOI. This technique is therefore useful for the determination of density and roughness of films and multilayers with a high precision. Holy, Pietsch, and Baumbach (Holy et al. 1999) provide an excellent discussion of the theoretical basis of the x-ray scattering measurements and model simulation. Recent developments in the three-dimensional micro XRF analysis, which employ two aligned confocal x-ray optics, provides interesting opportunities for three-dimensional imaging and quantitative analysis of a variety of samples, ranging from art pieces to geological and biological species (Malzer 2006). The visualization of major, minor, and trace constituents in these samples offers a powerful tool for nondestructive, *in situ* studies quantitatively (Janssens et al. 2004; Vincze et al. 2004). Illustrated in Figure 13.16 is an example by Briscoe et al. (2007) using a bending mica method to enable a proof-of-principle XRR measurements on three very different systems: (a) a Cr–Au nanofilm thermally evaporated in high vacuum on oxygen plasma-cleaned mica, (b) a surface-grown zwitterionic polymer brush made from poly(2-methacryloyloxyethyl phosphorylcholine) (pMPC) via atomic transfer radical polymerization from mica preinitiated with an adsorbed random copolymer, and (c) a LB monolayer of the phospholipid 1,2-dipalmitoylphosphatidylcholine (DPPC). The different film properties and morphologies gave distinct optical patterns in the measurements and analyses. By fitting the obtained reflectivity curves with the standard Parratt algorithm (Parratt 1954) they were able to extract the structural information of the nanofilms on thickness and apparent roughness.

For biofuel cells and biosensors, bioelectrodes require more complex designs than what LB films can offer. The seminal work by Decher and his coworkers (Lvov et al. 1993; Schmitt et al. 1993; Decher 1997) opened a new possibility of multilayer architecture and LBL fabrication for bioelectrodes. They show that multilayer structures can be tailored with careful control of polymeric or macromolecular assemblies in the architecture to fabricate bioelectrodes with unique properties. This approach has been widely adopted by many (e.g., Sun et al. 1999) to fabricate bioelectrodes for various applications. XRR, sometimes in combination with neutron reflectometry (NR), have been demonstrated to be quite useful in understanding the internal structure in the LBL adsorbed polyelectrolyte films (Schmitt et al. 1993; Lvov et al. 1993). Hollmann et al. used XRR in combination with NR and total internal reflection fluorescence (TIRF) to study the structure and protein bovine serum albumin (BSA) binding capacity on a planar polyacrylic acid (PAA) brush (Hollmann et al. 2007), which is an example illustrating a delicate control and optimization of protein immobilization for biological functions.

Another interesting approach is the design and synthesis of inorganic polymer hybrid materials and structures, most in thin films, that posses periodically organized nanoporosity in this class of so-called "periodically organized mesoporous materials and thin films" (POMMs and POMTFs). Sanchez and his coworkers recently presented an excellent review (Sanchez et al. 2008) on this class of materials and characterization techniques used to understand their formations and resulting properties. The tuning of the interface between the inorganic template and the polymerizing phase and the control over chemical and processing conditions are the key to producing tailor-made POMTFs with a high degree of reproducibility. XRR has been used with other modern analytical tools, including two-dimensional grazing incidence small-angle x-ray scattering (GISAXS), ellipsoporosimetry, high-resolution transmission electron microscopy (HRTEM), WAXS, time-resolved infrared spectroscopy, SAW, and optically polarized xenon NMR; to provide temporal and spatial resolutions to help researchers understand the film formation processes and the mesostructural properties. This class of materials utilizes the intrinsic physical and chemical properties of the inorganic or hybrid matrices in combination with a highly defined nanoporous network of a tunable pore size and connectivity, high surface area and accessibility, and a specific orientation with respect to the substrate. Therefore, POMMs and POMTFs are a promising class of advanced materials for a variety of future applications as biomaterials, bioelectrodes, and for biomicrofluidics, among others.

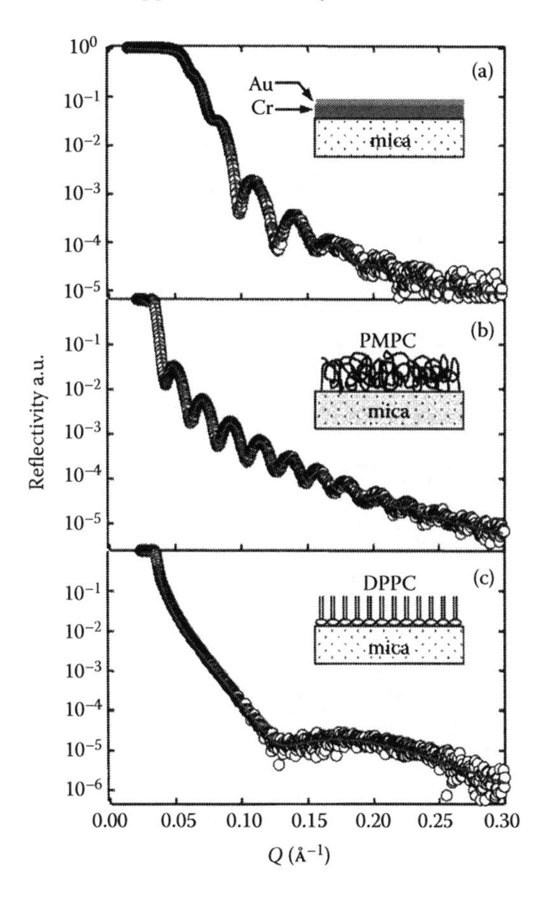

FIGURE 13.16
Reflectivity versus momentum transfer $Q = (4\pi/\lambda)\sin\theta$ for three nanofilms on mica in air, as schematically illustrated in the insets: (a) ~5 nm Au and ~20 nm Cr thermally evaporated on plasma treated mica in high vacuum; (b) ~28 nm surface grown zwitterionic polymer brush pMPC (poly (2-methacryloyloxyethyl phosphorylcholine)); and (c) a ~2.5 nm DPPC (1,2-dipalmitoylphosphatidylcholine) phospholipid Langmuir–Blodgett (monolayer) deposited on mica. The open circles are the experimental data, and the gray solid curves are computed using the fitting parameters reported by Briscoe et al. (Reprinted from Briscoe, W.H., Chen, M., Dunlop, I.E., Klein, J., Penfold, J., and Jacobs, R.M.J., *J. Coll. Inter. Sci.*, 306, 2007. With permission from Elsevier.)

The rapid advancement in the x-ray spectroscopic techniques can compliment ellipsometric techniques to provide a wide range of information regarding chemical and physical properties of thin films and porous media, offering a unique approach for *in situ*, noninvasive characterizations.

13.5.4 Other Spectroscopic Techniques

Confocal microscopy is an optical imaging technique employing in-focus processing through a pinhole to eliminate out-of-focus light in a specimen that is outside the focal plane. The ability to increase micrograph contrast and to reconstruct three-dimensional images makes this technique popular in the studies of life sciences and semiconductor industry. In a conventional wide-field fluorescence microscope, a light source illuminates the entire specimen. The resulting fluorescence from the specimen is detected by the photodetector or camera to construct the image. In a confocal microscope, a laser light source provides a sharp small-area illumination and a pinhole in an optically conjugate plane in front of the detector is used to eliminate out-of-focus light signals. Therefore, in a confocal configuration, only light produced by fluorescence very close to the focal plane is detected for image resolution, particularly in the sample depth direction. The image resolution of the confocal system is much better than that of a wide-field microscope. The tradeoff is a longer exposure time required to collect the light signals from the physical scanning process. There are three major types of confocal microscopes available commercially. LSCM usually provide better imaging quality than spinning-(Nipkow) disk confocal microscopes (SDCM) or programmable array microscopes (PAM) at a lower frame rate. SDCM can achieve higher frame rate for video, thus it is preferred for dynamic observations in live cells. To apply the fluorescence technique, a specimen is usually labeled with a fluorescent molecule called a fluorophore (e.g., green fluorescent protein or fluorescein) to enable light detection. A typical fluorescence microscope will comprise a light source, an excitation filter, a dichroic mirror (dichromatic beamsplitter), an emission filter, and a photodetector. The labeled specimen is illuminated with light of a specific wavelength (e.g., a laser source) or a range of wavelengths (e.g., a xenon arc lamp or mercury-vapor lamp). The fluorophore absorbs the light and then emits longer wavelengths of light (fluorescence). An emission filter will screen the spectrum for photo detection. Through proper control of the filters and dichroic mirror, different wavelengths of light can be detected and imaged with a distinct color. Different color images can be overlaid and reconstructed into a two-dimensional or three-dimensional image to display the specimen's unique property through fluorophore labeling. The best resolution below 10 nm can be achieved to date with a confocal system combining localization microscopy with spatially modulated illumination. Most of the fluorescence microscopes are epifluorescence types, in which excitation and observation of the fluorescence is from above (epi–) the specimen.

Figure 13.17 illustrates an experiment using an epifluorescence LSCM to understand the interaction of enzyme and porous polymer backbone during immobilization (Konash et al. 2006). In Figure 13.17a the distribution of the alcohol dehydrogenase labeled with Alexa 488 near the surface of an Eastman AQ 55 polymer film was shown. Using the advantage of depth profiling, the images of the distribution along the thickness of the film can be stacked to reconstruct the three-dimensional image for visualization (Figure 13.17b).

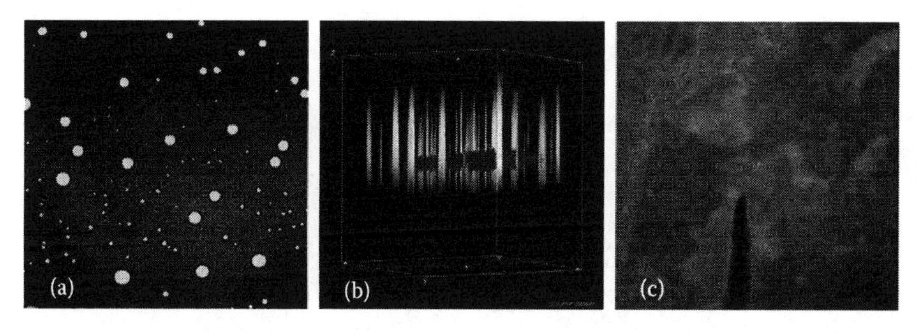

FIGURE 13.17
(a) A confocal microscopic images of alcohol dehydrogenase labeled with Alexa 488 and entrapped in Eastman AQ 55 polymer film, (b) a three-dimensional reconstructed confocal image showing the depth profile of the polymer film in (a), and (c) similar labeled enzymes entrapped in a modified Nafion film. (Konash, A., Cooney, M.J., Liaw, B.Y., and Jameson, D.M., *J. Mater. Chem.*, 16, 2006. Reproduced by permission of the Royal Society of Chemistry.)

Interestingly, columns of liquid-containing pore channels where the enzymes reside can be easily identified and visualized. These liquid columns open to the surface of the film as the circular openings (in white) on the image, as shown in Figure 13.17a. The same labeled enzymes were also entrapped in a modified Nafion film and their fluorescence image is shown in Figure 13.17c. In contrast, the Nafion film has network of micropores and water channels in small clusters, which is very different from the Eastman AQ backbone. Similarly, the LSCM technique has been used in the study of biofilms and their interactions with microorganisms in the production of electricity. Figure 13.18 shows examples of the investigations on *Geobacter sulfurreducens* biofilms attached to Au electrodes (Richter et al. 2008). In contrast to an early study (Bond and Lovley 2003) that used graphite as the electron accepting electrode, to which *G. sulfurreducens* can be attached directly and sustain current production; bare Au electrode seems to inhibit the electron transfer. On the other hand, with the growth of *G. sulfurreducens* biofilms up to 40 μm thick on the Au electrode surface, the current production was invigorated. No current was produced if *pilA*, the gene for the structural protein of the conductive pili of *G. sulfurreducens*, was deleted. The biofilm that formed after 10 days of growth (Figure 13.18a) covered most of the gold surface with pillars up to 12 μm high. Most of cells in the biofilm stained green on the microscope (medium to darker gray in the photo), indicating that most of the cells were metabolically active. With longer incubation (Figure 13.18b), the biofilm became thicker, ca. 40 μm, and more uniform. A high proportion of the cells in these older biofilms stained red on the micriscope (medium gray in the photo), suggesting that they had compromised membranes and might not be metabolically active. The utility revealed by the LSCM study of the biofilm property and the electron transfer

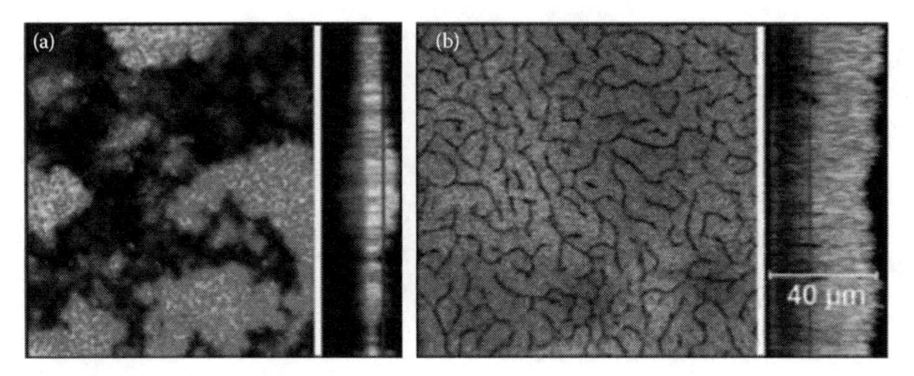

FIGURE 13.18
Confocal laser scanning microscope images of *Geobacter sulfurreducens* biofilms on gold electrodes. Magnification: 250×. The large images are top views on the biofilm. The smaller images are orthogonal cross-sections with the gold attached side of the biofilm on the left, the outer surface on the right. The vertical dark gray line is the plane in which the top view image was taken. (a) Biofilm from a gold electrode after 10 days. (b) Biofilm from another gold electrode after 18 days. (Reprinted from Richter, H., McCarthy, K., Nevin, K.P., Johnson, J.P., Rotello, V.M., and Lovley, D.R. *Langmuir,* 24, 2008. With permission from Elsevier.)

mechanism provides a good illustration of the increasing use of fluorophore-tagged imaging in the biofuel cell study.

13.6 Future Directions

Through a few interesting illustrations several spectroscopic techniques have been demonstrated and discussed. Properly employing these techniques one can explore, study, and characterize surface and interface properties and phenomena (such as adsorption and molecular interactions) in thin films or porous media. In combination with electrochemical and microgravimetric measurements, a fascinating *in situ,* noninvasive approach can be constituted to characterize thin films and porous structures and to monitor the evolution of physical and chemical changes from surface or interface to the bulk in details. In the wake of flourishing nano- and biotechnology developments, it is even more attractive to employ such an approach to characterize materials and biological species in relation to their functions and applications than ever.

As we have described earlier, more trendy developments and attentions have been directed to nanobiomaterials and controlled microstructures.

Advancements we observed from LB films to LBL polyelectrolyte multistructures (Lvov et al. 1993; Schmitt et al. 1993; Decher 1997), to composite inorganic polymer hybrids with periodically organized mesoporous materials (Sanchez et al. 2008), and to polymer brushes (Whiting et al. 2006; Hollmann et al. 2007) only indicate the desires and anticipations toward nanobiomaterial fabrications and optimizations for potential applications in biotechnology, biomedicine, bioenergy conversion and storage, and bioremediation of our polluted environment. To further such advancements, *in situ* nondestructive characterizations, syntheses, and controls of the fabrication processes are of paramount importance to the next generation of biotechnology and nanotechnology paradigm.

Emerging imaging capability and instrumentation could expand the horizon of the spectroscopy to provide fine temporal and spatial resolutions as a nondestructive tool for characterization. For instance, Gustafsson recently showed that structured illumination can enhance the lateral resolution of wide-field fluorescence microscopic imaging (Gustafsson 2000). The combined capability with confocal microscopy and other imaging techniques should be explored more in the porous media investigations for biofuel cell applications to help us understand the dynamic nature of the species interaction and transport involved in the system and to provide a vast amount of physical and chemical information about the porous media down to the nanometer scale. Such ability should be integrated for future applications in the control of porous media for biofuel cell applications, potentially through self-assembly, nanofabrication, smart microstructure, and precise surface modification. It should be cautioned though that some techniques such as the spectroscopic ones are more related to molecular and microscopic scale observations, while other techniques such as the electrochemical and microgravimetric ones are providing microscopic measurements. Therefore, the interpretation of the data and discussion of their relevance needs to be carefully evaluated to avoid confusions.

13.7 References

Akers, N. L., Moore, C. M., and Minteer, S. D. (2005). Development of alcohol/O_2 biofuel cells using salt-extracted tetrabutylammonium bromide/Nafion membranes to immobilize dehydrogenase enzymes. *Electrochimica Acta,* **50**:2521–2525.

Akkermans, R. P., Roberts, S. L., Marken, F., Coles, B. A., Wilkins, S. J., Cooper, J. A., Woodhouse, K. E., and Compton, R. G. (1999). Methylene green voltammetry in aqueous solution: studies using thermal, microwave, laser, or ultrasonic activation at platinum electrodes. *Journal of Physical Chemistry B,* **103**:9987–9995.

Allen, R. M. and Bennetto, H. P. (1993). Microbial fuel-cells: electricity production from carbohydrates. *Applied Biochemistry and Biotechnology,* **39/40**:27–40.

Arkles, B. (1977). Tailoring surfaces with silanes. *Chemtech,* **7**:766–778.

Arwin, H. (2005). *Ellipsometry in Life Sciences.* Chapter 12 in *Handbook of Ellipsometry,* William Andrew Pub, Norwich, NY, pp. 799–855.

Arwin, H., Gavutis, M., Gustafsson, J., Schultzberg, M., Zangooie, S., and Tengvall, P. (2000). Protein adsorption in thin porous silicon layers. *Physica Status Solidi A: Applications and Materials Science (a),* **182**:515–520.

Atanassov, P., Apblett, C., Banta, S., Brozik, S., Calabrese Barton, S., Cooney, M., Liaw, B. Y., Mukerjee, S., Minteer, S. D. (2007). Enzymatic biofuel cell. *Interface,* **16**:28–31.

Baklanov, M. R., Mogilnikov, K. P., Polovinkin, V. G., and Dultsev, F. N. (2000). Determination of pore size distribution in thin films by ellipsometric porosimetry. *Journal of Vacuum Science and Technology B,* **18**:1385–1391.

Benesch, J., Askendahl, A., and Tengvall, P., (2000). Quantification of adsorbed human serum albumin at solid interfaces: a comparison between radioimmunoassay (RIA) and simple null ellipsometry. *Colloids and Surfaces B: Biointerfaces,* **18**:71–81.

Benjamins, J-W., Jönsson, B., Thuresson, K., and Nylander, T. (2002). New experimental setup to use ellipsometry to study liquid–liquid and liquid–solid interfaces. *Langmuir,* **18**:6437–6444.

Bianco, P. (2002). Protein modified- and membrane electrodes: strategies for the development of biomolecular sensors. *Reviews in Molecular Biotechnology,* **82**:393–409.

Bond, D. R. and Lovley, D. R. (2003). Electricity production by *Geobacter sulfurreducens* attached to electrodes. *Applied and Environmental Microbiology,* **69**:1548–1555.

Bond, D. R., Holmes, D. E., Tender, L. M., and Lovley, D. R. (2002). Electrode-reducing microorganisms that harvest energy from marine sediments. *Science,* **295**:483–485.

Briscoe, W. H., Chen, M., Dunlop, I. E., Klein, J., Penfold, J., and Jacobs, R. M. J. (2007). Applying grazing incidence X-ray reflectometry (XRR) to characterising nanofilms on mica. *Journal of Colloid and Interface Science,* **306**:459–463.

Bruynooghe, S., Bertin, F., Chabli, A., Gay, J-C., Blanchard, B., and Couchaud, M. (1998). Infrared spectroscopic ellipsometry for residual water detection in annealed sol-gel thin layers. *Thin Solid Films*, **313–314**: 722–726.

Buttry, D. (1991). Application of the QCM to electrochemistry. In *Series of Advances in Electroanalytical Chemistry*, ed. A. J. Bard, Dekker, New York, pp. 23–33.

Calabrese-Barton, S., Gallaway, J., and Atanassov, P. (2004). Enzymatic biofuel cells for implantable and microscale devices. *Chemical Review*, **104**:4867–4886.

Chaudhuri, S. K. and Lovley, D. R. (2003). Electricity generation by direct oxidation of glucose in mediatorless microbial fuel cells. *Nature Biotechnology,* **21**:1229–1232.

Cheng, S., Liu, H., and Logan, B. E. (2006). Increased power generation in a continuous flow MFC with advective flow through the porous anode and reduced electrode spacing. *Environmental Science & Technology,* **40**:2426–2432.

Cooney, M. and Liaw, B. Y. (2008). In situ characterization techniques for design and evaluation of micro and nano enzyme-catalyzed power sources. In *Nano-scale Science and Technology in Biomolecular Catalysis*, eds., H. S. Kim, J. B, Kim, and P. Wang, ACS Symposium Series 986, Chapter 19, The American Chemical Soc., Cleveland, OH, pp. 289–333.

Cooney, M. J., Lau, C., Windmeisser, M., Liaw, B. Y., Klotzbach, T., and Minteer, S. D. (2008). Design of chitosan gel pore structure: towards enzyme catalyzed flow-through electrodes. *Journal of Materials Chemistry,* **18**: 667–674.

Cuypers, P. A., Corsel, J. W., Janssen, M. P., Kop, J. M. M., Hermens, W. Th., and Hemker, H. C. (1983). The Adsorption of prothrombin to phosphatidylserine multilayers quantitated by ellipsometry. *Journal of Biological Chemistry,* **258**:2426–2431.

Cuypers, P. A., Hermens, W. Th., and Hemker, H. C. (1978). Ellipsometry as a tool to study protein films at liquid–solid interfaces. *Analytical Biochemistry,* **84**:56–67.

de Feijter, J. A., Benjamins, J., and Veer, F. A. (1978). Ellipsometry as a tool to study the adsorption behavior of synthetic and biopolymers at the air-water interface. *Biopolymers,* **17**:1759–1772.

De Lacey, A. L., Detcheverry, M., Moiroux, J., and Bourdillon, C. (2000). Construction of multicomponent catalytic films based on avidin-biotin technology for the electroenzymatic oxidation of molecular hydrogen. *Biotechnology and Bioengineering,* **68**:1–10.

Decher, G. (1997). Fuzzy nanoassemblies: toward layered polymeric multicomposites. *Science,* **277**:1232–1237.

Delaney, G. M., Bennetto, H. P., Mason, J. R., Roller, S. D., Stirling, J. L., and Thurston, C. F. (1984). Electron-transfer coupling in microbial fuel cells. II. Performance of fuel cells containing selected microorganism-mediator combinations. *Journal of Chemical Technology and Biotechnology,* **34B**: 13–27.

den Engelsen, D. (1976). Optical anisotropy in ordered systems of lipids. *Surface Science,* **56**:272–280.

Diaz, A. F., Kanazawa, K. K., and Gardini, G. P. (1979). Electrochemical polymerization of pyrrole. *Journal of the Chemical Society, Chemical Communications,* **14**:635–636.

Dubarry, M., Gaubicher, J., Guyomard, D., Wallez, G., Quarton, M., and Baehtz, C. (2008). Uncommon potential hysteresis in the $Li/Li_{2x}VO(H_{2-x}PO_4)_2$ ($0 \leq x \leq 2$) system. *Electrochimica Acta,* **53**:4564–4572.

Dultsev, F. N. and Baklanov, M. R. (1999). Nondestructive determination of pore size distribution in thin films deposited on solid substrate. *Electrochemical and Solid State Letters,* **2**(4):192–194.

Gidley, D. W., Frieze, W. E., Dull, T. L., Sun, J., Yee, A. F., Nguyen, C. V., and Yoon, D. Y. (2000). Determination of pore-size distribution in low-dielectric thin films. *Applied Physics Letters,* **76**:1282–1284.

Gidley, D. W., Peng, H-G., Vallery, R., Soles, C. L., Lee, H-J., Vogt, B. D., Lin, E. K., Wu, W-L., and Baklanov, M. R. (2007). Porosity of low dielectric constant materials. In *Dielectric Films for Advanced Microelectronics*, eds. M. Baklanov, M. Green, and K. Maex, Chapter 3, 85–135. John Wiley & Sons, Ltd., New York.

Gorby, Y. A., Yanina, S., McLean, J. S., Rosso, K. M., Moyles, D., Dohnalkova, A., Beveridge, T. J., Chang, I. S., Kim, B. H., Kim, K. S., Culley, D. E., Reed, S. B., Romine, M. F., Saffarini, D. A., Hill, E. A., Shi, L., Elias, D. A., Kennedy, D. W., Pinchuk, G., Watanabe, K., Ishii, S., Logan, B., Nealson, K. H., and Fredrickson, J. K. (2006). Electrically conductive bacterial nanowires produced by *Shewanella oneidensis* strain MR-1 and other microorganisms. *Proceedings of the National Academy of Sciences of the United States of America,* **103**:11358–11363.

Gottesfeld, S., Kim, Y-T., and Redondo, A. (1995). Recent applications of ellipsometry and spectroellipsometry in electrochemical systems. *Physical Electrochemistry, Principles, Methods, and Applications*, ed. I. Rubinstein, Marcel Dekker, Inc., New York, pp. 393–467.

Greef, R., Kalaji, M., and Peter, L. M. (1989). Ellipsometric studies of polyaniline growth and redox cycling. *Faraday Discussions of Chemical Society,* **88**:277–289.

Gustafsson, M. G. L. (2000). Surpassing the lateral resolution limit by a factor of two using structured illumination microscopy. *Journal of Microscopy,* **198**:82–87.

Hamnett, A. and Hillman, A. R. (1985). An ellipsometric study of polymeric thionine films on platinum. *Journal of Electroanalytical Chemistry,* **195**:189–196.

Hamnett, A. and Hillman, A. R. (1987). A study of the electrochemical growth and optical properties of polymeric thionine films on platinum using ellipsometry. *Journal of Electroanalytical Chemistry,* **233**:125–146.

Heller, A. (2004). Miniature biofuel cells. *Physical Chemistry Chemical Physics,* **6**:209–216.

Hillman, A. R. and Mohamoud, M. A. (2006). Ion, solvent and polymer dynamics in polyaniline conducting polymer films. *Electrochimica Acta,* **51**:6018–6024.

Hollmann, O., Gutberlet, T., and Czeslik, C. (2007). Structure and protein binding capacity of a planar PAA brush. *Langmuir,* **23**:1347–1353.

Holy, V., Pietsch, U., and Baumbach, T. (1999). High-resolution X-ray scattering from thin films and multilayers. *Springer Tracts in Modern Physics,* Vol. 149. Berlin: Springer.

Huang, E., Toney, M. F., Volksen, W., Mecerreyes, D., Brock, P., Kim, H.-C., Hawker, C. J., Hedrick, J. L., Lee, V. Y., Magbitang, T., and Miller, R. D. (2002). Pore size distributions in nanoporous methyl silsesquioxane films as determined by small angle X-ray scattering. *Applied Physics Letters,* **81**:2232–2234.

Ikeda, T. and Kano, K. (2001). An electrochemical approach to the studies of biological redox reactions and their applications to biosensors, bioreactors, and biofuel cells. *Journal of Bioscience and Bioengineering,* **92**:9–18.

Janssens, K., Proost, K., and Falkenberg, G. (2004). Confocal microscopic X-ray fluorescence at the HASYLAB microfocus beamline: characteristics and possibilities. *Spectrochimica Acta Part B*, **59**:1637–1645.

Julbe, A. and Ramsay, D. J. (1996). Methods for the characterization of porous structure in membrane materials. In *Fundamentals of Inorganic Membrane Science and Technology,* eds. A. J. Burrgraaf and L. Cot, Chapter 4, Elsevier, Amsterdam, pp. 67–118.

Kanazawa, K. K. and Gordon II, J. G. (1985). Frequency of a quartz microbalance in contact with liquid. *Analytical Chemistry*, **57**:1770–1771.

Kanazawa, K. K. and Gordon II, J. G. (1985). The oscillation frequency of a quartz resonator in contact with liquid. *Analytica Chimica Acta,* **175**: 99–105.

Karyakin, A. A., Karyakina, E. E., and Schmidt, H-L. (1999). Electropolymerized azines: a new group of electroactive polymers. *Electroanalysis*, **11**: 149–155.

Karyakin, A. A., Morozov, S. V., Karyakina, E. E., Varfolomeyev, S. D., Zorin, N. A., and Cosnier, S. (2002). Hydrogen fuel electrode based on bioelectrocatalysis by the enzyme hydrogenase. *Electrochemistry Communications,* **4**:417–420.

Katz, E. and Willner, I. (2004). Integrated nanoparticle-biomolecule hybrid systems: synthesis, properties, and applications. *Angewandte Chemie, International Edition,* **43**:6042–6108.

Katz, E., Shipway, A. N., and Willner, I. (2003). Biochemical fuel cells. In *Handbook of Fuel Cells-Fundamentals, Technology, Applications*, eds. W. Vielstich, A. Gasteiger, and A. Lamm, Vol. **1**(4), Chapter 21, John Wiley & Sons, Chester, PA, pp. 355–381.

Kaynak, A. (1997). Effect of synthesis parameters on the surface morphology of conducting polypyrrole films. *Materials Research Bulletin,* **32**: 271–285.

Kerzenmacher, S., Ducrée, J., Zengerle, R., and von Stetten, F. (2008). An abiotically catalyzed glucose fuel cell for powering medical implants: reconstructed manufacturing protocol and analysis of performance. *Journal of Power Sources,* **182**:66–75.

Kim, B. H., Kim, H. J., Hyun, M. S., and Park, D. H. (1999). Direct electrode reaction of Fe (III) reducing bacterium, *Shewanella putrefacience. Journal of Microbiology and Biotechnology*, **9**:127–131.

Kim, H. J., Park, H. S., Hyun, M. S., Chang, I. S., Kim, M., and Kim, B. H. (2002). A mediator-less microbial fuel cell using a metal reducing bacterium, *Shewanella putrefaciens. Enzyme Microbial Technology*, **30**:145–152.

Konash, A., Cooney, M. J., Liaw, B. Y., and Jameson, D. M. (2006). Characterization of enzyme–polymer interaction using fluorescence. *Journal of Materials Chemistry,* **16**:4107–4109.

Lee, H. J., Lin, E. K., Bauer, B. J., Wu, W. L., Hwang, B. K., and Gray, W. D. (2003). Characterization of chemical-vapor-deposited low-k thin films using x-ray porosimetry. *Applied Physics Letters,* **82**:1084–1086.

Li, J., Wang, E., Green, M., and West, P. E. (1995). In situ AFM study of the surface morphology of polypyrrole film. *Synthetic Metals,* **74**:127–131.

Liaw, B. Y., Svoboda, V., Quinlan, F., and Cooney, M. J. (2005). Understanding conductive polypyrrole deposition via micro- and nano-scale observations. Presented in the 230*th* Am. Chem. Soc. National Meeting, Washington, DC, Aug. 28–Sept. 1, 2005.

Liu, H. and Logan, B. E. (2004). Electricity generation using an air cathode single chamber microbial fuel cell in the presence and absence of a proton exchange membrane. *Environmental Science & Technology,* **38**:4040–4046.

Lvov, Y., Decher, G., and Möhwald. H. (1993). Assembly, structural characterization, and thermal behavior of layer-by-layer deposited ultrathin films of poly(vinyl sulfate) and poly(allylamine). *Langmuir,* **9**:481–486.

Malzer, W. (2006). 3D Micro X-ray fluorescence analysis. *The Rigaku Journal,* **23**: 40–47.

Mano, N., Mao, F., and Heller, A. (2003). Characteristics of a miniature compartment-less glucose-O_2 biofuel cell and its operation in a living plant. *Journal of the American Chemical Society,* **125**:6588–6594.

Marx, K. A. (2003). Quartz crystal microbalance: a useful tool for studying thin polymer films and complex biomolecular systems at the solution-surface interface. *Biomacromolecules,* **4**:1099–1120.

McMillan, T., Rutledge, J. E., and Taborek, P. (2005). Ellipsometry of liquid helium films on gold, cesium, and graphite. *Journal of Low Temperature Physics,* **138**:995–1011.

Miles, M. J., Smith, W. T., and Shapiro, J. S. (2000). Morphological investigation by atomic force microscopy and light microscopy of electropolymerised polypyrrole films. *Polymer,* **41**:3349–3356.

Minteer, S. D., Liaw, B. Y., and Cooney, M. J. (2007). Enzyme-based biofuel cells. *Current Opinion in Biotechnology,* **18**:228–234.

Murray, C., Flannery, C., Streiter, I., et al. (2002). Comparison of techniques to characterise the density, porosity and elastic modulus of porous low-k SiO_2 xerogel films. *Microelectronic Engineering,* **60**:133–141.

Nivens, D. E., Chambers, J. Q., Anderson, T. R., and White, D. C. (1993). Long-term, on-line monitoring of microbial biofilms using a quartz crystal microbalance. *Analytical Chemistry,* **65**:65–69.

O'Sullivan, C. K. and Guilbault, G. G. (1999). Commercial quartz crystal microbalances—theory and applications. *Biosensors and Bioelectronics,* **14**:663–670.

Omote, K., Ito, Y., and Kawamura, S. (2003). Small angle X-ray scattering for measuring pore-size distributions in porous low-k films. *Applied Physics Letters,* **82**:544–546.

Palmore, G. T. R., Bertschy, H., Bergens, S. H., and Whitesides, G. M. (1998). A methanol/dioxygen biofuel cell that uses NAD^+-dependent dehydrogenases as catalysts: application of an electro-enzymatic method to regenerate nicotinamide adenine dinucleotide at low overpotentials. *Journal of Electroanalytical Chemistry,* **443**:155–161.

Parratt, L. G. (1954). Surface studies of solids by total reflection of X-rays. *Physical Review,* **95**:359–369.

Petkov, M. P., Weber, M. H., Lynn, K. G., Rodbell, K. P., and Cohen, S. A. (1999). Doppler broadening positron annihilation spectroscopy: a technique for measuring open-volume defects in silsesquioxane spin-on glass films. *Applied Physics Letters,* **74**:2146–2148.

Picart, C., Senger, B., Sengupta, K., Dubreuil, F., and Fery, A. (2007). Measuring mechanical properties of polyelectrolyte multilayer thin films: novel methods based on AFM and optical techniques. *Colloids and Surfaces A: Physicochemical Engineering Aspects,* **303**:30–36.

Pluddeman, E. P. (1980). Chemistry of silane coupling agents. In *Silylated Surfaces*, ed., D. E. Leyden, Gordon and Breach, New York, pp. 31–53.

Poksinski, M., Dzuo, H., Järrhed, J-O., and Arwin, H. (2000). Total internal reflection ellipsometry. In *Proc. of Eurosensors XIV.* Copenhagen, August 27–30, 2000.

Qian, D. J., Nakamura, C., Zorin, N., and Miyake, J. (2002). Hydrogenase–poly (viologen) complex monolayers and electrochemical properties in Langmuir–Blodgett films. *Colloids and Surfaces A: Physicochemical and Engineering Aspects,* **198–200**:663–669.

Rabaey, K., Lissens, G., Siciliano, S. D., Verstraete, W. (2003). A microbial fuel cell capable of converting glucose to electricity at high rate and efficiency. *Biotechnology Letters,* **25**:1531–1535.

Ramanavicius, A., K. Habermüller, E. Csoregi, V. Laurinavicius, and W. Schuhmann. (1999). Polypyrrole-entrapped quinohemoprotein alcohol dehydrogenase. Evidence for direct electron transfer via conducting-polymer chains. *Analtical Chemistry,* **71**:3581–3586.

Redondo, A., Ticianelli, E. A., and Gottesfeld, S. (1988). Ellipsometric measurements of the optical properties and dynamics of electrochemical conversion in films of polyaniline. *Molecular Crystals and Liquid Crystals,* **160**:185–203.

Reguera, G., McCarthy, K. D., Mehta, T., Nicoll, J. S., Tuominen, M. T., and Lovley, D. R. (2005). Extracellular electron transfer via microbial nanowires. *Nature,* **435**:1098–1101.

Rekveld, S. (1997). Ellipsometric studies of protein adsorption onto hard surface in a flow cell. *Fedobruk,* Enschede.

Richter, H., McCarthy, K., Nevin, K. P., Johnson, J. P., Rotello, V. M., and Lovley, D. R. (2008). Electricity generation by *Geobacter sulfurreducens* attached to gold electrodes. *Langmuir,* **24**:4376–4379.

Richter, R. P. and Brisson, A. R. (2005). Following the formation of supported lipid bilayers on mica: a study combining AFM, QCM-D, and ellipsometry. *Biophysical Journal,* **88**:3422–3433.

Ringeisen, B. R., et al. (2006). High power density from a miniature microbial fuel cell using *Shewanella oneidensis* DSP10. *Environmental Science & Technology,* **40**:2629–2634.

Rishpon, J. and Gottesfeld, S. (1991). Investigation of polypyrrole/glucose oxidase electrodes by ellipsometric, microgravimetric and electrochemical measurements. *Biosensors and Bioelectronics,* **6**:143–149.

Rishpon, J., Redondo, A., Derouin, C., and Gottesfeld, S. (1990). Simultaneous ellipsometric and microgravimetric measurements during the electrochemical growth of polyaniline. *Journal of Electroanalytical Chemistry,* **294**:73–85.

Rubinstein, I., Rishpon, J., Sabatani, E., Redondo, A., and Gottesfeld, S. (1990). Morphology control in electrochemically grown conducting polymer films. 1. Precoating the metal substrate with an organic monolayer. *Journal of the American Chemical Society,* **112**:6135–6136.

Rusling, J. F. and Suib, S. L. (1994). Characterizing materials with cyclic voltammetry. *Advanced Materials,* **6**:922–930.

Russell, A. E. and Rose, A. (2004). X-ray absorption spectroscopy of low temperature fuel cell catalysts. *Chemical Review,* **104**:4613–4635.

Sabatani, E., Redondo, A., Risphon, J., Rudge, A., Rubinstein, I., and Gottesfeld, S. (1993b). Morphology control in electrochemically grown conducting polymer films. Part 2: Effects of cathodic bias on anodically grown films studied by spectroscopic ellipsometry and quartz-crystal microbalance. *Journal of Chemical Society, Faraday, Transactions,* **89**:287–294.

Sabatani, E., Ticianelli, E., Redondo, A., Rubinstein, I., Rishpon, J., and Gottesfeld, S. (1993a). Morphological effects in conducting polymer films studied by combined OCM and spectroscopic ellipsometry. *Synthetic Metals,* **55**:1293–1298.

Sakai, H., Nakagawa, T., Tokita, Y., Hatazawa, T., Ikeda, T., Tsujimura, S., and Kano, K. (2009). A high-power glucose/oxygen biofuel cell operating under quiescent conditions. *Environmental Engineering Science*, **2**:133–138.

Salmon, M., Diaz, A. F., Logan, A. J., Krounbi, M., and Bargon, J. (1982). Chemical modification of conducting polypyrrole films. *Molecular Crystals and Liquid Crystals,* **83**:265–276.

Sanchez, C., Boissière, C., Grosso, D., Laberty, C., and Nicole, L. (2008). Design, synthesis, and properties of inorganic and hybrid thin films having periodically organized nanoporosity. *Chemistry of Materials,* **20**:682–737.

Sano, Y. (1988). Optical anisotropy of bovine serum albumin. *Journal of Colloid and Interface Science,* **124**:403–406.

Sauerbrey, G. (1959). Verwendung von schwingquarzen zur wägung dünner schichten und zur mikrowägung. *Zeitschrift für Physik,* **155**:206–222.

Schmitt, J., Grünewald, T., Decher, G., Pershan, P. S., Kjaer, K., and Lösche, M. (1993). Internal structure of layer-by-layer adsorbed polyelectrolyte films: a neutron and X-ray reflectivity study. *Macromolecules,* **26**:7058–7063.

Schubert, M. (1996). Polarization-dependent optical parameters of arbitrarily anisotropic homogeneous layered systems. *Physical Review B*, **53**: 4265–4274.

Scott, D., Cooney, M. J. and Liaw, B. Y. (2008). Sustainable current generation from the ammonia–polypyrrole interaction. *Journal of Materials Chemistry*, **18**:3216–3222.

Scott, D., and Liaw, B. Y. (2009). Harnessing electric power from monosaccharides—a carbohydrate–air alkaline fuel cell mediated by redox dyes. *Environmental Engineering Science*, **2**:965–969.

Severin, E. J. and Lewis, N. S. (2000). Relationships among resonant frequency changes on a coated quartz crystal microbalance, thickness changes, and resistance responses of polymer-carbon black composite chemiresistors. *Analytical Chemistry*, **72**:2008–2015.

Stenberg, M. and Nygren, H. (1983). The use of the isoscope ellipsometer in the study of adsorbed proteins and biospecific binding reactions. *Journal de Physique Colloque*, **C10**:83–86.

Stenberg, M., Sandström, T., and Stilbert, L. (1980). A new ellipsometric method for measurements of surfaces and surface layers. *Materials Science & Engineering, A: Structural Materials: Properties, Microstructure and Processing,* **42**:65–69.

Stoev, K. and Sakurai, K. (1997). Recent theoretical models in grazing incidence X-ray reflectometry. *The Rigaku Journal,* **14**:22–37.

Sun, C., Li, W., Sun, Y., Zhang, X., and Shen, J. (1999). Fabrication of multilayer films containing horseradish peroxidase based on electrostatic interaction and their application as a hydrogen peroxide sensor. *Electrochimica Acta,* **44**:3401–3407.

Sutton, S. J. and Vaughan, A. S. (1995). On the morphology and growth of electrochemically polymerized polypyrrole. *Polymer,* **36**:1849–1857.

Svoboda, V. and Liaw, B. Y. (2008). In-situ transient study of polymer nanofilm growth via simultaneous correlation of charge, mass, and ellipsometric measurements. *Pure and Applied Chemistry,* **80**:2439–2449.

Svoboda, V., Cooney, M. J., Liaw, B. Y., Minteer, S., Piles, E., Lehnert, D., Calabrese Barton, S., Rincon, R., Atanassov, P. (2008). Standardized characterization of electrocatalytic electrodes. *Electroanalysis,* **20**:1099–1109.

Svoboda, V., Cooney, M. J., Rippolz, C., and Liaw, B. Y. (2007). In situ characterization of electrochemical polymerization of methylene green on platinum electrodes. *Journal of the Electrochemical Society,* **154**: D113–D116.

Szalóki, I., Török, S. B., Ro, C-U., Injuk, J., and Van Grieken, R. E. (2000). X-ray spectrometry. *Analytical Chemistry,* **72**:211R-233R.

Tiwald, T. E., Thompson, D. W., Woollam, J. A., and Pepper, S. V. (1998). Determination of the mid-IR optical constants of water and lubricants using IR ellipsometry combined with an ATR cell. *Thin Solid Films,* **313–314**: 718–721.

Tjaernhage, T. and Sharp, M. (1994). The use of a quartz crystal microbalance combined with ellipsometry and cyclic voltammetry for determining some basic characteristics of an electroactive polymer film. *Electrochimica Acta,* **39**:623–628.

Tompkins, H. G. (1993). *A User's Guide to Ellipsometry.* Academic Press, Inc., San Diego, CA.

Tompkins, H. G. and McGahan, W. A. (1999). *Spectroscopic Ellipsometry and Reflectometry, a User's Guide.* John Wiley & Sons, Inc., New York.

Tronin, A. Y. and Konstantinova, A. F. (1989). Ellipsometric study of the optical anisotropy of lead arachidate Langmuir films. *Thin Solid Films,* **177**:305–314.

Vincze, L., Vekemans, B., Brenker, F. E., Rickers, K., Somogyi, A., Kersten, M., and Adams, F. (2004). Three-dimensional trace element analysis by confocal X-ray microfluorescence imaging. *Analytical Chemistry,* **76**:6786–6791.

von Bruggeman, D. A. G. (1935). Berechnung verschiedener physikalischer konstanten von heterogenen substanzen. *Annalen der Physik,* **5**:636–679.

Wang, J., Xu, Y., Chen, X., and Du, X. (2007). Electrochemical supercapacitor electrode material based on poly(3,4-ethylenedioxythiophene)/polypyrrole composite. *Journal of Power Sources,* **163**:1120–1125.

Weber, K., Achenbach, L., and Coates, J. (2006). Microbes pumping iron: anaerobic microbial iron oxidation and reduction. *Nature Reviews Microbiology,* **4**:752–764.

Welin-Klintström, S., Askendal, A., and Elwing, H. (1993a). Surfactant and protein interactions on wettability gradient surfaces. *Journal of Colloid and Interface Science,* **158**:188–194.

Welin-Klintström, S., Jansson, R., and Elwing, H. (1993b). An off-null ellipsometer with lateral scanning capability for kinetic studies at liquid-solid interfaces. *Journal of Colloid and Interface Science,* **157**:498–503.

Whiting, G. L., Snaith, H. J., Khodabakhsh, S., Andreasen, J. W., Breiby, D. W., Nielsen, M. M., Greenham, N. C., Friend, R. H., and Huck, W. T. S. (2006). Enhancement of charge-transport characteristics in polymeric films using polymer brushes. *Nano Letters,* **6**:573–578.

Willner, I., Basnar, B., and Willner, B. (2007). Nanoparticle–enzyme hybrid systems for nanobiotechnology. *The FEBS Journal,* **274**:302–309.

Wu, J., Yuan, X. Z., Martin, J. J., et al. (2008). A review of PEM fuel cell durability: degradation mechanisms and mitigation strategies. *Journal of Power Sources,* **184**:104–119.

Wu, W., Wallace, W. E., Lin, E., et al. (2000). Properties of nanoporous silica thin films determined by high-resolution x-ray reflectivity and small-angle neutron scattering. *Journal of Applied Physics,* **87**(1):1193–1200.

Xiao, Y., Patolsky, F., Katz, E., Hainfeld, J. F., and Willner, I. (2003). "Plugging into enzymes": nanowiring of redox enzymes by a gold nanoparticle. *Science,* **299**:1877–1881.

Xu, L-C. and Siedlecki, C. A. (2007). Effects of surface wettability and contact time on protein adhesion to biomaterial surfaces. *Biomaterials,* **28**:3273–3283.

Yang, R., Naoi, K., Evans, D. F., Smyrl, W. H., and Hendrickson, W. A. (1991). Scanning tunneling microscope study of electropolymerized polypyrrole with polymeric anion. *Langmuir,* **7**:556–558.

Yang, X. Q., Sun, X., McBreen, J. (2000). New phases and phase transitions observed in $Li_{1-x}CoO_2$ during charge: in situ synchrotron X-ray diffraction studies. *Electrochemistry Communications,* **2**:100–103.

Zangooie, S., Bjorklund, R., and Arwin, H. (1998). Protein adsorption in thermally oxidized porous silicon layers. *Thin Solid Films,* **313–314**:825–830.

Zayats, M., Katz, E., and Willner, I. (2002). Electrical contacting of glucose oxidase by surface-reconstitution of the apo-protein on a relay-boronic acid-FAD cofactor monolayer. *Journal of the American Chemical Society,* **124**:2120–2121.

Zhitomirsky, I. and Hashambhoy, A. (2007). Chitosan-mediated electrosynthesis of organic–inorganic nanocomposites. *Journal of Materials Processing Technology,* **191**:68–72.

Zhou, D-M., Fang, H-Q., Chen, H-Y., Ju, H-X., and Wang, Y. (1996). The electrochemical polymerization of methylene green and its electrocatalysis for the oxidation of NADH. *Analytica Chimica Acta,* **329**:41–48.

14

Spatial Pattern Formation of Motile Microorganisms: From Gravitactic Bioconvection to Protozoan Culture Dynamics

Tri Nguyen-Quang, Frederic Guichard

Department of Biology, McGill University, Montreal, Canada

The Hung Nguyen

Department of Mechanical Engineering, Polytechnic, University of Montreal, Montreal, Canada

CONTENTS

14.1 Description and Literature Review of Bioconvection

14.1.1 Overview

Studies of momentum, heat, and mass transports, originally developed as independent branches of classical physics, have grown into a unified field of fundamental engineering sciences with applications ranging from biotechnology, nanotechnology to environmental fluid mechanics. Within research fields of mass transport in aqueous environment, we might distinguish *diffusive substances* such as salt and sugar from *mobile microorganisms with self-driving ability* such as bacteria and algae playing a vital role in the nature. The mobility is an essential physiological activity of microorganisms seeking out new environment to colonize for continued growth and reproduction. Complex sensing and signaling mechanisms that allow microorganisms to move in response to external or internal cues have been described under the specific terms of "*taxis*" or "*kinesis*." The directional responses are called *taxis* while the nondirectional ones are *kinesis*. Once a signal is received and decoded, individual movement may occur via mechanical processes such as propulsion by flagella or cilia, or gliding mobility, which relies on the frictional movement of membrane proteins against a surface to propel the microorganism forward (Van Hamme et al. 2006). Generally speaking, the movement of swimming organisms may be classified according to their responses to various stimulating factors of the environment. Principal types of taxes are reported by Eisenbach (2001).

The interactions between the motile microorganisms and the surrounding fluid cannot be predicted by the well-established theory of classical mass transfer. Modeling the physiological activities of microorganisms constitutes the necessary background to understand, predict, and direct their activities as they are exposed to continually changing environmental conditions. Their responses may be either migration to a more favorable habitat or adaptation of their internal metabolism. This latter process is a relatively slow genetic evolution, while the migration to a more desirable environment, for example, swimming toward/away from increasing concentrations of nutrients/toxicants, is a more efficient short-term strategy. The complicated dynamics of this course of action is governed by the interplay between the laws of biology and those of fluid mechanics. One of the biggest challenges is therefore to develop a realistic mathematical model of microorganism population dynamics in aqueous media. A complete mathematical model should consist of two classes of equations. The spatiotemporal evolution equations of conserved quantities, such as momentum and cell concentration, constitute the first class of equations. The second class comprises the constitutive equations for bacterial fluxes, which are usually based on empirical observations, and thereby define the model. A detailed description of the individual and collective microorganisms behaviors may be found in the papers of Kessler (1986a), and Pedley and Kessler (1992a,b).

One of the most remarkable phenomena observed in the aqueous biosphere generated by the previously mentioned collective behavior of microorganisms is known under the name of bioconvection. This is the spontaneous pattern formation of motile microorganisms, which are denser than water and move randomly, but on average upwardly against gravity. The upswimming causes cells to accumulate in a thin layer near the upper surface, resulting in an unstable density distribution. The basic mechanism of bioconvection is qualitatively analogous to that of Rayleigh–Bénard convection in a fluid layer heated from below, in which an overturning instability develops when the upper region becomes sufficiently heavier than the lower region. The essential difference between bioconvection and thermoconvection lies in the fact that in the latter phenomenon the buoyancy force is due to heating at the boundary of the system, while the buoyancy force in bioconvection is due to the self-driving gradient, that is, upswimming ability of microorganisms.

The objective of this chapter is to review as well as to present some updated results on the onset and development states of gravitactic bioconvection in porous media via numerical and experimental approaches.

This chapter consists of four parts. The first one provides an overview of literature on the phenomenon of bioconvection. In the second one, we present a comprehensive study of the onset of bioconvection in porous media, using the linear stability theory and a simulation model. Attention will be focused on *gravitaxis.* The mathematical model is based on Darcy equation for the fluid flow within the porous medium, and a diffusion–advection equation for the conservation of the *gravitactic* microorganisms. The third part concerns the experimental study of the pattern formation in a suspension of gravitactic microorganisms. The last part presents some perspectives for future research on spontaneous pattern formation of living microorganisms, which is a central theme in population biology, discussing possible extensions of bioconvection theory to other systems involving similar nonlinear interactions between biological and physical processes.

14.1.2 Review of Literature

In 1911, Wager studied movement patterns of the flagellate *Euglena viridis*, among others, and showed that the tendency to swim against gravity was necessary for pattern formation among live organisms. In 1961, Platt coined the name bioconvection for this hydrodynamic phenomenon, and noted for the first time the analogy between thermoconvection (Bénard convection) and bioconvection. In 1970, Keller and Segel focused attention on this analogy in analyzing the aggregation process of *Amoebae,* as well as the instability problems in homogeneous mechanical systems subject to two or more forces of different types described by Chandrasekhar (1961). In the same year, Roberts suggested that the principal cause for geotaxis in ciliates such as *Paramecium* is a hydrodynamic interaction between the microorganisms and the medium, the magnitude of which is determined by the size and shape of the microorganisms. He developed a general mathematical theory of geotaxis to describe the motion of these organisms under gravity, and the predictions of the theory were compared with measurements in suspensions of *Paramecia* in long, vertical columns (Roberts 1970). In a later study, Roberts and Deacon (2002) examined in more details the movement of gravitactic ciliates such as *Paramecium* (Roberts and Deacon 2002). They found that the shape-dependent orientation plays an important role in the gravitactic responses of *Paramecium.* The first bioconvective model of pattern formation was developed by Plesset and Winet (1974) and Plesset and Whipple (1974) who explained the onset of bioconvection by applying the theory of Rayleigh–Taylor instability, in which the organism-rich sublayer is seen as a layer of dense fluid overlying a less dense, deeper layer of fluid. They found that the wavelength of the most rapidly growing mode agreed with the observed scale of the convection patterns. They concluded that the Rayleigh–Taylor instability is the mechanism for the observed bioconvection in *Tetrahymena pyriformis* culture (Figure 14.1) and may also explain the sedimentation patterns *(positive geotaxis)* in other microorganism swarms in so far as they are not disturbed by surface wind or by thermal gradients. In the Rayleigh–Taylor instability, the preferred horizontal wavenumber is basically predicted by the upper layer thickness and density difference regardless of the total depth of the container. The weakness of this model is due to neglecting cell diffusion between two layers with the result that the proposed basic state is not a solution of the governing equations. In the same year, Winet and Jahn (1974) proposed the gravity-propulsion theory of negative geotaxis of *T. pyriformis (hereafter referred to as TP)*, suggesting that the negative geotactic orientation was a physical consequence of the gyrational torque produced by geometrical asymmetry of the microorganisms.

Childress et al. (1975) and Levandowsky et al. (1975) developed the model of pattern formation of gravitactic microorganisms, based on the Navier–Stokes equation and a diffusion–convection equation for motile microorganisms, showing a remarkable analogy with Bénard convection. They

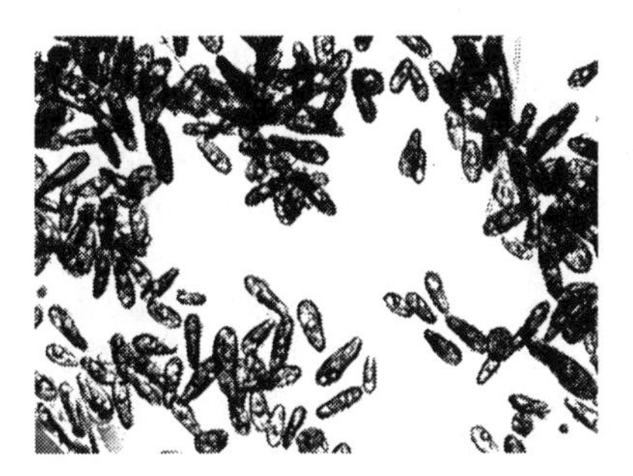

FIGURE 14.1
Tetrahymena pyriformis protozoan population. (Picture taken by T. Nguyen-Quang at the University of Minho, Braga, Portugal.)

then derived the equations for the onset of bioconvection in a fluid medium. The basis of their analysis is the equilibrium solution which is the balance between the upward swimming and downward diffusion in quiescent water. They obtained analytically the preferred wavenumber and growth rates of bioconvection, which qualitatively agree with experiments on both *TP* and algae. As noted by Pedley and Kessler, the model of Childress et al. was the first self-consistent *continuum model* of bioconvection (Pedley and Kessler 1992b). Subsequently, many researchers have extended the Childress et al. theory to suspensions of algae with gyrotactic behavior in fluid media (Hill et al. 1989; Bees and Hill 1998; Ghorai and Hill 2000).

One of the first simulations was performed by Childress and Peyret who, instead of using the continuum cell conservation equation, treated the cells as individual moving points each of which exerts a force on the fluid and move relative to the fluid by a superposition of upswimming and random walk. In this approach, each microorganism acts as a point mass in the Navier–Stokes equation. Inversely, the fluid flow affects the swimming of these organisms as each one is convected by the fluid (Childress and Peyret 1976).

In 1986, Fujita and Watanabe presented a numerical study of bioconvection based on the equations derived by Childress et al. (1975). They found a transition from steady flow to periodic and finally to nonperiodic motions through a sequence of period doubling bifurcations as the Rayleigh number increased. This route to chaos is quite similar to that of Bénard convection (Fujita and Watanabe 1986). A subsequent study by Harashima, Watanabe and Fujishiro was focused on the evolution of convection patterns in gravitactic suspensions (Harashima et al. 1988). Following the Childress et al. theory, they numerically solved the Navier–Stokes and cell conservation equations for upswimmers such as *TP*. They found that the evolution of the system proceeds in the direction to intensify the downward advection of microorganisms and to reduce the total potential energy of the system. In other words, the steady state pattern for a given Rayleigh number and for a given size medium seems to obey the principle of minimum total potential energy.

In contrast to the continuum model, Hopkins and Fauci presented a "discrete" mathematical model and a numerical method of solution for the fluid/microorganisms/nutrient system (Hopkins and Fauci 2002). Their approach was based on the method of Childress and Peyret (1976) in using a simplified representation of microbes as point particles, performing calculations with a large enough number of particles to model realistic cell concentrations and macroscopic fluid effects.

The aforementioned works were devoted exclusively to bioconvection in a fluid medium, while it has been noted that "*the fluid may be dynamically constrained so that convection is heavily damped. This constraint may be imposed by a porous medium..., which is located throughout the fluid or only near the top surface. The damping of fluid convection essentially eliminates downward re-mixing of the cells*" (Kessler 1986b). In fact, although bioconvection in porous media is frequently encountered in nature as well as in industrial applications, very few studies have been done to explicitly address bioconvective patterns in porous media.

According to Hill and Pedley (2005), a rational extension of the model of bioconvection from a fluid to a porous medium should take into account the fact that the local flow (through the pores) may cause the cells to tumble and strongly affect their ability to reorient if the pores are too small. So far, bioconvection in porous media has been described by the *continuum model* of Pedley et al. (1988), with the fluid flow being determined by Darcy equation, instead of the Navier–Stokes equation.

In 2001, Kuznetsov and Jiang presented a study of *gravitatic* bioconvection in a porous medium. The Darcy equation for the fluid flow and the diffusion–convection equation for the concentration of microorganisms were numerically solved to obtain the flow and concentration fields in a square cavity in term of the permeability of the porous medium. They concluded that there exists a critical value of permeability for the onset of convection and "*after numerous computations, it was determined that the critical value of permeability is approximately* $4 \times 10^{-7} m^2$*. If permeability is smaller than this value, then no convection develops. This, in turn, causes the cells to accumulate in the top layer and stay there.*" (Kuznetsov and Jiang 2001).

Later, Kuznetsov and Avramenko (2002) obtained a criterion for stability of a *gyrotactic* suspension in a porous medium using linear perturbation theory. This criterion gives the critical permeability in terms of the bacterial cell eccentricity, average swimming velocity, and other parameters. The onset of bioconvection has also been examined under the effect of cell deposition and declogging by Kuznetsov and Jiang (2003).

Nield et al. (2004) studied the onset of *gyrotactic* bioconvection in a horizontal saturated porous medium. The Darcy flow model was employed, and it was assumed that the *Peclet* number (dimensionless cell velocity) is not greater than unity. Critical conditions were obtained for various values of *Peclet* number, gyrotaxis number, and cell eccentricity.

Recently, the onset and development of bioconvection of *gravitactic* microorganisms within a confined fluid as well as a porous medium have been studied by Nguyen-Quang and coworkers (Bahloul et al. 2005; Nguyen-Quang et al. 2005; Nguyen-Quang 2006). Both linear stability analysis and numerical simulations based on the Navier–Stokes/Darcy and concentration conservation equations were performed to determine the onset and development of fluid flow and cell concentration in terms of the governing parameters (*Rayleigh, Peclet,* and *Schmidt* numbers). Attention was focused on effects of the swimming speed of the microorganisms and the aspect ratio of the cavity. For a more comprehensive bibliography, we would refer to the first excellent review of literature from 1911 to 1992 by Pedley and Kessler (1992b) and the second review by Hill and Pedley (2005) for the following years.

New developments of different types of bioconvection (oxytaxis, gyrotaxis, etc.) in porous media have been presented in two review chapters by Kuznetsov (2005, 2008).

14.2 Onset and Evolution of Gravitactic Bioconvection: Linear Stability Analysis and Numerical Simulation

14.2.1 Mathematical Formulation of Gravitactic Bioconvection in a Porous Medium

14.2.1.1 Description and Formulation of the Problem

Fluid flows and transport phenomena in porous media are frequently encountered in nature and industries (groundwater, building materials, composite materials, etc.) as well as in biosystems (human organs, aquifer ecosystems, etc.). An exact description of the flow dynamics and transport properties within the pore length scale is beyond the capabilities of both theoretical and experimental methods. Henry Darcy (1856) who successfully established an empirical relation between the filtration velocity and the pressure gradient in a porous medium has accomplished the first study of flow in porous media in 1856. The elegantly simple statement, that *the filtration velocity is proportional to the pressure gradient*, is now celebrated as the Darcy's law.

A rational derivation and extension of Darcy's law (which thereafter will be called Darcy equation) from the well-established Navier–Stokes equation was the subject of continuing research during the past 50 years. The method of volume averaging has provided a rational foundation for the derivation of spatially averaged equations for the fluid flow and heat and mass transfer in porous media. This approach consists essentially of a spatial smoothing process that leads to the governing equations for local volume average quantities (e.g., velocity, temperature, mass concentration). The next step is to develop and to solve the closure problem for the spatial deviations of the spatially smoothed quantities to determine the effective transport coefficients that appear in the spatially averaged equations. A comprehensive review of this method was presented in the monograph "*The Method of Volume Averaging*" by Whitaker (1999). In the present study, we consider a homogeneous, isotropic porous medium containing a dilute suspension of gravitactic microorganisms. The suspension is assumed to behave as a Newtonian incompressible fluid while the microorganisms swim through the porous medium randomly but with an average upward velocity, $\vec{V}_c^*$. Furthermore, the pore size should be significantly larger than the cell size so that the flow through the pores will not cause the cells to tumble and drastically affect their ability to reorient (Hill and Pedley 2005).

The objective of this part is to determine the critical conditions for the onset of bioconvection in term of the gravitactic velocity, $\vec{V}_c^*$, and to investigate the postonset status. The problem under consideration is schematically shown in Figure 14.2. Within the Boussinesq approximation (see Appendix), the fluid flow and microorganism concentration may be described by the following equations (Khaled and Vafai 2003; Nguyen-Quang et al. 2005; Nguyen-Quang 2006).

Continuity equation

$$\nabla \cdot \vec{V}^* = 0 \tag{14.1}$$

Darcy equation

$$-\vec{\nabla} P^* - \frac{\mu}{K}\vec{V}^* + \vec{g}\rho = 0 \tag{14.2}$$

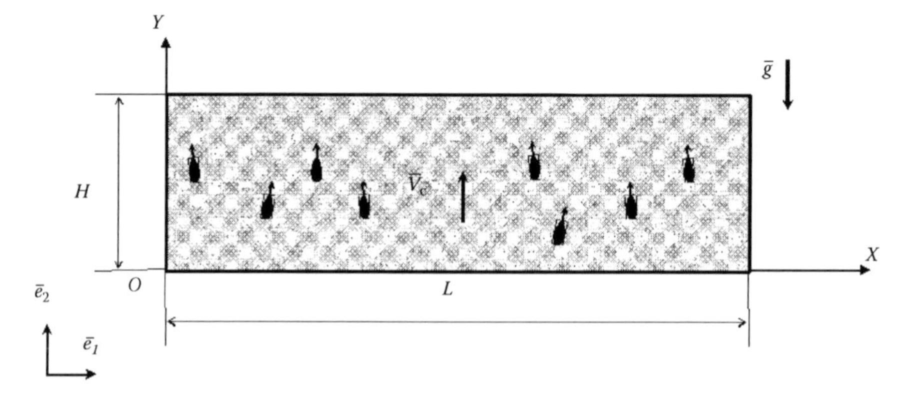

FIGURE 14.2
Geometry of the system.

Cell concentration equation

$$\frac{\partial n}{\partial t} + \nabla\cdot(n\vec{V}^*) + \nabla\cdot(n\vec{V}_c^*) = D_c\nabla^2 n \tag{14.3}$$

Boussinesq approximation

$$\rho = \rho_0(1+\beta(n-n_0)) \tag{14.4}$$

14.2.1.2 Initial and Boundary Conditions

At $t^* = 0$, it is supposed that the initial concentration is uniform, that is,

$$n(X,Y,t^*) = n(X,Y,0) = \overline{n} \tag{14.5}$$

At the impermeable boundaries the condition of zero-normal fluid velocity requires

$$\begin{cases} X = 0, L\colon & V_X^* = 0 \\ Y = 0, H\colon & V_Y^* = 0 \end{cases} \tag{14.6}$$

while the condition of zero-concentration flux is ($\vec{e}$ is the unit normal vector)

$$\vec{J}^* \bullet \vec{e} = \left[-D_c\nabla n + n(\vec{V}^* + \vec{V}_c^*)\right] \bullet \vec{e} = 0$$

$$\Leftrightarrow \begin{cases} X = 0, L\colon & J_X^* = -\partial n/\partial X = 0 \\ Y = 0, H\colon & J_Y^* = nV_c^* - D_c\partial n/\partial Y = 0 \end{cases} \tag{14.7}$$

14.2.2 Diffusion State

For $\vec{V}_c^* = (0, V_c^*)$, the system (14.1 through 14.4) under the initial and boundary conditions (14.5 through 14.7) admits the following steady state solution:

$$\vec{V}^* = 0 \quad \text{and} \quad n = \frac{\overline{n}\left(\dfrac{V_c^* H}{D_c}\right)}{\exp\left(\dfrac{V_c^* H}{D_c}\right) - 1}\exp\left(\frac{V_c^*}{D_c}Y\right) \tag{14.8}$$

which satisfies the conservation of microorganisms:

$$\overline{n} = \frac{1}{LH}\int_0^L dX \int_0^H n(X,Y,t^*)dY \tag{14.9}$$

Let $Pe = \frac{V_c^* H}{D_c}$ be the dimensionless cell velocity and rewrite equation (14.8) as $n = \frac{\overline{n} Pe e^{\frac{Pe}{H} Y}}{e^{Pe} - 1}$ from which we readily deduce the concentrations at the bottom ($Y = 0$) and the top ($Y = H$) of the cavity

$$n_0 = \frac{Pe}{e^{Pe} - 1}\overline{n};\ n_1 = \frac{Pe e^{Pe}}{e^{Pe} - 1}\overline{n}$$

$$\Delta n = (n_1 - n_0) = \overline{n} Pe \tag{14.10}$$

14.2.2.1 Nondimensional Equations

Nondimensional governing equations may be expressed in the following form, by choosing H and H^2/D_c as length and time scales, respectively.

$$\nabla \cdot \vec{V} = 0 \tag{14.11}$$

$$\vec{V} = -\frac{D_c K}{\nu H^2}\vec{\nabla} P - Ra N \vec{e}_2 \tag{14.12}$$

$$\frac{\partial N}{\partial t} + \nabla \cdot (N\vec{V}) + \nabla \cdot (N\overrightarrow{Pe}) = \nabla^2 N \tag{14.13}$$

In terms of the stream function ψ, they become

$$\nabla^2 \psi = Ra \frac{\partial N}{\partial x} \tag{14.14}$$

$$\frac{\partial N}{\partial t} + \nabla \cdot (N\vec{V}) + \nabla \cdot (N\overrightarrow{Pe}) = \nabla^2 N \tag{14.15}$$

where

$$\vec{V} = \left(\frac{\partial \psi}{\partial y}, -\frac{\partial \psi}{\partial x}\right)$$

$$N = \frac{n - n_0}{\Delta n}$$

and the Rayleigh number

$$Ra = \frac{gKH\beta \overline{n} Pe}{\nu D_c} \tag{14.16}$$

with initial and boundary conditions

$$N = \overline{N} = \frac{\overline{n} - n_0}{\Delta n} = \frac{e^{Pe} - Pe - 1}{(e^{Pe} - 1)Pe} \quad \text{at } t = 0 \tag{14.17}$$

$$\psi = 0,\ \partial N/\partial x = 0 \quad \text{at } x = 0,\ F \quad \text{and}$$

$$\psi = 0,\ \partial N/\partial y = PeN + Pe/(e^{Pe} - 1) \quad \text{at } y = 0 \tag{14.18}$$

The diffusion state (14.8) can be expressed in the nondimensional form

$$N_d = \frac{e^{Pey} - 1}{e^{Pe} - 1} \tag{14.19}$$

14.2.2.2 Linearized Equations

Let the diffusion state expressed by subscript "$_d$" be perturbed by quantities denoted by "$_1$", we will have

$$N(x,y,t) = N_d + N_1;\ \psi(x,y,t) = \psi_1;\ \vec{V}(x,y,t) = \vec{V}_1 \tag{14.20}$$

with N, ψ, $\vec{V}$ being the disturbed state and $N_1 \ll N_d = \frac{e^{Pey}-1}{e^{Pe}-1}$.

By substituting equation (14.20) in equations (14.14–14.15), neglecting second-order terms of perturbations, and then substituting the diffusion state (14.19) into the obtained system, we have

$$\frac{\partial^2 \psi_1}{\partial x^2} + \frac{\partial^2 \psi_1}{\partial y^2} = Ra \frac{\partial N_1}{\partial x} \tag{14.21}$$

$$\frac{\partial N_1}{\partial t} - G(y)\frac{\partial \psi_1}{\partial x} + Pe\frac{\partial N_1}{\partial y} = \nabla^2 N_1 \tag{14.22}$$

with

$$G(y) = Pe\frac{e^{V_c y}}{e^{Pe}-1} \tag{14.23}$$

The required boundary conditions are

$$\psi_1 = \frac{\partial N_1}{\partial x} = 0 \text{ at } x = 0,\ F \quad \text{and}$$
$$\psi_1 = \frac{\partial N_1}{\partial y} - PeN_1 = 0 \text{ at } y = 0, 1 \tag{14.24}$$

The linear system (14.21–14.22) with boundary conditions (14.24) determines the initial evolution of perturbations and the criterion for the onset of bioconvection. Numerical solutions are presented in the next section. Note that by making the change of variable $N_1 = -T_1$ and substituting it into equations (14.21–14.22), we readily obtain the following system:

$$\frac{\partial^2 \psi_1}{\partial x^2} + \frac{\partial^2 \psi_1}{\partial y^2} = -Ra \frac{\partial T_1}{\partial x} \tag{14.25}$$

$$\frac{\partial T_1}{\partial t} + G(y)\frac{\partial \psi_1}{\partial x} + Pe\frac{\partial T_1}{\partial y} = \nabla^2 T_1 \tag{14.26}$$

with the boundary conditions

$$\psi_1 = \frac{\partial T_1}{\partial x} = 0 \text{ at } x = 0, F \quad \text{and}$$
$$\psi_1 = \frac{\partial T_1}{\partial x} - PeT_1 = 0 \text{ at } y = 0, 1 \tag{14.27}$$

When the cell velocity $Pe \to 0$, $G(y) \to 1$, the above system of equations reduces formally to the equations governing the fixed-flux Bénard problem (Kimura et al. 1995; Prud'homme and Nguyen 2002).

14.2.3 Numerical Results

14.2.3.1 Linear Stability Analysis

It can be shown that the principle of exchange of stability is satisfied in the present problem as well as in the case of Bénard convection. We then look for solutions of the form

$$\psi_1(x,y,t) = \phi(y) \exp \gamma t \exp ikx$$

where γ and k are real numbers, representing the growth (or damping) rate and the wavenumber of the perturbations. The onset of convection corresponds to $\gamma = 0$. The stability boundary, *Ra(k,Pe),* is therefore determined by solving the eigenvalue problem of the system (14.21–14.24) for marginally stable perturbations with $\frac{\partial N_1}{\partial t} = \gamma = 0$. It should be noted that the results presented here apply to a rectangular cavity of aspect ratio F as well as to an infinite horizontal porous layer for a wavenumber $k = n\pi/F$ with n being the number of the convection cells in the considered domain. As the problem is governed by two parameters, *Ra* and *Pe*, the Rayleigh number at the onset of convection is a function of k and *Pe*.

Figure 14.3(a) shows a family of stability curves *Ra* versus k for values of *Pe* varying from 0.1 to 20. It divides the parameter space *(Ra, k)* into two regions: the region above the stability curve is unstable and the region below the curve is stable. For each value of *Pe* we get one stability curve. Each curve has a minimum at $Ra = Ra_m$, $k = k_m$. This is referred to as the critical point for onset of convection. Figure 14.3(d) for the critical conditions as functions of *Pe* shows that k_m is an increasing function of *Pe*, that is, the plow patterns change from elongated to slender shape as the swimming velocity is increased.

For *Pe* smaller than 1, the stability curves almost coincide as can be seen in Figure 14.3(b) and the critical points shown in Figure 14.3(e) may be approximated by

$$Ra_c = 12 + 2Pe \quad \text{and} \quad k_c = 0.9Pe^{1/2} \tag{14.28}$$

For *Pe* greater than 1, the stability curves are strongly dependent on *Pe* (Figure 14.3a). However, by setting the length scale to $h = D_c/V_c^*$ instead of H, they coalesce into a *unique stability curve with a critical point* $Ra_c^* = 10.2$, $k_c^* = 0.7$ for all *Pe* as shown in Figures 14.3(c) and 14.3(f). It should be noted that this renormalization leads to a renormalized Rayleigh number, $Ra^* = Ra/Pe$, and a wavenumber, $k^* = k/Pe$, based on the length scale, $h = D_c/V_c^*$.

On the basis of this result, the values of *Pe* chosen for the numerical simulations are varying between 0.1 and 5 while the Rayleigh numbers *Ra* may take the value several times as large as the critical one.

Also of interest is the growth rate of convection as predicted by the linear stability theory. For a given Rayleigh number above the critical value Ra_m, there exists a range of wavenumber for which the system is unstable. As shown

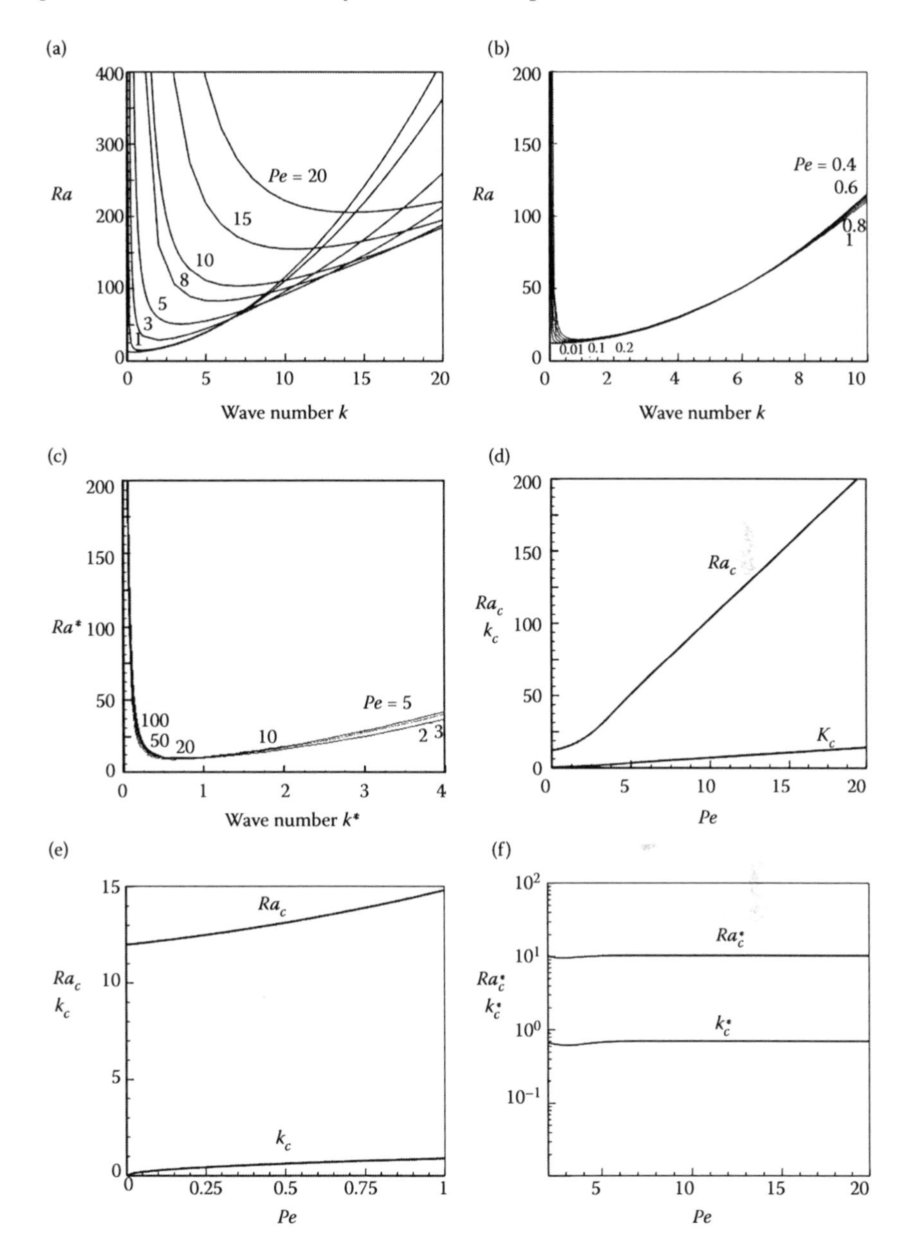

FIGURE 14.3
(a) Stability diagram for various *Pe*. (b) (Ra_c vs. k) for $Pe \leq 1$. (c) ($Ra^*{}_c$ *vs.* k^*) for $Pe \geq$ *1.* (d) (Ra_c vs. *Pe*) and (k_c vs. *Pe*) for various *Pe*. (e) (Ra_c vs. *Pe*) and (k_c vs. *Pe*) for $Pe \leq 1$ (F) ($Ra^*{}_c$ vs. *Pe*) and ($k^*{}_c$ *vs. Pe*) for $Pe \geq 2$. (Nguyen-Quang, T., Nguyen, T.H., Guichard, F., Nicolau, A., Smatzari, G., LePalec, et al., *Zoo. Sci.*, 26, 2009, doi:10.2108/zsj.26.54.)

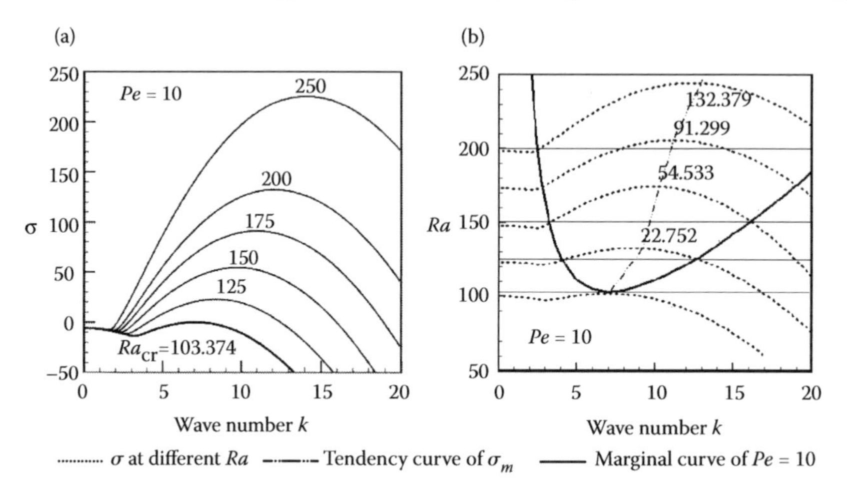

FIGURE 14.4
(a) Growth rate versus k for different Rayleigh number at $Pe = 10$. (b) Variation of growth rate in the stability diagram for $Pe = 10$.

in Figures 14.4(a) and 14.4(b), the growth rate varies with the wavenumber within this range to attain a maximum value σ_M at a wavenumber k_M, which is always greater than the critical wave number k_m. We also note from Figures 14.5(a) and 14.5(b) that k_M and σ_M increase as Ra is increased. Figure 14.5(c) shows a universal curve σ_M *versus* Ra when $Pe \leq 1$ while Figure 14.5(d) represents the variations of the renormalized growth rate σ_M^* for $Pe \geq 2$. It is noted that results presented here are obtained from the linear stability analysis of small perturbations. In other words, they are only valid at the very early development of convection. The subsequent evolution of the concentration and flow patterns are determined by nonlinear interactions and should be studied by solving the full set of governing equations.

14.2.3.2 Evolution of Bioconvection

14.2.3.2.1 Critical Threshold and Subcritical Regime

First simulations were presented for small values of swimming speed $Pe = 0.001$, 0.1, and 1. According to linear analysis, the critical Rayleigh numbers are calculated by (14.28) for $Pe \leq 1$ and by renormalization, established for $Pe \geq 2$ as mentioned earlier. In Table 14.1, we compare the two series of critical Rayleigh number values found by both methods (linear stability analysis and numerical simulation) for Peclet numbers from 0.1 to 10.

We recognize that for the case of $Pe = 0.001$, the results are very close to the ones obtained for thermoconvection under the constant flux heating condition (Figure 14.6(a)). When the swimming speed is increased, the unicellular convection core begins to shift (Figure 14.6(c)) toward the sidewalls.

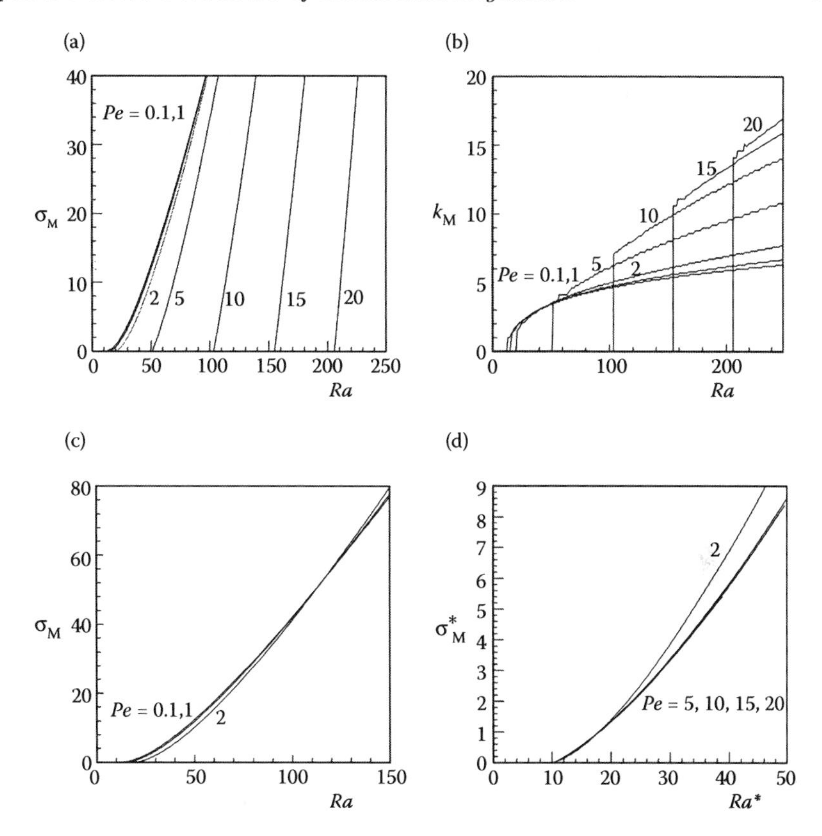

FIGURE 14.5
(a) σ_M versus Ra for various Pe = 0.1–20. (b) k_M versus Ra for various Pe = 0.1–20. (c) σ_M versus Ra for Pe = 0.1–2. (d) σ_M* versus Ra^* for Pe= 2–20.

Figure 14.7 illustrates the temporal evolution of ψ and N when $Pe = 5$ and $Ra = 52$, that is, just above the threshold value $Ra_{cr} = 51$. Contrary to thermoconvection, the iso-concentrations and streamlines in this case are strongly deformed (illustrated in Figure 14.6[d]). This indicates that we are dealing with a subcritical phenomenon in gravitactic bioconvection.

Bifurcation curves in Figure 14.8 for a range of swimming speed Pe from 0.1 to 10 illustrate the occurrence of this subcritical regime from the diffusion state, showing its dependence on the microorganism mobility. It can be seen that the higher swimming speed the stronger the subcriticity.

14.2.3.2.2 Supercritical State

In a cavity with an aspect ratio equal to half the critical wavelenghth $F = L_{cr}$, the bioconvection flow is unicellular, independent of the swimming speed and

TABLE 14.1
Critical Values of Rayleigh Number for a Porous Cavity with $F = L_{cr}$ Length Corresponding to Swimming Speed of Microorganisms

Swimming Speed or Peclet Number Pe	Critical Values of Rayleigh Number	
	By Linear Stability	By Simulation
0.1	12.188	12.22
1.	14.823	14.83
5.	51.028	51.437
10.	103.374	100

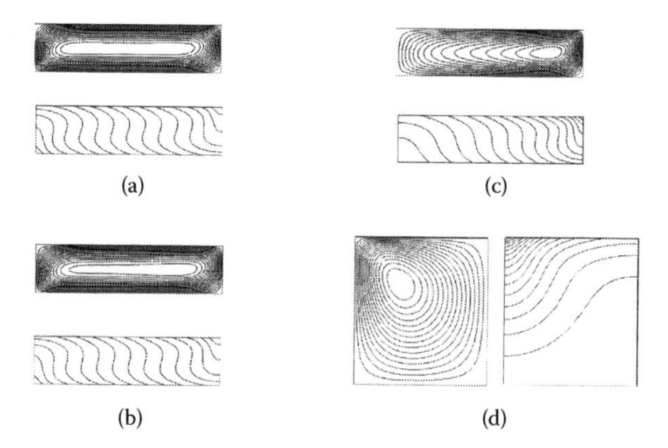

FIGURE 14.6
Streamlines ψ and isolines of concentration for the case of $F = 4$, $Ra = 100$, $n_{cell} = 1$ (a) $Pe = 0.001$, (b) $Pe = 0.1$, (c) $Pe = 1$, and (d) $F = L_{cr}$, $n_{cell} = 1$, $Pe = 5$, $Ra = 52$, close to $Ra_{cr} = 51$.

the initial conditions. This is also predicted by the linear stability analysis. Nevertheless, the growth rate and the convergence time for simulations are strongly affected by the initial conditions.

For Pe smaller than 1, the supercritical convection patterns at steady state are very similar to the ones obtained from thermoconvection (Rayleigh–Bénard convection) for the case of heating from below by a constant heat flux.

For $F > L_{cr}$ and $Pe \geq 2$, the morphology of steady state bioconvective cells depends strongly not only on the value of Rayleigh, but also on the initial condition and the aspect ratio of the cavity. When the Rayleigh number is close to the threshold, the steady state morphology predicted by linear stability

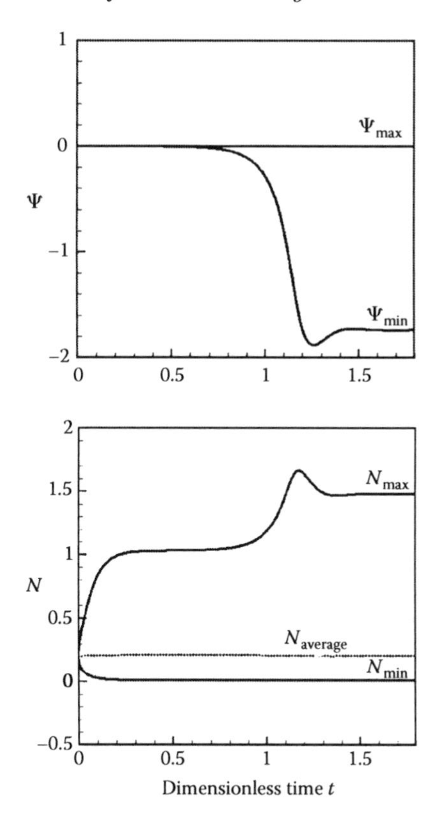

FIGURE 14.7
Evolution of ψ_{max}, ψ_{min}, and concentration N for $Pe = 5$ at $Ra = 52$, $F = L_{\text{cr}}, n_{\text{cell}} = 1$.

can appear, but only temporarily. In certain simulation cases, this "*pseudo-permanent*" state persists for a very long time and subsequently establish into one of two permanent states shown in Figure 14.9:

1. One cell, shifting toward the sidewall.
2. Two cells, symmetrical with respect to the vertical center line.

14.3 Experimental Study of the Pattern Formation in a Suspension of Gravitactic Microorganisms

14.3.1 Introduction

Many authors have revealed bioconvective patterns with different species of microorganisms. Wager studied pattern formation in populations of the

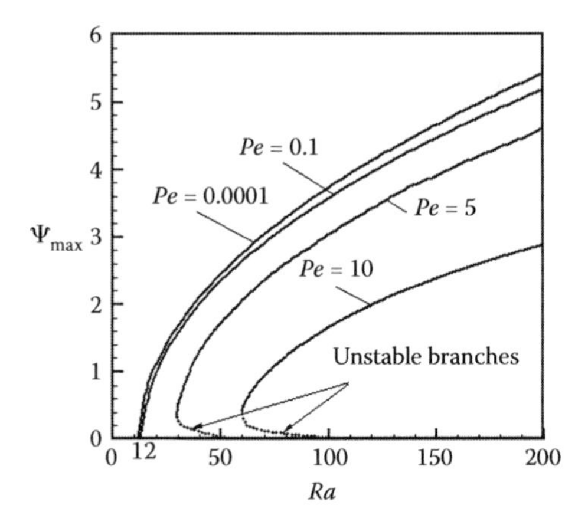

FIGURE 14.8
Bifurcation curves obtained with the $F = L_{cr}$ cavity.

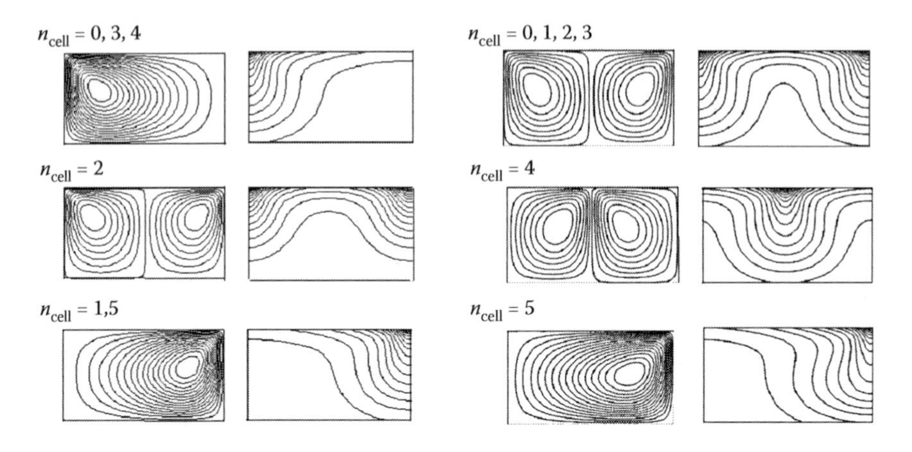

FIGURE 14.9
Streamlines and iso-concentration lines for $Pe = 5, F = 2L_{\text{cr}}$, $Ra = 52$ (left), and $Ra = 100$ (right), at different initial wavenumbers ($n_{\text{cell}} = 1, 2, 3, 4, 5$).

flagellate *Euglena viridis* (Wager 1911). Oxytactic bioconvection patterns in different strains of *Bacillus subtilis* at the stationary growth phase have also been observed (Czirok et al. 2000; Janosi et al. 2002). For eukaryotic unicellular microorganisms, Loefer and Meffert realized the anaerobiosis assay on *TP* (Loefer and Mefferd 1952). A quantitative description of bioconvection patterns given in terms of Rayleigh–Taylor instability has been proposed by Plesset and Winet (1974). Plesset et al. (1976) tried to provide an instructive justification for the configuration of upper layer *"granular nature"* of the *TP* suspensions. Assar (1988) tested the effects of *cadmium, pentachlorophenol,* and *light* on bioconvection patterns in *Chlamydomonas reinhardtii* and concluded that those factors had a significant effect on pattern formation. A study of *galvanotaxis* was published by Itoh and Toida (2001) who tried to control bioconvection of *Tetrahymena thermophila* by applying an electrical field. Other authors have also analyzed the spatiotemporal characteristics of *TP* bioconvection patterns under altered gravity acceleration (Mogami et al. 2004). Two-dimensional bioconvective plumes of *Tetrahymena pyriformis* have been recently reported from experiments by Nguyen-Quang et al. (2009) who noted that, two-dimensional systems are more amenable to experiment and simulation than three-dimensional systems as the reduction of dimension significantly reduces the amount of data required to specify the flow. Hence, experiments can straightforwardly determine an entire two-dimensional scalar field, and two-dimensional calculations can be performed at Rayleigh numbers much higher than those in three dimension (DeLuca et al. 1990). Nguyen-Quang et al. (2009) therefore considered a very thin cavity that constrains the fluid motion to a two-dimension vertical plane, making it possible to introduce a number of simplifications in the Navier–Stokes equation that lead to the so-called Hele-Shaw flow. The physical concept of this apparatus is well described by Hele-Shaw (1898) and Bear (1972).

14.3.2 Hele-Shaw Apparatus and Darcy's Law

A Hele-Shaw apparatus consists of two rigid parallel transparent plates (glass or Plexiglass), separated by a very thin spacer to create a two-dimensional fluid medium. The setup of Hele-Shaw apparatus is shown in Figure 14.10 with dimensions $H \times L$, separated by a thin spacer b. The Hele-Shaw apparatus is a well-known model of porous media where the law of Darcy is expressed as follows:

$$\vec{V} = -\frac{K}{\mu}\nabla P \tag{14.29}$$

The Hele-Shaw cell is equivalent to a "porous medium" having a permeability

$$K = \frac{b^2}{12} \tag{14.30}$$

which is well known under the name of "*Permeability of the Hele-Shaw cell.*" More details on the Hele-Shaw cell approach to gravitactic bioconvection can be found in Nguyen-Quang (2006).

14.3.3 Geometrical and Physicobiological Parameters

It is instructive to distinguish two groups of parameters that determine the development of bioconvection (Nguyen-Quang et al. 2009): one group is related to the geometry of the Hele-Shaw cell and the other one to the

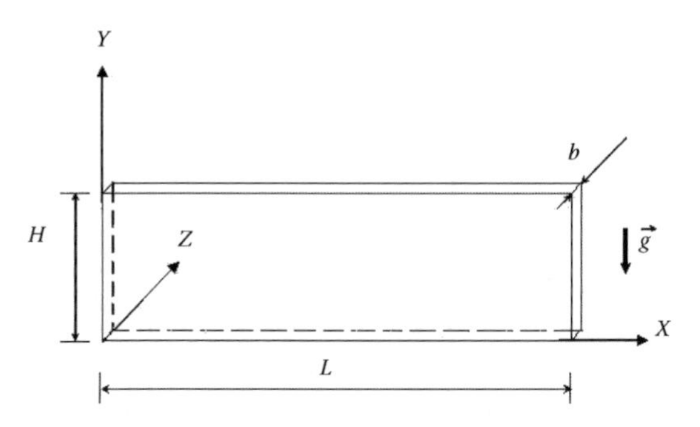

FIGURE 14.10
Conceptual model of the Hele-Shaw apparatus.

physicobiological characteristics of the system. The Rayleigh number is a combination of these two groups and is the parameter governing the development of the bioconvection patterns. From the linear stability analysis results, we arrive at two following parameters:

1. The geometrical parameter

$$C_1 = b^2 H^2 \text{ when } Pe \leq 1$$
$$C_1 = b^2 H \text{ when } Pe > 1 \tag{14.31}$$

2. The physicobiological parameter

$$C_2 = \frac{144 D_c^2 v}{g \bar{n} \beta V_c} \text{ when } Pe \leq 1$$
$$C_2 = \frac{122.4 D_c v}{g \bar{n} \beta} \text{ when } Pe > 1 \tag{14.32}$$

The condition for the onset of bioconvection may then expressed as

$$C_1 \geq C_2 \tag{14.33}$$

With a given species of microorganisms, C_2 can be a fixed constant. The geometry of the Hele-Shaw apparatus is then determined according to the criterion (14.33). On the other hand, it follows from (14.33) that *a minimum concentration*, $\bar{n}$, is required for convection to develop in a fixed-geometry Hele-Shaw cell. This *minimum concentration* is called *critical concentration* for the onset of convection.

The geometrical factors b, H, and L play an important role in the development of bioconvection patterns as the height H of the apparatus is directly proportional to the Peclet number and the Rayleigh number, while b represents the apparent permeability of the Hele-Shaw apparatus.

We should keep in mind that although there exists a certain mathematical analogy between Hele-Shaw and porous medium flows via the permeability term and general form following the Darcy's law, Hele-Shaw model can not be considered as an "exact" model for porous medium. The Hele-Shaw model can just reproduce certain behaviors of the fluid flow within two-dimensional porous medium. The results obtained from Hele-Shaw model might reflect qualitatively the nature of the pattern and behavior in the equivalent porous medium, and cannot evaluate exactly bioconvection patterning in real porous medium. The reasons are as follows:

- The flow in Hele-Shaw model is governed by an equation similar to the Darcy's law only for two-dimensional case.
- Physically, the real diffusion coefficients of the considered substance (herein the gravitactic microorganisms) are not the same for both media. In the real porous medium, microorganisms have to diffuse through different physical phases (real solid matrix, liquid, and pores), while in the Hele-Shaw model they disperse with a fluid medium diffusive coefficient.
- The thickness b in the formula (14.30) is considered as "analogous" to the pore size in porous medium, and should be larger than cell size, allowing microorganisms to move easily through the medium. Tortuosity, which is an important factor influencing significantly the convection and diffusion terms in real porous medium, does not, hence, exist in a Hele-Shaw model and microorganisms will interact with the inner surfaces of Hele-Shaw apparatus, not with solid matrix in a porous medium.

14.3.4 Key Results of Experimental Study

Bioconvection patterns of *T. pyriformis* are realized in various Hele-Shaw setups and recorded from a horizontal view. Regarding the materials and methods for experimental issues, we refer to Nguyen-Quang et al. (2009).

From these experiments, we have obtained three different regimes of pattern formation depending on experimental conditions.

- The diffusion regime
- The stationary convection regime
- The unsteady convection regime

14.3.4.1 The Diffusion Regime

A *diffusion state* as predicted by the mathematical model is shown in Figure 14.11(b). Beginning at $t = 0$, this state is established after *30 min to 2 h*, with cell concentration below the critical value for the onset of convection.

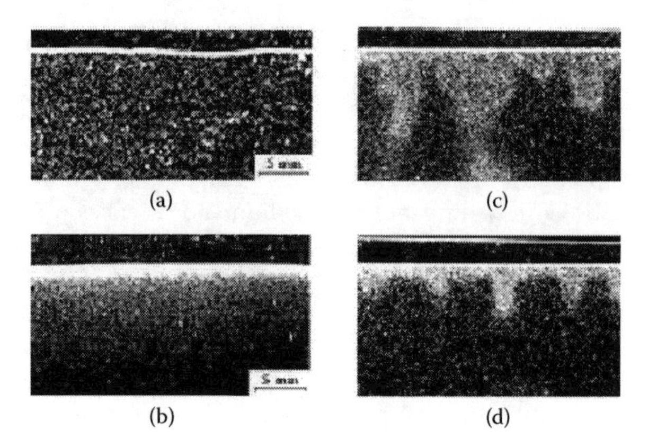

FIGURE 14.11
Different states of bioconvection pattern. (a) Uniform initial state; (b) diffusion state; (c) transient state; (d) steady state of the plumes. (Nguyen-Quang, T., Nguyen, T.H., Guichard, F., Nicolau, A., Smatzari, G., LePalec, et al., *Zoo. Sci.*, 26, 2009, doi:10.2108/zsj.26.54.)

Concentration values leading to this state range from 5×10^4 to 1×10^5 cells/cm^3, depending on the dimensions of the Hele-Shaw cell.

14.3.4.2 The Stationary Convection Regime

When the concentration is higher than a certain critical value, the diffusion state does not exist anymore and we observe the appearance of a *stationary convection regime* with many small plumes. Figure 14.11(d) illustrates the steady convection patterns obtained with a concentration $n = 1.27 \times 10^5$ cells/cm^3 (the Hele-Shaw apparatus dimensions are $H \times L \times b = 4 \times 15 \times 0.075\,\text{cm}$). The time t_c required for the steady state to establish varied from *30 minutes to 3 hours,* depending on each experimental parameter sets, but mostly on the initial filling concentration and the health status of *TP* cells. For this assay (Figure 14.11[d]) t_c is around 30 minutes. Figure 14.11(c) shows a transient state before reaching the stationary convection regime.

14.3.4.3 Unsteady Convection Regime

As the concentration is increased to higher values than those leading to the state described earlier (stationary convection), we observed an *unsteady convection* regime characterized by a collection of *cloudy* plumes. These plumes have nonstationary shapes, moving from side to side, dividing themselves into small plumes and then reassembling again (Figure 14.12). The motion of these plumes continues until the death of the *Tetrahymena* cells. The time for this *nonstationary* regime to appear is about *15–30 min,* depending mostly on the

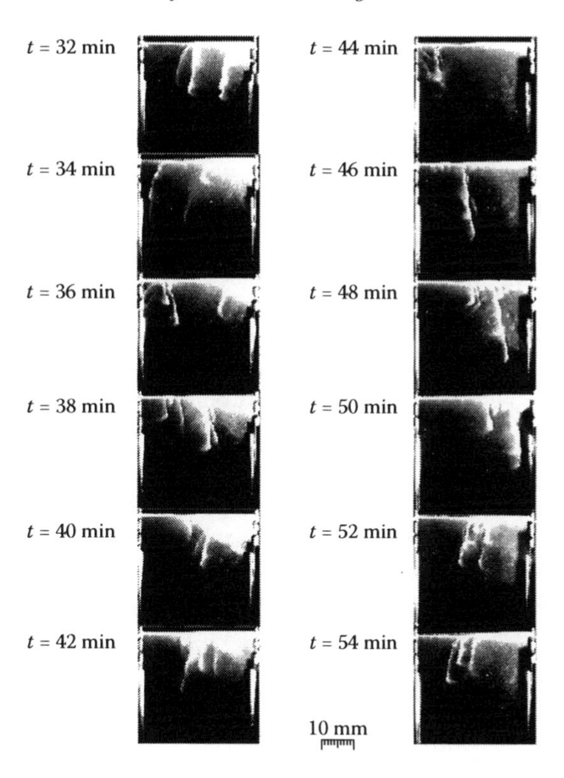

FIGURE 14.12
Evolution of unsteady convection patterns. (Nguyen-Quang, T., Nguyen, T.H., Guichard, F., Nicolau, A., Smatzari, G., LePalec, et al., *Zoo. Sci.*, 26, 2009, doi:10.2108/zsj.26.54.)

initial filling concentration. Plumes illustrated in Figure 14.12 were obtained in a Hele-Shaw cell of dimensions $H \times L \times b = 7.6 \times 8 \times 0.075$ cm, with a concentration of 3.08×10^5 cells/cm^3 after centrifugation, corresponding to a Rayleigh number of approximately $Ra^* = 52.7$.

14.3.4.4 Critical Threshold for the Transition

In the same Hele-Shaw apparatus of dimensions $H \times L \times b = 7.6 \times 8 \times 0.075$ cm, we have observed the unsteady flow for a range of concentrations from 2.29×10^5 to 2.45×10^5 cells/cm^3, corresponding to values of the Rayleigh number Ra^* varying from *39.2 to 41.9*, with a dimensionless swimming speed of 228 and a diffusion coefficient $D_c = 1.5 \times 10^{-3}$ cm^2/sec. We estimate that the critical Rayleigh number for the transition from steady state to unsteady state is around 40, that is, four times higher than the first critical Rayleigh number $Ra^* = 10.2$ predicted by the linear stability analysis.

It may be expected that as the Rayleigh number is gradually increased beyond a certain threshold, the convection patterns should change smoothly from a steady to an unsteady state with a well-defined moving flow. However, it is difficult to control the various parameters for this transition to occur. We have instead obtained a rather *cloudy* pattern, as shown in Figure 14.12, corresponding to a Rayleigh number $Ra^* = 52.7$, that is, rather far from the estimated critical value $Ra_c^* = 40$.

From these experiments, the following remarks are in order:

1. When H is varied, the Peclet and the Rayleigh numbers also vary proportionally. The variations of these two parameters could change the flow regime such that with a same initial concentration, there may establish a diffusion state in a Hele-Shaw cell with a height H_1; and a convection regime in another cell with a height $H_2 > H_1$.
2. The variation of the thickness b is related to the permeability K of the porous medium by the formula $K = b^2/12$ so that when b is increased, the Rayleigh number is increased accordingly and a small variation of b may strongly influence the convection pattern in a Hele-Shaw apparatus. It should be noted that b should not be too big to violate the two-dimensional Hele-Shaw approximation, and not too small to affect the motility of the microorganisms.
3. The number of plumes depends strongly on the length, L, of the apparatus. For steady convection regime, we observed regular plume patterns of 2.5–5.5 mm length and 2.0–2.6 mm width. The distance between two plumes is 0.4–0.8 cm in concentration of 1×10^5–3×10^5 cells/cm^3. These observations agree with those reported for *TP* (Plesset et al. 1976; wavelength $\lambda = 0.655$ cm and average culture concentration 2.7×10^5 cells/cm^3). The size and distance between plumes also depend on the initial concentration, which may lead to unsteady patterns beyond a certain value. In this unsteady regime, the shape and size of the unsteady patterns are not well defined.
4. Plesset et al. (1976) have underlined that a considerable amount of time is required before a cell turns around and resumes its normal upward swimming in the case of high concentration. Another considerable amount of time is also required to establish the convection state. The total time required for these cells to turn around and then resume its normal upward swimming and reach the convection state in our experiments is around *30 minutes to 3 hours* for steady convection regime, depending on the geometry and initial filling concentration (i.e., the Rayleigh and the Peclet numbers). In the case of unsteady regime (Figure 14.12), the amount of time for the cloudy patterns to appear is around *15–30 min,* depending on the Rayleigh and the Peclet numbers of the system.

14.4 Summary and Perspectives of Future Research

In this chapter, a numerical study has been made to predict the onset and evolution of bioconvection in porous media. A Hele-Shaw apparatus has been used as a model to study bioconvection phenomena in *Tetrahymena* culture.

Our experiments have reproduced stationary as well as time-varying two-dimensional bioconvection patterns.

It should be noted that in an experimental monoculture, the growth of *TP* is characterized by a logarithmic growth phase, a prestationary growth phase, and a stationary phase. In the logarithmic phase, which may last from a few hours to two days depending on inoculums, cell density increases logarithmically, the generation time being 3–7 h. In the prestationary phase, growth decreases for a few generations before entering the last stationary phase (Sauvant et al. 1999).

If the growth rate is taken into account, there will be a significant effect of density-dependent growth on pattern formation because the concentration and the Rayleigh number would both increase. We so far assumed that the coefficient of diffusion D_c and the swimming speed, V_c are independent constants and there is no interaction between the mobile *TP* cells. In the case of the population growth, there should be an important interaction between cells as density increases and both D_c and V_c should vary. Some fundamental questions naturally arise in the study of *TP* bioconvection with population growth: How is the dynamics of such populations affected by their motions? What regimes (stationary, oscillatory or chaotic…) can be established? Similar studies of diffusive instabilities and pattern formation in populations showed that the time scale of population growth relative to those of convection states are of critical importance to predict the maintenance of spatial heterogeneity in dynamic populations (Holmes et al. 1994; Malchow et al. 2001; Okubo and Levin 2002) and the nonlinear nature of biophysical systems is crucial in understanding self-organized phenomena over the ecological and evolutionary scales. Our results suggest that the *Rayleigh* and the *Peclet* numbers may be two important parameters in this kind of systems. Therefore, further studies at high *Rayleigh* and high *Peclet* numbers are necessary to better understand the development of time-dependent bioconvection.

It should be noted that the Hele-Shaw model is capable to qualitatively represent a system behaving as a two-dimensional porous medium, but surely with certain limitations. An "exact" analogy of Hele-Shaw with porous medium is impossible and *a Hele-Shaw model can never replace a real porous medium.* Therefore, although the study of gravitactic bioconvection in porous media might be simulated by a Hele-Shaw apparatus, further studies should be done in a real porous system. This remains one of the most challenging difficulties because the pattern visualization through a porous medium is not easy to realize.

Moreover, although some studies have been made of oxytaxis and gyrotaxis behaviors (Kuznetsov and Avramenko, 2003a,b, 2005; Kuznetsov 2003, 2008; Nield et al. 2004; Kuznetsov et al. 2004), more exhaustive works on the development of convection regimes beyond critical thresholds (laminar, turbulent, or chaotic...) should be done for microorganisms with various directional motions (e.g., phototaxis, galvanotaxis, chemotaxis, magnetotaxis...).

Another subject of interest is bioconvection in porous media with the double-diffusion effects, which is found in numerous applications in environmental sciences and engineering. Some works on this problem by Kuznetsov and coworkers (Nield and Kuznetsov, 2006; Kuznetsov, 2008) and Nguyen-Quang et al. (2008) have shown that the bioconvection in porous media with double-diffusion still remains an open field of research.

Finally, we should also consider the possibility of controlling bioconvection by/within porous media. In fact, the control of bioconvection by porous medium was first reported by Kessler (1986b) as mentioned in the previous part of literature review. Kuznetsov (2008) has realized a numerical investigation and indicated that vibration is an effective way of controlling the stability of bioconvection while Nguyen-Quang et al. (2008) have found that heating from above could strongly weaken the bioconvection flow. Further studies should be accomplished for this subject.

14.5 Appendix: Boussinesq Approximation for the Microorganism Suspension

Let us recall that from the Nomenclature below:

ρ_w is the specific mass of the fluid (N/m^3).
ρ is the total specific mass of the suspension (N/m^3).
ρ_c is the specific mass of the microorganism cells (N/m^3).
ϑ is the volume of 1 microorganism cell.

we have the following:

$N^*\vartheta$ is the total volume of cells in a unit volume of the suspension.
$1 - N^*\vartheta$ is the volume of fluid.
$\rho_c N^*\vartheta$ is the specific mass of cells.
$\rho_w(1 - N^*\vartheta)$ is the specific mass of fluid.
$\rho = \rho_w(1 - N^*\vartheta) + \rho_c N^*\vartheta$ is the specific mass of the suspension.

or

$$\rho = \rho_w + (\rho_c - \rho_w)N^*\vartheta = \rho_w + \Delta\rho\vartheta N^* \tag{14.34}$$

with

$$\Delta\rho = (\rho_c - \rho_w)$$

the term $N^*\vartheta$ is called "*volume fraction*" (Nguyen-Quang 2006). The expression (14.34), analogous to the Boussinesq approximation in thermoconvection, shows that the difference between the specific mass of microorganism cells and the water one could create a nonuniform relative mass density that depends directly on the cell concentration.

14.6 Nomenclature

b	Thickness space between two plates in Hele-Shaw cell, mm
D_c	Cell diffusivity, $\mathrm{m}^2/\mathrm{sec}$
$F = L/H$	Aspect ratio of *two-dimensional* porous cavity
$\vec{g}$	Gravitational acceleration, m^2/s
k	Wave number
K	Permeability of porous medium, m^2
k_c	Critical wave number
$\overline{n}$	Mean cell concentration in the cavity of length L, height H, and depth W, $\bar{n} = \frac{1}{WLH}\int_0^W dZ \int_0^L dX \int_0^H n(X,Y,Z,t^*)dY$, $\mathrm{cell/m}^3$.
$N = (n - n_0)/\Delta n$	Dimensionless cell concentration
$\overline{N} = (\bar{n} - n_0)/\Delta n$	Dimensionless mean cell concentration (at $t = 0$)
n_0	Cell concentration at the lower boundary $Y = 0$, $\mathrm{cell/m}^3$
n_1	Cell concentration at the upper boundary $Y = H$, $\mathrm{cell/m}^3$
N^*	Dimensional cell concentration, $\mathrm{cell/m}^3$.
$n(X,Y,Z,t^*)$	Cell concentration, $\mathrm{cell/m}^3$
$P = H^2P^*/\rho_0 D_c^2$	Dimensionless pressure
P^*	Dynamic pressure, Pa
$Pe = H\vec{V}_c^*/D_c$	Dimensionless cell velocity
$Ra = g\bar{n}b^2H\beta Pe/12vD_c$	Rayleigh number
$Ra^* = Ra/Pe = g\bar{n}b^2H\beta/12vD_c$	Renormalized Rayleigh number
Ra_c	Critical Rayleigh number
$\vec{V} = H\vec{V}^*/D_c$	Dimensionless Darcy velocity

$\vec{V}^*$	Darcy velocity, m/sec
$\vec{V}_c^*$	Gravitactic cell velocity, m/sec
(X, Y, t^*)	Cartesian coordinates, m, and time, sec
(x, y, t)	Dimensionless coordinates $x = X/H; y = Y/H$, and time $t = D_c t^*/H^2$
Greek symbols	
$\beta = \vartheta \Delta\rho/\rho_0$	Density variation coefficient of suspension
$\Delta\rho = \rho_c - \rho_w$	Difference of cell and water densities, kg/m^3
$\psi = \psi^*/D_c$	Dimensionless stream function
v	Kinematic viscosity of the suspension, m^2/sec
ρ	Density of suspension "fluid-cell," kg/m^3
ρ_c	Cell density, kg/m^3
ρ_w	Water density, kg/m^3
ρ_0	Suspension density at the bottom, kg/m^3
ϑ	Cell volume, m^3/cell

14.7 References

Assar, H. (1988). Effect of Cadmium, Pentachlorophenol and Light on bioconvection patterns in *Chlamydomonas reinhardtii.* Master Thesis, University of Montreal.

Bahloul, A., Nguyen-Quang, T., and Nguyen, T. H. (2005). Bioconvection of gravitactic microorganisms in a fluid layer. *International Communications in Heat and Mass Transfer,* **32**(1–2):64–71.

Bear, J. (1972). *Dynamics of Fluids in Porous Media.* Dovers.

Bees, M. A. and Hill, N. A. (1998). Linear bioconvection in a suspension of randomly swimming, gyrotactic micro-organisms. *Physics of Fluids,* **10**(8):1864–1881.

Chandrasekhar, S. (1961). *Hydrodynamic and Hydromagnetic Stability.* Clarendon Press, Oxford.

Childress, S. and Peyret, R. (1976). A numerical study of two-dimensional convection by motile particles. *Journal de Mecanique,* **15**(5):753.

Childress, S., Levandowsky, M., and Spiegel, E. A. (1975). Pattern formation in a suspension of swimming micro-organisms: equation and stability theory. *Journal of Fluid Mechanics,* **63**:591.

Czirok, A., Janosi, I. M., and Kessler, J. O. (2000). Bioconvective dynamics dependence on organism behaviour. *Journal of Experimental Biology,* **203**(21):3345–3355.

Darcy, H. P. (1856). *Les fontaines publiques de la ville de Dijon.* Vector Dalmont, Paris.

DeLuca, E. E., Werne, J., Rosner, R., and Cattaneo, F. (1990). Numerical simulations of soft and hard turbulence: Preliminary results for two-dimensional convection. *Physical Review Letters,* **64**:2370.

Eisenbach, M. (2001). Bacterial chemotaxis. *Encyclopedia of Life Sciences,* 1–14.

Fujita, S. and Watanabe, M. (1986). Transition from periodic to non-periodic oscillation observed in a mathematical model of bioconvection by motile micro-organisms. *Physica D: Nonlinear Phenomena,* **20**(2–3):435.

Ghorai, S. and Hill, N. A. (2000). Periodic arrays of gyrotactic plumes in bioconvection. *Physics of Fluids,* **12**(1):5–22.

Harashima, A., Watanabe, M., and Fujishiro, I. (1988). Evolution of bioconvection patterns in a culture of motile flagellates. *Physics of Fluids,* **31**(4):764.

Hele-Shaw, H. J. S. (1898). On the motion of a viscous fluid between two parallel plates. *Nature,* **58**:34–36.

Hill, N. A. and Pedley, T. J. (2005). Bioconvection. *Fluid Dynamics Research,* **37**:1–20.

Hill, N. A., Pedley, T. J., and Kessler, J. O. (1989). Growth of bioconvection patterns in a suspension of gyrotactic micro-organisms in a layer of finite depth. *Journal of Fluid Mechanics,* **208**:509–543.

Holmes, E. E., Lewis, M. A., Banks, J. E., and Veit, R. R. (1994). Partial differential equations in ecology: spatial interactions and population dynamics. *Ecology,* **75**:17–29.

Hopkins, M. M. and Fauci L. J. (2002). A computational model of the collective fluid dynamics of motile micro-organisms. *Journal of Fluid Mechanics,* **455**:149–174.

Itoh, A. and Toida, H. (2001). Control of Bioconvection and its mechanical application. 8–12 July; Como, Italy.

Janosi, I. M., Czirok, A., Silhavy, D., and Holczinger, A. (2002). Is bioconvection enhancing bacterial growth in quiescent environments? *Environmental Microbiology,* **4**(9):525–532.

Keller, E. F. and Segel, L. A. (1970). Initiation of slime mold aggregation viewed as an instability. *Journal of Theoretical Biology,* **26**:399–415.

Kessler, J. O. (1986a). Individual and collective fluid dynamics of swimming cells. *Journal of Fluid Mechanics,* **173**:191–205.

Kessler, J. O. (1986b). The external dynamics of swimming micro-organisms. In *Progress in Phycological Research,* ed, F. E. Round DJC, Bio Press, Bristol, pp. 257–307.

Khaled, A. R. A. and Vafai, K. (2003). The role of porous media in modelling flow and heat transfer in biological tissues. *International Journal of Heat and Mass Transfer,* **46**:4989–5003.

Kimura, S., Vynnycky, M., and Alavyoon, F. (1995). Unicellular natural circulation in a shallow horizontal porous layer heated from below by a constant flux. *Journal of Fluid Mechanics,* **294**:231–257.

Kuznetsov, A. V. (2005). Modeling bioconvection in porous media. In *Handbook of Porous Media*, ed, K. Vafai, 2nd ed., Taylor & Francis, New York, pp. 645–686.

Kuznetsov, A. V. (2008). New developments in bioconvection in porous media: bioconvection plumes, bio–thermal convection, and effects of vertical vibration. *Emerging Topics in Heat and Mass Transfer in Porous Media – From Bioengineering and Microelectronics to Nanotechnology.* Series: *Theory and Applications of Transport in Porous Media*, Vol. 22, ed, P. Vadasz, Springer, Dordrecht, pp. 181–217.

Kuznetsov, A. V. and Jiang, N. (2003). Bioconvection of negatively geotactic microorganisms in a porous medium: the effect of cell deposition and declogging. *International Journal of Numerical Methods for Heat and Fluid Flow,* **13**(2–3):341–364.

Kuznetsov, A. V. and Avramenko, A. A. (2003a). Analysis of stability of bioconvection of motile oxytactic bacteria in a horizontal fluid saturated porous layer. *International Communications in Heat and Mass Transfer,* **30**(5):593–602.

Kuznetsov, A. V. and Avramenko, A. A. (2003b). Stability analysis of bioconvection of gyrotactic motile microorganisms in a fluid saturated porous medium. *Transport in Porous Media,* **53**(1):95–104.

Kuznetsov, A. V., Avramenko, A. A., and Geng, P. (2004). Analytical investigation of a falling plume caused by bioconvection of oxytactic bacteria in a fluid saturated porous medium. *International Journal of Engineering Science,* **42**(5–6):557–569.

Kuznetsov, A. V. and Jiang, N. (2001). Numerical investigation of bioconvection of gravitactic microorganisms in an isotropic porous medium. *International Communications in Heat and Mass Transfer,* **28**(7): 877–886.

Kuznetsov, A. V. and Avramenko, A. A. (2002). A 2D analysis of stability of bioconvection in a fluid saturated porous medium—Estimation of the critical permeability value. *International Communications in Heat and Mass Transfer,* **29**(2):175–184.

Kuznetsov, A. V. and Avramenko, A. A. (2005). Effect of fouling on stability of bioconvection of gyrotactic microorganisms in a porous medium. *Journal of Porous Media,* **8**(1):45–53.

Levandowsky, M., Childress, W. S., Spiegel, E. A., and Huthner, S. H. (1975). A mathematical model of pattern formation by swimming micro-organisms. *Journal of Protozoology,* **22**:296.

Loefer, J. B. and Mefferd, R. B. J. (1952). Concerning pattern formation by free swimming microorganisms. *The American naturalist,* **86**(830): 325–329.

Malchow, H., Petrovskii, S., and Medvinsky A. (2001). Pattern formation in models of plankton dynamics. A synthesis. *Oceanologica Acta,* **24**(5): 479–487.

Mogami, Y., Yamane, A., Gino, A., and Baba, S. A. (2004). Bioconvective pattern formation of Tetrahymena under altered gravity. *Journal of Experimental Biology,* **207**:3349–3359.

Nguyen-Quang, T. (2006). Gravitactic bioconvection study in porous media - Etude de la bioconvection gravitactique en milieux poreux. Ph.D. diss. Ecole Polytechnique de Montréal, University of Montreal.

Nguyen-Quang, T., Bahloul, A., and Nguyen, T. H. (2005). Stability of gravitactic micro-organisms in a fluid-saturated porous medium. *International Communications in Heat and Mass Transfer,* **32**(1–2):54–63.

Nguyen-Quang, T., Nguyen, T. H., and Le Palec G. (2008). Gravitactic bioconvection in a fluid-saturated porous medium with double diffusion. *Journal of Porous Media,* **11**(8):751–764.

Nguyen-Quang, T., Nguyen, T. H., Guichard, F., Nicolau, A., Smatzari, G., LePalec, et al. (2009). Two dimensional gravitactic bioconvection in protozoan (*Tetrahymena pyriformis)* culture. *Zoological Science,* **26**: 54–65, doi:10.2108/zsj.26.54.

Nield, D. A., Kuznetsov, A. V., and Avramenko, A. A. (2004). The onset of bioconvection in a horizontal porous-medium layer. *Transport in Porous Media,* **54**(3):335–344.

Okubo, A. and Levin, S. (2002). *Diffusion and Ecological Problems: Modern Perspectives.* 2nd ed, Springer.

Pedley, T. J. and Kessler, J. O. (1992a). Bioconvection. *Science Progress,* **76**(299):105–125.

Pedley, T. J. and Kessler, J. O. (1992b). Hydrodynamic phenomena in suspensions of swimming micro-organisms. *Annual Review of Fluid Mechanics,* **24**:313–358.

Pedley, T. J., Hill, N. A., and Kessler, J. O. (1988). The growth of bioconvection patterns in a uniform suspension of gyrotactic micro-organisms. *Journal of Fluid Mechanics,* **195**:223–238.

Platt, J. R. (1961). Bioconvection patterns in cultures of free-swimming organisms. *Science,* **133**:1766.

Plesset, M. S. and Whipple, C. G. (1974). Viscous effects in Rayleigh–Taylor instability. *Physics of Fluids,* **17**:1–7.

Plesset, M. S., Whipple, C. G., and Winet, H. (1976). Rayleigh–Taylor instability of surface layers as the mechanism for bioconvection in cell cultures. *Journal of Theoretical Biology,* **59**:331–351.

Plesset, M. S. and Winet, H. (1974). Bioconvection patterns in swimming micro-organism cultures as an example of Rayleigh–Taylor instability. *Nature,* **248**:441–443.

Prud'homme, M. and Nguyen, T. H. (2002). Parallel flow stability under a uniform heat flux: effect of the Prandtl number. *Int. Comm. Heat and Mass Transfer,* **29**(6):749–756.

Roberts, A. M. (1970). Geotaxis in motile micro-organisms. *Journal of Experimental Biology,* **53**:687–699.

Roberts, A. M. and Deacon F. M. (2002). Gravitaxis in motile micro-organisms: the role of fore-aft body asymmetry. *Journal of Fluid Mechanics,* **452**(452):405–423.

Sauvant, M. P., Pepin, D., and Piccini, E. (1999). *Tetrahymena pyriformis,* a tool for toxicological studies-a review. *Chemosphere,* **38**(7):1631–1669.

Van Hamme, D. J., Singh, A., and Ward, O. P. (2006). Physiological aspects Part 1 in a series of papers devoted to surfactants in microbiology and biotechnology. *Biotechnology Advances,* **24**:604–620.

Wager, H. (1911). On the effect of gravity upon the movements and aggregation of *Euglena Viridis,* Ehrb., and other micro-organisms. *Philos. Trans. R. Soc*, London Ser. B, **201**:333–390.

Whitaker, S. (1999). *The Method of Volume Averaging.* Kluwer Academic, Dordrecht, Boston.

Winet, H. and Jahn, T. L. (1974). Geotaxis in protozoa. *Journal of Theoretical Biology,* **46**:449–465.

Index

Note: Page numbers in *italics* refer to figures and tables.

H

I